SPECTRAL LINE SHAPES

SPECTRAL LINE SHAPES

Volume 10
14th ICSLS

State College, Pennsylvania June 1998

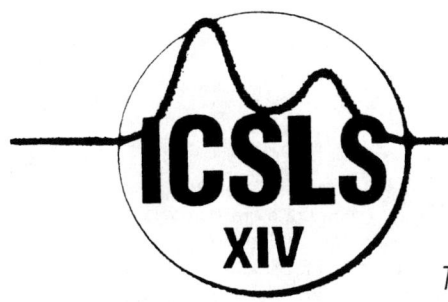

EDITOR
Roger M. Herman
The Pennsylvania State University

AIP CONFERENCE
PROCEEDINGS 467

American Institute of Physics   Woodbury, New York

Editor:

Roger M. Herman
Department of Physics
104 Davey Laboratory
The Pennsylvania State University
University Park, PA 16802
U.S.A.

Email: rmh@phys.psu.edu

The articles on pp. 49–63 and pp. 275–285 were authored by U.S. Government employees and are not covered by the below mentioned copyright.

Authorization to photocopy items for internal or personal use, beyond the free copying permitted under the 1978 U.S. Copyright Law (see statement below), is granted by the American Institute of Physics for users registered with the Copyright Clearance Center (CCC) Transactional Reporting Service, provided that the base fee of $15.00 per copy is paid directly to CCC, 222 Rosewood Drive, Danvers, MA 01923. For those organizations that have been granted a photocopy license by CCC, a separate system of payment has been arranged. The fee code for users of the Transactional Reporting Service is: 1-56396-754-5/ 99 /$15.00.

© 1999 American Institute of Physics

Individual readers of this volume and nonprofit libraries, acting for them, are permitted to make fair use of the material in it, such as copying an article for use in teaching or research. Permission is granted to quote from this volume in scientific work with the customary acknowledgment of the source. To reprint a figure, table, or other excerpt requires the consent of one of the original authors and notification to AIP. Republication or systematic or multiple reproduction of any material in this volume is permitted only under license from AIP. Address inquiries to Office of Rights and Permissions, 500 Sunnyside Boulevard, Woodbury, NY 11797-2999; phone: 516-576-2268; fax: 516-576-2499; e-mail: rights@aip.org.

ISBN 1-56396-754-5
ISSN 0094-243X
DOE CONF- 9806180

Printed in the United States of America

Contents

Preface xiii
International and Local Committees xv

PLASMA LINESHAPES OVERVIEW

Review of Plasma Broadening of Spectral Lines from Multiply
Ionized Non-Hydrogenic Atoms 3
    H. R. Griem
Spectroscopic Methods for the Measurement of Electric Fields in Plasma ... 14
    V. P. Gavrilenko
Progress in Spectral Line Shapes and Shifts Evaluation of Experimental
Stark Broadening Parameters 27
    A. Lesage, N. Konjević, and J. R. Fuhr

HIGH TEMPERATURE/DENSITY PLASMAS

Line Shape Measurements Employing Independently Diagnosed
Pinch Plasmas 39
    S. Büscher, T. Wrubel, I. Ahmad, and H.-J. Kunze
X-Ray Spectroscopy from Fusion Plasmas 49
    S. H. Glenzer
Alternative Treatment of Line Broadening in Dense and Hot Plasmas 64
    E. Leboucher-Dalimier, P. Sauvan, P. Angelo, H. Derfoul, T. Ceccotti,
    P. Gauthier, A. Poquérusse, A. Calisti, and B. Talin
Measurements of Hydrogen Isotope Densities by Two-Photon Induced
Lyman-α Fluorescence as a Proof for Tokamak Diagnostics 77
    A. Steiger, K. Grützmacher, C. Seiser, M. I. de la Rosa, and U. Johannsen
Radiative Redistribution Modeling for Hot and Dense Plasmas 88
    C. Mossé, A. Calisti, B. Talin, R. Stamm, R. W. Lee, and L. Klein
Exact Analytical Solution for the Ion Impact Broadening of Hydrogen
Lines in Plasmas at High Densities or High Principal Quantum Numbers 99
    E. Oks
Polarization Capture of Electrons by Ions as a New Channel
of Recombination: Shapes and Total Recombination Rates 111
    L. Bureyeva and V. Lisitsa
Improved Calculations on High-n Balmer and Paschen Series Profiles 115
    S. Ferri, A. Calisti, R. Stamm, B. Talin, and R. W. Lee
High Density Effects on Hydrogen Stark Profiles 119
    A. Calisti, S. Ferri, M. Koubiti, L. Mouret, R. Stamm, B. Talin,
    M. Gigosos, and V. Cardenoso
A New Spectroscopic Effect Depressing the Electron Impact
Broadening in Dense Plasmas 123
    E. Oks

**New Advances in Redistribution of Resonance Radiation
in Dense Plasmas** 127
    A. V. Demura, N. Feautrier, I. N. Kosarev, V. S. Lisitsa, and C. Stehlé

**The Influence of Plasma Microfields on Spectral Properties
of Lithium-like Ions in Non-ideal Plasma** 132
    A. N. Starostin and I. I. Yakunin

**Measurement of the D/T Fuel Mixture in MCF Plasmas by Doppler-Free
Two-Photon Spectroscopy** 134
    J. Seidel and D. Voslamber

Dense Plasma Line Shifts: Theory and Experiment 136
    G. C. Junkel, M. A. Gunderson, D. A. Haynes, Jr., C. F. Hooper, Jr.,
    D. K. Bradley, J. A. Delettrez, P. A. Jaanimagi, and S. Regan

**Charge Exchange Between Two Nearest Neighbour Ions Immersed
in a Dense Plasma** 139
    P. Sauvan, P. Angelo, H. Derfoul, E. Leboucher-Dalimier, A. Devdariani,
    A. Calisti, and B. Talin

**Time Resolved FLy β Spectroscopic Measurements in Colliding
Foil Experiments.** 141
    T. Ceccotti, H. Derfoul, P. Sauvan, P. Angelo, P. Gauthier, A. Poquérusse,
    C. Chenais-Popovics, E. Leboucher-Dalimier, M. Vollbrecht, and E. Förster

**Space-Resolved XUV Spectra of C VI and B V Lines from a 10 ps KrF
Laser-Produced Plasma.** 143
    E. J. Iglesias, H. R. Griem, R. C. Elton, and H. Scott

**Shapes of Polarization-Induced Resonances in Stimulated
Bremsstrahlung Spectra** 146
    V. A. Astapenko

**Exact Solution for the Impact Broadening of the Hydrogen Lines
Lyman-beta and Lyman-gamma** 148
    A. Derevianko and E. Oks

**Distinctive Features of the Advanced Generalized Theory of Stark
Broadening of Hydrogen Lines in Plasmas.** 150
    E. Oks and J. Touma

LOWER TEMPERATURE/DENSITY PLASMAS

**Stark Broadening of the 2S-Level of Hydrogen at Low Electron Densities
Measured by Doppler-Free Two-Photon Polarization Spectroscopy** 155
    M. Schmidt, K. Grützmacher, and A. Steiger

Radiative Collisions and the Sonoluminescence Continua 167
    L. Frommhold

Estimation of the Spectral Line Profiles in the High-Frequency Discharge 179
    G. Revalde and A. Skudra

**Profiles of Neutral Lines Emitted from Weakly Non-Ideal Helium
Plasmas Produced in Flashlamps.** 183
    Y. Vitel, M. El Bezzari, L. G. D'yachkov, and Y. K. Kurilenkov

The Hydrogen Balmer α Emission Line in Underwater Laser Plasmas 185
    A. Escarguel, A. Lesage, and J. Richou

Interpretation of Measured Stark Broadened Profiles of Highly Excited Hydrogen Lines Emitted from a Low-Density, Low-Temperature Plasma 187
    R. D. Bengtson, E. Oks, and J. Touma

Systematic Experimental Study of the Stark Broadening of C II, C III, N II, N III, O II and O III Spectral Lines 189
    B. Blagojević, M. V. Popović, and N. Konjević

Plasma Broadened 419.07 nm and 419.10 nm Neutral Argon Lines 191
    D. Nikolić, S. Djurović, Z. Mijatović, R. Kobilarov, and N. Konjević

Deconvolution Procedure for Plasma Broadened Neutral Atom Lines 193
    D. Nikolić, Z. Mijatović, R. Kobilarov, S. Djurović, and N. Konjević

Anomalous Asymmetry of a Helium Spectral Line—Anomalous Electric Fields in a Current Sheet Plasma 195
    A. G. Frank, V. P. Gavrilenko, and N. P. Kyrie

Modified Helium Microwave—Induced Plasma Discharge Chamber 197
    M. M. Mohamed and Z. F. Ghatass

New Three Phase Double Arc Plasma for Spectrochemical Analysis of Solid and Powder Samples 199
    M. M. Mohamed, Z. F. Ghatass, E. A. Shalaby, M. M. Kotb, and M. El-Raey

Line Shapes Obtained by Polarization Spectroscopy 201
    A. Steiger and K. Grützmacher

Transition Probability Measurement in a NeI Plasma 203
    J. A. del Val, J. A. Aparicio, and S. Mar

ASTROPHYSICAL AND ATMOSPHERIC APPLICATIONS

Collision-Induced Absorption in Dense Atmospheres of Cool Stars 207
    A. Borysow and U. G. Jørgensen

Lyman Series Profiles: From Laser-Plasmas to White Dwarf Stars 228
    J. F. Kielkopf and N. F. Allard

Spectral Line Shapes for Atmospheric Work: Problems and Solutions 240
    J. R. Drummond, A. D. May, R. Berman, and S. Dolbeau

Accurate Determination of Electron Densities from Hydrogen Line Widths in Active and Quiescent Prominences: The Mid-Infrared Advantage 252
    E. S. Chang, D. W. Blair, and D. Deming

Polarized Lines of Hydrogen in Magnetized Stellar Atmospheres 256
    S. Brillant, G. Mathys, and C. Stehlé

The Search for the Bound-Free Emission from the LiH $C\ ^1\Sigma^+$ State 259
    H. Skenderović, T. Ban, and G. Pichler

Transition Probabilities of Si II Lines from Laser Induced Breakdown Spectroscopy on Solid Target 260
    P. Matheron, A. Lesage, and J. Richou

Quasi-Molecular Satellites in the Lyman β Profile: Application to the White Dwarf Stars 263
    N. F. Allard, I. Drira, and J. F. Kielkopf

Effect of the Variation of Electronic Dipole Moment on Theoretical
Spectra: Application to the λ Bootis Stars.............................. 264
    N. F. Allard, I. Drira, R. Faraggiana, M. Gerbaldi, J. F. Kielkopf,
    and R. Kurucz

INNOVATIVE TECHNIQUE FOR LINESHAPE MEASUREMENT AND ANALYSIS

Spectroscopy with Non-Classical Light and Non-Classical Atoms 267
    E. S. Polzik, J. Hald, and J. L. Sørensen
Single-Mode Cavity Ring-Down Spectroscopy for Line Shape
Measurements.. 275
    J. P. Looney, R. D. van Zee, and J. T. Hodges
Quantum Oscillations: Collisional Broadening of Rydberg States
by Alkali Perturbers 286
    M. E. Henry and R. M. Herman
Spectral Boltzmann Distribution: an "Infrared Catastrophe" 301
    A. N. Starostin, A. G. Leonov, D. I. Chekhov, A. Y. Sechin,
    and Y. K. Zemtsov
Theory of Optical Shielding in Cold Atom Beams 313
    V. A. Yurovsky and A. Ben-Reuven
Coherent Transients as an Effective Technique to Distinguish Different
Collisional Relaxational Channels 317
    N. N. Rubtsova, L. S. Vasilenko, and E. B. Khvorostov
Ionic Spectra under Strong Laser Field in Plasma......................... 321
    D. A. Shapiro and M. G. Stepanov
Towards a 'Rule of Thumb' for the Wings of Forbidden Transitions......... 325
    A. Devdariani
Asymmetric Rautian-Sobelman Profile 328
    R. Ciuryło and J. Szudy
Moyal Quantum Dynamics: Atomic Scattering and Line Shapes................ 332
    B. R. McQuarrie, T. A. Osborn, M. F. Kondrat'eva, and G. C. Tabisz
The "On the Energy Shell" Simplification of the Impulse
Approximation in Rydberg Atom-Neutral Collisions 336
    D. Hoang-Binh and H. Van Regemorter
Quantum Interference and Thermodynamic Equilibrium Between
the Gas of Three-Level Atoms and the Photon Gas 339
    V. I. Savchenko, N. J. Fisch, A. A. Panteleev, and A. N. Starostin
Relative Intensity of Brillouin Lines Resulting from Quasi-Transverse
Hypersonic Acoustic Waves 341
    T. Błachowicz

COLD ATOM INTERACTIONS

Optical Spectra of Atoms in Liquid Helium and Cold Helium Gas............ 345
    M. Takami

Far-Wing Excitation Studies on the Quasimolecular Transitions in the Hg—Rare-Gas, Simple-Molecule Half-Collisions 352
 K. Ohmori

Probing a Bose-Einstein Condensate by Near-Resonant Light Scattering 364
 C. A. Sackett, J. M. Gerton, M. Welling, and R. G. Hulet

Determination of Long-Range Interactions from Photoassociative Spectroscopy of Ultracold Atoms 377
 W. C. Stwalley

Far-Wing Line Shape Study of the Collision-Induced $c \leftarrow X$ Transition in Hg—Rare-Gas Quasimolecules 388
 Y. Sato, T. Kurosawa, K. Ohmori, H. Chiba, M. Okunishi, K. Ueda, A. Z. Devdariani, and E. E. Nikitin

Combined Half-Collision Approach to the Nonadiabatic Transitions in the $Hg(6s6p^3P_2) - N_2$, CO Cold and Thermal Quasimolecules 389
 K. Ohmori, T. Kurosawa, K. Amano, H. Chiba, M. Okunishi, K. Ueda, Y. Sato, A. Z. Devdariani, and E. E. Nikitin

First Observation of the Bound Hg—Rare-Gas Complex in the Dark c-state Using Free—Bound—Bound 2-Step Laser Excitation 390
 K. Amano, K. Ohmori, T. Kurosawa, H. Chiba, M. Okunishi, K. Ueda, Y. Sato, A. Z. Devdariani, and E. E. Nikitin

Far-Wing Line-Shape Study of the Inter-Excited-State Transitions of the Hg-Ar and Hg-Ne Collisional Quasimolecules 391
 K. Amano, K. Ohmori, M. Okunishi, H. Chiba, K. Ueda, and Y. Sato

NEUTRAL ATOM LINESHAPES

Recent Progress in the Determination of Interatomic Potentials of Alkali—Argon Systems 395
 D. Zimmermann, M. Braune, and D. Schwarzhans

Experimental Study of Thermal Radiation in Dense Sodium Vapor for Broad Visible and Infrared Range 400
 A. G. Leonov, A. N. Starostin, D. I. Chekhov, A. A. Rudenko, A. Y. Sechin, and Y. K. Zemtsov

Quantum Width and Shift of Ar-Perturbed K Spectral Lines in Non-Impact Regime Based on a Pseudo-Potential for the Na/Ar 3s S State 402
 W. C. Kreye and J. F. Kielkopf

Influence of Excitation Processes on the Shape of Argon Lines 404
 A. Bielski, S. Brym, R. Ciuryło, and J. Szudy

Speed-Dependent Narrowing of the 687.1 nm Argon Line Perturbed by Neon 406
 A. Bielski, S. Brym, R. Ciuryło, and J. Szudy

The Effect of Perturber on the Pressure Broadening and Shift of Spectral Lines 408
 G. D. Roston and M. S. Helmi

Excitation Transfer Li(3D→3P) Occuring in Optical Collisions
with Rare Gas Atoms at Thermal Energies 411
    G. Lindenblatt, W. Behmenburg, F. Rebentrost, M. Jungen, M. Smit,
    and W. Meyer

COLLISION-INDUCED SPECTRA: ATOMIC AND MOLECULAR INTERACTIONS

Mixed Vibrational and Rotational Excitations in Liquid and Solid
Para-Hydrogen 415
    M. Zoppi, L. Ulivi, M. Moraldi, and M. Santoro

Interaction Properties of Hg Probed by Collision-Induced Raman
Scattering.. 427
    A. Bonechi and M. Moraldi

New Formulation for Far-Wing Line Shapes: Application to CO_2
and H_2O.. 439
    Q. Ma and R. H. Tipping

Interatomic Potentials of Cd-Kr from Far-Wing Line Profiles 444
    G. D. Roston and T. Grycuk

Absolute Interaction-Induced Light Scattering Spectral Intensities
from Helium Diatoms over a Large Frequency Domain 449
    F. Rachet, C. Guillot-Noël, M. Chrysos, and Y. Le Duff

Analysis of the Collision-Induced Absorption Spectra of Double
Vibrational Transitions in H_2-N_2 453
    C. Stamp, R. D. G. Prasad, P. G. Gillard, and S. P. Reddy

Collision-Induced Absorption in the Fundamental of O_2 and N_2:
Comparison Between Experimental and Theoretical Results 457
    G. Moreau, J. Boissoles, R. Le Doucen, R. H. Tipping, and C. Boulet

Collision-Induced Absorption by H_2 Pairs in the Second Overtone Band:
Comparison between Experimental and Theoretical Results................ 459
    A. Borysow

The Internuclear Distance of Molecular Hydrogen in the Fluid
and Solid Phases at Room Temperature................................. 460
    L. Ulivi, M. Zoppi, and G. Pratesi

Determination of the Ground State Well Depth Position R_m of Cd-inert
Gases Van der Waals Molecules Experimentally.......................... 462
    G. D. Roston and M. S. Helmi

Collision-Induced Emission Spectra of H_2-He and D_2-He Pairs
at Temperatures of Thousands of Kelvin 465
    D. Hammer, L. Frommhold, and W. Meyer

MOLECULAR LINESHAPES

Line Mixing in CO_2 Infrared Q-Branches. A Test of the Energy
Corrected Sudden Approximation 469
    J.-M. Hartmann

Egelstaff Time and the Birnbaum-Cohen Line Shape 482
    J. C. Lewis and C. Stamp
Energy Corrected Sudden Calculations of Line Widths and Line Shapes Based on Coupled States Cross Sections: The Test Case of CO_2-Argon.. 485
    F. Thibault, J. Boissoles, C. Boulet, L. Ozanne, J. P. Bouanich, C. F. Roche, and J. M. Hutson
Pressure Broadening and Saturation Lineshapes of CO at Pressures between 10^{-2} Pa and 10^2 Pa... 487
    P. Palm, D. Hanke, M. Mürtz, B. Frech, and W. Urban
The 2093 and 2130 cm^{-1} CO_2 Q-Branches Revisited: Line Mixing Effects..... 489
    A. Predoi-Cross, R. Berman, J. R. Drummond, and A. D. May
Evidence of Inhomogeneous Broadening and Shifting in the Raman Q Branch of D_2 and D_2—He at Low Temperatures 491
    S. H. Fakhr-Eslam, G. D. Sheldon, J. R. Drummond, and A. D. May
Direct Measurements of Line Mixing in Pure CO_2........................ 493
    R. Berman, P. M. Sinclair, J. R. Drummond, and A. D. May
Line Mixing in HD: Bridging Density Regimes........................... 495
    G. D. Sheldon, J. R. Drummond, and A. D. May
On the Form of Rotational Relaxation Matrix in the Infinite-Order Sudden Approximation Corrected for Energy and Frequency 497
    A. P. Kouzov
Line Mixing Effect on IR Line Clusters and Line Wings: Relaxation Matrix and Applications 499
    M. V. Tonkov and N. N. Filippov
Spectrally Resolved Determination of the Linear Dipole-Polarizability of Molecular Iodine in the Range of the B←X Transition Between 11500 cm^{-1} and 17800 cm^{-1}.. 501
    U. Hohm
Investigation of Line Broadening in the $\nu_1+3\nu_3$ Band of Acetylene.......... 503
    H. Valipour and D. Zimmermann

SYMPOSIUM IN HONOR OF THE 100TH ANNIVERSARY OF THE BIRTH OF ALEKSANDER JABŁOŃSKI

Aleksander Jabłoński and the Atomic and Molecular Physics 509
    J. Szudy
Polarization Processes in Electron-Heavy Ions Collisions 520
    V. A. Astapenko, L. A. Bureyeva, and V. S. Lisitsa
Optical Transitions in Excited Lithium+Rare Gas Collision Molecules and Related Interatomic Potentials 532
    W. Behmenburg
Collision Perturbations in the Spectra and Incremental Polarizabilities of Inert Gas Atoms.. 546
    M. O. Bulanin

Following in the Footsteps of A. Jabłoński: Some Considerations
on Collisional Interference in the HD Fundamental Band................... 552
    R. M. Herman

APPENDIX

Minutes of the Meeting of the ICSLS Program Committee.................. 559
List of Participants by Nationality.. 561
Author Index.. 563

PREFACE

The 14th International Conference on *Spectral Line Shapes* was held in State College, Pennsylvania, USA, from June 22 to 26, 1998. It was attended by 97 participants from 15 nations. These conferences are held every other year, alternately in Europe and North America. The Proceedings of the past nine conferences (Berlin 1980, Boulder 1982, Aussois 1984, Williamsburg 1986, Torun 1988, Austin 1990, Marseille 1992, Toronto 1994, and Firenze 1996) have been published under the title *Spectral Line Shapes* by de Gruyter, Berlin (Volumes 1-3); A. Deepak Publishing, Hampton, Virginia (Volume 4); Ossolineum Publishing, Warsaw, Poland (Volume 5); American Institute of Physics Press, New York (Volumes 6, 8, 9); and Nova Science Publishers, Commack New York (Volume 7).

The Conference covered a wide range of subjects emphasizing the physical processes associated with the formation of line profiles. This included invited lectures, oral contributions and contributed papers presented in three poster sessions. A symposium in honor of the 100th anniversary of the birth of A. Jabłoński consisted of three invited and one oral contribution in areas related to the pioneering work of Professor Jabłoński. The program of invited speakers was suggested primarily by the International Committee, with particular emphasis on newly emerging results and techniques. Some of the contributed papers were chosen for an oral presentation on the same basis. All contributions to the conference, to the extent possible, have been included in the present volume. The minutes of the meeting of the International Committee, and other information related to conference organization, are included in the Appendix.

A quantitative determination and analysis of the line profiles that are observed in absorption, emission, or scattering of electromagnetic radiation by plasmas and neutral fluids are powerful tools for studying the fundamental physics of atomic and molecular interactions in any of the phases of matter. Line shape analysis is also a powerful diagnostic tool for many media in extreme conditions of temperature and pressure. Indeed some of the hottest/highest pressure conditions known to mankind (magnetically confined and laser produced plasmas; stellar atmospheres) and coldest/lowest pressure conditions (cryogenic situations, magnetooptically trapped atoms including Bose-Einstein condensates, cold-atom collisions) are natural candidates for line shape analysis, which is among the few non-invasive means of determining the conditions characterizing the systems under study. In the Conference held this year, a special attempt was made to incorporate novel techniques for studying lineshapes, as well as a full discussion of lineshape phenomena in the ultracold regime. The proceedings are intended to record the latest advances in the field and to stimulate new collaborations among scientists of different countries for novel and interesting applications of the line shape analysis technique.

The first three sections of the present volume are devoted to plasma lineshapes. These include (part 1) overview of plasma lineshapes, (part 2) high temperature/density plasmas, and (part 3) lower temperature/density plasmas. We then include the section on astrophysical and planetary (including Earth) atmospheric applications (part 4). Following this is (part 5) a section on innovative techniques. The problems of neutral atomic and molecular lineshapes are then taken up in four sections: (part 6) cold atom/molecule lineshapes, (part 7) neutral atoms, (part 8) collision induced spectroscopy and (part 9) neutral molecules. The volume concludes with the papers included in (part 10) the Jablonsky symposium.

The Appendix includes the minutes of the meeting of the International Committee, a listing of the present Committee membership, the list of conference attendees by nationality (together with e-mail addresses) and an author index of the Conference Proceedings.

We wish to thank those organizations and individuals whose time, effort and financial support insured the success of the Conference. Supporting organizations included The Pennsylvania State University, Corning, Inc. Lambda-Physik, Inc., Thorlabs, Inc. and the International Union of Pure and Applied Physics (IUPAP). The sponsorship by the latter organization is contingent upon adherence to its statement of principles, which appears below. We are grateful to Xerox, Inc. for making some equipment available to the Conference.

The assistance of the Eberly College of Science, and especially the Office of Conferences and Institutes of The Pennsylvania State University were invaluable. Particularly noteworthy were the efforts of J. Hall and C. Anskis of this office. E. Olbrich and the staff of the Nittany Lion Inn (the Conference site) are commended for their fine service and attention to detail.

Outstanding efforts in program organization were made by W. Ernst and C. Back along with the rest of the International Program Committee. M. Zoppi, R. Berheim and R. Bruehl also lent helpful assistance. A special thanks is extended to J. Roberts for clerical assistance. The efforts of N. Herman in ensuring the success of the program for accompanying persons is gratefully acknowledged. Finally, we are grateful to all the participants for the top-level science presented at the conference and for the lively discussions during the various sessions and during informal activities.

The 15th International Conference on Spectral Line Shapes will be hosted by Professor Joachim Seidel of the Physikalisch-Technische Bundesanstalt (PTB), Berlin, Germany, in the summer of 2000. We express our best wishes to him and to his colleagues for a successful conference.

Roger M. Herman
The Pennsylvania State University

Organizing Chairman

Roger Herman
Department of Physics
The Pennsylvania State University
104 Davey Laboratory
University Park, PA USA 16802
rmh@phys.psu.edu

International Program Committee

Christina Back
Lawrence Livermore National Lab
tinaback@llnl.gov

Alexander Z. Devdariani
Department of Optics and
    Spectroscopy
St. Petersburg University
ponik@devdar.spb.su

M. S. Dimitrijević
Astronomical Observatory
Belgrade
mdimitrijevic@aob.bg.ac.yu

Nicole Feautrier
Observatoire de Paris-Meudon
Nicole.Feautrier@obspm.fr

John Kielkopf
University of Louisville
jfkiel01@nimbus.physics.louisville.edu

Valery S. Lisitsa
Kurchatov Institute-Moscow
LISITSA@qq.nfi.kiae.su

A. David May
University of Toronto
dmay@physics.utoronto.ca

Massimo Moraldi
Universita di Firenze
moraldi@firenze.infn.it

Hoe Hguyen
Universite Pierre et Marie Curie
Paris
hoe.nguyen@lsai.u-psud.fr

Eugene Oks
Auburn University
goks@physics.auburn.edu

Gillian Peach
University College - London
g.peach@ucl.ac.uk

Goran Pichler
Institute of Physics
Zagreb
pichler@its.hr

Joachim Seidel
Physikalisch-Technische Bundesanstalt
Berlin
seidel@chbrb.berlin.ptb.de

Roland Stamm
Universite de Provence
Marseille
rstamm@piima1.univ-mrs.fr

Josef Szudy
Nicolas Copernicus University
Torun
szudy@phys.uni.torun.pl

George C. Tabisz
University of Manitoba
Winnepeg
tabisz@cc.umanitoba.ca

Richard Tipping
University of Alabama
rtipping@bama.ua.edu

LOCAL ORGANIZING COMMITTEE
(THE PENNSYLVANIA STATE UNIVERSITY)

Robert Bernheim
Department of Chemistry
r5b@psuvm.psu.edu

William Steele
Department of Chemistry
was@chem.psu.edu

Wolfgang Ernst
Department of Physics
wee1@psuvm.psu.edu

John Yeazell
Department of Physics
yeazell@phys.psu.edu

Megan Henry
Clinch Valley College
Wise, VA, USA
m_henry@clinch.edu

The Pennsylvania State University is committed to the policy that all persons shall have equal access to programs, facilities, admission, and employment without regard to personal characteristics not related to ability, performance, or qualifications as determined by University policy or by state or federal authorities. The Pennsylvania State University does not discriminate against any person because of age, ancestry, color, disability or handicap, national origin, race, religious creed, sex, sexual orientation, or veteran status.

In accordance with IUPAP rules, no bona fide scientist is excluded from participation in the Conference on the grounds of national origin or political considerations unrelated to science.

PLASMA LINESHAPES

OVERVIEW

Review of Plasma Broadening of Spectral Lines from Multiply Ionized Non-Hydrogenic Atoms

H.R. Griem

Institute for Plasma Research, University of Maryland, College Park, MD 20742-3511

Abstract. There has been substantial progress made in our quantitative understanding of the widths, shifts and asymmetries of lines from neutral atoms and low charge-state ions due to (mostly) dipole interactions with plasma electrons and ions, i.e., of Stark broadening proper. Nevertheless, recent measurements and calculations of plasma broadening of higher charge-state ions continue to yield surprising and sometimes contradictory results. Experimentally, both line profile measurements and independent measurements of plasma conditions become increasingly difficult as one proceeds to higher charge-states, i.e., to higher temperatures (and densities). In the calculations, especially of the broadening by electrons, some of the approximations used with good success for atoms and low charge-state ions may be questionable (e.g., the dipole and classical-path approximations). Suitably improved quantum-mechanical calculations give both significant shifts for ion lines due to penetrating monopole interactions, and for certain lines widths smaller by a factor of about 2 than predicted by some semiclassical calculations, because of quantum effects. In some experiments, the problem of distinguishing Doppler broadening from nonthermal motions remains unresolved.

INTRODUCTION

For almost forty years semiclassical calculations [1-5] have been rather successful in evaluations of the broadening of lines from one-electron atoms [1] and ions [2] and of lines from two- and more-electron atoms [3] and ions [4,5]. Especially for "isolated" lines, (i.e., for lines which are narrower than separations of unperturbed energy levels) from non-hydrogenic levels of two- and more-electron systems, there is a close relation between electron-collisional cross sections and line widths. This was emphasized by Van Regemorter [6], and quantitatively formulated by Baranger [7] in quantum-mechanical language even earlier than the semiclassical calculations just cited.

It always was realized that completely quantum-mechanical calculations would be necessary in cases where significant contributions came from collisions with impact parameters ρ which are comparable to or even smaller than the reduced de

Broglie wavelength, $\bar{\lambda} = \lambda/2\pi = \hbar/mv$. Moreover, to avoid violations of unitarity, a minimum impact parameter was introduced, which for hydrogenic radiators and assuming straight classical paths was estimated [1,2] to be $\rho_{\min} \approx n^2\hbar/Zmv = (n^2/Z)\bar{\lambda}$, n and Z being the principal quantum number and effective (spectroscopic) charge, respectively. Since calculated line widths are mostly due to collisions well beyond $\rho_{\min}$, the condition $n^2/Z \gtrsim 1$ is therefore a reasonable one for the validity of the semiclassical approximation, provided the use of hyperbolic orbits [2,4,5] and allowing for $\Delta n = 0$ level splittings do not cause significant changes in the relative importance of so-called strong collisions. Corresponding effects will be discussed in the following section.

On the experimental side, much progress was made in the development of plasma light sources suitable for line profile measurements and of diagnostic methods, especially of those for the determination of the most important plasma property for the evaluation of Stark broadening data, namely, the electron density. A series of critical reviews of such data [8-12] has been instrumental in establishing a balanced view of experiment-theory comparisons of widths and shifts of isolated lines from many neutral atoms and low charge-state ions. Not surprisingly [13], the agreement was generally much better for widths than for shifts. This situation was further discussed in corresponding chapters of two books [14,15], but experiment-theory comparisons have remained less satisfactory for higher charge-state ions. (For some discussion of asymmetries of isolated lines, see Sec. 4.10 of Ref. 15.) Nevertheless, there were substantial advances toward the measurement of line profiles of lines from such ions through the development of plasma light sources capable of reaching higher densities and temperatures, e.g., of the gas-liner pinch [16] and of appropriate diagnostic methods, e.g., collective Thomson scattering [17]. However, as will be discussed in Sec. III, some conclusions [18] concerning the accuracy and validity of improved semiclassical calculations [19] must be revised [20].

THEORETICAL CONSIDERATIONS

Before discussing possible modifications of the simple criterion for the validity of the semiclassical calculations, namely $n^2/Z \gtrsim 1$, it is worth mentioning that fully quantum-mechanical calculations based on Baranger's [7] general formula for the collision broadening of isolated lines have been made for almost thirty years [21-23]. Such calculations are an integral part of the Opacity Project [24-26] and use the close-coupling (CC) method, with the total electron wavefunction expanded in terms of a set of target states and optimized perturber wavefunctions.

Similar calculations were made for hydrogen and ionized helium resonance lines [27,28] and can be used to check the $n^2/Z \gtrsim 1$ criterion. For higher Z members of the one-electron [29,30] and two-electron [31,32] sequences it was later realized that strong-coupling effects were rather small, justifying the use of Coulomb-Born or distorted-wave approximations. These calculations could not only be used to improve the semiclassical calculations by adjusting the so-called strong-collision constant [29] but also through their use of more (rather than only dipole) terms in the multipole expansion of the perturber-radiator interactions; they yielded significant red-shifts, mostly due to the (penetrating) monopole ($\Lambda = 0$) term [30-32]. These shifts therefore correspond to the earlier "plasma polarization shift" [14,33].

Returning to close-coupling (CC) calculations, care must be exercised not to use them for lines from high-lying levels, unless the important perturbing levels are included in the basis set, and also not for electron energies above the thresholds for the excitation of such levels. Extension of the expansion basis [26] and inclusion of "pseudo-states," which can, e.g., account for the target ion continuum, in corresponding R-matrix with pseudo-states (RMPS) calculations [34] largely removed these restrictions. Calculated elastic and inelastic scattering cross sections, e.g., of Be atoms in the ground state [35] were found to be in good agreement with convergent close-coupling calculations (CCC) [36]. The latter method involves the use of a large Laguerre basis and also takes the target continuum into account. It treats highly excited states and can be used at any energy. Assuming a sufficiently complete basis, both RMPS and CCC results for electron collisional broadening should therefore be correct within the numerical accuracy of the very large calculations and assuming the impact approximation to be valid.

Also semiclassical calculations have recently been improved, in particular by refining the definition of strong collisions by using the time-dependent unitarity-condition [18,19], rather than its time-integrated version [3], and by including higher than second-order terms in the Dyson series for the solution of the time-dependent Schrödinger equation. However, the electron-ion interaction is still replaced by its long-range dipole plus quadrupole approximation, and only the reduced de Broglie wavelength $\bar{\lambda}$ and impact parameter ρ were used in assessing the validity of the semiclassical approximation [37]. Because during an electron-ion collision the velocity increases, say, to $v_{\max}$ and the distance of closest approach, $r_{\min}$, is smaller than the impact parameter ρ, a more appropriate set of criteria would be

$$\lambda_{\min} = 2\pi \bar{\lambda}_{\min} = 2\pi \hbar / m v_{\max} \lesssim r_{\min}, \tag{1}$$

and

$$r_n \lesssim r_{\min}, \tag{2}$$

where $r_n \approx n^2 a_0 / Z$ is the excited-state atomic radius. If the second criterion is violated, both semiclassical and long-range interaction approximations are questionable. Including the factor 2π in the first criterion may be conservative, so that using the symbol $\lesssim$ rather than $\ll$ seems reasonable. Omitting this factor might seem consistent with $\Delta \rho / \rho \approx 1/L$ from the uncertainty principle, but would correspond to significant errors if small L-contributions to the line width are important.

Because of the conservation of angular momentum L (in units of $\hbar$), the first criterion leads to

$$\frac{\lambda_{\min}}{r_{\min}} = \frac{2\pi \hbar}{m r_{\min} v_{\max}} = \frac{\lambda}{\rho} = \frac{2\pi}{L} \lesssim 1. \tag{3}$$

With the help of Eqs. (116), (117) and (118) of [14], the second criterion can also be expressed in terms of L, namely using also the Coulomb parameter $\eta = (Z-1)e^2/\hbar v$,

$$\frac{r_n}{r_{min}} = \frac{n^2 a_0/Z}{r_{min}} = \frac{(Z-1)}{Z}n^2\frac{[1+(L/\eta)^2]^{1/2}+1}{L^2} \lesssim 1. \qquad (4)$$

The remaining problem in assessing the validity of the semiclassical long-range interaction approximation is therefore the determination of the range of significant L-values. In quantum-mechanical calculations, this is naturally accomplished by the partial wave expansion; whereas in semiclassical calculations only internal consistency checks are possible, but often not easily extracted from the literature. The provisional $n^2/Z \gtrsim 1$ condition, as mentioned earlier, is not always sufficient. Moreover, it corresponds to omitting the factor 2π in Eq. (1).

According to the recent CCC calculations [20,38] of Li-like B III and Be-like Ne VII, $\Delta n = 0$, $n = 2$ and 3 lines, respectively, widths were mostly (within 90%) due to $L \le 6$ or $L \le 15$ partial waves, respectively. For the B III lines, criterion (3) is therefore clearly violated; whereas for the Ne VII lines, semiclassical calculations could be valid, say, for $L \ge 10$ partial waves, which contribute about 30% of the line width. As already discussed in [20], for the B III lines also the criterion (4) is at best fulfilled only marginally. For the Ne VII lines, $\eta \approx 4$ to 8 is larger, because of the much larger ionic charge, as are contributing L-values. Table I was calculated from Eq. (4) for an effective principal quantum number $n = 3$ (although r_n values from Hartree-Fock calculations are larger by a factor 1.4); and it shows that typical values of r_n/r_{min} are probably sufficiently small to justify the use of the long-range interaction approximation in this case for $L \gtrsim 6$, which contribute about 75% of the line width in the CCC calculations. This leaves criterion (3) as the most restrictive for the Ne VII line.

To summarize, in both cases the semiclassical method as such is less reliable than had been inferred from estimated bounds of excluded quantal contributions [18,19,37,39]. (See also the discussion of [58] below.)

TABLE I. Ratio of atomic radius and perihelion (r_n/r_{min}) for electron–Ne VII ion collisions according to Eq. (4)

$L\backslash\eta =$	4	6	8
4	1.16	1.06	1.02
8	0.39	0.32	0.29
12	0.22	0.17	0.15

For both the B III and Ne VII lines recent quantum-mechanical (CCC) results for the line widths [20,38] are smaller than the recent semiclassical results [18,19] by a factor of 2, but agree with earlier close-coupling calculations [25] for B III to within 15% at the temperatures of interest. Any numerical uncertainty of this order originates probably in the strong cancellations in the elastic scattering contribution to the line width [20]. This contribution is about 30% in the temperature range of interest for B III and 10% for Ne VII. Especially in the latter case, electron-collisional widths can thus be used to make comparisons with the sums of excitation and de-excitation rates of ions in upper and lower levels from calculated or measured electron-ion cross sections. While for some Li-like ions some related data were available to support the accuracy of the B III CCC-calculation [20], and recent measurements [40] of the most important 2s-2p cross section for B III have confirmed the RMPS and CCC calculations, no measurements of cross sections seem to be available for Be-like ions in support of the Ne VII calculations. Only some excitation rate coefficients inferred from measurements on time-dependent plasmas [41,42] are available for comparisons. Since these measurements involved ground state and metastable target ions, they can be used only to check some $\Delta n = 1$ collisional rates, which are not very important here. However, one can say that measured [41] rates involving the $n = 3$ states at an electron temperature of 260 eV are 2 to 3 times smaller than those used in [38].

Another difference is in the elastic scattering contribution, which for the B III lines is mostly due to $L \lesssim 4$ partial waves [20], i.e., well in the region where both criteria (3) and (4) are violated. To the extent that any account is taken of the elastic scattering term [7] in semiclassical calculations, it would normally have been included in the strong-collision term; or in semiempirical calculations [14] been allowed for by extrapolating inelastic cross sections below threshold energies. A notable exception was the early work of Roberts [43]. Another indication of the very different treatment of small-L effects in quantum-mechanical and semiclassical calculations is the relatively large inelastic quadrupole ($\Lambda = 2$) collisional rate for 3s-3d in case of Ne VII [18,19], which is almost an order of magnitude larger than the CCC result [38].

A detailed comparison of trends along iso-electronic sequences, e.g., for the 2s-2p Li-like lines, between the non-perturbative semiclassical calculations [18] and fully quantum-mechanical calculations must wait for more calculations, e.g., CCC. However, given the dominance of inelastic contributions and the good agreement between CCC and CBE (Coulomb-Born with exchange) calculations [20,38], one expects theoretically that widths (in frequency or energy units) will more closely scale as $1/Z^2$ than as $1/Z$. For low Z-members of a given sequence, elastic scattering will be relatively more important, but the corresponding steepening of the Z-dependence is compensated by the well-known [15] increase of dipole excitation cross sections over the $1/Z^2$ scaling inferred from that of the dipole matrix elements. This increase is fairly weak (see below), so that we may conjecture that the two methods of calculation will differ also in regard to the Z-dependence.

EXPERIMENT-THEORY COMPARISONS

Measurements of electron-collisional broadening of lines from multiply ionized atoms have been relatively scarce until gas liner pinch [16] and linear-pinch-discharge experiments [44,45] became available for line profile measurements at sufficiently high temperatures (for the ions to be produced and excited) and electron densities (for Stark broadening to dominate). However, these measurements remain difficult, not only because it is not possible to meet all requirements for ideal Stark broadening experiments [9], but also because additional diagnostics were needed, e.g., collective Thomson scattering [17]. As always in new experiments, possible systematic errors are easily underestimated, as indicated by the different ratios of measured widths of the B III 2s-2p lines and electron densities in a low-pressure pulsed arc [46] and in the gas liner pinch [47], the former being larger by a factor of 3 than the latter. This disagreement is probably due to an underestimate of Doppler broadening in the arc experiment, in which the measured electron density was lower by a factor of 7. Even in the high density ($N_e = 1.8 \times 10^{18}\,\text{cm}^{-3}$), gas liner experiment, Doppler broadening may also have been underestimated [20], because the implosion velocities correspond to well above critical Reynolds numbers for gas-dynamic (fluid) turbulence to develop. In that case, test gas ions would move with the turbulent eddies of the hydrogen fill gas at random velocities approaching thermal velocities of the hydrogen atoms, thus increasing Doppler widths by a factor of the order of the square root of the test gas and hydrogen atomic weights. This effect could increase calculated widths by a factor of 2 for the conditions of the B III gas liner experiment [47].

Besides the B III 2s-2p (Li-like ion) and Ne VII 2s3s–2s3p (Be-like) measurements, both of which agreed with improved semiclassical calculations [18,19] but yielded twice the widths calculated quantum-mechanically [20,25,38], there are two other types of experiments of interest here. Of primary interest are measurements of widths for low principal quantum number transitions along iso-electronic sequences of Li- [48-51], Be- [51] and B- [52,53] like ions. Since the temperature of necessity also increases along the sequences, the temperature dependence must be measured as well, which has been done, e.g., for Li- [49] and B- [52,54] like ions. Most of these measurements are consistent with the simple $T^{-1/2}$ dependence; although in the case of Li-like $3s^2S$-$3p^2P$ lines of C IV and N V, deviations from earlier measurements in the same device [48,55] are larger than the changes of measured widths with temperature [49]. Similarly, the temperature-dependence of calculated widths does not depart from the $T^{-1/2}$ dependence by more than deviations, e.g., of specific close-coupling results for various $n = 3$, $\Delta n = 0$ transitions in Be-like ions, from a fit to an empirical relation [25] with temperature-independent Gaunt factors. However, the fit corresponds to an effective Gaunt factor [56] which scales for multiply ionized atoms as $1.30-(1.22/Z)$, for a spectroscopic charge Z. (One reason for not using this relation for low charge-state ions was already mentioned, namely, the larger role of elastic collisions which require other-than-dipole matrix elements.)

Recent experiments, on the other hand, have consistently yielded line widths for highly charged ions in excess of the $1/Z^2$ (dipole) scaling prediction [49,50], although an earlier observation regarding C IV [48] was not confirmed. While the improved semiclassical calculations [19] indeed reach agreement with the measured Ne VII 2s3s-2s3p line widths [57], provided quadrupole interactions are taken into account, this agreement may nevertheless be spurious. According to the considerations in the preceding section, both the semiclassical and long-range interaction approximations are questionable in this case, which can be seen by reinterpreting Fig. 11 of [58]. The "excluded region," to avoid quantum corrections, corresponds to $L \lesssim 3$, for impact parameters less than 1.05 Å at electron velocities $v = 3 \times 10^8$ cm/sec. The corresponding Coulomb parameter is $\eta \approx 4$, so that the criteria (3) and (4) are clearly violated, which is especially serious for the quadrupole terms. As already suggested in connection with the B III 2s-2p lines [20], the wave-mechanical averaging over particle positions and less steep increase of the exact Coulomb interaction relative to the long-range approximation are evidently responsible for the smaller widths, by a factor of 2, of the CCC calculations [38], relative to both measured [57] and semiclassically calculated widths. The situation for the Ne VIII 3s-3p line seems very similar. Its measured width [49] also exceeds the width obtained from $1/Z^2$ scaling, etc., but CCC calculations have not yet been made. However, the CC results of [25] suggest that in this case quantum-mechanically calculated widths will be smaller than measured widths by more than a factor of 2. This continues a trend already noticed [59] when the first CC-calculations had just been published, of a factor 1.5–2 smaller widths than semiclassical results. (See also [60].)

For the lines discussed so far ion-ion collisions are not too important, contributing, e.g., less than 10% to the collisional widths of the B III [20] and Ne VII [38] lines from $\Delta n = 0$ transitions. For $\Delta n \geq 1$ transitions, especially from $n \geq 3$ levels, the relative importance of ion collisions increases rapidly, especially if separations from perturbing levels are small enough that ion-produced, quasistatic fields would cause nearly linear Stark effects and give rise to forbidden components. Examples here are the relatively broad Li-like C IV, N V and O VI lines due to $n = 4$ to 5 transitions [61] for which agreement with calculated profiles is very good if ion-dynamical effects are included. For more highly charged ions, experiments still suffer from difficulties with independent density measurements, a challenge remaining especially for densities approaching $N_e \approx 10^{25}$ cm^{-3} in laser-fusion experiments. In these experiments, lines due to $n = 1$ to 3 transitions in hydrogen- and helium-like ions of argon, added as a seed to the fuel, are especially useful [62]; and there is no question that broadening and (red) shifts (mostly due to $\Lambda = 0$ monopole interactions) must be calculated quantum-mechanically [29-32]. However, in these cases broadening by ions is dominant, requiring allowance for ion-dynamics. Moreover, at lower temperatures, below or near 1 keV, various satellites due to inner-shell transitions in lower charge-state ions (mostly on the red wing), must also be allowed for in detailed calculations [63-65]. The situation is simpler for lower-Z ions, such as C VI [66] and Aℓ XIII [67], for which red-shifts were observed without noticeable interference with satellites.

SUMMARY

The intended emphasis of this brief review of Stark broadening experiments and calculations was on low n, $\Delta n = 0$ lines from multiply ionized, three- and more-electron ions for which plasma conditions can still be measured independently, and wavelengths are relatively long for the given ion. This is not only convenient experimentally, but for given n also favors Stark broadening over Doppler broadening, the only competing broadening mechanism in fully ionized, optically thin plasmas. Moreover, the theoretical problem is also relatively simple in that electron-ion collisions dominate over ion-ion collisions, and that the impact approximation is definitely applicable to the broadening by plasma electrons.

Particularly interesting have been the experimental investigations of line widths along iso-electronic sequences. As already discussed, there is still substantial scatter of data from various experiments. Nevertheless a trend exists toward some excess broadening of high series members relative to the theoretically expected Z-scaling. This trend has been attributed to ion-ion collisions, underestimates in the strong-collision term in perturbative semiclassical calculations and, more recently and again here, also to additional Doppler broadening caused by turbulence or nonthermal (flow) velocities in the highly dynamical experiments.

As to differences between various methods of calculations, there is a persistent excess broadening (by factors of 1.5 to 2) obtained by nonperturbative semiclassical calculations compared to fully quantum-mechanical calculations. This excess broadening can be attributed to the quantum-mechanical smoothing of the interaction and to the erroneous use of only long-range interactions in the semiclassical calculations; it is therefore unphysical.

More precision measurements of line widths and, perhaps, also shifts of lines from multiply charged ions are clearly needed to resolve the remaining differences from quantum calculations. Should they persist, this might indicate to some readers that collisional rate coefficients presently used in plasma modeling would be too small by the same factor of 1.5 to 2. However, given the good agreement between measured excitation cross sections and the best quantum (RMPS, CCC) calculations, this conclusion seems untenable. This would leave the possibility of other line broadening mechanisms than electron-ion (mostly inelastic) collisions, e.g., of unresolved Doppler line splitting associated with the radial implosion velocities in gas-liner pinches [68,69]. These Doppler effects are important only for the resonance lines and some other lines from high-Z ions, which are relatively insensitive to Stark broadening.

ACKNOWLEDGMENTS

This work was partially supported by the National Science Foundation.

REFERENCES

1. Griem H.R., Kolb A.C. and Shen K.Y., *Phys. Rev.* **116**, 4 (1959).
2. Griem H.R., and Shen K.Y., *Phys. Rev.* **122**, 1490 (1961).
3. Griem H.R., Baranger M., Kolb A.C., and Oertel G.K., *Phys. Rev.* **125**, 177 (1962).
4. Bréchot S., and Van Regemorter H., *Ann. Astrophys.* **27**, 432 (1964).
5. Sahal-Bréchot S., *Astron. Astrophys.* **1**, 91 (1969); **2**, 322 (1969).
6. Van Regemorter H., *Ann. Review Astron. Astrophys.* **3**, 71 (1965).
7. Baranger H., *Phys. Rev.* **112**, 855 (1958).
8. Konjević H., and Roberts J.R., *J. Phys. Chem. Ref. Data* **5**, 209 (1976).
9. Konjević N., and Wiese W.L., *J. Phys. Chem. Ref. Data* **5**, 259 (1976).
10. Konjević N., Dimitrijević M.S., and Wiese W.L., *J. Phys. Chem. Ref. Data* **13**, 619 (1984); **13**, 649 (1984).
11. Konjević N., and Wiese W.L., *J. Phys. Chem. Ref. Data* **19**, 1307 (1990).
12. Wiese W.L., and Konjević N., *J. Quant. Spectrosc. Radiat. Transfer* **47**, 185 (1992).
13. Griem H.R., and Shen C.S., *Phys. Rev.* **125**, 196 (1962).
14. Griem H.R., *Spectral Line Broadening by Plasmas*, Academic Press (1974).
15. Griem H.R., *Principles of Plasma Spectroscopy*, Cambridge University Press (1997).
16. Kunze H.-J., in *Spectral Line Shapes, vol. 4*, ed. R.J. Exton, Deepak, Hampton, VA (1987).
17. Glenzer S.H., in *Atomic Processes in High Temperature Plasmas*, ed. A.L. Osterheld and W.H. Goldstein, AIP Conf. Proc. 381, Woodbury, NY (1996).
18. Alexiou S., in *Spectral Line Shapes, vol. 9*, ed. M. Zoppi and L. Ulivi, AIP Conf. Proc. 386, Woodbury, NY (1997).
19. Alexiou S., *Phys. Rev. Lett.* **75**, 3406 (1995).
20. Griem H.R., Ralchenko Yu. V., and Bray I., *Phys. Rev. E* **56**, 7186 (1997).
21. Bely O., and Griem H.R., *Phys. Rev. A* **1**, 97 (1970).
22. Barnes K.S., and Peach G., *J. Phys. B* **3**, 350 (1970).
23. Peach G., and Barnes K.S., *J. Phys. B* **4**, 1377 (1971).
24. Seaton M.J., *J. Phys. B* **20**, 6431 (1987).
25. Seaton M.J., *J. Phys. B* **21**, 3033 (1988).
26. Burke V.M., *J. Phys. B* **25**, 4917 (1992).
27. Griem H.R., *Comments At. Mol. Phys.* **2**, 53 (1970).
28. Tsuji A., and Narumi H., *Prog. Theor. Phys.* **44**, 1557 (1970).
29. Griem H.R., Blaha M., and Kepple P.C., *Phys. Rev. A* **19**, 2421 (1979).
30. Nguen-Hoe H., Koenig M., Benredjem D., Caby M., and Coulaud G., *Phys. Rev. A* **33**, 1279 (1986).
31. Koenig M., Malnoult P., and Nguen H., *Phys. Rev. A* **38**, 2089 (1988).
32. Griem H.R., Blaha M., and Kepple P.C., *Phys. Rev. A* **41**, 5600 (1990).
33. Berg H.F., Ali A.W., Lincke R., and Griem H.R., *Phys. Rev.* **125**, 199 (1962).
34. Bartschat K., Hudson E.T., Scott M.P., Burke P.G., and Burke V.M., *J. Phys. B* **29**, 115 (1996).
35. Bartschat K., Burke P.G., and Scott M.P., *J. Phys. B* **30**, 5915 (1997).
36. Fursa D.V., and Bray I., *J. Phys. B* **30**, 5895 (1997).
37. Alexiou S., Phys. Rev. A **49**, 106 (1994).

38. Ralchenko Yu. V., Griem H.R., Bray I., and Fursa D., to be published.
39. Lee R.W., Glenzer S., Nash J., Osterheld A., Alexiou S., and Ralchenko Y., Proc. 11 APS Topical Conf. on Atomic Processes in Plasmas (1998).
40. Bannister M.E., private communication.
41. Johnston W.D., and Kunze H.-J., *Phys. Rev. A* **4**, 962 (1971).
42. Lang J., *J. Phys. B* **16**, 3907 (1983).
43. Roberts D.E., *Astron. Astrophys.* **6**, 1 (1970).
44. Purić J., Djeniže S., Sreckovič A., Platiša M., and Labat J., *Phys. Rev. A*, **37**, 498 (1988).
45. Djeniže S., Sreckovič A., Milosavljević M., Labat O., Platiša M., and Purić J., *Z. Physik D* **9**, 129 (1988).
46. Sreckovič S., Djeniže S., and Platiša M., Proc. XVI Int. Symp. Phys. Ionized Gases, ed. M. Milosavljević, Inst. Nucl. Sci., Beograd, 201 (1993).
47. Glenzer S., and Kunze H.-J., *Phys. Rev. A* **53**, 2225 (1996).
48. Böttcher F., Breger P., Hey J.D., and Kunze H.-J., *Phys. Rev. A* **38**, 2690 (1988).
49. Glenzer S., Uzelac N.I., and Kunze H.-J., *Phys. Rev. A* **45**, 8795 (1992).
50. Glenzer S., in *Spectral Line Shapes, vol. 8*, eds. A.D. May, J.R. Drummond and E. Oks, AIP Conf. Proc. **328**, 134 (1995).
51. Blagojević B., Popović M.V., Konjević N., and Dimitrijević M.S., in *Spectral Lines, vol. 9*, eds. M. Zoppi and L. Ulivi, AIP Conf. Proc. **386**, 143 (1997); see also Wrubel Th., Ahmad I., Büscher S., Kunze H.-J., and Glenzer S.H., *Phys. Rev. E* **57**, 5972 (1998).
52. Blagojević B., Popović M.V., Konjević N., and Dimitrijević M.S., *Phys. Rev. E* **50**, 2986 (1994).
53. Glenzer S., Hey J.D., and Kunze H.-J., *J. Phys. B* **27**, 413 (1994).
54. Blagojević B., Popović M.V., Konjević N., and Dimitrijević M.S., *Phys. Rev. E* **54**, 743 (1996).
55. Ackermann U., Fincken K.H., and Musielok J., *Phys. Rev. A* **31**, 2597 (1985).
56. Griem H.R., *Phys. Rev.* **165**, 258 (1968).
57. Wrubel Th., Glenzer S., Büscher S., Kunze H.-J., and Alexiou S., *Astron. Astrophys.* **306**, 1023 (1996).
58. Alexiou S., Calisti A., Gauthier P., Klein L., Leboucher-Dalimier E., Lee R.W., Stamm R., and Talin B., *J. Quant. Spectrosc. Radiat. Transfer* **58**, 399 (1997).
59. Sahal-Brechot S., and Segre E.R., *Astron. Astrophys.* **13**, 161 (1971).
60. Dimitrijević M.S., and Sahal-Bréchot S., *Ann. Phys.* (Paris), Colloque No. 3, Suppl. **15**, 77 (1990).
61. Glenzer S., Wrubel Th., Büscher S., Kunze H.-J., Godbert L., Calisti A., Stamm R., Talin B., Nash J., Lee R.W., and Klein L., *J. Phys. B* **27**, 5507 (1994).
62. Griem H.R., *Phys. Fluids B* **4**, 2346 (1992).
63. Woolsey N.C., Asfaw A., Hammel B., Keane C., Back C.A., Calisti A., Mossé C., Stamm R., Talin B., Wark J.S., Lee R.W. and Klein L., *Phys. Rev. E* **53**, 6396 (1996).
64. Haynes D.A., Jr., Garber D.T., Hooper C.F., Jr., Mancini , Lee Y.T., Bradley D.K., Delettrez J., Epstein R., and Jaanimagi P.A., *Phys. Rev. E* **53**, 1042 (1996).
65. Hooper, C.F. Jr., Junkel G.C., Gunderson M.A., Haynes D.A. Jr., Mancini R.C.,

Bradley D., Delettrez J., and Jaanimagi P., in Proceedings of the Int. Conf. on Strongly Coupled Coulomb Systems, Boston (1997).
66. Leng Y., Goldhar J., Griem H.R., and Lee R.W., *Phys. Rev. E* **52**, 4328 (1995).
67. Renner O., Salzmann D., Sondhauss P., Foerster E., Djaoui A., and Krovský E., in *Spectral Line Shapes, vol. 9*, eds. M. Zoppi and L. Ulivi, AIP Conf. Proc. **386**, 57 (1997); Renner O., Sondhauss P., Salzmann D., Djaoui A., Koenig M., and Förster E., *J. Quant. Spectrosc. Radiat. Transfer* **58**, 851 (1997).
68. Kunze H.-J., private communication.
69. Büscher S., Wrubel Th., Ahmad I., and Kunze H.-J., these proceedings.

SPECTROSCOPIC METHODS FOR THE MEASUREMENT OF ELECTRIC FIELDS IN PLASMA

V. P. Gavrilenko

Center for Surface and Vacuum Research of the Russian State Committee for Standards, Andreevskaya nab. 2, Moscow 117334, Russia

Development of methods for the measurement of electric fields (EFs) in plasmas is of great practical importance. Firstly, these EFs may represent electromagnetic waves penetrating into a plasma. Secondly, these are EFs of natural plasma oscillations (e. g., the EFs of plasma turbulence). Thirdly, such EFs may be EFs existing in discharges. Many of spectroscopic methods for diagnostics of EFs in a plasma are considered, e.g. in [1,2]. This paper reviews some new spectroscopic methods for the measurement of the parameters of EFs in a plasma.

METHODS BASED ON EMISSION SPECTROSCOPY

1. SPECTRAL LINES OF HYDROGEN IN OSCILLATING ELECTRIC FIELD: IMPACT BROADENING AND DYNAMIC STARK EFFECT

Oscillating EFs of the form $\vec{E}(t) = \vec{E}_0 \cos(\omega t + \phi)$ play a decisive role in many physical processes in a plasma. One of the most commonly used methods for diagnostics of oscillating EFs in a plasma is the spectroscopic method based on a modification of the emission spectra of atomic hydrogen under the influence of such fields. In a plasma, the interaction of a hydrogen atom with the field $\vec{E}(t)$ occurs against a "background" of chaotic collisions of the atom with charged particles. Let us consider a situation when microfields of charged particles exert impact influence on the formation of profile of hydrogen spectral line corresponding to the transition $a \to b$. We analyse below two different cases. In the first case (the case of rare collisions: $\omega \tau_0 \gg 1$, τ_0 is the mean free path time for the broadening collisions) the profile of hydrogen spectral line has the form

$$I(\Delta\omega) = \frac{1}{\pi \tau_0} \sum_{\alpha \in a, \beta \in b} \sum_{p=-\infty}^{+\infty} J_p^2\left(\frac{\gamma_{\alpha\beta} E_0}{\omega}\right) \frac{1}{(\Delta\omega - p\omega)^2 + \tau_0^{-2}}, \quad (1)$$

where the frequency $\Delta\omega$ is measured from the unperturbed frequency of the transition $a \to b$. In (1) $J_p(x)$ is the Bessel function, $\gamma_{\alpha\beta}$ is the Stark constant: $\gamma_{\alpha\beta} = <\alpha|z|\alpha> - <\beta|z|\beta>$, axis $z \parallel \vec{E}_0$ (here and below we are using atomic units, where $\hbar = m_e = e = 1$). In fact, spectrum (1) is the Blochinzew-type spectrum [3], in which impact broadening of all satellites is taken into account. In the second case (the case of often collisions: $\omega \tau_0 \ll 1$) the profile of hydrogen spectral line is the following

$$I(\Delta\omega) = \sum_{\alpha \in a, \beta \in b} I_{\alpha\beta}(\Delta\omega), \quad (2)$$

where

$$I_{\alpha\beta}(\Delta\omega) = \frac{1}{2\pi^2\tau_0}\int_0^{2\pi}\frac{d\Phi}{(\Delta\omega - \gamma_{\alpha\beta}E_0\cos\Phi)^2 + \tau_0^{-2}}$$

$$= \frac{1}{\sqrt{2}\pi\tau_0(\gamma_{\alpha\beta}E_0)^2}\left\{\left[\left(\frac{1}{\tau_0\gamma_{\alpha\beta}E_0}\right)^2 + \left(1 - \frac{\Delta\omega}{\gamma_{\alpha\beta}E_0}\right)^2\right]^{-1/2}\right.$$

$$+ \left[\left(\frac{1}{\tau_0\gamma_{\alpha\beta}E_0}\right)^2 + \left(1 + \frac{\Delta\omega}{\gamma_{\alpha\beta}E_0}\right)^2\right]^{-1/2}\right\}\left\{\left[\left(\frac{1}{\tau_0\gamma_{\alpha\beta}E_0}\right)^2 + \left(\frac{\Delta\omega}{\gamma_{\alpha\beta}E_0}\right)^2 - 1\right]^2\right.$$

$$\left.+ \left(\frac{2}{\tau_0\gamma_{\alpha\beta}E_0}\right)^2\right\}^{1/2} + \left(\frac{1}{\tau_0\gamma_{\alpha\beta}E_0}\right)^2 + \left(\frac{\Delta\omega}{\gamma_{\alpha\beta}E_0}\right)^2 - 1\right\}^{-1/2}.$$

(3)

The dependence of the spectrum $I_{\alpha\beta}(\Delta\omega)$ in (3) on the parameter $(\tau_0|\gamma_{\alpha\beta}|E_0)^{-1}$ is the following. While $(\tau_0|\gamma_{\alpha\beta}|E_0)^{-1} \ll 1$ the profile $I_{\alpha\beta}(\Delta\omega)$ has two sharp lateral symmetrical peaks (when $\Delta\omega = \pm(|\gamma_{\alpha\beta}|E_0 - 3^{-1/2}\tau_0^{-1}))$ and a dip in the line center (when $\Delta\omega = 0$). When the parameter $(\tau_0|\gamma_{\alpha\beta}|E_0)^{-1}$ increases, the lateral peaks are moving to the center with the decrease of their intensities. When $(\tau_0|\gamma_{\alpha\beta}|E_0)^{-1} \approx 0.71$ the profile $I_{\alpha\beta}(\Delta\omega)$ turns from a "two-humped" profile into a "one-humped" profile with the peak in the line center $\Delta\omega = 0$. When $(\tau_0|\gamma_{\alpha\beta}|E_0)^{-1}$ grows further, the profile $I_{\alpha\beta}(\Delta\omega)$ turns into Lorentzian. The "reduced" profiles $\tilde{I}(x)$, corresponding to the profiles $I_{\alpha\beta}(\Delta\omega)$, are present in Fig. 1 for several values of the parameter $P = [1 + (\tau_0\gamma_{\alpha\beta}E_0)^2]^{-1/2}$. The relation between the profiles $\tilde{I}(x)$ and $I_{\alpha\beta}(\Delta\omega)$ is the following:

$$\tilde{I}(x) = I_{\alpha\beta}(\Delta\omega)[\tau_0^{-2} + (\gamma_{\alpha\beta}E_0)^2]^{1/2}, \quad \Delta\omega = x[\tau_0^{-2} + (\gamma_{\alpha\beta}E_0)^2]^{1/2}.$$

By using (1) – (3), it is possible to measure the field $\vec{E_0}\cos(\omega t + \phi)$ in a plasma for the wide range of frequencies ω.

2. STARK PROFILES OF HYDROGEN SPECTRAL LINES IN ELECTRIC FIELDS OF PLASMA TURBULENCE

In case of a turbulent plasma, the typical situation is that the oscillating EF of plasma turbulence is a superposition of a considerable number of independent harmonics with random phases ϕ_j and different frequencies ω_j:

$$\vec{E}(t) = \sum_{j=1}^{N}\vec{E}_j\cos(\omega_j t + \phi_j), \quad N\to\infty. \qquad (4)$$

We assume further that vectors $\vec{E}_j$ in (4) are parallel to each other, and thus, the field $\vec{E}(t)$ in (4) is linearly polarized. In accordance with the central limit theorem the electric field $\vec{E}(t)$ in (4) is a stochastic Gaussian process.

Let the field $\vec{E}(t)$ in (4) be a stationary Gaussian process ($\{\vec{E}(t)\}_{av} = 0$) with the correlation function

$$\{\vec{E}(t)\vec{E}(t+\tau)\}_{av} = B \cdot G(\tau). \tag{5}$$

Here $\{\ldots\}_{av}$ means averaging, $B = \{E^2\}_{av}$ is the mean intensity of the stochastic field $\vec{E}(t)$, and the correlation coefficient $G(\tau)$ describes the time behaviour of the correlation function [$G(0)=1$].

The result of the calculation of the emission spectrum $S(\Delta\omega)$ of a hydrogen atom for the transition $a \to b$ in the "Gaussian" field $\vec{E}(t)$ is the following:

$$\begin{aligned}
S(\Delta\omega) &= \pi^{-1} \sum_{\alpha \in a, \beta \in b} \mathrm{Re} \int_0^\infty \exp(-i\Delta\omega\tau)\Phi_{\alpha\beta}(\tau)d\tau, \\
\Phi_{\alpha\beta}(\tau) &= |<\varphi_\alpha|\vec{r}\cdot\vec{e}|\varphi_\beta>|^2 P_{\alpha\beta}(\tau), \\
P_{\alpha\beta}(\tau) &= \exp[-\gamma_{\alpha\beta}^2 B \int_0^\tau (\tau - t) G(t)\, dt], \quad \tau \geq 0.
\end{aligned} \tag{6}$$

In (6) φ_α and φ_β are "parabolic" wave functions (WFs) of a hydrogen atom, $\vec{e}$ is the unit vector of the emitted photons.

The calculations were performed for the model in which the field $\vec{E}(t)$ represents a band noise with Lorentzian spectrum. In the model of this type the correlation coefficient can be represented in the form

$$G(\tau) = \exp(-\nu|\tau|)\cos(\omega_0 \tau). \tag{7}$$

The value 2ν is the halfwidth of the spectrum corresponding to the function $G(\tau)$. Substituting (7) in (6) we obtain

$$P_{\alpha\beta}(\tau) = \exp\{-\gamma_{\alpha\beta}^2 B\{[(\nu^2 - \omega_0^2)\cos(\omega_0\tau) - 2\nu\omega_0 \sin(\omega_0\tau)]\exp(-\nu\tau) \\ + \nu\tau(\omega_0^2 + \nu^2) + \omega_0^2 - \nu^2\}(\omega_0^2 + \nu^2)^{-2}\}. \tag{8}$$

In the case when $\nu/\omega_0 \to 0$ the result of [4] can be derived from (8):

$$P_{\alpha\beta} = \exp\{-(\gamma_{\alpha\beta}^2 B/\omega_0^2)[1 - \cos(\omega_0\tau)]\},$$

which leads to the following emission spectrum

$$S_{\alpha\beta}(\Delta\omega) = \exp(-\kappa_{\alpha\beta}) \sum_{p=-\infty}^{+\infty} I_p(\kappa_{\alpha\beta})\delta(\Delta\omega - p\omega_0), \quad \kappa_{\alpha\beta} = (\gamma_{\alpha\beta}B^{1/2}/\omega_0)^2. \tag{9}$$

In (9) $I_p(\kappa_{\alpha\beta})$ is the modified Bessel function. From (8) it follows that for $\nu/\omega_0 \ll 1$ each satellite in the spectrum (9), excepting a few satellites near $\Delta\omega = 0$, has a Lorentzian profile of the halfwidth $\Delta_{1/2} = 4\nu\gamma_{\alpha\beta}^2 B/\omega_0^2$. In the case $\Delta_{1/2} > \omega_0$ the profiles of satellites are overlapping, and the resultant profile is described by the Gaussian function

$$S_{\alpha\beta}^{(G)}(\Delta\omega) = \exp[-\Delta\omega^2/(2\gamma_{\alpha\beta}^2 B)]. \tag{10}$$

Now let us assume that the following conditions are valid

$$\nu/\omega_0 \geq 1, \quad \gamma_{\alpha\beta}^2 B/\omega_0^2 \gg 1, \quad \gamma_{\alpha\beta}^2 B/\nu^2 \gg 1. \tag{11}$$

Under these conditions the profile of the spectral component $\alpha \to \beta$ is also Gaussian (10). A simple result can be also derived from (8) for the case

$$\gamma_{\alpha\beta}^2 B/(\omega_0^2 + \nu^2) \ll 1. \tag{12}$$

Under the condition (12) the profile of the spectral component is Lorentzian of the halfwidth $2\gamma_{\alpha\beta}^2 B\nu/(\omega_0^2 + \nu^2)$. Figs. 2,3 show the spectra of hydrogen spectral lines H_β and H_δ with the polarization of the radiation parallel to the vector $\vec{E}(t)$ (Figs. 2,a; 3,a) and orthogonal to the vector $\vec{E}(t)$ (Figs. 2,b; 3,b). The parameters of the stochastic field $\vec{E}(t)$ are the following: $B^{1/2} = 6$ kV/cm, $\omega_0 = 1.88 \cdot 10^{11}$ s^{-1}, $\nu/\omega_0 = 0.02$. Let γ_1 and γ_2 be mean Stark constants for a hydrogen spectral line with polarizations parallel $\vec{E}(t)$ and orthogonal $\vec{E}(t)$ respectively. For the line H_β the satellites are present for both polarizations (cf. Fig. 2), although due to the condition $\gamma_2 < \gamma_1$ the halfwidth of the satellites in Fig. 2,a is greater than in Fig. 2,b. For the line H_δ the satellites are visible clearly for the polarization orthogonal to the field $\vec{E}(t)$ (cf. Fig. 3,b). The disappearance of the most of satellites in Fig. 3,a (excepting a few satellites near the center of the line H_δ) is due to the fact that the formally calculated halfwidths of such satellites are greater than the separation between them ω_0.

The obtained results give an opportunity to determine simultaneously two parameters of the turbulent electric field $\vec{E}(t)$ in a plasma: the root mean square strength $B^{1/2}$ and the spectral width ν. To do it we need to register spectral line profiles of hydrogen with two polarizations: parallel to the vector $\vec{E}(t)$ (the profile $S_1(\Delta\omega)$) and perpendicular to the vector $\vec{E}(t)$ (the profile $S_2(\Delta\omega)$). The profiles $S_1(\Delta\omega)$ and $S_2(\Delta\omega)$ may differ from each other not only in their halfwidths, but also qualitatively (as is shown, e.g., in Figs. 2,3). Let us assume that the frequency ω_0 is known beforhand (or is measured by using the separation between satellites of a hydrogen spectral line). Then the values $B^{1/2}$ and ν may be obtained by comparing the experimental profiles $S_1(\Delta\omega)$ and $S_2(\Delta\omega)$ with a set of the theoretical profiles calculated for different values of B and ν by using the formulas (6), (8).

3. SATELLITES OF FORBIDDEN SPECTRAL LINES. METHOD FOR DETERMINING THE POLARIZATION STATE OF OSCILLATING ELECTRIC FIELDS IN PLASMAS

In the energy spectrum of nonhydrogen-like atoms in many cases one can distinguish a relatively isolated system of three levels 0,1, and 2. Levels 1 and 2 constitute a pair of closely spaced upper levels, while 0 is a distant lower level, to which a radiative transition occurs (Fig. 4). We assume that the only nonzero matrix elements for the electric dipole moment $\vec{d}$ are those between the 1 and 2 states and between the 0 and 2 states: $<1|\vec{d}|2> \neq 0$, $<0|\vec{d}|2> \neq 0$. Oscillating EF (having the frequency ω) can mix the WFs of the levels 1 and 2. As a result, the emission spectrum will contain, in addition to the allowed spectral line with intensity I_{20}, two satellites of the forbidden line $1 \to 0$ with intensities $S_{1,2}$, spaced a frequency distance of $\pm\omega$ from the position of this line (see Fig. 5). It was suggested in [5] that the ratio $S_{1,2}/I_{20}$ could be used for the diagnostics of oscillating EFs in a plasma. Since then this "satellite method" was used with various modifications in many experiments, the first of them was the experiment in a low-density θ pinch in helium [6]. We note that the "satellite method" primarily aims at measurements of the strength of oscillating EFs, and it requires certain *a priori* information on the polarization state of these fields. For example, by using the results of Ref.[5] one can measure the mean square strength of oscillating EFs in plasmas under the assumption that the oscillatin EFs are an isotropic high-frequency noise with uncorrelated components in different directions.

In this paper we examine the polarization of satellites of forbidden spectral lines of nonhydrogen-like atoms as a function of the polarization characteristics of the oscillating EF acting on these atoms. On the base of the analytical expressions presented below, a method is proposed for determining the polarization state of an oscillating EF in a plasma.

In the present paper we take the oscillating EF to be a quasimonochromatic EF of arbitrary form

$$\vec{E}(t) = \vec{E}_0(t) \exp(-i\omega t) + \vec{E}_0^*(t) \exp(i\omega t), \qquad (13)$$

where the asterisk means complex conjugation. In Eq. (13) it is assumed that the complex amplitude $\vec{E}_0(t)$ varies much more slowly than $\exp(\pm i\omega t)$ as a function of time. We denote by $\varphi_{pJ_pM_p}$ the WF of the state corresponding to level p, while ϵ_p is the energy of level p. We assume that the field $\vec{E}(t)$ as in (13) acts on the two-level system 1,2. Then from the Schrödinger equation in first-order time-varying pertubation theory, we can derive the following expression for the WF $\Psi_{1J_1M_1}(t)$ of level 1 in the field $\vec{E}(t)$:

$$\Psi_{1J_1M_1}(t) = \varphi_{1J_1M_1} \exp(-i\epsilon_1 t) + \sum_{M_2=-J_2}^{J_2} \varphi_{2J_2M_2} \left\{ \frac{<\varphi_{2J_2M_2}|\vec{dE}_0(t)|\varphi_{1J_1M_1}>}{\epsilon_{21} - \omega} \right.$$

$$\left. \times \exp[-i(\epsilon_1 + \omega)t] + \frac{<\varphi_{2J_2M_2}|\vec{dE}_0^*(t)|\varphi_{1J_1M_1}>}{\epsilon_{21} + \omega} \exp[-i(\epsilon_1 - \omega)t] \right\}, \qquad (14)$$

where $\epsilon_{kk'} = \epsilon_k - \epsilon_{k'}$. We assume that the vector $\vec{E}_0(t)$ and the unit polarization vector of the emitted photons $\vec{e}$ are of the form

$$\vec{E}_0 = \sum_{k=1}^{3} E_{0k}(t)\vec{e}_k, \quad \vec{e} = \sum_{k=1}^{3} \beta_k \vec{e}_k,$$

where $\vec{e}_1$, $\vec{e}_2$, $\vec{e}_3$ are unit vectors along the x, y, z axes, respectively. By using expression (14) we find the formula for the intensity $S_1^{(e)}$ of the satellite of the forbidden spectral line $1 \to 0$ (below $\{...\}_{av}$ means time averaging):

$$S_1^{(e)} = \frac{|(1, J_1 \| d \| 2, J_2)(2, J_2 \| d \| 0, J_0)|^2}{30(\epsilon_{21} - \omega)^2 \Delta} \sum_{k,k'=1}^{3} \{\{|E_{0k}(t)|^2\}_{av}[a_1|\beta_k|^2 \delta_{kk'} \qquad (15)$$

$$+a_2|\beta_{k'}|^2(1-\delta_{kk'})] - \{E_{0k}(t)E_{0k'}^*(t)\}_{av}(1-\delta_{kk'})(a_3\beta_k\beta_{k'}^* + a_4\beta_k^*\beta_{k'})\},$$

where $\delta_{kk'}$ is the Kronecker delta, $(pJ \| d \| p'J')$ is the reduced matrix element. We note that the coefficients Δ, a_1, a_2, a_3, a_4 are given in [7]. The formula (15) is valid for all possible relations among the quantum numbers J_0, J_1, J_2.

In order to measure the absolute intensity of the oscillating EF we need to compare the satellite intensities $S_{1,2}^{(d)}$ with the intensity of the dipole-allowed spectral lines. These dipole-allowed lines may have as upper level either level 2 (the $2 \to 0$ transition is shown in Fig. 4) or level 1 (an arrow in Fig. 4 shows the transition $1 \to 0'$, where $0'$ is a distant lower energy level which is not level 0). Assuming that the Zeeman states of levels 2 and 1 are equally populated, we find the following expressions for the intensities of the allowed $2 \to 0$ and $1 \to 0'$ spectral lines, respectively:

$$I_{20} = 3^{-1}|(2, J_2 \| d \| 0, J_0)|^2, \quad I_{10'} = 3^{-1}|(1, J_1 \| d \| 0', J'_{0'})|^2. \qquad (16)$$

By using expressions (15), (16), and the data for Δ, a_1, a_2, a_3, a_4 presented in [7], it is not difficult to obtain the results of the work [5], in which it is assumed that the oscillating EF has an isotropic directional patten with uncorrelated components in different directions. In the latter case it is necessary to take into account the following expression

$$\{E_{0k} E^*_{0k'}(t)\}_{av} = 6^{-1} < E_p^2 > \delta_{kk'},$$

where $< E_p^2 >$ is the time average of the square strength of the oscillating EF.

From (15) – (16) we can find all nine elements of the tensor

$$\sigma_{kk'} = \{E_{0k}(t) E_{0k'}(t)\}_{av}$$

that determines the polarization state and energy density of the oscillating field. For this purpose we must measure the intensity ratio of a satellite to one of the allowed spectral lines ($1 \to 0'$ or $2 \to 0$) for nine different polarization vectors $\vec{e}$ of emitted photons, recording the emission spectrum along three noncomplanar axes. Here is one of the possible set of polarization vectors $\vec{e}$:

$$\vec{e}_I = \vec{e}_1, \quad \vec{e}_{II} = \vec{e}_2, \quad \vec{e}_{III} = \vec{e}_3,$$
$$\vec{e}_{IV} = 2^{-1/2}(\vec{e}_1 + \vec{e}_2), \quad \vec{e}_V = 2^{-1/2}(\vec{e}_1 + i\vec{e}_2), \quad \vec{e}_{VI} = 2^{-1/2}(\vec{e}_1 + \vec{e}_3),$$
$$\vec{e}_{VII} = 2^{-1/2}(\vec{e}_1 + i\vec{e}_3), \quad \vec{e}_{VIII} = 2^{-1/2}(\vec{e}_2 + \vec{e}_3), \quad \vec{e}_{IX} = 2^{-1/2}(\vec{e}_2 + i\vec{e}_3),$$

where i is the square root of -1. We note that the vectors $\vec{e}_I$, $\vec{e}_{II}$, $\vec{e}_{III}$, $\vec{e}_{IV}$, $\vec{e}_{VI}$, $\vec{e}_{VIII}$ correspond to the linear polarization of the emitted photons, while the vectors $\vec{e}_V$, $\vec{e}_{VII}$, $\vec{e}_{IX}$ correspond to the circular polarization.

METHODS BASED ON LASER SPECTROSCOPY

1. LASER-INDUCED FLUORESCENCE OF POLAR DIATOMIC MOLECULES

Diagnostics of relatively weak electric fields ~ 100 V/cm is of great importance for various gas discharges and plasmas. In [8] a novel diagnostic method was proposed, which is based on laser-induced fluorescence (LIF) of polar diatomic molecules, and was implemented in a number of experiments after publication of the work [8]. The primary characteristic feature of this method is its high sensitivity with respect to electric fields that can be explained as follows.

Laser pumping populates an electronic state $^1\Pi$ of a polar diatomic molecule. Each rotational level of the molecule in the $^1\Pi$ state is split into two close sublevels

(Λ-doubling sublevels, see Fig. 6) coupled by a dipole matrix element. Therefore, even not very strong electric field $\vec{F}$ can significantly intermix the WFs of the Λ-doubling sublevels. As a result, intensities of spectral components of the LIF originating from those Λ-doubling sublevels may be considerably modified. As it was shown in [9,10], these intensities can be determined not only by the strength of the EF and the laser pumping of the Λ-doubling sublevels, but also by the process of the collision equalization of the populations of the Zeeman states within each of the Λ-doubling sublevels.

Let us assume that the laser radiation exitates the $R(J)$-transition in the polar diatomic molecules. Then for diagnostics of the EF it is convenient to use the ratio of the intensities of the components in the LIF spectrum corresponding to the Q- and P-transitions. Such intensities were calculated in [9] in the assumption that collisions completely equalize the populations of the Zeeman states within each of the Λ-doubling sublevels:

$$I_Q = [(J+1)^2(2J+1)(2J+3)^2]^{-1} \sum_{M=-(J+1)}^{J+1} \sum_{M'=-J}^{J} [(J+1)^2 - M'^2]$$
$$\times [(J+1)(J+2) + M^2](1 - \cos\beta_M \cos\beta_{M'}),$$
$$I_P = [(2J+1)(2J+3)^3(2J+5)]^{-1} \sum_{M=-(J+1)}^{J+1} \sum_{M'=-J}^{J} [(J+1)^2 - M'^2]$$
$$\times [(J+2)(3J+7) - M^2](1 + \cos\beta_M \cos\beta_{M'}),$$
(17)

where
$$\beta_M = \tan^{-1} g_M, \qquad g_M = \{2\mu FM/[(J+1)(J+2)\Delta_{ef}]\}. \tag{18}$$

In (17) μ is the dipole moment of the molecule in the $^1\Pi$ state, Δ_{ef} is the zero field Λ-doublet splitting, J is the quantum number of the total angular momentum of the molecule, M is the quantum number of the projection of the total angular momentum onto the z axis (the z axis is directed along the vector $\vec{F}$). It is assumed in formula (17) that laser radiation is polarized along the vector $\vec{F}$, and that the direction of the observation of the fluorescence radiation is orthogonal to the vector $\vec{F}$.

The theoretical dependence of the ratio I_Q/I_P versus the field strength F, calculated by formula (17) for the cases of $R(2)$- and $R(3)$-pumping of the transition $(X^1\Sigma^+, v=0) \to (A^1\Pi, v=0)$ in CS molecules, is represented by the solid line in Fig. 7,a and Fig. 7,b, respectively (Figs. 7,a and 7,b are taken from [9]). The dashed line in these figures shows the corresponding theoretical dependences of the ratio I_Q/I_P versus F, calculated by using the results of [8]. The points in Fig. 7,a and Fig. 7,b show the experimental ratio I_Q/I_P versus F, obtained in [9]. It is seen that both in Fig. 7,a ($R(2)$-pumping) and in Fig. 7,b ($R(3)$-pumping) the experimental points fall between the dashed curve, corresponding to the complete neglect of collisions, and the solid curve, corresponding to the collision-induced equalization of populations of Zeeman states within each of the Λ-doubling sublevels. It can also be seen that at relatively weak fields, that is $F \leq 2$ kV/cm for $R(2)$-pumping and $F \leq 4$ kV/cm for $R(3)$-pumping, the experimental points are closer to the solid curve than to the dashed curve.

2. NONLINEAR OPTICAL METHODS FOR DIAGNOSTICS OF ELECTRIC FIELDS

2.1. We consider firstly a nonlinear optical method for measurements of the strength of the static EFs in a plasma [11]. This method is based on the generation of coherent infrared (IR) radiation as a result of biharmonic laser pumping in a static EF.

The typical scheme of four-wave mixing process is presented in Fig. 8,a. The interaction of three electromagnetic waves (with frequencies ω_1, ω_2, and ω_3) with the medium results in the generation of coherent light at the frequency $\tilde{\omega} = \omega_1 - \omega_2 + \omega_3$.

Let us assume that there are molecules in a plasma, the vibrations of which are active for the Raman scattering. In this case, if the frequency difference ($\omega_1 - \omega_2$) of two waves is close to the vibration frequency Ω of these molecules, there will be a resonant growth of the cubic susceptibility of the molecules $\chi^{(3)}(\tilde{\omega} = \omega_1 - \omega_2 + \omega_3)$, and respectively, a resonant growth of the intensity of the generated radiation $I(\tilde{\omega}) \propto |\chi^{(3)}(\tilde{\omega})|^2 I_1 I_2 I_3$. (Here I_k is the intensity of the electromagnetic wave, $k = 1, 2, 3$.) A static EF can be used in place of one of the interacting waves (see Fig. 8,b). This corresponds to the situation $\omega_3 = 0$. In the latter case the intensity of the generated radiation at the frequency $\tilde{\omega}$ can be written

$$I(\tilde{\omega} = \omega_1 - \omega_2) \propto |\chi^{(3)}|^2 F^2 I_1 I_2, \qquad (19)$$

where F is the strength of the static EF. The expression (19) can be considered as the basis for the measurements of the strength F of the static EFs in gaseous or plasma media.

In our experiment (see [11]) we have detected the IR radiation (at the frequency $\tilde{\omega} = \omega_1 - \omega_2$), emitted by molecular hydrogen in the presense of the static EF F. The wave with frequency ω_1 ($\lambda_1 = 532$ nm) was the second harmonic from a pulsed ($\tau_i = 10$ ns) Nd:YAG laser. A tunable dye laser (the dye was pyridine-1) generated the light with ω_2 ($\lambda_2 \sim 683$ nm), which satisfied the Raman resonance condition $\omega_1 - \omega_2 = \Omega$ for $v = 0, v = 1$ transition of the $H_2(X^1\Sigma_g^+)$ molecule.

The intensity of the generated light at the frequency $\tilde{\omega} = \omega_1 - \omega_2$ (with $\lambda \sim 2.4\ \mu m$) had a quadratic dependence on the strength of the static electric field F (in accordance with (19)). This intensity was also a quadratic function of the pressure of molecular hydrogen (the latter was in accordance with the fact that the cubic susceptibility $\chi^{(3)}$ is proportional to the density of H_2).

We note that in order to measure the EF F, the working molecules need not be H_2; they could also be such molecules as N_2, CO_2, H_2O, and CO.

2.2. Let us assume that in plasma containing Raman-active molecules, there is an oscillating EF $E_p(t)$, which is the field of the Langmuir wave. In this case the frequency spectrum of the field $E(t)$ is localized in the narrow region near the plasma frequency $\omega_{pe} = (4\pi e^2 N_e/m_e)^{1/2}$ (we assume $\omega_{pe} \ll \Omega$), and the spectrum of wave vectors $\vec{k}_p$ is restricted from above by the inverse Debye length: $|\vec{k}_p| \lesssim r_D^{-1} = (8\pi N_e e^2/T_e)^{1/2}$. For the diagnostics of the field $E_p(t)$ it is necessary to direct into the plasma volume two laser beams with frequencies ω_1 and ω_2, and with wave vectors $\vec{k}_1$ and $\vec{k}_2$, having the frequency difference $\omega_1 - \omega_2$ close to the frequency Ω. Fig. 9 shows two possible schemes, for which there occurs a resonant increase of the intensity of the generated light at frequency $\omega' = \omega_1 - \omega_2 + (-1)^n \omega_p$, where $n = 1, 2$, and ω_p is the frequency of the Langmuir wave. In order to make the generation of the light at the frequency ω' appreciable, the phase matching condition should also be satisfied:

$$\vec{k}' = \vec{k}_1 - \vec{k}_2 + (-1)^n \vec{k}_p + \Delta \vec{k}, \quad |\Delta \vec{k}| \ll \min(|\vec{k}'|, |\vec{k}_1 - \vec{k}_2 + (-1)^n \vec{k}_p|), \quad n = 1, 2,$$

where $\vec{k}_p$ and $\vec{k}'$ are the wave vectors of Langmuir wave and generated wave respectively, and $\Delta \vec{k}$ is the phase mismatch. Changing the directions of the laser beams at frequencies ω_1 and ω_2 (and hence changing the difference $\vec{k}_1 - \vec{k}_2$) and/or changing the direction of the observation of light at the frequency ω', it is possible to satisfy the phase matching condition for the interacting waves. It provides an opportunity to determine simultaneously both the frequency ω_p and the wave vector $\vec{k}_p$ of the plasma wave. It could allow us to obtain the information on the dispersion relation for waves in plasmas.

This work was sponsored by the Russian Foundation for Basic Research (project No. 97-02-17657) and by INTAS (project No. 96-456).

REFERENCES

1. H.R. Griem, "Spectral Line Broadening by Plasmas", Academic Press, New York (1974)
2. E. Oks. "Plasma Spectroscopy: The Influence of Microwave and Laser Fields", Springer Series on Atoms and Plasmas, V. 9, Springer-Verlag, New York (1995)
3. D.I. Blochinzew, Phys. Z. Sow. Union 4, 501 (1933)
4. E.V. Lifshitz, Sov. Phys. – JETP 26, 570 (1968)
5. M. Baranger, and B. Mozer, Phys. Rev. 123, 25 (1961)
6. H.-J. Kunze, and H.R. Griem, Phys. Rev. Lett. 21, 1048 (1968)
7. V.P. Gavrilenko, JETP 76, 236 (1993)
8. C.A. Moore, G.P. Davis, and R.A. Gottscho, Phys. Rev. Lett. 52, 538 (1984)
9. S. Maurmann, H.-J. Kunze, V.P. Gavrilenko, and E. Oks, J. Phys. B: At. Mol. Opt. Phys. 29, 25 (1996)
10. S. Maurmann, V.P. Gavrilenko, H.-J. Kunze, and E. Oks, J. Phys. D: Appl. Phys. 29, 1525 (1996)
11. V.P. Gavrilenko, E.B. Kupriyanova, D.P. Okolokulak, V.N. Ochkin, S.Yu. Savinov, S.N. Tskhai, and A.N. Yarashev, JETP Lett. 56, 1 (1992)

Fig. 1. Reduced profile $\tilde{I}(x)$ of one Stark component ($\alpha \to \beta$) of a spectral line of a hydrogen atom interacting in plasma with a regular electric field $\vec{E}_0 \cos \omega t$ and with an electric field $\vec{e}_p$ of charged particles at different values of P. It is assumed that the effect of a field $\vec{E}_0 \cos \omega t$ results in the quasistatic Stark broadening of the hydrogen spectral line, whereas the effect of a field $\vec{e}_p$ results in the impact Stark broadening of the line. The parameter $P = [1 + (\tau_0 \gamma_{\alpha\beta} E_0)^2]^{-1/2}$, where $\gamma_{\alpha\beta}$ is the Stark constant for the transition $\alpha \to \beta$, and the value τ_0^{-1} characterizes the impact width of the hydrogen spectral line.

Fig. 2. Spectral line profile H_β of a hydrogen atom interacting with a stochastic electric field of plasma turbulence $\vec{E}(t)$ as in (4). The parameters of the field $\vec{E}(t)$ are the following: the root mean square strength $B^{1/2} = 6$ kV/cm, the mean frequency $\omega_0 = 1.88 \cdot 10^{11}$ s^{-1}, $\nu = 0.02\omega_0$, where 2ν is the halfwidth of spectrum corresponding to the correlation function for the field $\vec{E}(t)$ (cf. (7)).
a – polarization of the radiation is parallel to the vector $\vec{E}(t)$; b – polarization of the radiation is orthogonal to the vector $\vec{E}(t)$.

Fig. 3. The same as in Fig. 2, but for H_δ.

Fig. 4. Energy level diagram of a non-hydrogen-like atom. The two closely spaced upper levels, 1 and 2, are coupled by a dipole matrix element. The solid arrows show electric dipole transitions to the distant lower levels 0 and 0'. The dashed arrow shows a transition which is forbidden in the dipole approximation in the absence of an external electric field.

Fig. 5. Emission spectrum of a non-hydrogen-like atom in a quasimonochromatic electric field $\vec{E}(t)$ as in (13). The spectral components I_{20} and $I_{10'}$ correspond to allowed electric dipole transitions $2 \to 0$ and $1 \to 0'$. The satellites S_1 and S_2 arise at frequencies $\pm \omega$ with respect to the dipole-forbidden spectral line $1 \to 0$ as a result of the field $\vec{E}(t)$ in (13).

Fig. 6. Energy level diagram for the $^1\Pi - ^1\Sigma$ transition in a polar diatomic molecule. Laser radiation pumps the R-branch transition. The Q- and P-branch transitions are observed in the fluorescence spectrum.

Fig. 7. Ratio of intensities Q and P lines in the laser-induced fluorescence spectrum of CS molecules in the (0,0) band of the CS $A^1\Pi - X^1\Sigma^+$ system: ϕ, experimental points; dashed curves, calculations by using the results of [8]; solid curves, calculations by formula (17).
a – R(2)-transition laser exitation; b – R(3)-transition laser exitation. From [9].

Fig. 8. Possible schemes of processes involving the cubic nonlinearity of a medium: a – generation of light at the frequency $\tilde{\omega} = \omega_1 - \omega_2 + \omega_3$; b – generation of light at the difference frequency $\tilde{\omega} = \omega_1 - \omega_2$ in a static electric field $\vec{F}$. The frequency Ω is the frequency of Raman-active molecular vibrations.

Fig. 9. Possible schemes for active laser diagnostics of a Langmuir wave in plasma. Here ω_1 and ω_2 are the frequencies of two laser beams, ω_p is the frequency of the Langmuir wave, Ω is the frequency of Raman-active molecular vibrations, and $\omega' = \omega_1 - \omega_2 \pm \omega_p$ is the frequency of the light generated in the process of wave mixing.

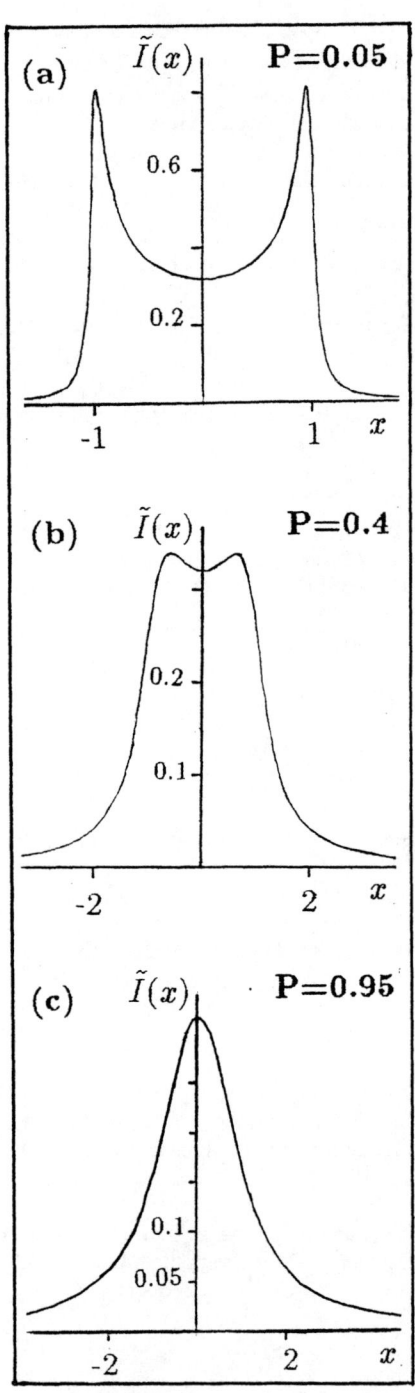

Fig. 1

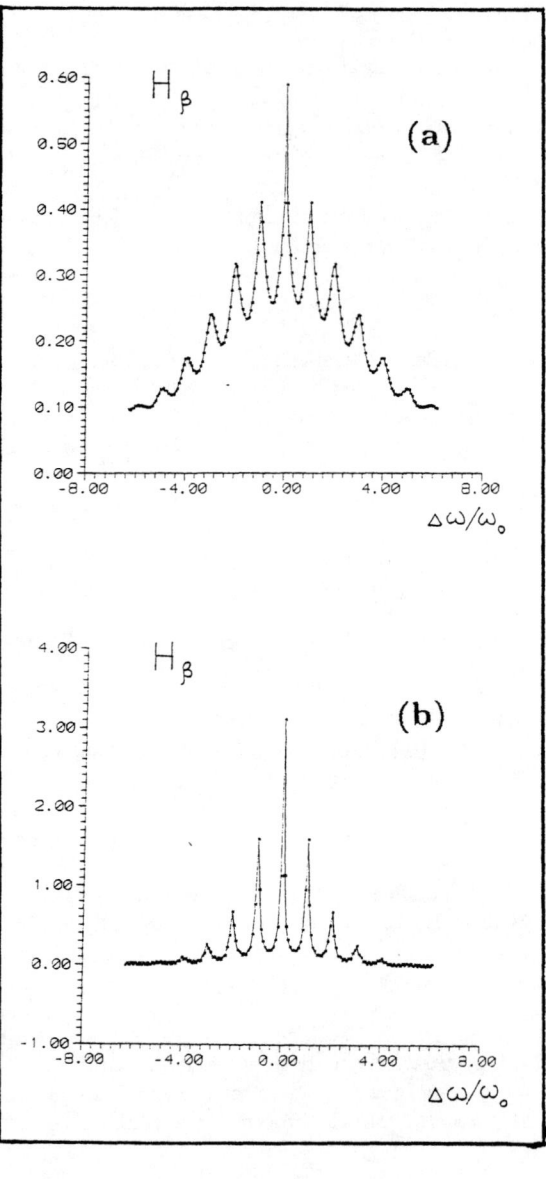

Fig. 2

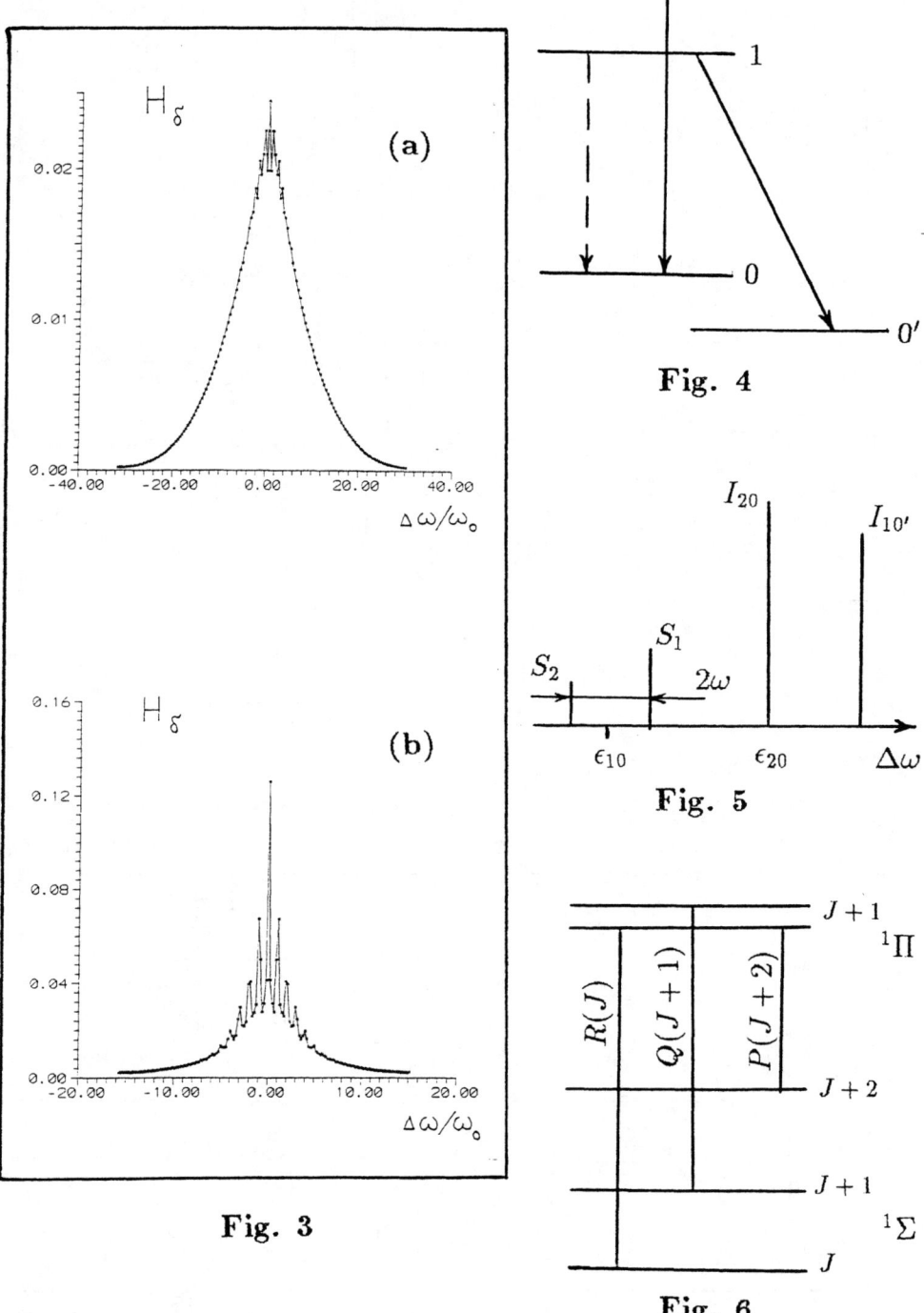

Fig. 3

Fig. 4

Fig. 5

Fig. 6

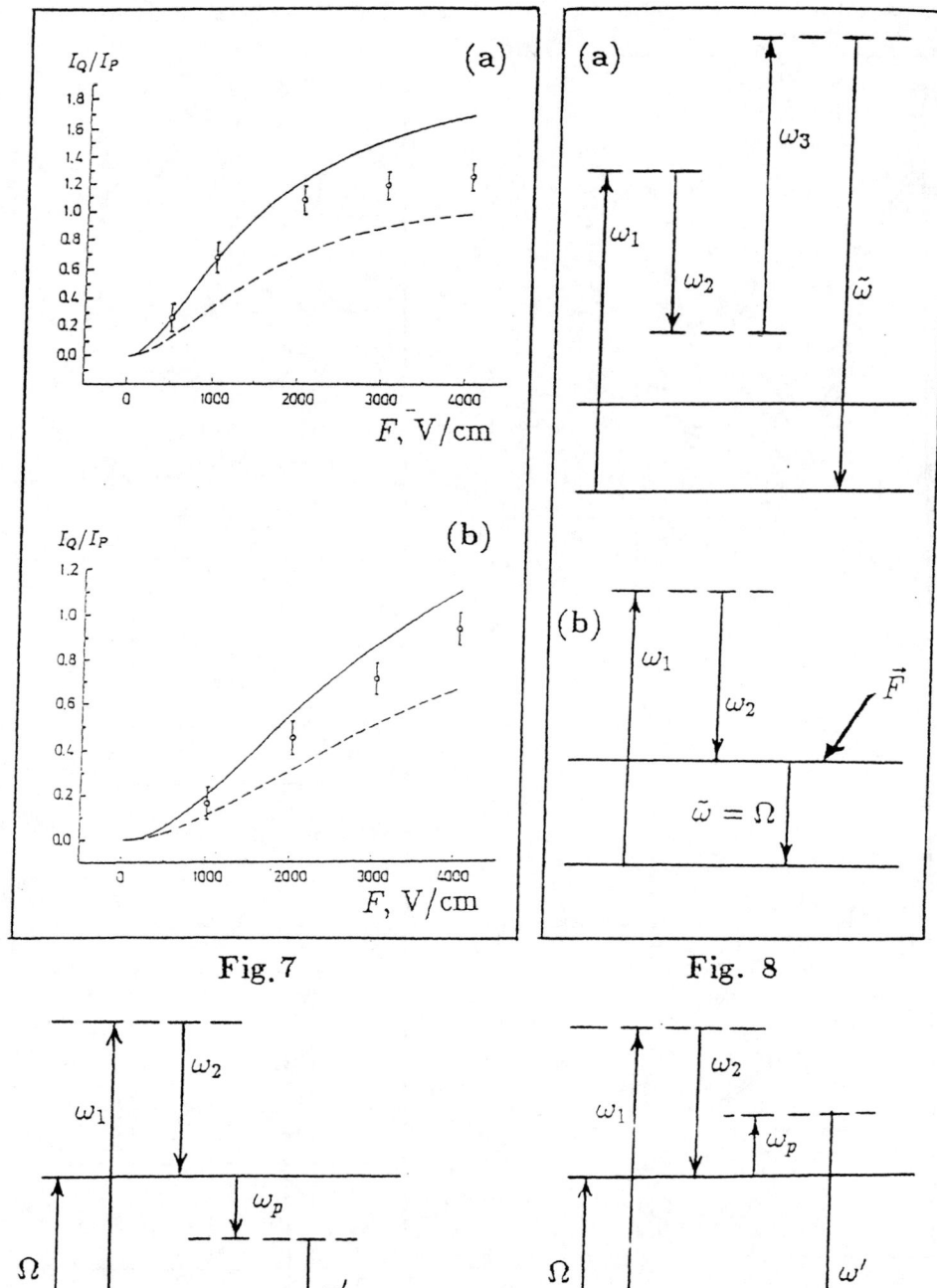

Fig. 7

Fig. 8

Fig. 9

Progress in Spectral Line Shapes and Shifts Evaluation of Experimental Stark Broadening Parameters

A. Lesage[*], N. Konjevic[†] and J. R. Fuhr[‡]

[*]Département d'Astrophysique Stellaire et Galactique (URA 335)
Observatoire de Paris 92195 Meudon CEDEX.
[†]Institut of Physics, 11080 Beograd P. O. Box 68 Yugoslavia
[‡]National Institute of Standards and Technology, Gaithersburg, Maryland 20899 USA

SUMMARY

Two volumes of the Bibliography on Atomic Line Shapes and Shifts [1,2] have been published, in 1993 and 1998, by J. R. Fuhr and A. Lesage, covering the period from July 1978 up to now. Here, we will describe the progress made in the field during the past twenty years. In particular, we will focus, more particularly, on Stark broadening and on the reason why such parameters are important for stellar atmosphere opacity calculations [3].

For recent work, we present an example, of the evaluation of experimental Stark widths and shifts of spectral lines of non-hydrogenic atoms and ions.

1. INTRODUCTION

The first article on spectral line shapes was published by Lord Rayleigh in 1889 : « On the limit to interference when light is radiated from moving molecules »[4]. It is a basic article on Doppler broadening. From that time up to the end of the word war I, 2-5 articles have been published every ten years. They were pioneering works, which opened that field of activity to the next generation of scientists, as is the case with the paper by Michelson[5] , « On the broadening of spectral line », which is the first article published on that topic.

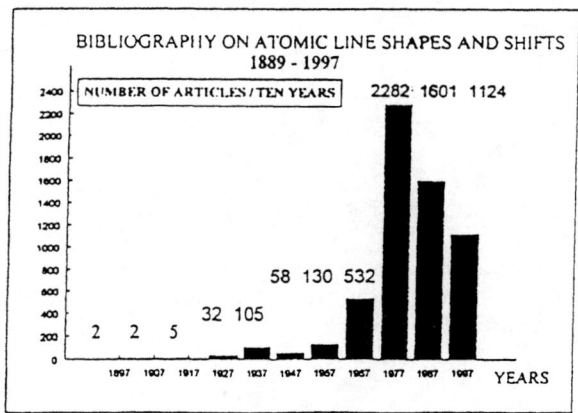

2. PROGRESS IN LINE SHAPES AND SHIFTS

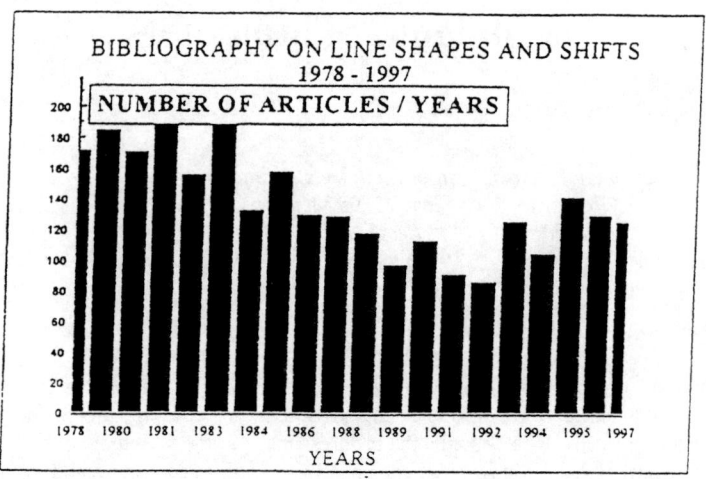

Fig. 1

The number of articles published by year, during the period 1978-1997 corresponding to supplements 4 and 5 of the bibliography is shown on Fig. 1. It is a proof of the intensity of the activity in that field. The main sections are: pressure broadening; Stark and van der Waals broadening; resonance, Doppler and natural broadening. There are specialized sections, such as instrumental broadening and applications, such as plasma diagnostics and astrophysics. The new sections which have been opened since Supplement 4 give a better idea of the advances being made in line broadening. They are:

- Dielectronic satellite
- Rydberg atoms
- Doppler free spectroscopy
- Plasma chemistry
- Redistribution of radiation
- Laser field induced broadening
- Charge exchange effects
- Line narrowing mechanisms
- Studies of regularities (important for astrophysical applications)
- Ion dynamic effects

Note that " non ideal plasmas" is not included in the list of general categories, though it is of great interest, especially for technical applications. However, this area is usually covered by other, above-mentioned categories.

3. STARK BROADENING

Reviews on Stark broadening are regularly published: in 1987, Wiese et al. have published " Ion broadening of heavy elements lines in plasmas"[6] , in 1988 H. Griem " Stark broadening in multielectron atoms and ions"[7], and Kelleher et al. in 1993 " Advances in plasma broadening of atomic hydrogen"[8]; Oks in 1993 " New ideas for Stark broadening and Stark narrowing in high density plasmas"[9]; Babin and Shapiro (1994) " Spectral line broadening due to the Coulomb interaction in plasmas"[10]; Redmer (1997) " Physical properties of dense, low temperature plasmas"[11], and Alexiou in 1997 reviewed the topic of " Stark broadening in weakly coupled plasmas" [12]. The most recent contribution to that field is Griem's Chapter IV in " Principle of Plasma Spectroscopy " [13].

4. ASTROPHYSICAL APPLICATIONS

Looking at the actual number of articles published during the past five years, we find that after « plasma diagnostics », « astrophysical applications » is, the next most common category. That is understandable, as astrophysical spectra has always been a major motivation for the study of spectral line broadening.

There are two review articles particularly devoted to astrophysical applications, which have been published during the period of time covered by Supplements 4 and 5 of the bibliography. M. Dimitrijevic wrote (in Russian) on " Stark broadening parameters of spectral lines needed for stellar and laboratory plasma investigations"[14]. Chantal Stehle wrote another on " Line shapes in astrophysics"[15], where is treated the case of spectra emitted by atoms (mainly hydrogen) or ions in stars' atmospheres and envelopes. She treats the case when the plasma temperature and density are such that the broadening is essentially attributed to interactions between the radiating species and the free plasma charges. She shows how the knowledge of stellar emergent Balmer line shapes, helps to deduce gravity, when other parameters, such as luminosity, are known.

The variety of plasma sources in the universe is very large and the range of temperature, density and gravity extend from the very faint H II region up to the very dense white dwarfs and neutron stars. Then there are articles which are not cited in the above mentioned reviews, which, nevertheless, represent significant progress and promising developments. This is the case of new theoretical calculations done by: N. F. Allard and J. Kielkopf[16] for Lyman α profiles in DA white dwarfs; by A. Beauchamp et al. [17] for helium lines in DB white dwarfs; by F. Paerels[18] in order to estimate if the pressure broadening of heavy ions could be observed in the atmosphere of neutron stars; by Van Regemorter et al. [19], on the broadening and shift of solar Rydberg lines in solar atmosphere.

<u>DA white dwarf</u> : These have a hydrogen-rich atmospheric composition whose spectra are characterized by broad band hydrogen lines. They have been the subject of several studies by B. Grabowsky et al.[20-22] by F. Allard and R. Wehrse [23], N. F. Allard and D. Koester[24] and by N. F. Allard and J. Kielkopf [16].

In their 1991 article, N. F. Allard and J. Kielkopf [16] have presented new theoretical calculations for the red wing of the Lyα profile. Close collisions with neutral and ionized hydrogen lead to the formation of pseudo-molecules H-H and H-H$^-$ with the appearance of satellite features near 1600 and 1400 Å. The calculations include multiperturber effects, which are responsible for the formation of H_3^- and H_3, with features near 1950 and 2600 Å. These calculations are used to determine the effective temperatures and surface gravity for the ZZ Ceti stars.

In their most recent (1995) article[25] the authors have experimentally confirmed the existence of satellites due to collision broadening of Ly_α and the validity of the unified theory line profile as a diagnostic for stellar atmospheres.

<u>DB white dwarfs</u>: These have a helium-rich composition, whose spectra are dominated by the lines of neutral Helium. In their 1997 article A. Beauchamp, F. Wesemael and P. Bergeron[17] have calculated 21 new He I Stark profiles and their forbidden components for electron density corresponding to the 10,000 to 40,000 K temperature range. Their calculations provide a very good match to the observed profiles of a large sample of white dwarfs, provided Stark broadening dominates other broadening mechanisms.

Neutron stars

F. Paerels[18] has suggested that at the very high density encountered in the neutron star's atmosphere, the pressure broadening of heavy ion absorption lines, due to the Stark effect, should be appreciable. He believes that these line widths should be measurable in the near future, with the next generation of X-ray spectrometers, if the heavy-element abundance is high enough to produce absorption features. In such a case, a new technique for measuring neutron star masses and radii, derived from that observation, could be used. That technique combines measurements of the gravitational red shift and of the acceleration of gravity at the surface, to produce separate estimations of the stellar mass and radius. (Paerels' calculations are nevertheless restricted to the case where the magnetic field is negligible.) From his pressure broadening calculation, he estimates that the O, and perhaps Fe ion absorption lines, should be resolved in the near future, (thanks to the increasing sensitivity of the X-Ray spectrometers) and then could provide the expected parameters.

Rydberg atoms

It is a new field of activity included in the Bibliography in the eighties. I. L. Beigman and V. S. Lebedev have written, in Physics Reports, a review on « Collision theory of Rydberg atoms with neutral and charged particles »[26].

The observations of many infrared lines in the solar spectrum (Kitt Peak; ATMOS, in spacelab) have stimulated the interest in the broadening theory of these kind of transitions between high Rydberg states. As shown by Chang and Schoenfeld [27] both the widths and the shifts of such lines are important as diagnostic tools.

Van Regemorter and Hoang-Binh, in their 1993 article have developed a « Stark broadening theory of solar Rydberg lines in the far infrared spectrum »[19]. They have shown that inelastic electron collisions usually dominate the width and that adiabatic ion collision always dominates the shift, giving the correct signs of the observed lines. Hoang-Binh[28] calculated the shifts of magnesium Rydberg lines in the solar infrared spectrum to reanalyze the shifts measured by Chang and Schoenfeld. He found (on the assumption that the shifts due to interaction with neutral hydrogen, 10^4 more abundant than electrons in the photosphere, are negligible) that the lines are formed in a photospheric layer between 130 and 350 Km at $1.5\ 10^2 < Ne < 6.6 \times 10^{12}$ cm^{-3}.

Neutral atom line broadening

A subject very close to the last topic (and fundamental for astrophysicists) is neutral line broadening. Lewis did a review of the « Astrophysical aspects of neutral atom line broadening »[29]. He has discussed several aspects of line shapes, which are relevant to astrophysical spectra: the impact broadening of strong resonance lines by atomic hydrogen; the effect on line profile of collisional redistribution of radiation in the impact region; the influence of velocity-changing collisions in the impact region.

Opacity and Stark broadening

Another astrophysical application of Stark broadening is the relationship between broadening parameters and stellar opacity. It has important consequences on the abundance of elements in the stars' atmospheres.

If we consider the opacity due to spectral lines in stellar spectra, the total intensity of an absorption line is related to different broadening mechanisms: the radiative and the collisionnal damping (Lorentz profile) and the Doppler (Gaussian profile).

The radiative width is related to the lifetime of the two levels of the transitions; its value varies over a very wide range around the « classical » value of 10^{-4} and can be as large as 10^{-3} or 10^{-2}. When levels above the ionization limit are auto ionizing, they are strongly broadened, with natural widths as large as several Angströms.

The collisional width is due to the interaction between atoms or ions and other particles, free electron in general; it is proportional to the perturber density and varies from one transitions to the other, increasing with upper level energy.

The fine and hyperfine structure, the isotopic shifts and the Zeeman effect (which splits one line into several components) have to be taken in account, since a blend of unresolved components behave as a profile for radiative transfer.

An example taken from T. Lanz, M. S. Dimitrijevic and M. C. Artru[30] shows: why Stark broadening parameters are important for stellar atmosphere opacity calculations.

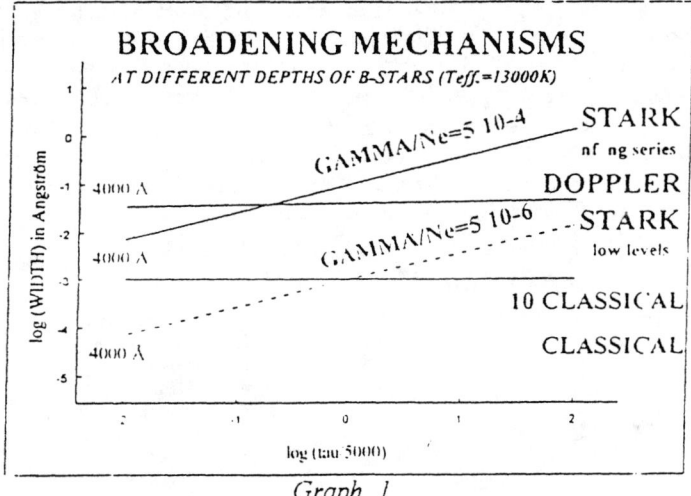

Graph. 1

On Graph 1 are shown the « classical » (10^{-3} Å) and ten times the « classical » (10^{-2} Å) values sometimes used for synthetic spectra to represent the Stark effect in the $\tau_{5000} = 1$ range of depth in stellar interior.

These values are compared to the real values calculated for the Si II transitions at 4000 Å in two cases: the $\gamma / Ne = 5 \times 10^{-6}$ value corresponds to the low excitation levels and $\gamma / Ne = 5 \times 10^{-4}$ corresponds to the highly excited levels.

Then it can be shown that the classical damping roughly represents the Stark effect in the first case and does not reproduce its real variation versus electron density along the atmosphere depth. In the second case, for lines which are more sensitive to the electric field fluctuation, it underestimates the width by about two orders of magnitude. In this case (for example Si II 4201 Å) the Stark width is comparable or even larger than the Doppler width around $\tau_{5000} = 1$.

The calculated value of the equivalent widths becomes very sensitive to the adopted Stark broadening as soon as the core begins to saturate. Then an accurate determination of the Stark width is essential, when it involves the high-excitation levels.

In the very particular case of the Stark broadening parameters of the Si II Mult. (1) lines, T. Lanz et al. have shown that, depending on the adopted value (available in the literature), the equivalent width could be reduced from 10% to 35%. The consequence on the abundance evaluation of Silicon in the stars atmosphere is an uncertainty of the order of 0.3 dex (a factor of two).

It is that fact and also the evaluation done by N. Konjevic and W. L. Wiese[31] in the 1990 Critical Review (they have shown that the scatter in the width data, published in the literature available at that time, was still quite large), which leads a group of scientists from Düsseldorf and Toulon University and the Observatory of Paris to plan new measurements of the astrophysically important silicon spectra. It is also for that reason that the example of Si II Stark broadening parameters will be chosen to show how the data are evaluated in the Critical Review.

5. EVALUATION OF EXPERIMENTAL DATA

The situation for the experiments, (up to those performed at Düsseldorf and including them), is shown on Graph. 2. Values measured by M. H. Miller[32] et al. and W. T. Chiang et al.[33] are almost a factor of two larger than those of N. Konjevic et al.[34], J. Puric et al. [35-37] and A. Lesage et al. [38]. Notice on Graph. 4 that the weakest line of the multiplet (λ 3854 Å) has been measured for the first time at Düsseldorf and that the λ 3862 Å line has been measured independently twice and simultaneously with the same experimental setup, using two lines of sight. All measured[39] values (widths of weak or stronger lines and width of the λ 3862 Å line) agree with each other, within the appropriate error bars, independently of the method of measurement.

Si II Mult. (1) Stark Width
EXPERIMENTS (1970-1997)

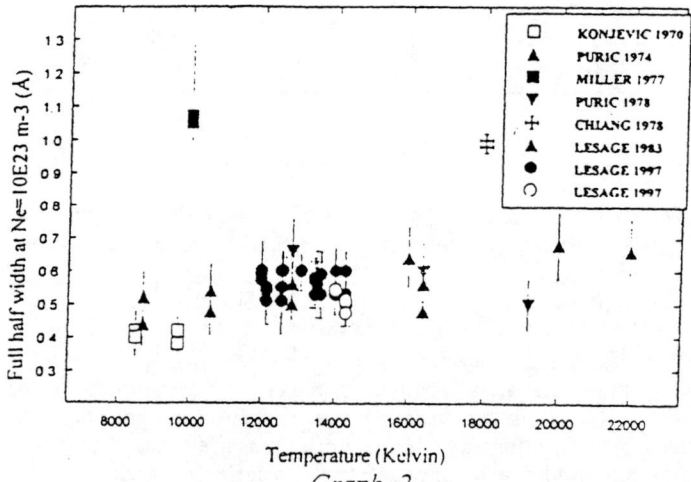

Temperature (Kelvin)

Graph. 2

Konjevic 1970, Ref. 34; Puric 1974, Ref. 36; Miller 1977, Ref. 32; Puric 1978, Ref. 37
Chiang 1978, Ref. 33; Lesage 1983, Ref. 38; Lesage 1997, Ref. 39.

Si II Mult. (1) Stark Width

CALCULATIONS:

Semi-Classical - Quant. Mech. - Semi-Empirical

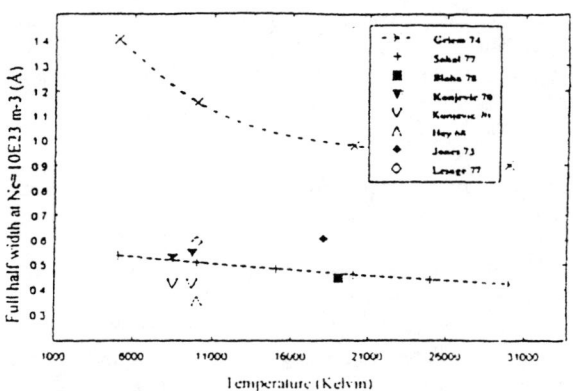

Graph. 3
Griem 74, Ref. 40; Sahal 77, Ref. 32; Blaha 78, Ref. 33; Konjevic 70, Ref. 34;
Hey 68, Ref. 41; Jones 73, Ref. 42; Lesage 77, Ref. 32.

Calculated values by H. R. Griem[40] and by S. Sahal[30,38] using the semi-classical theory, but with different kind of approaches, are shown on Graph. 3 along with semi-empirical calculated values by N. Konjevic et al.[31], J. D. Hey[41], W. Jones[42] and A. Lesage et al.[32]. All of them agree with the Sahal values rather than with those calculated by H. R. Griem. Notice that Blaha's value calculated with the fully quantum mechanical theory (mentioned in Chiang's article) agrees perfectly with that of S. Sahal.

Si II Mult. (1) Stark Width

EXPERIMENTS (1997) compared with CALCULATIONS

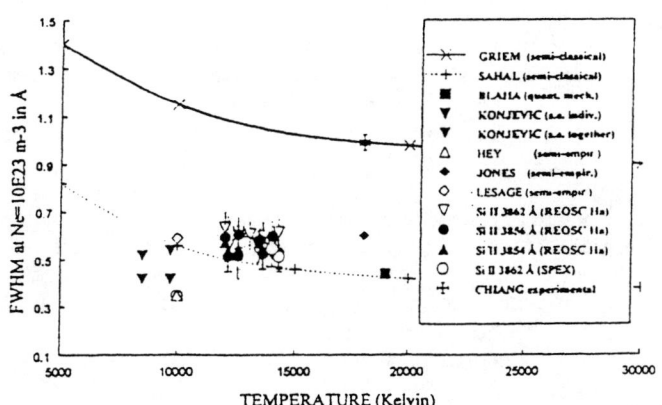

Graph. 4
References as for Graph. 2 & 3

Si II Mult. (1) line widths measured at Düsseldorf (Ref. 39) are detailed on Graph. 4 and compared with all available calculations.

The main criteria which are used to evaluate such kinds of experiments are: the plasma source homogeneity and reproducibility; the electron density and temperature methods of measurement; how are the different contributions to the observed line shape estimated and unfolded; how the profile is corrected for self absorption.

In the case of the above mentioned experiments for Si II Stark broadening parameters, our comments are as follows:

Plasma source and optical depth:

The kind of plasma source used in the above mentioned experiments, cannot explain the large discrepancy between high and low measured line width values for the Si II mult. (1) lines. For the high (M. H. Miller and W. T. Chiang), classical and electromagnetic shock tubes have been used, but these are the same kind of sources which have been used to measure the low values. Nevertheless, what can be noticed is that great care has been taken by N. Konjevic et al. to create the physical conditions which minimize self absorption. The optical depth has even been measured. The plasma homogeneity also has been tested carefully. J. Puric et al.(1978) and A. Lesage et al.(1983) have used the same source as N. Konjevic et al.. In the later case an alumina tube replaced the Pyrex tube and sapphire windows were used to avoid silicon evaporation from the walls, and silicon has been introduced in gaseous phase.

Before the experiments at Düsseldorf were performed, the data situation for Si II was not satisfactory. Compared to the previous experiments, those performed at Düsseldorf are the only ones which provide a line width value for the faintest Mult. (1) λ 3854 Å line, and this value is in close agreement with those of the two stronger lines. Consequently it provides the proof of a pertinent self absorption correction. Then it is likely, that the lower values have to be retained for the Si II mult. (1) Stark parameter, as the correct values.

Electron density:

M. H. Miller used Hβ and W. T. Chiang used the He I λ 3889 and 5016 Å line widths (separately calibrated with laser interferometry for Ne), while N. Konjevic and J. Puric used independent measurement with laser interferometers. But at Düsseldorf, a double wavelength interferometer was used.

Temperature:

It is not a critical factor, as line width varies slowly with temperature (except for temperatures below 10,000 K). The method of measurement varies according to the experiment from the line reversal technique applied to Hα line, to the absolute intensity of Ne I λ 5852 Å line and to the ratio of intensity of He I λ 3839 and 5016 Å lines to the continuum.

Unfolding procedure:

A priori, the method used by the various authors to obtain the Stark component from the observed line profile is not likely to introduce scatter among the measured values. But in fact, Biraud's [43,44] method, which introduces the concept of positivity and constraints in the deconvolution process, enables the measurement of faint lines, as the Mult. (1) λ 4854 Å line. Then it permits the extraction of the profile from the noise, when it is not possible by the traditional methods.

By the evaluation methods given in this Critical Review, one should be able to assess the reliability of Stark parameters.

6. REFERENCES

1. J. R. Fuhr and A. Lesage, Bibliography on Atomic Line Shapes and Shifts (July 1978 through March 1992) NIST Special Publication 366 Supplement 4, (1993). U. S. Government Printing Office, Washington DC.
2. A. Lesage and J. R. Fuhr, Bibliography on Atomic Line Shapes and Shifts (April 1992 through March 1998) Publication de l'Observatoire de Paris, (1998).
3. A. Lesage, Why Stark Broadening Parameters are Important for Stars Atmospheres Opacity Calculations. IAU XXIInd General Assembly (1994). Poster Abstracts JD 16.18. Hugo van Woerden editor.
4. Lord Rayleigh, Phil. Mag. 27; 298, (1889).
5. A. A. Michelson, Astrophys.J. 2, 251 (1895).
6. W. L. Wiese, D. W. Jones, "Spectral Line Shapes" Vol. 4, 3-21 (Ed. R. J. Exton, A. Deepak Publishing,Hampton, VA, (1987).
7. H. R. Griem, " Spectral Line Shapes" Vol. 5, 17-33 (Ed., J. Szudy, Elsevier, New York (1988).
8. D. E. Kelleher, W. L. Wiese, V. Helbig, R. L. Greene, D. H. Oza, Phys. Scr. T47, 75 (1993).
9. E. Oks, "Spectral Line Shapes", Vol. 7, 65-85 (Eds., R. Stamm & B. Talin, Nova Science Publishers, Inc., Commack, NY (1993).
10. S. A. Babin, D. A. Shapiro, Phys. Rep. 241, 119 (1994).
11. R. Redmer, Phys. Rep., 282, 35 (1997).
12. SD. Alexiou, "Spectral Line Shapes", Vol. 9, AIP Conf. Proc. 386, 79-98 (Eds., M. Zoppi & L. Ulivi, AIP Press, Woodbury, NY, (1997).
13. H. R. Griem, "Priciple of Plasma Spectroscopy", Chap. 4, Spectral line broadening, pages 54-131. Cambridge University Press NY (1997).
14. M. S. Dimitrijevic, Publ. Obs. Astron. Belgrade N 48, 127 (1995).
15. C. Stehlé « Line shapes in Astrophysics », Spectral Line Shapes. Vol. 8, AIP Conf. Proc. 328, 36-57 (Eds. A.David May, J. R. Drummond, Eugene Oks, AIP Press, New York 1995.
16. N. F. Allard and J. Kielkopf, Astron. Astrophys., 242, 133 (1991).
17. A. Beauchamp, F. Wesemael and P. Bergeron, Astrophys. J. Suppl. Series, 108, 559 (1997).
18. F. Paerels, Astrophys. J. 476, L47 (1997).
19. H. Van Regemorter, D. Hoang-Binh, Astron. Astrophys. 277, 623 (1993).

20. B. Grabowsky, J. Halenka and J. Madej « Gravitational and pressure redshift in white-dwarf stars « Proceeding of the Sixteenth Int. Cof. on Phenmena in Ionized Gases, Vol. 4, Contributed Papers, » 104-105 (Eds. W. Butticher, H. Wenk & E. Schulz-Gulde, Organ. Comm. ICPIG XVI, Dusseldorf, 1983).
21. B. Grabowsky, J. Halenka and J. Madej « Pressure shift in core and wings of α and ß Balmer lines and its relevance to the gravitationnal-redshift measurements in white dwarfs « Proc. 17th Int. Conf. Phenom. Ionized Gases » 987-988 (Budapest, 1985).
22. B. Grabowsky, J. Madej and J. Halenka, Astrophys. J. 313, 750 (1987)
23. F. Allard and R. Wehrse, J. Quant. Spectrosc. Radiat. Transfer 44, 209 (1990)
24. N. F. Allard and D. Koester, Astron. Astrophys., 258, 464, (1992).
25. J. F. Kielkopf and N. F. Allard, Astrophys. J., 450, L75 (1995).
26. I. L. Beigman and V. S. Lebedev, Phys. Rep. 250, 95 (1995).
27. E. S. Chang, W. G. Schoenfeld, Astrophys. J, 383, 450 (1991).

28. D. Hoang-Binh, Astron. Astrophys. **286**, 607 (1994).
29. E. L. Lewis "Lline Shapes, Vol. 6, AIP Conf. Proc. 216, 541-562 (Eds.,L. Frommhold & J. W. Keto, AIP Press, New York 1990).
30. T. Lanz, M. S. Dimitrijevic, and M. C. Artru, Astron. Astrophys., **192**, 249 (1988).
31. N. Konjevic and W. L. Wiese, J. Phys. Chem. Ref. Data, **19**, 1307 (1990).
32. A. Lesage, S. Sahal-Brechot, and M. H. Miller, Phys. Rev. A **16**, 1617 (1977).
33. W. T. Chiang and H. R. Griem, Phys. Rev. A **18**, 1169 (1978).
34. N. Konjevic, J. Puric, Lj. Circovic, and J. Labat, J. Phys. **B 3**, 999 (1970).
35. J. Puric, S. Djenize, J. Labat, and Lj. Cirkovic, Phys. Lett. **45A**, 97 (1973).
36. J. Puric, S. Djeñize, J. Labat, and Lj. Cirkovic, Z. Phys. **267**, 71 (1974).
37. J. Puric, A. Lesage, and V. Knezevic, in *Proceeding of the Ninth Summer School-and Symposium on the Physics of Ionizied Gases, Dubrovnik*, Edited by R. K. Janev (Institut of Physic, Beograd, 1978).
38. A. Lesage, B. A. Rathore, I. S. Lakicevic and J. Puric, Phys. Rev. A **28**, 2264 (1983).
39. F. Wollschlager, J. Mitsching, D. Meiners, M. Depiesse, J. Richou and A. Lesage, J. Quant. Spectrosc. Radiat. Transfer **58**, 135 (1997).
40. H. R. Griem, *Spectral Line Broadening by Plasmas*, Academic, New York (1974).
41. J. D. Hey, J. Quant. Spectrosc. Radiat. Transfer, **18**, 425, (1977).
42. W. Jones, Phys. Rev., **A7**, 1826, (1973).
43. Y. Biraud, Astron. Astrophys., **1**, 124 (1969)
44. M. Depiesse, Y. Biraud, A. Lesage and J. Richou, J. Quant. Spectrosc. Radiat. Transfer, **54**, 539 (1995).

HIGH TEMPERATURE/DENSITY PLASMAS

Line Shape Measurements Employing Independently Diagnosed Pinch Plasmas

S. Büscher*, Th. Wrubel*, I. Ahmad[†], and H.-J. Kunze*

*Institute for Experimental Physics, Ruhr-University-Bochum
44780 Bochum, Germany
[†]Department of Physics, Quaid-i-Azam University, Islamabad, Pakistan

Abstract. Line shape studies for comparison with theoretical calculations rely on well-diagnosed plasmas. At Bochum we have utilized for some time the gas-liner pinch facility, where the plasma parameters are obtained independently from spectroscopic measurements by collective Thomson scattering. Both spectroscopy and laser scattering allow spatially resolved investigations. Characteristics and potentials, but also deficiencies, of the device are discussed for conditions where the plasma is stable or unstable.

Specific attention is paid to recent investigations. The profile of the HeII Balmer-alpha line was studied at densities from 0.2×10^{18} to $3 \times 10^{18} \mathrm{cm}^{-3}$ and electron temperatures from 5 to 35 eV, where two different plasma formation regimes could be created: a high density and a high temperature mode. The Stark width shows a $n_e^{2/3}$ dependence, and it is compared with different theoretical approximations. Attempts are reported to study the broadening of the 2s-2p doublet of CIV, a possible benchmark experiment, where semiclassical and quantum-mechanical calculations differ by a factor of two. In order to achieve the necessary resolution a visible spectrograph flushed with argon with a suitable grating of 3600 lines/mm is employed. The detector is an ICCD camera with a MgF_2 window.

INTRODUCTION

The contributions to this bi-annual conference series vividly reflect the advances in plasma broadening of spectral lines made both in theory and experiment. Improved calculations specifically for lines of higher charge states on the one hand, and new calculations in the pure quantal approach on the other hand, challenge the experimentalists to improve the reliability of their measurements. This is certainly important where discrepancies between theories and experiment still exist, and it is crucial where large discrepancies between theoretical calculations show up. Benchmark experiments are asked for. Our present knowledge, and existing problems in this field, are critically analyzed and discussed by Griem in his review at this conference [1].

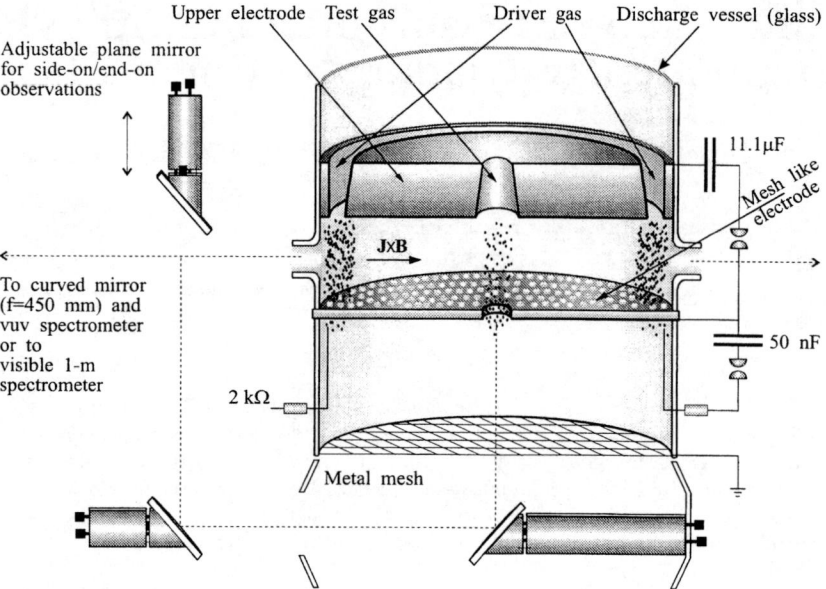

FIGURE 1. Schematic of the discharge vessel. The mirrors shown give the opportunity to investigate the plasma also end-on.

A useful plasma source for such studies is the gas-liner pinch first introduced at the conference in Williamsburg in 1986 [2] and described in detail in, e.g., Ref. [3]. The present device is discussed briefly, and attention is paid specifically to the independent diagnostic by Thomson scattering and spatial resolution of the measurements. The characteristics of the setup and current measurements of the Balmer-α line of HeII are presented. The discussion of problems in the context of some recent measurements of the resonance transition in CIV concludes the presentation.

EXPERIMENTAL SETUP AND DIAGNOSTICS

The device. In principle, the device is a large-aspect-ratio z-pinch with two fast acting electromagnetic valves as gas inlet system. The first valve injects a gas stream of hydrogen (so-called driver gas) into the vacuum vessel as a concentric hollow gas cylinder near the wall. The second valve injects the gas of spectroscopic interest (so called test-gas) along the z-axis of the discharge (Fig. 1). The main capacitor bank (11.1 μF) is charged to about 35 kV. A preionization system below the lower electrode reduces macroscopic instabilities and thus improves the reproducibility of the plasma, which is checked by the continuum emission at 520 nm with a 1/4 monochromator. After maximum compression a stable plasma column of about 1 to 2 cm in diameter and 5 cm in length is formed, if the test-gas concentration is kept at levels of about 1% or lower.

One spectrograph each is used for scattering diagnostics and spectroscopy, both viewing the plasma column through opposite ports. For the spectroscopic investigation a vuv-spectrograph (McPherson model 225) covering the spectral range down to 30 nm and a 1-m spectrograph (Spex M1000) for the visible spectral range can be chosen. The spectral range of the visible spectrograph was extended down to at least 150 nm, because the transitions under investigation are in the vacuum ultraviolet, and the grating of the vuv-spectrograph did not provide the necessary resolution. Therefore, the spectrograph for the visible spectral range was equipped with a holographic grating (3600 l/mm) which is optimized for the wavelength region of 150 nm $< \lambda < 450$ nm. The radiation is imaged with a 1:1 MgF_2 optics onto the entrance slit of the spectrograph (50 μm), which is equipped with an ICCD-camera with MgF_2 entrance window. The whole line of sight including the spectrograph was flushed with argon in order to prevent absorption by oxygen and nitrogen molecules of air.

Thomson scattering. The plasma is diagnosed by Thomson scattering employing a ruby laser system of 50 MW with a pulse duration (FWHM) of 25 ns. A fast FND-100 photodiode monitors the time evolution of the laser pulse. The laser beam is focused to a cross section of 0.7 mm^2 at the center of the discharge vessel. A 1-m spectrograph (Spex model 1704) detects the scattered radiation at an angle of 90°. The linear reciprocal dispersion is 5.83 pm/pixel in second order with a 1200 lines/mm grating blazed at 1000 nm and employing an ICCD camera (Princeton Instruments 578×384 pixel). The apparatus profile has a width of 2.5 pixels (FWHM). With an OMA system (EG&G model 1456-990G) instead of the ICCD camera, the apparatus profile has a width of 3.5 pixel (FWHM), the dispersion being 6.3 pm/pixel.

The standard application with the OMA system as detector yields the plasma parameters (i. e. density and temperature of the electrons, test-gas ions and driver-gas ions) in the center of the device as described previously [4]. Employing the ICCD camera instead of the OMA system gives the plasma parameters radially resolved [5,6]. The spatial resolution is 60 μm due to the pixel size, taking into account the apparatus profile. However, to obtain a usable scattering signal-to-noise ratio 20 neighboring channels are averaged.

Radial resolution. This setup was used to check the radial homogeneity of the discharge over the central part of the plasma of 1.4 cm in diameter [5,6]. The good localization of the test-gas particles in the center of the discharge was verified again in this way: With argon as test-gas, the respective ions are confined to a region smaller than 3 mm in a homogeneous plasma region (diameter of 1 cm), so that no cold boundary layers are present, thus avoiding self-absorption.

Axial resolution. Using krypton as test-gas the dependence of the rise of instabilities on the amount of injected test-gas was investigated [7]. In this investigation the electron density along the z-axis was determined spatially-resolved via Stark width measurements of three transitions of the 5s $^5S^o$ - 5p ^5P Kr III multiplet. An amount of less than 1% test-gas particles ensures the absence of instabilities and turbulences, especially when the investigation takes place after maximum pinch compression. An unstable plasma condition could be cultivated if injecting 5% or more test-gas. One goal of those studies was to look for modifications of line shapes by turbulence and instabilities as suggested in [8].

HEII BALMER-ALPHA LINE

For the fundamental theoretical understanding of Stark broadening, hydrogen-like ions are still of interest. Especially for ionized helium the situation is not satisfying, because Stark widths of some transitions fit very well with calculations while others fail [9]. Recent progress has been made in the treatment of the HeII P_α line by including an improved electron collision operator into the FFM (frequency fluctuation model) calculations [10,11].

We measured the HeII Balmer-α line at the wavelength $\lambda = 164.0$ nm. Former measurements of this line by Böddeker employed a vuv-spectrograph with an open MCP [12]. In order to avoid uncertainties regarding ion feedback and to increase the resolution we employed the argon flushed 1-m spectrograph originally designed for the visible spectral range. The linear reciprocal dispersion is 5.88 pm/pixel in first order, with an apparatus profile of 2.5 pixels width (FWHM). The amount of helium (test-gas) was kept below 1% to avoid inhomogeneities and to keep the optical thickness low.

We achieved two different plasma confinements by changing the gas pressure of the driver gas and the voltage of the main capacitor bank. The first condition is rather cold (5-15 eV) and dense ($0.5\text{-}3\times 10^{18}$cm^{-3}), while the other is hotter (up to 36 eV) but with lower density (up to 1.3×10^{18}cm^{-3}). Table 1 gives electron density and temperature for each discharge for times when line profiles were recorded. They represent average values of 10 to 15 measurements. For both conditions the line

TABLE 1. Electron density and temperature together with the experimental Stark width (FWHM) of the HeII Balmer-α line.

$n_e/10^{18}$cm^{-3}	$k_B T_e/eV$	Stark width/Å
0.46	5.17	1.20
0.73	6.74	1.61
1.05	10.86	2.02
1.64	13.18	2.53
2.30	11.67	3.17
2.81	14.97	3.32
0.22	10.94	1.20
0.39	16.63	1.44
0.61	21.16	1.75
0.96	30.78	2.38
1.34	35.75	2.58

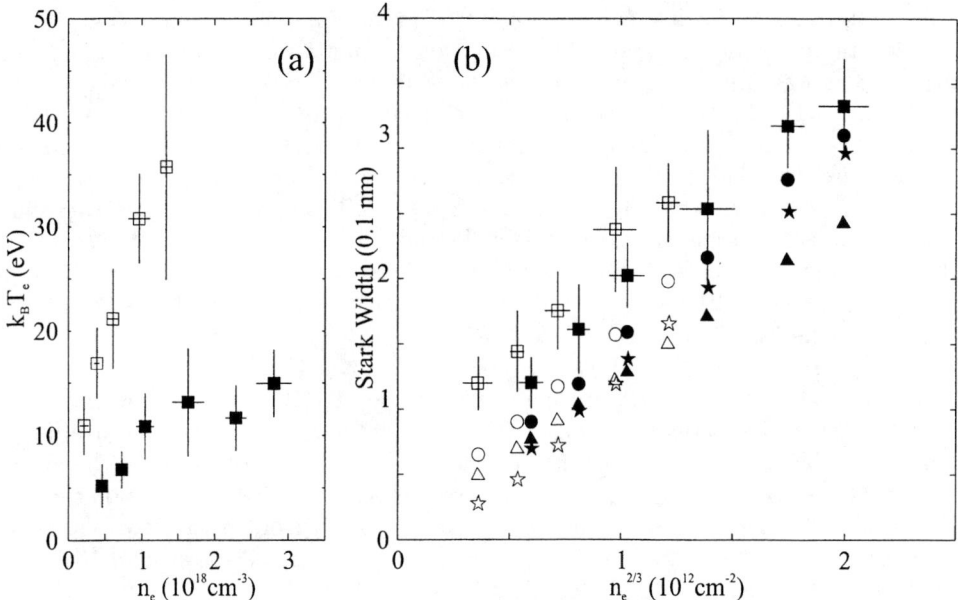

FIGURE 2. (a) mirrors the difference between the hot (hollow symbols) and the cold (filled symbols) plasma conditions. (b) gives the corresponding experimental Stark width (*hollow and filled squares*) vs. electron density (i.e. $n_e^{2/3}$) together with theoretical calculations for the experimental plasma parameter according to Godbert (FFM, *hollow and full circles*, [13]), Könies (MMM, *hollow and full triangle*, [15]) and Kepple (*hollow and full stars* [17]).

emission was homogeneous along the z-axis of the discharge (at least over 8 mm), as could be seen from each ICCD spectrum. Since the Stark width does not change along the z-axis, the spectrum was averaged over the height to reduce the noise. The resulting spectra were fitted with Voigt profiles consisting of numerical profiles to account for Stark- and Doppler broadening and the apparatus function. The experimental Stark widths are also given in Tab. 1.

The experimental data are plotted in Fig. 2. For both plasma conditions they show a $n_e^{2/3}$ dependence, as theoretically expected. However, the widths of the hotter condition are systematically larger than the widths measured at the lower temperatures, but overlap in the range of their error bars. In contrast to the cold condition, a fit through the data of the hot condition does not pass through the origin, which might be attributed to a temperature effect.

Figure 2 also displays preliminary theoretical calculations [13] following the FFM model [14] and the Model Microfield Method (MMM [15,16]). Both calculations were performed for the specific experimental plasma parameters and predict a smaller width, the widths of [15] differing somewhat more from the experimental data. For both theories the agreement is better in the low temperature regime (full symbols, 5-25 % discrepancy for [13]) but becomes more than 30-40 % for the high temperatures. The MMM calculation accounts for ion dynamics and fine structure and assumes hydrogen ions as perturbers. The preliminary calculations according to FFM also includes ion dynamic effects and fine structure splitting; for details see [10,11]. While the widths of [15] show almost no change due to the temperature, the widths of [13] slightly increase with increasing temperature. In Fig. 2 also the results of calculations by Kepple [17] are shown, which treat the electron impact contribution semiclassically and the ions quasistatically. The tabulated theoretical data were linearly scaled to the experimental plasma parameters and underestimate the experimental Stark width, especially for low densities. In contrast to the FFM model, the width of Kepple [17] becomes smaller with higher temperatures.

Measurements of the shift are currently prepared. However, a reference line for a direct wavelength calibration in the range of HeII Balmer-α is not available. Another problem is the relatively low intensity of the line emission using the argon flushed visible-spectrograph. Therefore, measurements in second order suffer under the continuum radiation at 328 nm in first order. This problem can not be solved by interference filters (which have a transmission of less than 30 % around 164 nm), due to the low intensity of the HeII Balmer-α line.

STARK BROADENING OF 2S-2P IN CIV

Recent measurements of the resonance transition in BIII by Glenzer [18] showed good agreement with some semi-classical calculations [19–21]. However, this experimental value differs from close-coupling calculations of Sanchez et al. [22] and Seaton [23] by a factor of about 2. Since for resonance transitions broadening by electrons is mostly due to contributions of small angular momentum L in a partial wave expansion — where the semiclassical treatment is near to break down — close-coupling calculations are thought to give a priori the most accurate values. This situation gave rise to many discussions (see e.g. [24,8]) and encouraged new close-coupling calculations which confirmed the former ones [8]. A more detailed analysis of this problem is given by Griem in these proceedings [1].

We attempted to study the broadening of the analogous resonance transition in CIV, since the introduction of gaseous boron compounds into the pinch always lead to serious contaminations. This benchmark experiment, however, is faced with several experimental problems: The transitions at wavelengths of $\lambda = 154.813$ (J=3/2-1/2) and $\lambda = 155.071$ nm (J=1/2-1/2) are in the vacuum ultraviolet. We used the setup described above and recorded a linear reciprocal dispersion of 2.68 pm/pixel in second order. This high spectroscopic resolution is necessary because of the small expected width of the resonance line for the plasma parameters of the gas liner pinch plasma. Furthermore, one has to resolve the profile to a high degree in order to separate Stark broadening from contributions by Doppler broadening. Another intrinsic problem is the optical thickness of this resonance transition. In

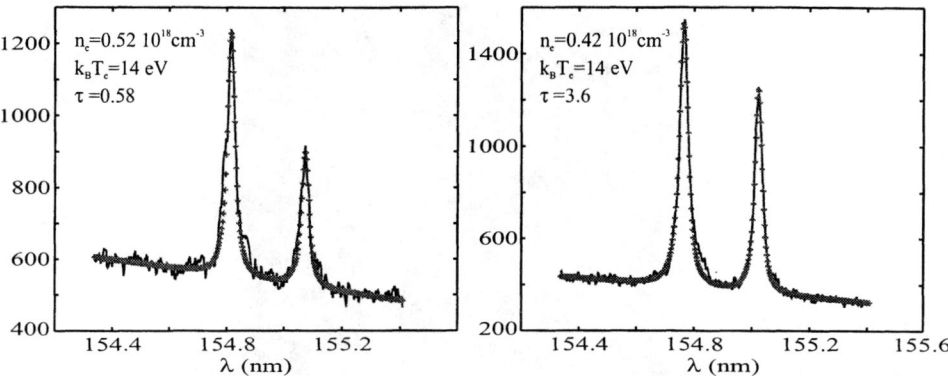

FIGURE 3. Examples of two spectra of the 2s-2p CIV resonance transition which are averaged over a distance of 5 mm along the z-axis. The simultaneously measured plasma parameters are given in the figure.

order to achieve low optical thickness we installed a new discharge vessel and used very small amounts of CH_4 as test-gas.

Employing this equipment we succeeded in detecting the resonance line doublet with good intensity, examples of spectra in second order averaged over an axial distance of about 5 mm are given in Fig. 3. The doublet transition is fitted with a theoretical profile accounting for a small optical thickness, Doppler- and Stark-broadening and the apparatus profile [25]. This fitting procedure also yields the optical thickness of the transition. Depending on the discharge condition we found varying optical thickness of the transition from discharge to discharge. Although the theoretical profile matches the experimental line shape satisfactorily, a closer look shows a deformation of the line shape: It seems to consist of a superposition of slightly frequency shifted spectra. In fact, some spectra recorded before and at maximum compression clearly show two such shifted components, in some cases even more. At a time of 100 ns after maximum pinch compression this shift is unresolvable, but every multiplet component shows up again twice at late time in the decompression phase of the plasma. In addition, inspection of the spatially resolved ICCD spectra reveals that the shift of the spectral line is connected with the axial position of the pinch column. Figure 4 shows an example of an ICCD spectrum along the axis recorded shortly after maximum compression. Taking the mean of this ICCD spectrum along the z-axis, the averaged spectrum consists of two parts shifted slightly against each other.

These shifts are due to radial motion of the carbon ions caused by plasma compression. It seems that injecting carbon as test-gas leads to a particular plasma condition. Carbon ions are always present in our plasma, since carbon is an unavoidable contamination of the chamber wall. Therefore, the CIV line is still detectable without injecting methane as test-gas. Nearly all such spectra measured without puffing in test-gas show Doppler shifts. In this situation the carbon ions are

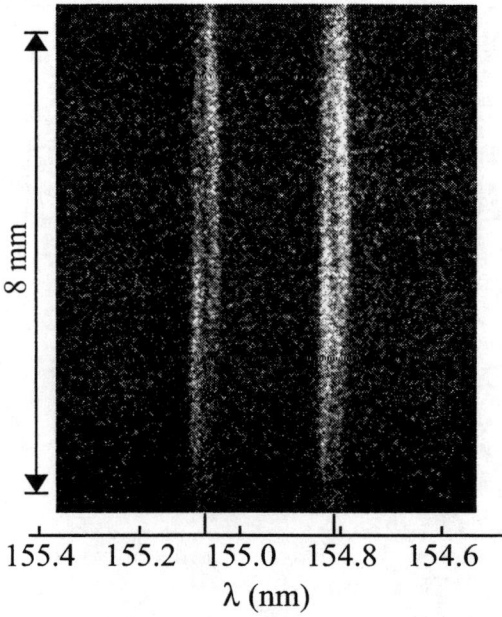

FIGURE 4. The ICCD camera provides spatial resolution along the z-axis. The plasma forms two spatially separated regions which give different frequency shifts of the transition.

localized in the outer regions of the compressing (or decompressing) plasma shell, where the compression velocity is high. Therefore we conclude that the injected test-gas diffuses before the incoming plasma ring sweeps up the carbon ions which are, therefore, not localized in the resting center of the plasma. In fact, analyzing the Doppler shifts of the spectra reveals that the corresponding velocity is equal to the velocities measured when studying the pinch dynamics by Thomson scattering [6].

Since the measured spectra suffer from these additional Doppler shifts, we unfortunately so far cannot give the Stark width of the 2s-2p transition in CIV. Nevertheless, we like to point out that we are capable to recognize if additional Doppler shifts are present or not in the experiment. This is made possible by the improved diagnostic [5] and the spatial resolution when performing the spectroscopic observations simultaneously. In general, a linear density dependence of the Stark width matching zero when extrapolating to zero density, is a strong hint that the plasma is at rest and that no additional Doppler shifts, originating from plasma motions due to turbulence and instabilities, are present. In addition, shifts occur mostly in combination with inhomogenous line emission along the z-axis, which can be recognized easily in the ICCD spectra. Situations where inhomogeneities occur can be intentionally produced when preparing special discharge conditions [7].

In order to obtain a Stark width, a discharge condition has to be found where no shifts occur. This can be done by changing the way of injecting the test-gas or changing the pressure of the gas or even the gas itself. Measurements along the axis (end-on observation) should also be done, in which case the component of the decompression velocity in the direction of the observer is zero.

CONCLUSION

The gas-liner pinch is a well suited plasma source to study Stark broadening of spectral lines. The density of the plasma covers an intermediate density regime so that on the one hand Stark broadening dominates other broadening mechanisms and on the other hand the plasma can be diagnosed independently by Thomson scattering. Stark broadening of transitions in hydrogenlike ions and of isolated lines have been studied, encouraging new calculations and refining theory [1].

In the present investigation we studied the profile of the HeII Balmer-α line and compared it with some theoretical calculations, showing fair agreement with preliminary FFM calculations. In another investigation of the broadening of the narrow 2s-2p doublet in CIV the experiment was driven to its limit with respect to spectroscopic resolution and characteristics of the device. When injecting CIV ions, motions originating from the compression velocity of the plasma give additional Doppler shifts deforming the narrow line shape of this line so that no Stark width could be obtained. Nevertheless, these additional Doppler shifts — which are only crucial for very narrow spectral lines — can be recognized employing refined scattering diagnostics and spatially resolved spectroscopic observations.

ACKNOWLEDGEMENT

These studies were supported by the Sonderforschungsbereich 191 (A6) of the DFG. One of us (I.A.) thanks the DAAD for support.

REFERENCES

1. H. R. Griem, *this Proceedings*, ICSLS Pennsylvania, USA, 1998
2. H. J. Kunze, in *Proc. 8th Int. Conf. on Spectral Lineshapes*, Williamsburg, Virginia, 1986, Vol. 4, ed. R. J. Exton, A. Deepak Publishing, pp. 23-35 (1987)
3. K. H. Finken and U. Ackermann, J. Phys. D: Appl. Phys. **16**, 773 (1983)
4. Th. Wrubel, S. Glenzer, S. Büscher, and H.-J. Kunze, Astron. Astrophys. **306**, 1023 (1996)
5. Th. Wrubel, I. Ahmad, S. Büscher, and H.-J. Kunze, in *Proceedings of the 4th International Conference on Dense Z-Pinches, May 28-30, 1997, Vancouver, CP409*, eds. N.R. Pereira, J.Davis, and P.E. Pulsifer (AIP Press 1997), p. 455
6. Th. Wrubel, S. Büscher, I. Ahmad, and H.-J. Kunze, *Proceedings of 11th APS Topical Conference on Atomic Processes in Plasmas*, Auburn (AIP press 1998), in print

7. I. Ahmad, S. Büscher, Th. Wrubel, and H.-J. Kunze, submitted to Phys. Rev. E
8. H. R. Griem, Yu. Ralchenko and I. Bray, Phys. Rev. E **56**, 7186 (1997)
9. S. Büscher, S. Glenzer, Th. Wrubel, and H.-J. Kunze, J. Phys. B: At. Mol. Opt. Phys. **29**, 4107 (1996)
10. T. Meftah, S. Alexiou, A. Calisti, L. Godbert, R. Stamm, and B. Talin, AIP Conference Proceedings # **386**, Spectral Line Shapes, eds. M. Zoppi and L. Ulivi (AIP Press, New York 1998) Florenz, Italy, Vol. 9, p. 31 (1997)
11. L. Godbert, A. Calisti, R. Stamm, and B. Talin, *Proc. 11th ICSLS*, Marseilles, 1992, eds. R. Stamm and B. Talin, p. 155 (1994)
12. St. Böddeker, Dissertation, unpublished, Bochum (1995)
13. L. Godbert-Mouret, private communication, (1998)
14. A. Calisti, F. Khelfaoui, R. Stamm, B. Talin, and R. W. Lee, Phys. Rev. A **42**, 5433 (1990)
15. A. Könies, private communication, (1998)
16. A. Könies, S. Günter et al., J. Quant. Spectrosc. Radiat. Transfer, in print (1998)
17. P. C. Kepple, Phys. Rev. A **6**, 1 (1972)
18. S. Glenzer and H.-J. Kunze, Phys. Rev. A **53**, 2225 (1996)
19. H. R. Griem, *Principles of Plasma Spectroscopy*, Cambridge Univ. Press, Cambridge (1997)
20. J. D. Hey and P. Breger, J. Quant. Spectrosc. Radiat. Transfer **24**, 349 (1980)
21. J. D. Hey and P. Breger, J. Quant. Spectrosc. Radiat. Transfer **24**, 427 (1980)
22. A. Sanchez, M. Blaha, and W. W. Jones, Phys. Rev. A **8**, 774 (1973)
23. M. J. Seaton, J. Phys. B **21**, 3033 (1988)
24. S. Alexiou, S. Glenzer, and R. W. Lee, private communication
25. Th. Wrubel, S. Glenzer, S. Büscher and H.-J. Kunze, AIP Conference Proceedings # **386**, Spectral Line Shapes, eds. M. Zoppi and L. Ulivi (AIP Press, New York 1997) Florenz, Italy, Vol. 9, p. 71

X-ray Spectroscopy from Fusion Plasmas

S. H. Glenzer

Lawrence Livermore National Laboratory, L-399, P.O. Box 808, Livermore, CA 94551

Our understanding of laser energy coupling into laser-driven inertial confinement fusion targets largely depends on our ability to accurately measure and simulate the plasma conditions in the underdense corona and in high density capsule implosions. X-ray spectroscopy is an important technique which has been applied to measure the total absorption of laser energy into the fusion target, the fraction of laser energy absorbed by hot electrons, and the conditions in the fusion capsule in terms of density and temperature. These parameters provide critical benchmarking data for performance studies of the fusion target and for radiation-hydrodynamic and laser-plasma interaction simulations. Using x-ray spectroscopic techniques for these tasks has required its application to non-standard conditions where kinetics models have not been extensively tested. In particular, for the conditions in high density implosions, where electron temperatures achieve 1 - 2 keV and electron densities reach 10^{24} cm^{-3} evolving on time scales of < 1 ns, no independent non-spectroscopic measurements of plasma parameters are available. For these reasons, we have performed experiments in open-geometry gas bag plasmas at densities of 10^{21} cm^{-3} and which are independently diagnosed with Thomson scattering and stimulated Raman scattering. We find that kinetics modeling is in good agreement with measured intensities of the dielectronic satellites of the He-β line (n=1-3) of Ar XVII. Applying these findings to the experimental results of capsule implosions provides additional evidence of temperature gradients at peak compression.

INTRODUCTION

In the indirect drive approach to inertial confinement fusion [1], gas-filled hohlraums are used as radiation enclosures converting high-power laser energy into soft x rays to achieve a symmetric ablation pressure on the fusion capsule and a symmetric high convergence capsule implosion [2]. Present ignition designs for the future National Ignition Facility (NIF) show a significant yield of ~16MJ when absorbing more than 1MJ of laser energy into the hohlraum [2,3]. For this reason, it is crucial to understand backscattering losses by stimulated Brillouin (SBS) and stimulated Raman (SRS) scattering [4,5] since these processes can scatter a significant fraction of the laser energy away from the fusion plasma. Present experiments therefore measure the absolute scattering losses by these processes in addition to the total x-ray production in hohlraums applying various beam smoothing techniques. Smoothing of the laser beams have been shown to reduce laser scattering losses by suppressing filamentation and hot spots. This result has led to more than 95% absorption of the laser energy in gas-filled hohlraums [6]. Absolute x-ray spectroscopic measurements have shown that 80-90% of the laser light is then converted into x rays which drive the implosion of the fusion capsule.

A small fraction of the laser energy, however, is absorbed into hot electrons [7,8,9]. These electrons have energies in the range of 20 - 40 keV preheating the fusion capsule which in turn results in a reduced capsule compressibility. Calculations show that the time-integrated fraction of hot electrons in an ignition hohlraum must be smaller than 15% (6%) for 20 keV (40 keV) electrons to avoid a reduction of the capsule gain [2]. Applying x-ray spectroscopic measurements of line radiation [10-15] from small tracers have shown that the local hot electron fraction in gas-filled hohlraums is strongly correlated with stimulated Raman scattering [16]. Electron Landau damping of the electron plasma waves which are excited by the stimulated Raman process [17] is the physical mechanism by which these hot electrons have been produced. A peak hot electron fraction of 10% (2% time integrated) has been observed with unsmoothed laser beams indicating the importance to suppress stimulated Raman scattering in fusion targets [16] through the use of beam smoothing techniques as planned on the NIF.

Capsule implosions driven by x rays from high-Z hohlraums converge to a final radius ~10 smaller than the original radius of about 275 μm in present indirect drive experiments. Consequently plasma conditions similar to those of stars of extremely high densities 10^{24} cm^{-3} and temperatures 1 - 2 keV have been produced [18]. The characterization of these dense plasmas requires x-ray or neutron diagnostics. In particular, the spectrum of the He-β line (n=1-3) of Ar XVII with its dielectronic satellites has been found to be a valuable diagnostic of electron densities and temperatures. This line is Stark broadened so that densities can be inferred from the width of the spectral line, and the upper levels of the dielectronic satellites on the red wing of the He-β line are populated by dielectronic recombination so that their relative intensity is sensitive to the electron temperature. The contribution of the satellites is especially important at earlier times in the implosion where temperatures are below 1 keV so that the satellites need to be self-consistently included in the fit of the whole line shape with a Stark-broadening code coupled to a kinetics model. Recent temporally resolved measurements [19] of the spectrum of the He-β line plus satellites have resulted in a density of 10^{24} cm^{-3} and peak temperatures of ~1 keV which are somewhat lower than calculated with two-dimensional hydrodynamic modeling using the code LASNEX. For this reason we have performed new experiments in open-geometry gas bag plasmas [20-22] at densities of 7×10^{20} cm^{-3} and 10^{21} cm^{-3}. These gas bag plasmas are independently diagnosed with spectroscopy [22], Thomson scattering [23], and stimulated Raman scattering [20,24]. In particular, the Thomson scattering diagnostics has been developed to a high degree of accuracy in these fusion plasmas. It is now possible to measure electron temperatures T_e and densities n_e, ion temperatures T_i, macroscopic flow velocities $\mathbf{v}$, and the averaged ionization stage Z. Possible effects due to turbulence as suggested in Ref. [25] can be clearly ruled out from the experimental Thomson scattering spectra [26]. We find that the kinetics modeling is in good agreement with the measured intensities of the dielectronic satellites of the He-β line (n=1-3) of Ar XVII for the two different electron densities. This result verifies the kinetics modeling used for the interpretation of implosion spectra. It further indicates that the somewhat lower temperature from the satellites might be a combined effect of spatial gradients in the fusion capsule and of the weak intensity of the satellites at peak temperature where Stark broadening introduces great difficulties in determining their intensity.

HOHLRAUM COUPLING EXPERIMENTS

Cylindrical gold hohlraums of 2,750 μm length and 800 μm radius were heated in experiments at the Nova Laser Facility at the Lawrence Livermore National Laboratory [27] applying various beam smoothing techniques. These hohlraums are of the standard size for capsule implosions [18,28] and present benchmarking experiments [29-31] at Nova. On each side five laser beams enter the hohlraum through laser entrance holes (LEH). Besides regular empty hohlraums we shot hohlraums filled with 1 atm of methane (CH_4) and used 0.35 μm thick polyimide to cover holes. A total of ten shaped laser beams were applied with a duration of 2.4 ns rising from a 7 TW foot to 17 TW peak power (pulse shape no. 22: PS22). The total energy supplied to the target was 29 kJ. Experiments were performed with standard unsmoothed Nova beams and with two types of beam smoothing conditions. The beams were smoothed by using Kinoform Phase Plates (KPP) [32-34] which break each beam into many several-mm-scale beamlets whose diffraction limited focal spots are then superposed in the target plane producing an intensity envelope without large scale-length inhomogeneities but consisting of fine scale hot spots (speckles).

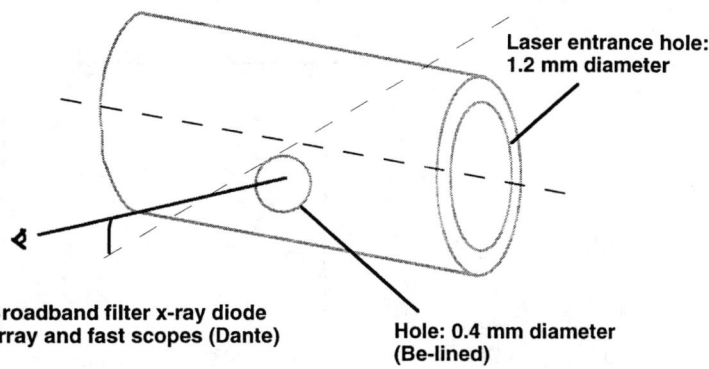

Figure 1. Schematic of a scale-1 Nova hohlraum. The holes are covered with polyimide to contain the gas. The hohlraum temperature is measured through the holes.

The focal spot is further smoothed through use of Smoothing by Spectral Dispersion (SSD) [35,36], where the combination of bandwidth of 0.22 nm at 1ω

together with a dispersive grating in the beam line serves to move the speckles in the focal plane on time scales ~5 ps short compared with that required for hot spots to form a filament in the plasma (~10 ps - 20 ps). The present configuration results in peak laser intensities of 2×10^{15} W cm^{-2} at the LEH which are comparable to those anticipated in NIF hohlraums.

We measure the x ray production in the hohlraums with a broadband filter x-ray diode array and fast scopes [37]. This instrument detects the radiation flux per steradian temporally and spectrally resolved $\phi(t,\nu)$ emitted along the collimated line of sight from the indirectly heated wall opposite a diagnostic hole (450 μm diameter) in the side of the hohlraum (Fig. 1). The hohlraum wall temperature is then obtained by $\Phi(t) = \sigma T_{wall}^4(t) A_d$, where $\Phi(t)$ is the frequency-integrated radiation flux (0-2 keV) emitted from the wall through a diagnostic hole (area A_d) and σ is the Stefan-Boltzmann constant. Figure 2(a) shows examples of temporally resolved emission spectra of the gold hohlraum wall in the energy range 1 keV - 2 keV. The spectra are integrated over 100 ps. From the absolute power we deduce the hohlraum wall temperature as function of time as plotted in Fig. 2(b). The temperature measurements compare well to the detailed radiation hydrodynamic modeling [6,38-40]. In particular, at the peak of the x-ray drive at $t = t_0 + 1.8$ ns measured and calculated temperatures match well. The calculations use corrected pulse shapes with SBS and SRS losses subtracted. From these wall temperature data it is possible to arrive at an estimate of the hohlraum radiation temperature T_{RAD} [41-44] which is a measure of the total x-ray production in hohlraums and is important for the purpose of simulating capsule implosion dynamics because the capsule is irradiated by x rays from both, the indirectly and directly heated hohlraum wall. This correction, called albedo correction, is performed on calculational basis [40] and corroborated by witness plate measurements [44] and x-ray emission measurements through the laser entrance hole [6,45,46].

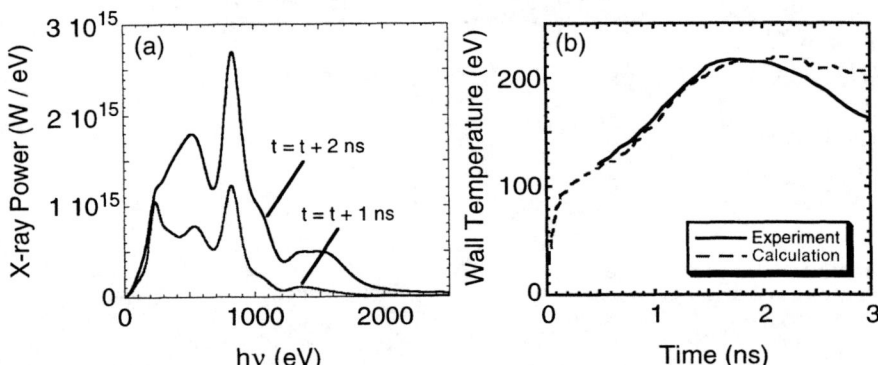

Figure 2. Temporal resolved absolute broadband spectra from the hohlraum gold wall (a) and hohlraum wall temperature as function of time together with LASNEX simulations (b).

Simultaneously, three independent detection systems have been applied for a complete measurement of the scattering losses. Backscattering into the lens of one of the Nova laser beams has been detected with a full aperture backscattering diagnostic [20]. The light was imaged onto a frosted silica plate and detected temporally and spectrally resolved with filtered diodes, spectrometers, and optical streak cameras. The whole detection system was absolutely calibrated *in situ* by retro-reflecting 8% of a full-power laser shot into the detector. Light scattered at larger angles up to 22° was measured with a near backscattering imager [47] consisting of a calibrated aluminum scatter plate mounted around the lens and two-dimensional imaging detectors for SBS and SRS. Light scattered at larger angles was collected with calibrated diodes.

Figure 3 shows the time-integrated SBS and SRS losses and the experimental wall temperatures of methane-filled and empty hohlraums for various beam smoothing conditions. For gas-filled hohlraums we find that the total scattering losses are reduced from $(18 \pm 3)\%$ for unsmoothed laser beams to $(3 \pm 1)\%$ when applying KPPs plus 0.22 nm SSD. Simultaneously, we observe that the temperatures increase by 15 eV. This increase of the hohlraum temperatures for smoothed laser beams is consistent with the reduction of backscatter losses and is clear evidence of improved energy coupling into the hohlraum

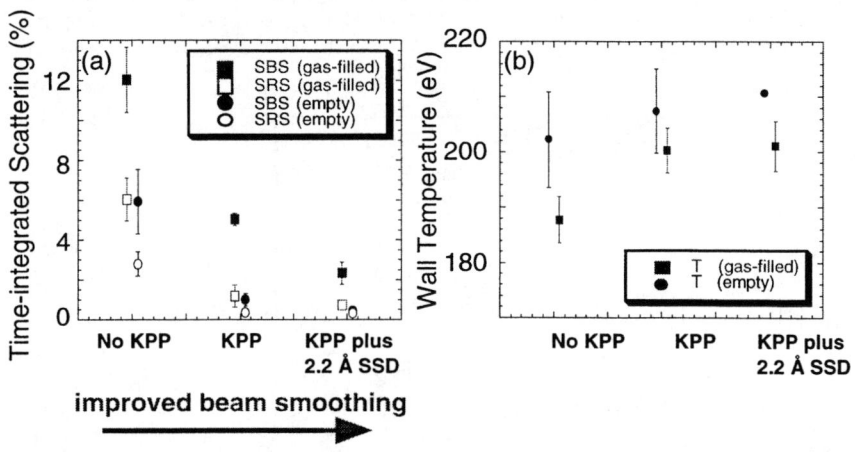

Figure 3. Stimulated scattering losses by SRS and SBS (a) and peak wall temperatures (b) for gas-filled and empty hohlraums.

These experimental observations are consistent with calculations of the filamentation threshold [48,49] and with measured Raman spectra. We observe that short wavelength Raman scattering is gradually suppressed with improved beam smoothing consistent with a reduction of energy in hot spots which produce filamentation (KPP only) and the reduction of the filamentation rate of hot spots (with additional SSD). For our plasma conditions [31], calculations show that the laser beams filament for intensities of $I > 5 - 8 \times 10^{15}$ W cm^{-2}. For unsmoothed laser beams more than 30% of the laser beam exceeds this threshold in the LEH region where Raman scattering occurs. On the other hand, a smoothed laser beam shows a significantly smaller fraction of high intensity spots. Typically less than

5% of a Nova beam smoothed with a KPP is above the filamentation threshold and therefore such beams are more stable against filamentation.

The comparison of the experimental wall temperatures with simulations [6,40] and scalings [2,50] indicate a laser conversion efficiency into x rays of 80% - 90%. This estimate is based on the measured absorbed energy into the hohlraum where SBS and SRS losses are already subtracted. The suppression of these parametric scattering processes is therefore important for the energetics of hohlraums avoiding laser energy losses and consequently maximizing the x-ray production. In addition, it is particularly critical to suppress stimulated Raman scattering because of the production of hot electrons. Stimulated Raman scattering is a three wave process where an incident light wave decays into a scattered light wave of smaller energy and an electron plasma (Langmuir) wave. The electron plasma wave is damped by electron Landau damping where electrons can gain significant energy from the wave turning into so-called hot electrons. The process obeys the Manley-Rowe relations (energy and momentum conservation) from which we can calculate energies of 20-40 keV of the hot electrons for typical parameters of inertial confinement fusion plasmas. These electrons can preheat the fusion capsule reducing its compressibility. Calculations show that for future ignition experiments, e.g., at the National Ignition Facility (NIF), the time-integrated hot electron fraction must be smaller than 15% (6%) for 20 keV (40 keV) electrons to avoid a reduction of the capsule gain.

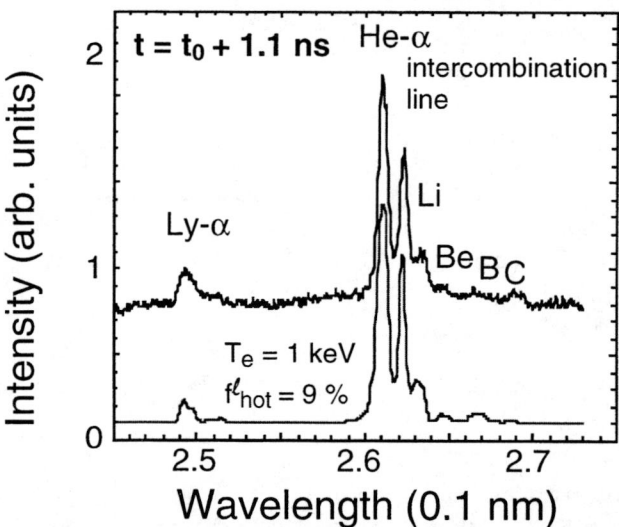

Figure 4. Example of a temporally resolved titanium emission spectrum coated on a tracer foil in a gas-filled Nova hohlraum. A peak hot electron fraction of 9 % is inferred.

We have measured the *local* hot electron fraction in gas-filled hohlraums by x-ray emission spectroscopy using titanium-chromium tracers mounted at various locations in a gas-filled hohlraum. These hohlraums were heated with unsmoothed Nova beams showing a significant fraction of the total laser energy ($E_{tot} = 27$ kJ) scattered by stimulated Raman scattering ($E_{SRS} \cong 1$ kJ). These experiments show a clear correlation between the time histories of stimulated Raman scattered light and of the hot electron fraction measured with x-ray spectroscopy. The presence of highly energetic electrons in these plasmas gives rise to characteristic x-ray emission features showing hydrogen-like resonance emission simultaneously with the emission of Li-, Be-, B-, and C-like satellite lines on temporally resolved spectra. An example of a spectrum measured 1.1 ns after the beginning of the 2.4 ns long laser pulse is shown in Figure 4. The intensity of the hydrogen-like resonance transition in Ti (1s ^{2}S - 2p ^{2}P^o or Ly-α) is found to be more than one order of magnitude larger than calculated for a Maxwellian electron velocity distribution function with a temperature obtained from independent measurements. This spectral line together with the emission from lower ionized charged stages can be used to quantify the hot electron fraction.

For our conditions, we can approximate the electron velocity distribution function by a cold thermal Maxwellian distribution (T_e) plus a Maxwellian tail of hot electrons (T_{hot}) produced by SRS [8,9]. A standard spectroscopic tool to infer T_e is the line intensity ratio of the He-α line to the lithium-like dielectronic satellites which are dominated by the $jk\,l$ dielectronic satellites [51]. Extending the analysis of Ref. [15] to the temperature and density regime encountered in this experiment shows that T_e is high enough so that this method is not sensitive to the presence of hot electrons in the plasma. For example, at $t = t_0 + 1.1$ ns this ratio gives $T_e \cong 1$ keV for the titanium-chromium foil plasma. Complete spectra simulations as shown in Figure 4 and additional measurements of the intensity ratio of the He-β lines from titanium and chromium [52] verify this result. This value is slightly smaller than 1.5 keV of the surrounding CH-plasma as measured with Thomson scattering [31], because the heating of the foil lags behind that of the gas.

The comparison of the experimental spectra with synthetic spectra applying collisional-radiative model calculations shows that the experimental spectra are influenced by hot electron excitation. Including a non-Maxwellian electron velocity distribution function into the modeling gives a quantitative estimate of the plasma conditions. On the other hand, assuming a Maxwellian velocity distribution with temperatures of 1 - 1.5 keV results in a calculated relative intensity of the Ly - α transition which is at least one order of magnitude smaller than experimentally observed (Fig. 4). Using time-dependent non-Maxwellian calculations which include hot electrons with an energy of 20 keV (based on x-ray continuum measurements [10] results in a wider range of ionization stages showing the Ly - α lines, the He - α line, intercombination line, and the Li-, Be-, B-, and C-like $n = 2$ satellites as observed in the experiments. The calculated spectrum shown in Fig. 4 matches the observed intensities rather well and has been calculated for a hot electron fraction of 9%. This result is valid for unsmoothed Nova beams and is significantly larger than the experimental values obtained for smoothed laser beams. X-ray continuum measurements from gas-filled hohlraums which have been heated with beams smoothed by KPPs plus SSD show that the (time-integrated) hot electron fraction is reduced by a factor of about 10. This finding is consistent with the factor of 6 drop of SRS losses seen in Fig. 3(a) for gas-filled hohlraums.

Figure 5 shows the temporal evolution of the hot electron fraction inferred from the spectra synthesis for spectra measured at various times during the hohlraum heating. The independently measured SRS losses are shown for comparison. A correlation between the hot electron fraction and the SRS signal is apparent. The error bar for the hot electron fraction is in the range of 25% - 50%. Other parameters such as the electron density and opacity are obtained self consistently from the spectra. They compare well with the results of the hydrodynamic modeling and with calculations based on the initial conditions of the experiments. From the temporally integrated spectroscopic hot electron fraction

$$1/T \int_0^{2.4} f_{hot}(t)\, dt \cong 0.2$$

we find an averaged hot electron energy of $E_{hot} = 500$ J with $E_{hot}/E = 0.019$ ($E = 27$ kJ the total laser energy delivered to the hohlraum) assuming that the local value of the hot electron fraction is valid for a large part of the hohlraum plasma and further using an averaged electron density of 10^{21} cm^{-3} from hydrodynamic calculations.

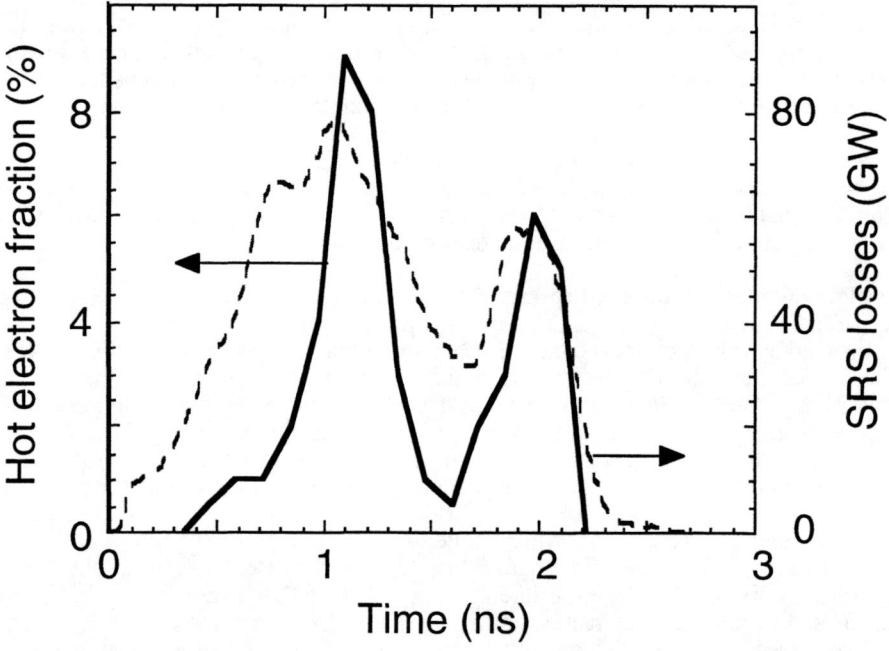

Figure 5. Hot electron fraction from x-ray emission spectroscopy and stimulated Raman scattering losses as function of time. A clear correlation can be observed.

This value for the time-integrated hot electron fraction is in close agreement with the temporally and spatially integrating continuum measurements which give $E_{hot} \cong 400$ J; $E_{hot} / E = 0.015$ as well as with the hot electron fraction and energy estimated from the measurements of the Raman scattered light. From the Manley-Rowe relations we can calculate the energy in suprathermal electrons from SRS [8,9]. For these experiments we measured simultaneously with x-ray spectroscopy a total energy loss into SRS of 1 kJ. The energy of the scattered photons by SRS is found from the measured spectra so that we can infer the energy into hot electrons of $E_{hot} \cong 410$ J which compares well with the above results.

GAS BAG BENCHMARKING EXPERIMENTS

To validate kinetic code calculations (or collisional-radiative modeling) of atomic spectra, well diagnosed plasmas which can serve as test beds are a necessary prerequisite. At Nova, spherical gas bag plasmas [20] have been developed for laser plasma interaction studies in large scale-length plasmas with typical electron densities of 10^{21} cm^{-3} and peak electron temperatures of 3 keV. The targets consist of an aluminum washer covered with two thin polyimide membranes on either side. By filling the target with 1 atm of propane (C_3H_8) or neopentane (C_5H_{12}) with a small amount of argon (1%) for spectroscopy, an almost spherical gas balloon with a known electron density and of 2.75 mm diameter is created. Figure 6 shows a picture of the target. In the figure, vertical slits can be seen which have been mounted for spatially resolving x-ray spectroscopic measurements.

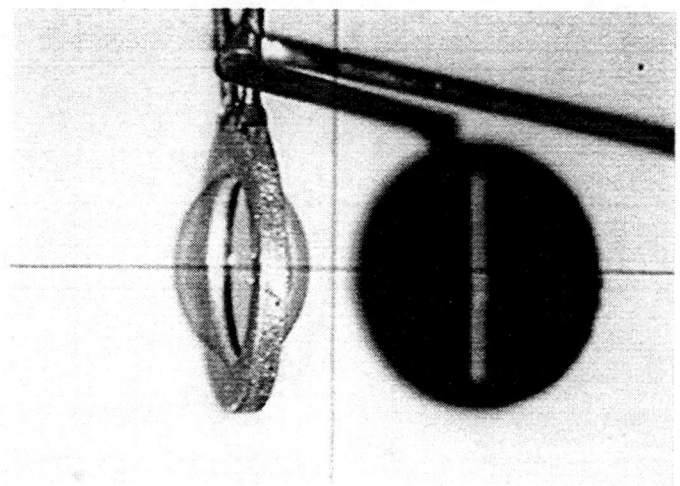

Figure 6. Gas bag target of 2.75 mm diameter together with vertical slits for spatially resolving x-ray spectroscopic measurements.

The gas bags are heated with nine laser beams with a total of 22 kJ for a duration of 1 ns. X-ray spectroscopy from argon and chlorine impurities has been applied to determine the electron temperature in these plasmas and a peak value of 3 keV [22] was measured consistent with radiation hydrodynamic modeling.

Recently, we have performed Thomson scattering and we verified the plasma temperatures measured with x-ray spectroscopy. Thomson scattering observes the scattering of electromagnetic radiation from a probe laser by the electrons of the plasma. It is a spatially and temporally resolving technique which interpretation is based on theory. In the collective scattering regime, where the wavelength of the probe laser is larger than the Debye screening length, the collective behavior of the plasma is observed. In this case, the light is scattered at the ion acoustic and electron plasma wave resonances, and from the measured frequencies and damping of these features - called ion feature and electron feature - one obtains the electron temperatures T_e and densities n_e, ion temperatures T_i, and the averaged ionization stage Z. The macroscopic flow velocities v can be inferred from the Doppler shift of the experimental spectra. We applied this powerful diagnostic at Nova using a 4ω probe laser of 50 J energy focused into the gas bag plasma at a distance of 800 μm from the target center. Figure 7 shows an experimental Thomson scattering spectrum of the ion feature. The spectrum was measured temporally resolved with an optical streak camera and a 1m spectrometer. A fit to the experimental data using the theory for multi-ion species [23,53] gives T_e = 2 keV and T_i = 0.6 keV.

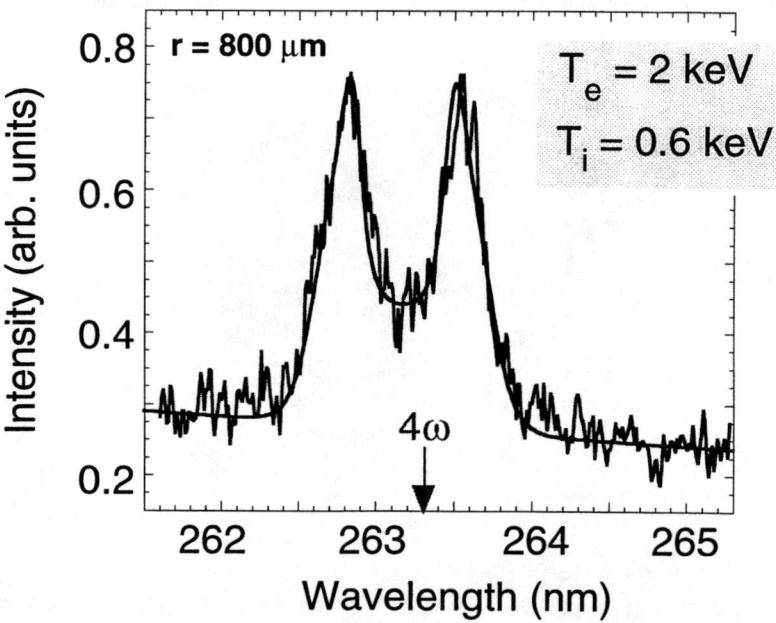

Figure 7. Experimental Thomson scattering spectrum from a gas bag plasma measured at $t = t_0 + 1.2$ ns after the beginning of the 1 ns heater pulse of squared intensity shape.

The experimental electron temperatures show mutual agreement between the results from the Thomson scattering technique and with temporally and spatially resolved x-ray spectroscopy using the intensity ratio of the He-α line to the lithium-like $jk\ell$ dielectronic satellites. This intensity ratio has many advantages over other spectroscopic techniques: the lines can easily be made optically thin by adjusting the impurity concentrations; the lines are separated only by a small wavelength interval so that the sensitivity of the instrument does not change drastically reducing errors due to the relative calibration; and the upper states of the dielectronic satellite transitions are populated on time scales of ~100 ps so that there are only small differences in the time-resolved and steady state analysis of the spectra reducing possible errors even further. The two techniques are compared in Fig. 8 together with LASNEX simulations. The error bars for the spectroscopic data are in the range of 20% and Thomson scattering data are accurate to within 10%. The experimental data are also in reasonable agreement with the simulations when including heater beam backscattering losses by SBS and SRS. The latter show slightly higher peak temperatures and also a faster rise of the electron temperature than experimentally observed. At present, we are working on improved simulations to take better into account the geometry of the heater beams.

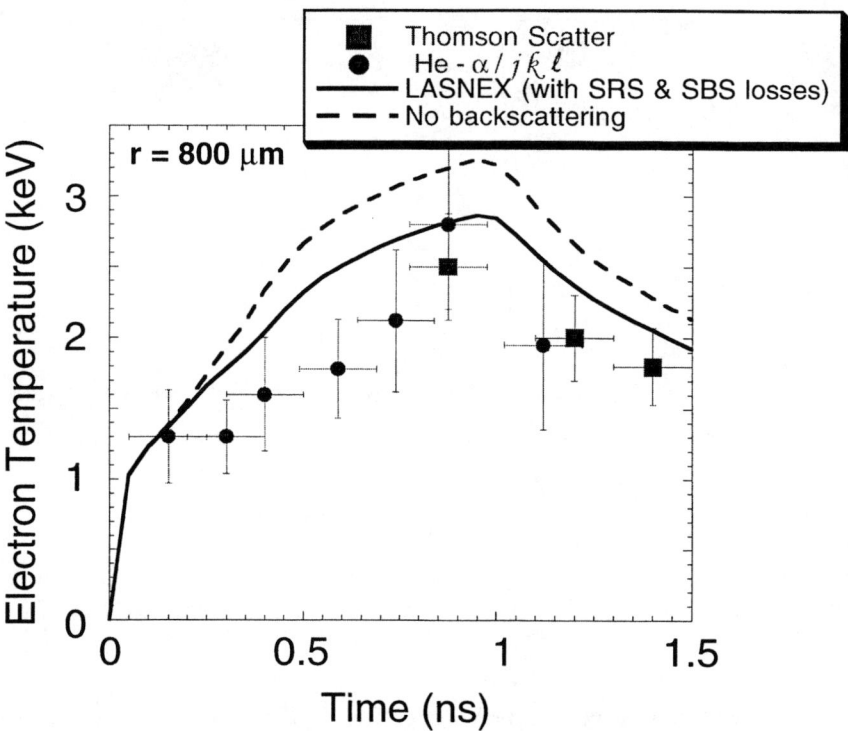

Figure 8. Comparison of simulated and experimental electron temperatures measured temporally resolved with Thomson scattering and x-ray spectroscopy.

While the electron and ion temperature in these gas bag plasmas is well known from the measurements described above the electron density is principally known by the density of the gas fill. Measurements of the wavelength of the Raman scattered light which occurs at the frequency of the electron plasma or Langmuir wave give a value for the electron density which is consistent with the gas fill density [20,24]. In addition, x-ray spectroscopic measurements of the resonance and intercombination line of helium-like argon have also been shown to be in agreement with the expected densities [22]. The detailed measurements and the simulations provide a good understanding of the plasma conditions in these gas bag target and it can therefore be used as a spectroscopic source to benchmark kinetics with experimental spectra. These type of experiments begun two years ago investigating Xe L-shell spectra [54].

Recently, we have measured the dielectronic satellites on the red wing of the He-β line of argon which relative intensity is sensitive to the electron temperature and therefore important for the characterization of capsule implosions [18]. Figure 9 shows an example of a measured spectrum of the He-β line from a gas bag plasma where the satellites are labeled following Ref.[55].

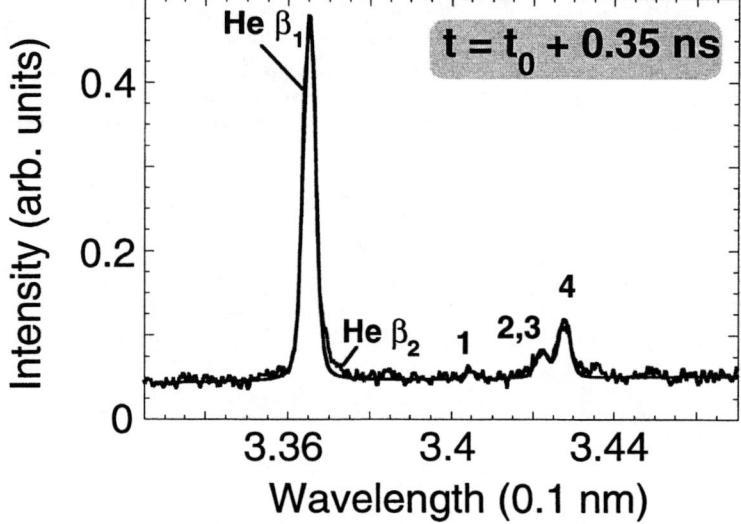

Figure 9. Experimental spectrum of the He-β line of argon from a gas bag plasma where n_e and T_e is determined by independent measurements such as Thomson scattering.

The spectrum has been fit by Voigt profiles to determine the relative intensities of the lines (the He-β_2 component is excluded from the fit to show its small contribution to the intensity of the He-β line but it is included when comparing with calculated line ratios). For example the experimental ratio between the satellite feature 4 and the He-β line is 0.15 ± 0.01 for $n_e = 10^{21}$ cm^{-3} and $T_e = 1.3$ keV. This result is in excellent agreement with collisional radiative modeling which predict a ratio of 0.17. The complete spectrum is presently under investigation. In particular, the consistency of the various line ratios with the code predictions for a single temperature and density are being studied. In addition, data have been taken at various times during the gas bag heating and for different densities. These results will be published elsewhere. The present results verify the collisional radiative modeling which is applied for the interpretation of implosion spectra [18,54]. It indicates that the smaller temperatures (when compared to simulations) obtained for the capsule conditions with this method are not due to the kinetic models. Spatial temperature gradients and/or difficulties with the satellite technique at small intensities and large widths are probably responsible for the discrepancies.

CONCLUSIONS

The present laser absorption measurements at the Nova laser facility are important to understand the physics of closed-geometry gas-filled hohlraums. They were designed to study plasma conditions similar to those anticipated at the National Ignition Facility. For example, laser beam smoothing by kinoform phase plates and by spectral dispersion has been adopted in the NIF design because of its beneficial effects seen in the Nova experiments. Its will be particularly important to suppress stimulated Raman scattering in the large scale plasmas encountered at the NIF because of the production of hot electrons. Our spectroscopic measurements will be a useful tool for future experiments to determine the local fraction of the hot electrons near the fusion capsule to estimate possible preheat effects.

Experiments in well-diagnosed gas bag plasmas have shown that collisional radiative (kinetics) modeling of the intensities of the He-β line and its satellites is in excellent agreement with the measured spectra. This result increases our confidence in the characterization of implosion capsules which requires Stark broadening calculations coupled to a kinetics model to calculate the detailed line shapes for density and temperature diagnostics.

ACKNOWLEDGMENTS

I would like to thank C. A. Back, C. Decker, B. A. Hammel, R. M. Herman, R. W. Lee, B. J. MacGowan, A. Osterheld, L. J. Suter, and R. E. Turner for helpful discussions. This work was performed under the auspices of the U. S. Department of Energy by the Lawrence Livermore National Laboratory under Contract No. W-7405-ENG-48.

REFERENCES

[1] J. H. Nuckolls, Phys. Today **35**, No. 9, 24 (1982).
[2] J. D. Lindl, Phys. Plasmas **2**, 3933 (1995).
[3] S. W. Haan *et al.*, Phys. Plasmas **2**, 2480 (1995).

[4] W. L. Kruer, *The Physics of Laser Plasma Interactions* (Addison-Wesley, New York, 1988).
[5] T. P. Hughes, *Plasmas and Laser Light* (John Wiley and Sons, New York, 1975).
[6] S. H. Glenzer et al., Phys. Rev. Lett. **80**, 2845 (1998).
[7] J. W. Shaerer et al., Phys. Rev. A **6**, 764 (1972).
[8] K. G. Estabrook et al., Phys. Rev. Lett. **45**, 1399 (1980).
[9] K. G. Estabrook and W. L. Kruer, Phys. Fluids **26**, 1892 (1983).
[10] R. L. Kauffman, in *Handbook of Plasma Physics*, edited by M. N. Rosenbluth and R. Z. Sagdeev, *Physics of Laser Plasma*, edited by A. Rubenchik, and S. Witkowski (North-Holland, Amsterdam, 1991), pp. 111-162.
[11] R. W. Lee et al., J. Quant. Spectrosc. Radiat. Transfer **32**, 91 (1984).
R. W. Lee *private communication* (1997).
[12] H. R. Griem, Phys Fluids B **4**, 2346 (1992).
[13] J. P. Matte et al., Phys. Rev. Lett. **72**, 1208 (1994).
[14] J. Abdallah et al., Phys. Scripta **53**, 705 (1996).
[15] F. B. Rosmej, J. Phys. B **28**, L747 (1995).
F. B. Rosmej, J. Quant. Spectrosc. Radiat. Transfer **51**, 319 (1994).
[16] S. H. Glenzer et al., Phys. Rev. Lett. **81**, 365 (1998).
[17] A. A. Offenberger et al., Phys. Rev. Lett. **49**, 371 (1982).
[18] B. A. Hammel et al., Phys. Rev. Lett. **70**, 1263 (1993).
[19] N. Woolsey et al., Phys. Rev. E **57**, 4650 (1998).
[20] B. J. MacGowan et al., Phys. Plasmas **3**, 2020 (1996).
[21] D. H. Kalanter et al., Phys. Plasmas **2**, 3161 (1995).
[22] S. H. Glenzer et al., Phys. Rev. E **55**, 927 (1997).
[23] S. H. Glenzer et al., Phys. Rev. Lett. **77**, 1496 (1996).
[24] D. S. Montgomery et al., Phys. Plasmas **5**, 1935 (1998).
[25] H. R. Griem et al., Phys. Rev. E **56**, 7186 (1997). (See also these proceedings).
[26] S. Alexiou et al., Phys. Rev. E (Comment), in press (1998).
[27] E. M. Campbell et al., Rev. Sci. Intrum. **57**, 2101 (1986).
[28] M. D. Cable et al., Phys. Rev. Lett. **73**, 2316 (1994).
[29] A. A. Hauer et al., Phys. Plasmas **2**, 2488 (1995).
[30] N. D. Delamater et al., Phys. Plasmas **3**, 2022 (1996).
[31] S. H. Glenzer et al., Phys. Rev. Lett. **79**, 1277 (1997).
[32] S. N. Dixit et al., Optics Lett. **21**, 1715 (1996).
[33] Y. Kato et al., Phys. Rev. Lett. **53**, 1057 (1985).
[34] D. M. Pennington et al., Proc. SPIE **1870**, 175 (1993).
[35] R. H. Lehmberg and S. P. Obenschein, Opt. Commun. **46**, 27 (1983).
[36] S. Skupsky et al., J. Appl. Phys. **66**, 3456 (1989).
[37] H. Kornblum et al., Rev. Sci. Instrum. **57**, 2179 (1986).
[38] G. B. Zimmerman and W. L. Kruer, Comments Plasma Phys. Controlled Fusion **2**, 85 (1975).
[39] L. J. Suter et al., Phys. Rev. Lett. **73**, 2328 (1994).
[40] L. J. Suter et al., Phys. Plasmas **3**, 2057 (1996).
[41] R. E. Marshak, Phys. Fluids **1**, 24 (1958).
[42] R. Sigel et al., Phys. Rev. Lett. **65**, 587 (1990).
[43] H. Nishimura et al., Phys. Rev. A **44**, 8323 (1991).
[44] R. L. Kauffman et al., Phys. Rev. Lett. **73**, 2320 (1994).

[45] C. Decker *et al.*, Phys. Rev. Lett. **79**, 1491 (1997).
[46] R. E. Turner *et al.*, *submitted to* Phys. Rev. Lett. (1998).
[47] R. K. Kirkwood *et al.*, Rev. Sci. Instrum. **68**, 636 (1997).
[48] E. M. Epperlein and R. Short, Phys. Fluids B **4**, 2211 (1992).
[49] R. L. Berger *et al.*, Fluids B **5**, 2243, (1993).
[50] R. Sigel *et al.*, Phys. Rev. A **38**, 5779 (1988).
[51] A. H. Gabriel, Mon. Not. R. Astron. Soc. **160**, 99 (1972).
[52] C. A. Back *et al.*, Phys. Rev. Lett. **77**, 4350 (1996).
[53] D. E. Evans, Plasma Phys. **12**, 573 (1970).
[54] C. A. Back *et al.*, in *Spectral Line Shapes*, edited by M. Zoppi and L. Ulivi, Vol. 9 (AIP Conference Proceedings 386, New York, 1997), pp. 35-44.
[55] P. Beiersdorfer *et al.*, Phys. Rev. E **52**, 1980 (1995).

Alternative treatment of line broadening in dense and hot plasmas

E.Leboucher-Dalimier, P.Sauvan, P.Angelo, H.Derfoul, T.Ceccotti,
P.Gauthier, A.Poquérusse, A.Calisti*, B.Talin*

*Physique Atomique dans les Plasmas Denses,
LULI CNRS-Ecole Polytechnique-CEA-Université Paris VI
Université Paris VI, 4 Place Jussieu, 75252 Paris Cedex 05, France
and Ecole Polytechnique, 91128, Palaiseau Cedex, France.*

**Physique des Interactions Ioniques et Moléculaires, Université de Provence, Centre de St Jérôme,
13397, Marseille, France*

in collaboration with S.J.Rose (RAL Didcot), E.Förster (XROP Jena),
A.Devdariani (University of St Petersburg), R.W.Lee (LLNL)

Abstract - A spectral line shape code using the "quasimolecular model" has been developped for dense and hot plasmas. First this alternative treatment is justified. Then the importance of using a two-centre basis and the effects of the plasma screening are discriminated. Specific hydrogen-like and helium-like transitions are studied for the exhibition of dense plasma effects (PPS, asymmetries, satellite features) and for a comparison with experimental results. It is shown that satellites due to extrema are enhanced by the ion dynamics correction.

INTRODUCTION

This work aims at studying the effects of high material densities on atomic processes and the spectral line shapes characterizing dense and hot plasmas. This subject gives rise to new basic problems to be solved for understanding the transition between plasma physics and solid state physics. The plasmas of interest are *highly correlated* ($N_e > 10^{23} cm^{-3}$) : the mean interionic spacing $R_i = (3/4\pi N_i)^{1/3}$ is comparable to the orbital extent of the outer orbitals $r_n = n^2 a_0/Z$. Because of the resulting overlapping of neighboring ion orbitals, we can use a N centre basis. This emitting structure is confined with free electrons in a bounded volume. As the temperature is high ($T_e > 300 eV$), the plasmas are *moderately coupled* (ionic coupling parameter Γ about 1). Therefore *the N centre states are not established* (i.e. as in a molecule bound by a chemical bond). The emission merging from those plasmas is characterized by asymmetries, satellite-like features and forbidden transitions (1). These features are attributed to two-centre structures (2) and they give new diagnostics interesting for laser-driven implosion experiments (ICF). Similar diagnostics had yet been pointed out in astrophysics (3) and that is why laboratory experiments reproducing astrophysical situations such as the centre of the sun, the λ Bootis stars or the white or brown dwarfs are expected. Such studies will contribute to the knowledge of the stellar evolutions and the energy transfer from stars.

In the first part, the two-centre model is justified and then used for the computation of the electronic structures of hydrogen-like and helium-like quasi-molecules immersed in a dense plasma. Then in a second part the spectral line shapes of transitions of a particular interest for experiments are calculated within the quasistatic approximation and the results are compared with those given by standard codes using an atomic basis. It is shown that the dynamics correction must be introduced for the central part of the profiles and for the singularities. Finally, the last section is devoted to the discussion of a possible exhibition of satellite-like features in connection with experimental results.

THE ELECTRONIC STRUCTURE OF THE QUASIMOLECULE

Validity conditions of the Two-Centre Model

The formation of bound molecules is plainly not possible for the plasma conditions probed here because the ion kinetic energy is much higher than any binding energy. But nevertheless the N centre states representation is justified because the radiative transitions are instantaneous on the ionic motion time scale.

Moreover it can be shown that the *Two-Centre Model* is an alternative approach for moderately coupled plasmas. To support this assumption we have first demonstrated that the Nearest Neighbor (NN) interaction is a good approximation for the microfield distribution. Molecular Dynamics (MD) simulations made for argon in Ar-filled implosion conditions ($N_e=5.10^{24}\text{cm}^{-3}$, $T_e=800\text{eV}$) show that this approximation gives the exact microfield (all neighbor ions involved) within 20% (see Fig.1).

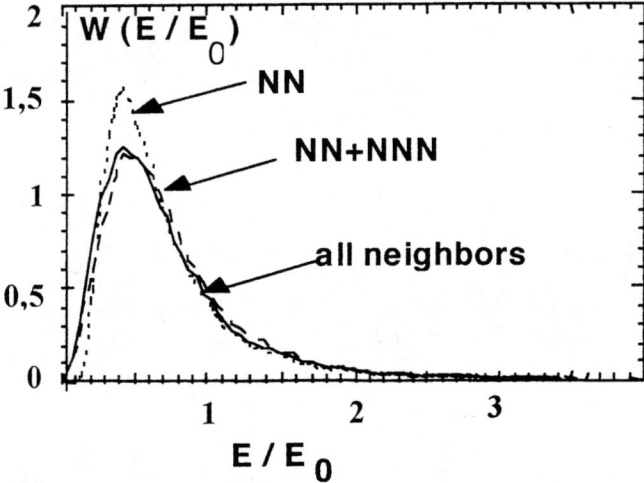

Figure 1. Molecular Dynamics microfield distributions for argon plasma from the nearest neighbor contribution only (NN), from the nearest and next nearest neighbor only (NN+NNN) and from all neighbors. E_o is the microfield for the mean interionic separation.

Secondly MD simulations for a $F^{8+}-F^{9+}$ transient molecule embedded in a hot and dense plasma ($N_e = 2 \; 10^{23}$ cm^{-3}, $T_e = 300$ eV) reveal a very small change of the interionic distance R(t) on the ionic motion time scale ; so that the two-centre formation is highly probable during the time of interest (w_{ip}^{-1} the inverse of the ionic plasma frequency, estimated at about $3 \; 10^{-1}$ ps) (see Fig 2). More precisely, the lifetime of the molecule can be estimated at 30 fs (see Fig 3). As a consequence we chose a transient dicenter model to describe the emitting structure. The probability distributions needed for the emission were computed also by MD simulations using an effective electron screening. We present in Fig 4 Nearest Neighbor NN probability densities $P_{NN}(R)$ in a fluorine plasma for different electron densities ranging from 10^{22}cm^{-3} to 10^{25} cm^{-3} and the same temperature 300 eV. They give the radii of interest and we remark that the radius connected with the maximum decreases strongly with the density. It has been shown that it increases with Z (1).

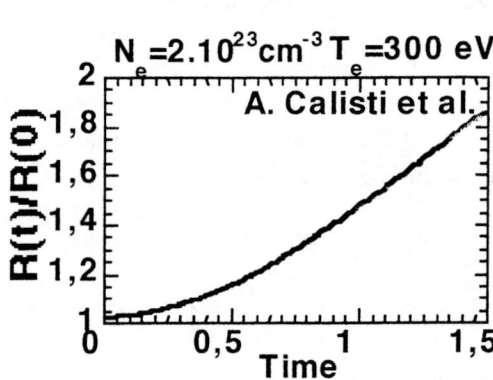
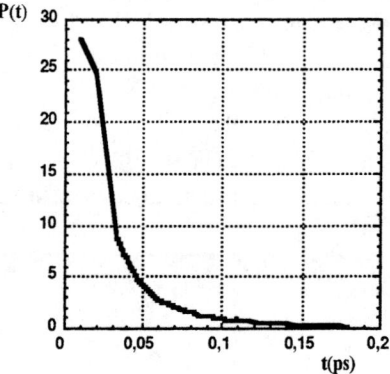

S. Alexiou

Figure 2. Interionic distance evolution with time (w_{ip}^{-1} time unit) for $F^{8+}-F^{9+}$ transient molecule.

Figure 3. $F^{8+}-F^{9+}$ quasimolecule lifetime distribution for 10^{23}cm^{-3} and 300eV.

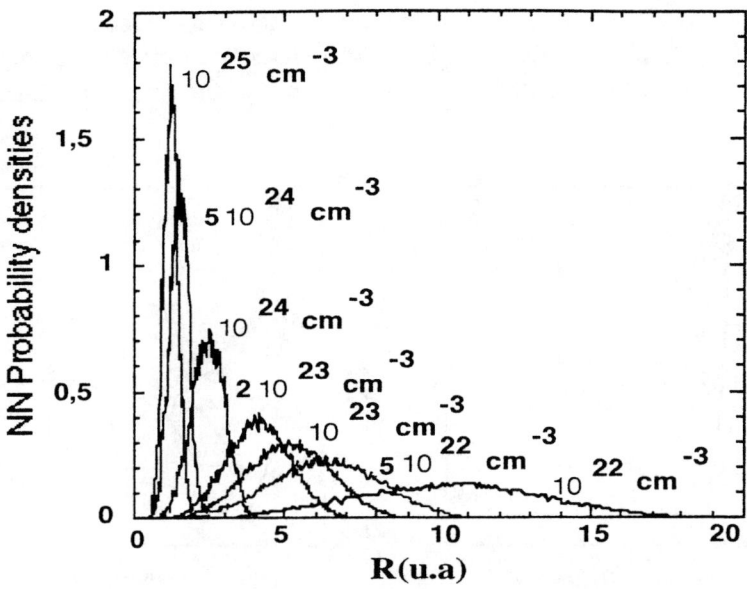

Figure 4. Nearest Neighbor probability densities $P_{NN}(R)$ for a fluorine plasma (Te=300eV)

The electronic structure

The emitting structure, two NN ions (with one or two bound electrons) embedded in a non uniform electron gas, is confined in a volume satisfying the quasi-neutrality condition. Using a self-consistent treatment to solve the coupling between the Schrödinger and the Poisson equations, we determine with the code IDEFIX (2) the adiabatic energies, the transition energies and the oscillator strengths. The model is an exact treatment of multipolar interactions to all orders and the outputs for the previous quantities are parametered with the interionic spacing R. The code has been designed for hydrogen-like and helium-like ionization stages, for homonuclear and heteronuclear molecules and for a large range of Z and plasma conditions (N_e, T_e).
We show in the following the efficacity of the code. The line FLyβ has been chosen as an example.

i) *The Two-Centre model is adapted to exhibit at short interionic spacing the splitting of the transition energies, the possible extrema in the transition energies and the possible strong variation of the oscillator strengths.* Figure 5 shows for the F^{8+}-F^{9+} system the transition energies involved for FLyβ and their splitting due to the NN interaction of the n=3 energy level for a mean electron density $N_e = 2.10^{23} cm^{-3}$ and an electron temperature T_e=300eV.

ii) *The self consistent treatment is efficient for solving the shielding of the NN interaction by plasma free electrons.* Two major effects in the transition energies at large internuclear separations result from the free electron screening : a reduction of the level splitting and a Plasma Polarisation Shift (PPS). These two effects can be seen for FLyβ from the comparison between Fig.5a (without screening) and Fig.5b (with screening).

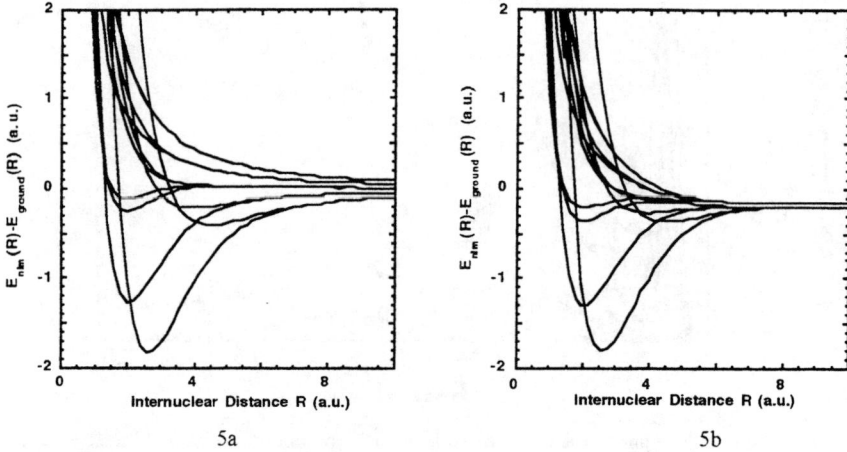

5a   5b

Figure 5. Variation of the transition energies involved for FLyβ, as a function of the distance from its nearest neighbor. Electron screening is considered in 5b ($N_e=2.10^{23}$ cm^{-3}) but not in 5a. The calculations were performed for the same electron temperature $T_e=300$eV.

In Fig.6 we give the dipolar matrix elements for the line FLyβ for $N_e = 0$ (solid lines) and $N_e = 2 \cdot 10^{23}$ cm^{-3} (dotted lines), $T_e = 300$ eV as a function of the NN distance. They vary strongly at short interionic spacing and the plasma screening effect is important at large interionic spacing.

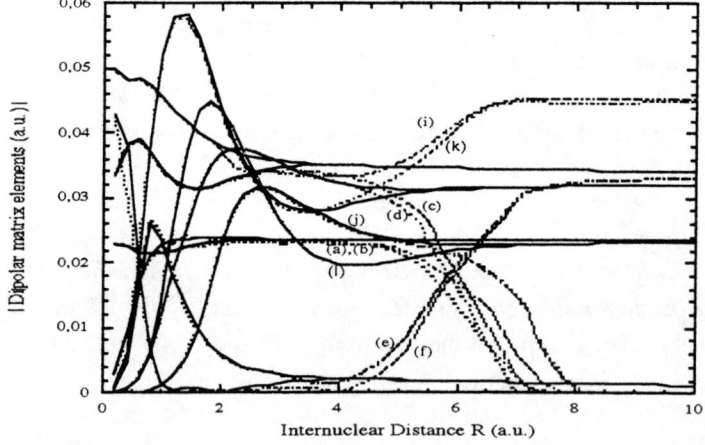

Figure 6. Variation of the dipolar matrix elements for the line FLyβ as a function of the internuclear distance R for $N_e = 0$ (solid lines) and $N_e = 2 \cdot 10^{23}$ cm^{-3} (dotted lines), $T_e = 300$ eV.

The previous treatment has been extended for two bound electron states (i.e. helium-like ionization stages) using the Molecular Orbital (MO) approach. A perturbation theory is employed to account for the two bound electron repulsion.

Transition energies involved in the ArHeα spectral range are displayed in Fig.7 for $N_e = 6.10^{24} cm^{-3}$ and $T_e = 600 eV$. At large internuclear distances it is seen that the two-centre energy levels go with a very good accuracy over the Heα atomic states. At short NN distances, two well separated energy levels arise from the ^1S - ^1P resonant component ((a) in Fig.7), the splitting of which increasing rapidly as the NN distance decreases.

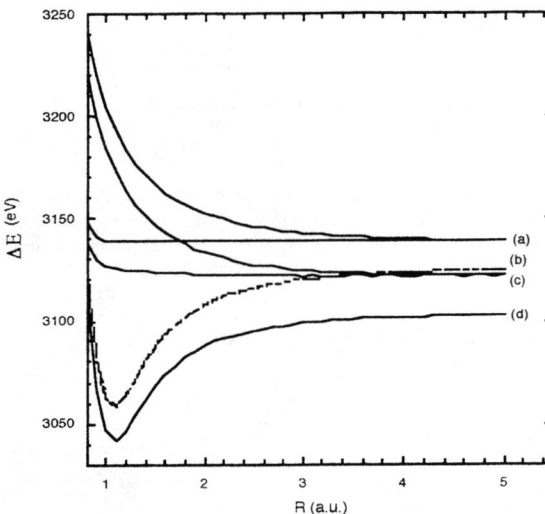

Figure 7. Transition energies involved in ArHeα as a function of the nearest neighbor separation. At large distances, the two-centre transition energies go over to the Ar^{16+} atomic energies corresponding to the transitions (a) ^1S - ^1P, (b) ^1S - ^1S, (c) ^1S - ^3P, (d) ^1S - ^3S. Calculations were performed for $N_e = 6\ 10^{24}\ cm^{-3}$ and $T_e = 600$ eV

The relative intensity of the high-energy (blue) ^1S - ^1P sub-component to the unshifted ^1S - ^1P sub-component is plotted in Fig.8 as a function of the NN separation.

The interionic spacing corresponding to the maximum of the NN probability for the plasma conditions of interest being 1.67 a.u., the two-centre description is expected.

THE SPECTRAL LINE SHAPE WITHIN THE TWO-CENTRE MODEL

In this section dense plasma effects on spectral line shapes are analysed. We use at first the quasistatic approach for the emitting ionic configuration. A dynamics treatment is then developed for line broadening in the vicinity of singularities where the time of interest is important.

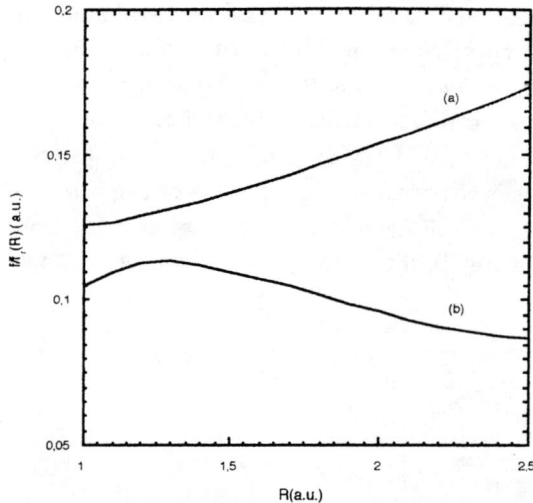

Figure 8. Reduced oscillator strengths f/f_0 for the two resonant components as a function of the NN distance ((a) high-energy ^1S - ^1P sub-component (b) ^1S - ^1S component). Calculations were performed for $N_e = 6 \, 10^{24}$ cm^{-3} and $T_e = 600$ eV. f_0 corresponds to the unshifted ^1S - ^1P sub-component. At large R transition (b) is forbidden.

Emission from a quasistatic transient molecule

The Stark-broadened theoretical line profiles are obtained in the QMSPECTRA code by an average over the transient molecule weighted by its probability density $P_{NN}(R)$, namely

$$I(\omega) \propto \sum_{i,f} \int_0^\infty P_{NN}(R) |f_{if}(R)|^2 \frac{\phi_{if}}{(\omega - \Delta E_{if}(R))^2 + \phi_{if}^2} dR \qquad (1)$$

For given plasma parameters (N_e, T_e), $\Delta E_{if}(R)$ represents the transition energy between sublevels i and f and $f_{if}(R)$ is the related oscillator strengths, both computed by the code IDEFIX. The electron collisional broadening ϕ_{if} of the sublevels was included in the usual impact approximation using the second-order perturbation theory (4).

In the interionic distance window open by $P_{NN}(R)$, the transition energies are widely splitted and shifted and the oscillator strengths vary strongly. These are arguments for the *shifts, asymmetries and satellite-like features* exhibited in dense plasmas.

The following results show *the importance of using a molecular basis* (2). We present in Fig. 9 FLyβ line profiles at 300eV for two densities (a) 10^{22}cm^{-3} and (b) 2.10^{23}cm^{-3}. The profiles are computed with the standard code Pim Pam Poum (PIIM Marseille), the dotted lines, and with the Two-Centre Model QMSPECTRA, the solid lines.

While at low densities (curves (a)) the spectra come together, at high densities there is a discrepancy (curves (b)).

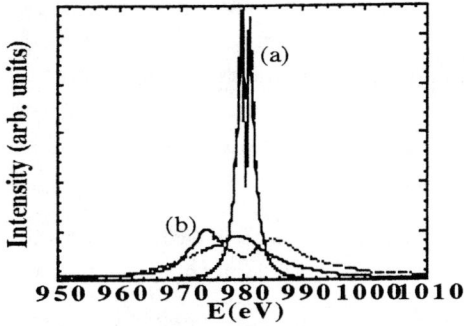

Figure 9. FLyβ line shapes at $T_e=300eV$ for $N_e=10^{22}cm^{-3}$ (curves (a)) and $N_e=2.10^{23}cm^{-3}$ (curves (b)) : Standard Pim Pam Poum simulations (dotted lines) and QMSPECTRA simulations (solid lines).

The comparison between the two code results is relevant as long as the quasi-static approximation is used for ions and the only NN microfied distribution is taken into account in the Pim Pam Poum simulations.

Calculations performed at various densities show *the sensitivity of the line shapes to electron screening* (2). In figure 10 we present FLyβ line profiles at $T_e = 300eV$ for $N_e = 5.10^{22}cm^{-3}$ (curves (a) and (b)) and $N_e = 2.10^{23}cm^{-3}$ (curves (c) and (d)). Curves (a) and (d) include the electron screening whereas this influence has not been included in curves (b) and (c). Plasma screening leads to a reduction of the Stark effect (i.e. a reduction of the Full Width at Half Maximum and a reduction of the splitting of the components) and a global Plasma Polarization Shift (PPS) of the line in the direction of the lower energies.

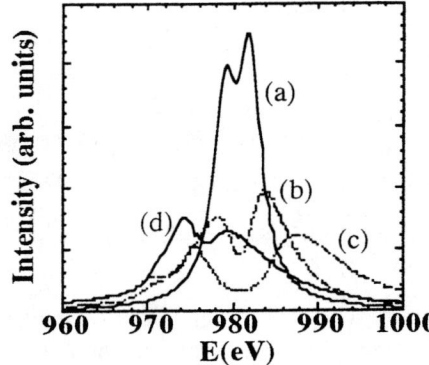

Figure 10. FLyβ line profiles for $N_e=5.10^{22}cm^{-3}$ ((a) and (b)) and $N_e=2.10^{23}cm^{-3}$ ((c) and (d)) ($T_e=300eV$ for all cases). Calculations have been performed with QMSPECTRA. For profiles (b) and (c) electron screening effects have not been included.

At high densities the line profiles may exhibit satellite-like features (5). There are several reasons for such exhibitions. First, the splitting at short R of transitions degenerated at large R can lead to a satellite when the oscillator strengths are comparable in the window open by $P_{NN}(R)$. *Secondly, a satellite-like feature may correspond to the excitation at short R of a transition forbidden at large R. Another possibility to get a satellite is related to the exhibition of an extremum in a transition energy simultaneously with an oscillator strength of importance in the NN probability window.*

We present in Fig. 11 Stark-broadened ArHeα line profiles at 600eV for various electron densities in the range $\left[2.10^{24} \text{cm}^{-3} - 10^{25} \text{cm}^{-3}\right]$. The MD simulations indicate that the corresponding NN most probable distances are in the range $\left[2.3 \text{a.u} - 1.08 \text{a.u.}\right]$. As singlet-triplet transitions ($^1S-^3P$ and $^1S-^3S$) were expected to not significantly contribute to the line shape (6), we neglected them in our computations.

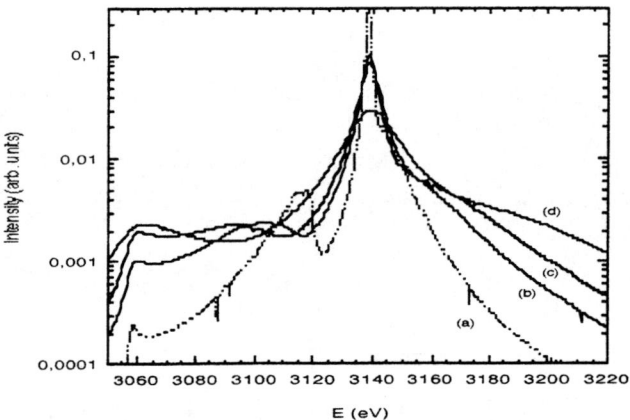

Figure 11. Stark-broradened Ar-Heα line profiles at 600eV for the density range $\left[2.10^{24} \text{cm}^{-3} - 10^{25} \text{cm}^{-3}\right]$. The low-energy side is dominated by the $^1S-^1S$ transition and the high-energy side by the $^1S-^1P$ transition. The line profiles are area-normalised.

As the density increases, the high-energy component of $^1S-^1P$ results in the amplification of the emission at energies greater than 3160eV and in the appearance of a blue satellite. The low-energy side of ArHeα is dominated by the $^1S-^1S$ transition and the extremum in the related transition energy at short R leads to a red satellite feature : this satellite does not depend on the density.

Emission from a dynamics transient molecule

The dynamics approach for the Two-Centre Model had been tested analytically with simplifying assumptions for dense plasma (7). In the present work the ion dynamics is treated selfconsistently with the plasma.

As explained before, the molecule formation is basically transient and the resulting dynamics correction is not negligible in the vicinity of profile singularities ω_S (satellite-like features) giving a new reference for the time of interest $(\omega - \omega_S)^{-1}$.

Let us consider an ion moving on a classical trajectory depending on the potential energy $E_p(R, R_c)$ due to the NN fixed ion and a non uniform free electron gas. This trajectory is truncated because of the cut off R_c corresponding to the interionic distance of the molecule at the limit of the confined quasineutral volume. The line profile within this approach is given by :

$$I(\omega) \propto \sum_{if} \int d\bar{R}_c d\bar{v}_{R_c} P(\bar{R}_c, \bar{v}_c) \left| \int_{R_{tp}}^{R_c} \frac{dR}{v} D_{if}(R) \cos\left(\int_{R_{tp}}^{R} (\omega - \Delta E_{if}(R')) \frac{dR'}{v} \right) \right|^2 \quad (2)$$

It depends on the history of the collision controlled through the interionic distance R and the radial velocity v. R_{tp} corresponds to the turning point and v_c to the radial velocity for the cut off. The ensemble average involves the conditional probability $P(\bar{R}_c, \bar{v}_c)$ computed by Molecular Dynamics. The transition energies $\Delta E_{if}(R)$, the dipole moments $D_{if}(R)$ and the potential energy $E_p(R, R_c)$ are computed by the code IDEFIX. Electron collisional broadening, not expressed in (2), has also been included in our calculations.

The enhancement with the dynamics correction of satellite-like features due to extrema in $\Delta E_{if}(R)$ is a highlight result. In figure 12 a comparison between the two versions (quasistatic / dynamics) of the Two-Centre model (QMSPECTRA) is given for FLyβ at high density conditions ($N_e = 4.10^{23}$ cm^{-3}, $T_e = 300$eV). The dynamics correction smoothes the central part of the profile but enhances the satellites. These satellites involve interionic distances shorter than 3 a.u. (see Fig.5) which can be achieved for all trajectories such as $R_{tp} < 3$ a.u. In the quasistatic approach the observability of these satellites was conditionned by the more restrictive NN probability window $P_{NN}(R_c \approx 3 \text{a.u.}, v_c = 0)$.

The profiles computed with the dynamics version of QMSPECTRA have been compared to standard profiles issued from the ion dynamics version of Pim Pam

Poum. As seen in figure 13 for FLyβ ($N_e = 2.10^{23}$ cm^{-3}, T_e = 300eV), our model gives the same near wings shape as Pim Pam Poum and the molecular satellites in addition.

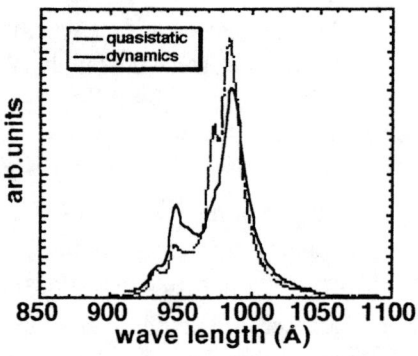

Figure 12. Comparison between quasistatic (the dotted line) and dynamics (the solid line) FLyβ profiles computed with QMSPECTRA.

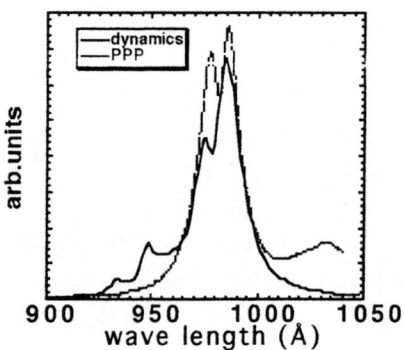

Figure 13 Comparison between FLyβ profiles computed with Pim Pam Poum (the dotted line) and the dynamics version of QMSPECTRA (the solid line). For the comparison the PPS shift (not taken into account in Pim Pam Poum) has been eliminated in our computation.

ANALYSIS OF EXPERIMENTAL RESULTS WITHIN THE TWO-CENTRE MODEL

Some experimental spectra emitted from high density plasmas are reported. Their comparisons with synthetic theoretical profiles computed with QMSPECTRA provide a density diagnostic. However the exhibition of satellite-like features may be questionable because of the possible smoothings due to the electron collisional broadening and to the space/time integrations over different plasma conditions (N_e, T_e).

In both experiments reported, the NN approximation is shown to be relevant to reproduce asymmetries and satellite-like features.

Fluorine Lyman β at high densities

Recently, the colliding foil system known for its capability to generate dense and hot plasma has been complemented with a time-resolved spectroscopic diagnosis. On the LULI nanosecond chain, focal spots of 60μm provided a 4ω laser intensity of $\approx 4.10^{14} \text{Wcm}^{-2}$ on target. Two thin (5μm) Teflon foils initially separated by 100μm collide approximately at the pulse maximum. A TℓAP crystal coupled to a X Streak camera recorded a FLyβ spectra evolution for the 25μm central plasma. The time resolution ($\approx$ 50ps) was sufficient to show a red plasma polarisation shift (PPS) and a varying broadening due to strong density gradients. At peak density, the spectrum exhibits a PPS induced asymmetry on the red wing and a forbidden 3d-1s transition on the blue wing. We present in figure 14 a synthetic theoretical profile which fits the experimental data. This profile uses the Two-Centre Model (QMSPECTRA) and takes into account for the electronic broadening and the spatial integration (25μm) over strong density and emissivity gradients. These gradients were determined previously from a 1D1/2 hydrocode (FILM).

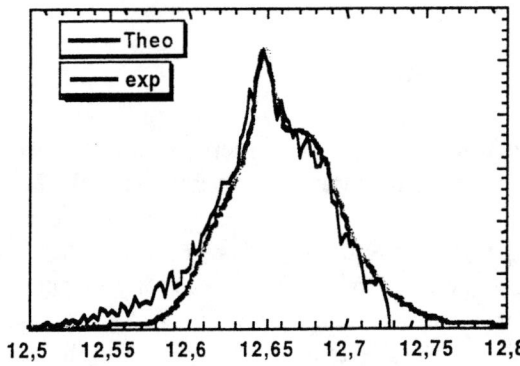

Figure 14. FLyβ time resolved spectrum from a colliding foil experiment performed at LULI. A forbidden transition and a PPS induced asymmetry are clearly exhibited.

Argon Helium α at high densities

The time-resolved (60ps) ArHeα spectra merging from implosions of Argon-filled plastic shell targets (experiments made at Rochester (8) were analysed. A blue-satellite feature is exhibited at the peak compression. A theoretical profile computed by a standard model using an atomic basis (9) was not able to explain the experimental results. On the contrary, the Two-Centre Model, in its quasistatic version, has been shown to be efficient to reproduce the asymmetry and the satellite feature. This blue satellite is due to the splitting of the resonant component ($^1S-^1P, 1s^2 - 1s2p$) increasing rapidly as the NN distance decreases (Fig.15). For the comparison, electronic and opacity broadenings have been included. As the blue satellite is very sensitive to the density (here 6.10^{24}cm^{-3}), it potentially provides a

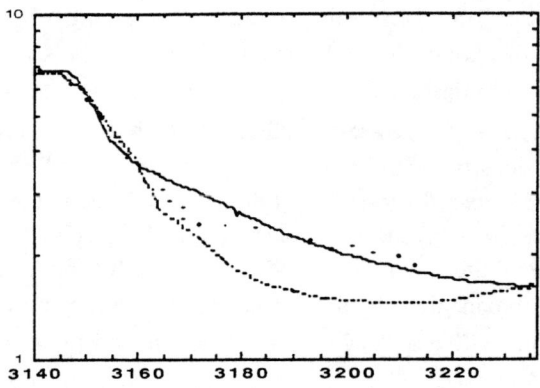

Figure 15. ArHeα high energy wing at the peak compression. The Two-Centre Model (QMSPECTRA) is in agreement with the experimental data. This agreement was obtained for $N_e = 6.10^{24} \text{cm}^{-3}$ and $T_e = 800\text{eV}$. There is a discrepancy with the standard model (Woltz and Hooper).

density diagnosis. The possible overlapping of the excited states orbitals taken into account in the Two-Centre Model explaining the anomalous splitting of components may indicate the initiation of a plasma-solid phase transition.

CONCLUSION

This paper shows the efficiency of the Two-Centre Model to reproduce dense plasma effects involving short interionic distances for which transition energies are splitted and oscillator strengths vary strongly. This treatment is of importance for the understanding of radiative properties in high energy density plasma and we show that the exhibition of satellite-like features gives a new possibility for density diagnosis in Laboratory dense plasmas reproducing astrophysical conditions (1,8). The same alternative treatment has yet been used for the calculation of the opacity at the center of the sun (3), and it should be considered for the study of the hydrogen emission (with transient $H-H^+$ and $H-H^+-H^+$ molecules) from the white and brown dwarfs (3).

REFERENCES

1. Leboucher-Dalimier, E. et al., J.Quant.Spectrosc.Radiat.Transfer **58**, 721 (1997)
2. Gauthier, P. et al., J.Quant.Spectrosc.Radiat.Transfer **58**, 597 (1997) ; Phys. Rev. E **58**, n°1 (1998)
3. Allard, N.F. et al., Astron.Astrophys. **330**, 782 (1998) ; 14th ICSLS PennState Univ., June 1998 ; Rose, S. J. et al., J. Phys. B **31**, L127 (1998)
4. H. R. Griem, *Spectro line broadening by plasma*, Academic Press, NY (1974)
5. Leboucher-Dalimier, E. et al., AIP, 11th Atomic Processes in Plasmas Auburn, March 1998
6. Drake, G.W.F., Phys.Rev. A **19**, 1387 (1979)
7. Devdariani, A. et al., Sauvan, P. et al., *LULI Reports*, NTIS PB97- 170963 (1997)
8. Hooper, C. F. et al., Phys. Rev. Lett. **63**, 267 (1989)
9. Woltz, I. A. et al., Phys. Rev. A **38**, 4766 (1989).

Measurements of Hydrogen Isotope Densities by Two-Photon Induced Lyman-α Fluorescence as a Proof for Tokamak Diagnostics

A. Steiger, K. Grützmacher, Ch. Seiser, M.I. de la Rosa and U. Johannsen

Physikalisch-Technische Bundesanstalt, Abbestr. 2-12, D-10587 Berlin, Germany

Abstract. Future application of two-photon induced Lyman-α fluorescence (2γ-Lα) measurements as an isotope selective determination of small neutral densities (H, D, T) in magnetically confined fusion plasmas requires quantitative studies of this novel technique. A first feasibility study has been terminated recently at a plasma beam (Ne $\approx 5 \cdot 10^{12}$ cm^{-3}, Te ≈ 10 eV), dedicated to the development of fusion plasma diagnostics. The 2γ-Lα signals were obtained from a measurement volume of 0.1 cm^3 only, although the background radiation was as expected for tokamak plasmas. Good detection sensitivity was achieved for atomic number densities down to 10^{10} cm^{-3}.

INTRODUCTION

The basic theory of Doppler-free 2γ-Lα fluorescence for hydrogen atoms in magnetically confined fusion plasmas was developed by Voslamber [1]. The two-photon absorption line profile was found to consist of two broad side components (due to the motional Stark effect in the electric field $E = v \times B$ as seen by an atom moving with velocity v through a magnetic field B), but also has a sharp, unshifted central component. Recent theoretical investigations revealed [2,3,4], that under fusion plasma conditions, regardless of temperatures from 0.1 keV to 10 keV, the spectral width of the central component is of the order of 1 GHz for two-photon excitation of all isotopes (H, D, T). These widths are very small compared with respect to the isotope separation which is 335.5 GHz for H - D and 111.6 GHz for D - T. Furthermore, the widths are fitting well with the spectral resolution of advanced UV-laser spectrometers, i.e. 500 MHz at 243 nm, and allow therefore for selective measurements of hydrogen isotopes by Doppler-free two-photon excitation.

Although the principle of 2γ-Lα fluorescence is very simple, measurements of neutral densities of hydrogen isotopes in magnetically confined fusion plasmas require extensive experimental developments in order to achieve the required single shot detection sensitivity at isotope densities as small as possible. The performance of 2γ-Lα measurements is depending basically on three parameters:
- the quality of the laser-spectrometer generating pulsed tunable narrowband radiation around 243 nm required for two-photon excitation,
- the properties of the detection system for measuring the 2γ-Lα fluorescence,
- the background emission of the plasma in the VUV spectral region around 121 nm.

During the last five years PTB and the Max-Planck-Institute of Plasma Physics (IPP) had a close cooperation in order to demonstrate first quantitative 2γ-Lα measurements at the plasma generator PSI-1 of IPP, Berlin Branch [5]. IPP has been responsible for the determination of plasma parameters by various diagnostic techniques [5] and PTB has been responsible for the 2γ-Lα measurements. The experimental setup and the results obtained are described in this report.

EXPERIMENTAL SETUP AND CONDITIONS

Plasma Parameters

The plasma generator PSI-1 is providing a magnetically confined plasma beam, which is generated by a high dc-current between a cathode and a hollow anode and expands horizontally through a differential pumping system into a diagnostic chamber. Finally the beam is stopped by a plasma dump. The diagnostic chamber is about 1 m long, and the plasma beam has a diameter of about 100 mm. For the 2γ-Lα measurements, the discharge was operated in pure deuterium and various hydrogen-deuterium mixtures at currents from 70 A to 400 A. The corresponding parameters of the plasma beam were: Ne in the range of 10^{12} cm^{-3} to $5 \cdot 10^{12}$ cm^{-3}, Te in the range of 5 eV to 10 eV, magnetic field strength about 0.1 T, total pressure in the target chamber about 0.1 Pa.

Lyman-α Detection System

It should be mentioned, that one real challenge of 2γ-Lα measurements at tokamak plasmas is the detection of the 2γ-Lα fluorescence, which appears on the nanosecond scale, with respect to the strong continuos VUV-background radiation. The situation at PSI-1 was even more difficult, because the flux of VUV-radiation from the stationary PSI-1 plasma was even higher than expected for tokamak plasmas. The detection system required therefore a flexible design, which allowed for adaptation to fairly unknown radiation flux conditions. The final detection system consisted of MgF$_2$ optics, a special photomultiplier, an oxygen gas filter, and could be equipped additionally with Lα-interference filters. It was optimized to achieve a high and selective sensitivity for Lα-radiation at 121 nm and high temporal resolution.

The Lα detection system had the following properties:

imaging: 1:1, f = 125 mm MgF$_2$ lens, Ø = 50 mm
photomultiplier: particle detector + MgF$_2$ window (special design, custom made)
 temporal resolution: 2,5 ns, gain of about 10^6
 electronic gate: 8 µs on, 60 times suppression of the dc signal while gate off
spatial filter: aperture in the image plane : 3 mm · 15 mm
spectral filters: the photomultiplier itself, sensitivity range: 115 nm-140 nm (figure 1)
 Lα-interference filters (6% transmittance) for stepwise attenuation,
 O$_2$ gas-filter, pressure dependent attenuation (see figure 2).

The O_2 filter was operated in a pressure range from $4 \cdot 10^4$ Pa to 10^5 Pa and had an absorption length of about 25 mm. This provided excellent suppression (i.e. two orders of magnitude) of the strong molecular background radiation emitted by the PSI-1 plasma, while the $L\alpha$-radiation was attenuated only by a factor of about 3.

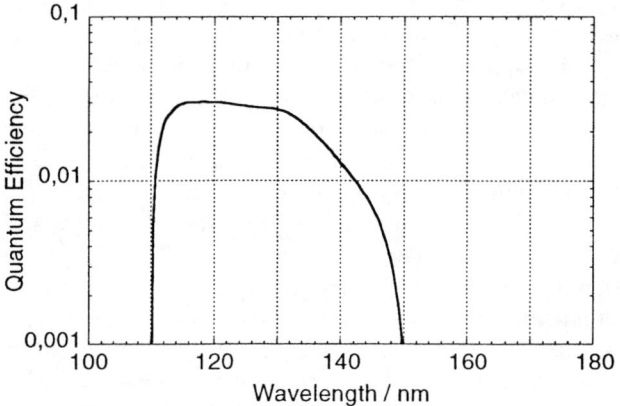

FIGURE 1. Quantum Efficiency (QE) of the custom made photomultiplier. The cut-off at shorter wavelength is due to the transmission cut-off of the MgF_2 entrance window, while the QE of the photocathode material is seen to decay rapidly for larger wavelength.

The absolute detection sensitivity of the actual configuration of the entire detection system was determined several times, and was found to be sufficiently stable and reproducible. For this, a special $L\alpha$-lamp [7] was used, which has been calibrated as $L\alpha$-radiation transfer-standard by the PTB laboratory for synchrotron radiometry at the electron storage ring BESSY I, Berlin.

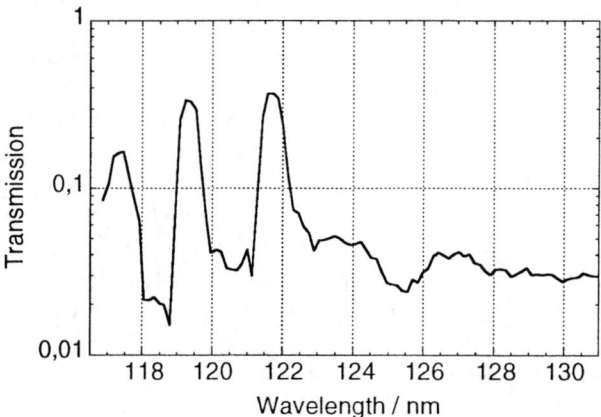

FIGURE 2. Typical transmission characteristics of the O_2 gas-filter - corresponding to 40 mm absorption length and $5 \cdot 10^4$ Pa at room temperature - measured with a spectral resolution of 1 nm.

UV-Laser Spectrometer

Because of the small two-photon absorption probability, the method requires pulsed tunable laser radiation of high beam quality and high spectral quality at 243 nm with much more power than commercial laser systems can presently provide. A new type of solid-state laser spectrometer [6] has been developed based on a commercial injection seeded Q-switched Nd:YAG laser. A single mode optical parametric oscillator and amplification by Ti:Sapphire crystals, both pumped by the second harmonic of the Nd:YAG, generate tunable pulsed radiation at 772 nm. This output is finally converted to 243 nm radiation by frequency mixing with the third harmonic of the Nd:YAG in a BBO crystal. At present, this table-top system delivers up to 50 mJ pulse energy in 2,5 ns at 243 nm with a bandwidth of about 300 MHz at a repetition rate of 10 Hz. The pointing stability (better than 100 µrad) and shot to shot reproducibility are excellent, and the spatial laser beam profile is sufficiently good.

Frequency control of the laser output at 243 nm is provided by an optogalvanic reference cell containing hydrogen or deuterium at a pressure of about 5 mbar. Partial thermal dissociation of the molecules is obtained by a resistively heated tantalum wire of about 2000 K. Less than 50 µJ are sufficient for two-photon excitation in a retroreflected beam. Subsequent photo-ionization, i.e. absorption of a third photon is producing ions which are finally measured by a pickup electrode as a current pulse. This optogalvanic detection allows for Doppler-free measurements and reliable frequency control.

Beam Preparation for Two-Photon Excitation

The laser-spectrometer was set up in a nearby laboratory and the 243 nm radiation was directed to the PSI-1 underneath the ceiling. There, the laser beam came down to a horizontal optical plane covering both sides of the diagnostic chamber and was prepared for two-photon excitation. Finally, two counter-propagating circularly polarized beams of equal pulse energy were directed into the PSI-1. The measurement volume, at a distance of 20 m from the laser, was as small as 0.1 cm^3 (3 mm in diameter, 15 mm long). It is defined by the diameter of the laser beams at the excitation zone and the length of the spatial filter of the $L\alpha$-detection system. The detection system was situated on top of the target chamber viewing downwards and a translator allowed for precise vertical positioning. Precise alignment of the lenses, used for directing the laser beams into the measurement volume, was provided by stepping motor driven translators.

Quantitative 2γ-$L\alpha$ measurements require the knowledge of the beam profile and the pulse energy present in the measurement volume. Beam profiles were recorded with a CCD camera using wedged plates for attenuation. Starting with typically 30 mJ pulse energy from the laser, about 10 mJ were available for each beam, having 3 mm in diameter. This is fitting well with the conditions required for an optimum 2γ-$L\alpha$ fluorescence yield (see below) in the measurement volume. Smaller beam diameters would lead to higher irradiances, which would start to quench the 2γ-$L\alpha$ fluorescence yield by photo-ionization losses due to absorption of a third photon.

DATA RECORDING AND EVALUATION

Data Recording and Measurements

During two campaigns, measurements were performed mainly in order to determine atomic number densities N_D of deuterium at different plasma conditions. In November 1996 the dependence on discharge currents was studied in pure deuterium plasmas, and in June 1997 N_D was determined at low discharge currents, mainly at 100 A, for various gas mixtures of deuterium and hydrogen.

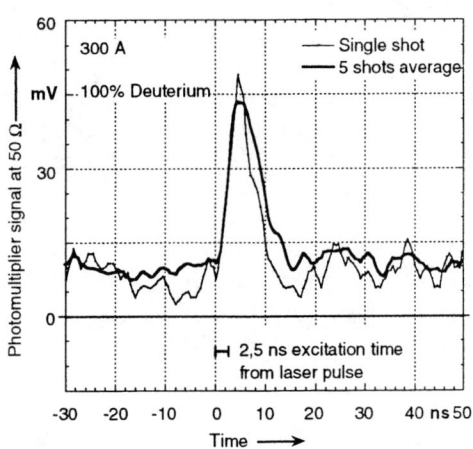 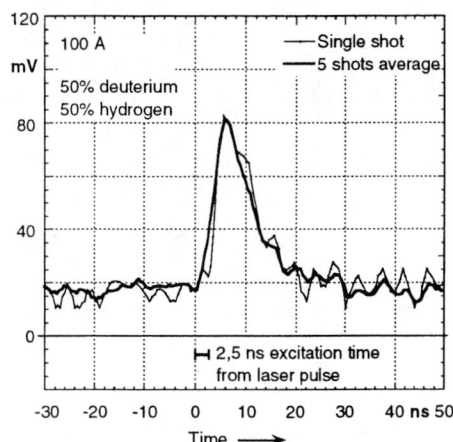

FIGURE 3. Single shot signals compared to 5 shot averages, obtained for deuterium at different discharge conditions and detector configurations. Left side: deuterium discharge at 300 A, detection system equipped with one Lα-interference filter. Right side: discharge with 50% hydrogen and 50% deuterium at 100 A, detection system without interference filter. Immediately with the laser excitation appears the 2γ-Lα fluorescence, well pronounced compared to the background signal of the plasma. The example at the right side corresponds to tokamak background radiance and an atomic number density of about 10^{11} cm^{-3}.

Typical measurements consisted of 600 Lα-detector signals recorded at the repetition rate of the laser. Such a measurement was performed with the laser frequency either fixed at the center of the resonance or tuned across the resonance in order to obtain spectral profiles. For each laser pulse the Lα-detector signal at 50 Ω was recorded with a fast digital storage oscilloscope for a time interval of 100 ns at 2 GHz sample rate, see figure 3. Additionally the optogalvanic signal was recorded as frequency monitor. The 2γ-Lα signal could be measured with good signal to noise ratio, immediately with the laser excitation appears the 2γ-Lα fluorescence, lasting about 10 ns, well pronounced compared to the background signal of the plasma.

The calibrated sensitivity of the detection system allowed for the determination of the background radiance emitted from the stationary PSI-1 plasma. The example at the right side in figure 3 corresponds to a background signal as expected from tokamak plasmas, i.e. to a photon flux of about 10^{15} $(s \cdot sr \cdot cm^2)^{-1}$ and to an atomic number density of about $8 \cdot 10^{10}$ cm^{-3}. However, at the highest discharge current of 400 A investigated, the radiance of PSI-1 is twenty times larger compared to tokamak plasmas.

Statistical Analysis of the 2γ-Lα Signals

The statistical analysis of the data was performed in order to investigate the single shot detection noise of the 2γ-Lα signals. For this, it is necessary to relate the photomultiplier currents to an average number of detected 2γ-Lα photons per time interval, e.g. 10 ns. This requires the knowledge of the photomultiplier gain, which was determined from the single photon pulse response. The gain was found to be about $8.5 \cdot 10^5$ (at 2.75 kV photomultiplier voltage), i.e. in good agreement with the nominal gain of 10^6. A photomultiplier signal of 20 mV at 50 Ω integrated over 10 ns corresponds to 30 photons therefore. Based on this relation, the integral leads to a number of detected 2γ-Lα photons $N_{2\gamma\text{-L}\alpha}$, while the background signal corresponds to a number of background photons N_{BG}. According to Poisson statistic, the expression

$$\Delta N/N = (N_{2\gamma\text{-L}\alpha} + N_{BG})^{1/2} / N_{2\gamma\text{-L}\alpha} \qquad (1)$$

gives the theoretical limit of the standard deviation $\Delta N/N$ for the detected 2γ-Lα photons. The statistical analysis of the recorded signals exhibited, that the standard deviations were found close to the limit of Poisson statistic, e.g. see Table 1.

TABLE 1. Statistical analysis of 2γ-Lα signals measured for deuterium. The deuterium concentration in a hydrogen discharge of 100 A was varied from 3% to 50%. The atomic number density at a deuterium concentration of 50% was about 10^{11} cm^{-3}

deuterium concentration / %	3	10	50
single shot standard deviation / %	55	35	28
theoretical limit of Poisson statistic / %	60	28	14

2γ-Lα Yield at PSI-1

The determination of absolute atomic number densities requires beside the calibrated detection system the knowledge of the 2γ-Lα fluorescence yield Y_α, i.e. the probability per atom for the emission of a two-photon induced Lα-photon.

In general, the yield depends on spatial and temporal distribution of the irradiance E provided by the laser beams and on further experimental parameters as well, i.e.:
- spectral properties of the laser radiation and the two-photon absorption profile,
- the fluorescence decay time for Lα emission after two-photon excitation,

and on atomic constants, i.e. the cross-section for two-photon excitation and the cross-section for photo-ionization due to a third photon.

A reasonable treatment for the determination of the temporal and spatial average of the yield present in the measurement volume is achieved by a two step procedure:

In a first step the irradiance dependence of Y_α is calculated for the smooth temporal pulse shape of the laser by numerical integration of rate equations which account for two-photon excitation and de-excitation between the 1S and 2S levels, the photo-ionization as a loss channel, and the 2γ-Lα fluorescence decay via collisional excitation transfer from the 2S state to the 2P state [8].

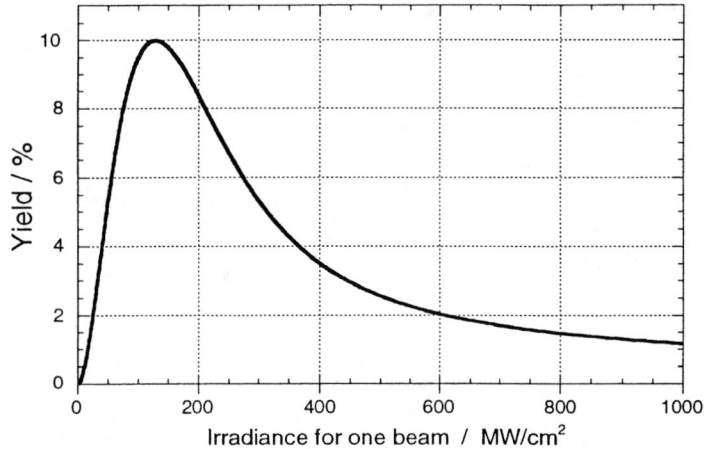

FIGURE 4. 2γ-Lα fluorescence yield Y_α curve for PSI-1 versus the mean irradiances $<E_{(t)}>$

This integration results in the dependence of Y_α on the temporal mean value $<E_{(t)}>$ of the irradiance as shown in figure 4. The experimental inputs for the determination of the yield curve are:
- temporal pulse shape of the laser pulse: FWHM = 2.5 ns,
- spectral widths of the laser obtained from the reference cell: Δv_L = 360 MHz,
- spectral widths of the 2γ-absorption profile as measured at PSI-1: $\Delta v_{2\gamma}$ = 450 MHz,
- 2γ-Lα fluorescence decay time as measured at PSI-1: τ_{21} = 8 ns.

Concerning the shape of the yield curve, it should be noted, that the photo-ionization is negligible for small irradiances (up to about 20 MW/cm^2) and Y_α is increasing therefore with the square of $<E_{(t)}>$. For larger $<E_{(t)}>$ values the ionization starts to dominate rapidly the Lα yield. Finally, the photo-ionization causes remarkable losses for the two-photon excited level and diminishes strongly the Lα yield.

The second step accounts for the measured spatial distribution of the irradiance in the measurement volume as given in figures 5 and the volume average $<Y_\alpha>$ of the yield is calculated as follows.

As already mentioned, intensity distributions of the beam profile were recorded in between 2γ-Lα measurements via a beam splitter with a CCD camera outside the PSI-1. Together with the measured pulse energy and temporal FWHM of the laser pulses, the CCD intensity picture were transformed into the spatial distribution of the irradiance of the laser beam in the measurement volume. An example is given in figure 5. Combined with the calculated yield, the irradiance distribution is transformed into the yield distribution figure 6. Please note, that the maximum irradiance areas in the beam profile lie already beyond the maximum yield in figure 4. Therefore, the corresponding areas in the yield profile (figure 6) are not the ones with the maximum yield. However, they are surrounded by large areas with higher yield values. These areas are belonging to irradiances, which lie closer to the optimum irradiance of 125 MW/cm^2.

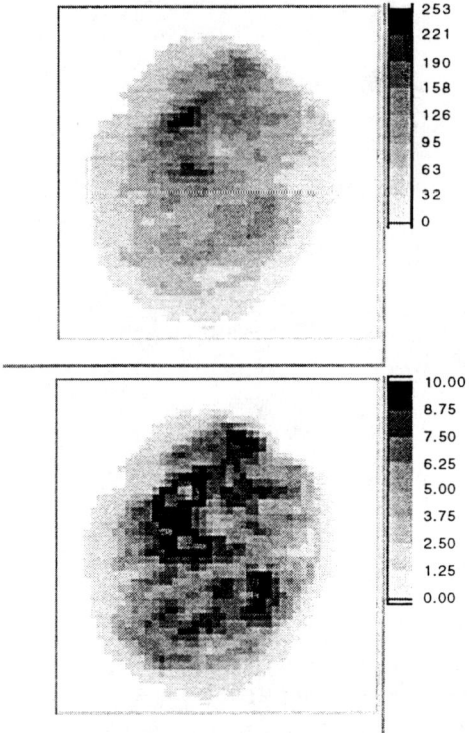

FIGURE 5. A two-dimensional plot of a typical laser beam irradiance distribution at a wavelength of 243 nm is shown, as present in the measurement volume. With a typical pulse energy of 10 mJ within 2.5 ns for each beam, the maximum local irradiance is 253 MW/cm^2, whereas most of the pixels have an irradiance of about 100 MW/cm^2.

FIGURE 6. The corresponding spatial distribution of the Lα yield is displayed. This is obtained by assigning the calculated value (see figure 4) to each pixel of figure 5

Finally the yield profile was used to determine $<Y_\alpha>$, i.e. the volume average of the 2γ-Lα yield over the whole measurement volume. For typical conditions, the mean value $<Y_\alpha>$ was about 4%. This is about 90 % of the corresponding value to be achieved by a perfect Gaussian beam profile.

Determination of Atomic Densities

Finally, the number densities N_D of deuterium ground state atoms in the measurement volume V can be determined, using the following relation:

$$S_\alpha = <Y_\alpha> \cdot V \cdot N_D \cdot DS \qquad (2)$$

S_α is the time integrated 2γ-Lα fluorescence signal, $<Y_\alpha>$ the volume average of the 2γ-Lα yield and DS the sensitivity of the calibrated detection system.

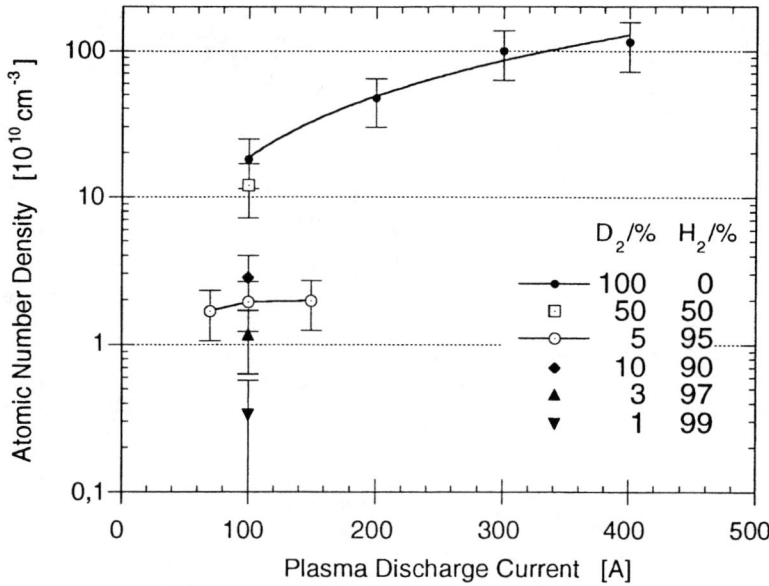

FIGURE 7. Atomic deuterium number density (varying over more than two orders of magnitude of) versus the discharge current in a pure deuterium plasma, and for various input gas mixtures at a fixed discharge current of 100 A

Figure 7 shows the main results obtained for atomic deuterium densities from the two measurement campaigns: the dependence of N_D on discharge currents in pure deuterium and the variation of N_D at low discharge currents, mainly at 100 A, with various gas mixtures of deuterium and hydrogen.

The total relative uncertainty for N_D was less than 40%. This results from uncertainty contributions (typically below 15%) related to the calculated 2γ-Lα yield Y_α, the positioning of the detection system, the calibration procedure of the detection system and the determination of the laser beam irradiance distribution. The standard deviation, which was normally achieved for the integrated signals averaged over 10 laser pulses, was usually well below 15%, e.g. for deuterium densities larger than $5 \cdot 10^{10} cm^{-3}$.

Kinetic Temperature of Deuterium

In addition to the standard optical setup where the two laser beams entered the diagnostic chamber exactly counter-propagating, a second arrangement was realized for measuring the kinetic temperature of the deuterium atoms: a small crossing angle of 52 mrad for the two laser beams was used causing a residual Doppler width of 2.6 %. This resulted in a spectral profile of Gaussian line shape with a half width well above the laser bandwidth but small enough to be detected, as can be seen in figure 8. From this

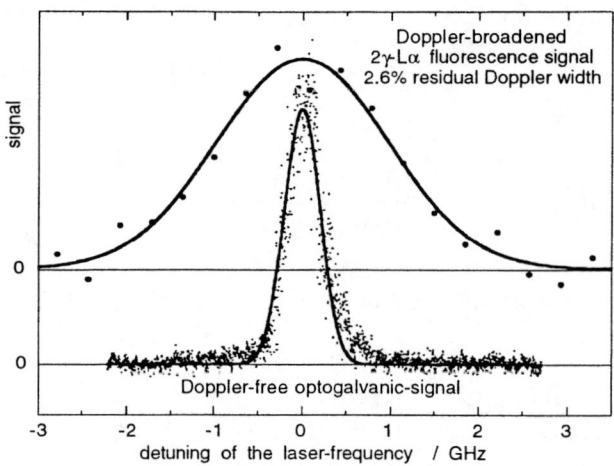

FIGURE 8. Spectral profile obtained from a pure deuterium plasma at a discharge current of 300 A. The fluorescence signal was recorded together with the Doppler-free optogalvanic signal from the reference cell, while slowly scanning the laser across the resonance. Each point for the 2γ-Lα fluorescence signal is the mean value over 100 laser pulses while the stability of the laser can be seen from the optogalvanic signal, plotted for each laser pulse.

half width, a kinetic temperature of the deuterium atoms of 1.6 eV could be derived after subtraction of the laser bandwidth.

CONCLUSION

First quantitative 2γ-Lα measurements at the plasma generator PSI-1 were performed and the feasibility of isotope selective diagnostic was demonstrated successfully. Although the measurement volume was only 0.1 cm^{-3} in a plasma beam of 100 mm in diameter, atomic deuterium number densities as small as 10^{10} cm^{-3} were determined close to theoretical detection limit predicted for PSI-1 plasma conditions and experimental setup. Typical pulse energies provided in the measurement volume were 10 mJ for each of the two counter-propagating laser beams. The measurements were performed at plasma conditions, were the perturbing background radiance was in a range as expected for tokamak plasmas, and up to twenty times higher.

Based on these results and on the experience gained so far, it is possible now to define an experimental setup for a feasibility study of 2γ-Lα measurements in edge plasma regions of existing tokamaks. However, the verification of basic theoretical predictions and the demonstration of a single shot detection limit for isotope concentrations (as small as 10^8 cm^{-3} for temperatures not exceeding 1 keV) requires the realization of especially suited laser and detection systems.

ACKNOWLEDGMENT

This work was carried out in a close cooperation with the Max-Planck-Institute of Plasma Physics (IPP), Ass. EURATOM, in order to develop 2γ-Lα measurements for tokamak diagnostic. We are very grateful to the Berlin Branch of IPP, headed by G. Fußmann and especially to W. Bohmeyer, H. Grote and H.D. Rainer operating the plasma generator PSI-1.

REFERENCES

1. D. Voslamber, Determination of neutral particle density and magnetic field direction from laser-induced Lyman-α fluorescence, II - Two-photon excitation, Report EUR-CEA-FC-1387 (1990)
2. D. Voslamber, *Spectral Line Shapes, Vol. 8* (AIP Conf. Proc. 328), Eds. A.D. May, J.R. Drummond, E. Oks, American Institute of Physics, New York (1995), p.3
3. J. Seidel and D. Voslamber, to be published
4. D. Voslamber, W. Mandel, Determination of the D/T fuel mixture using two-photon laser induced fluorescence in combination with neutral beam injection, *23rd EPS Conference on Controlled Fusion and Plasma Physics*, Vol. 20 C, 1996, p. 987
5. H. Behrendt, W. Bohmeyer, L. Dietrich, G. Fussmann, H. Greuner, H. Grote, M. Kammeyer, P. Kornejew, M. Laux, E. Pasch, *21st EPS Conference on Controlled Fusion and Plasma Physics*, E. Eds. Joffrin et al., Europhysics Conference Abstracts Vol. 18 B, part III, 1994, p. 1328
6. A. Steiger, K. Grützmacher and M.I. de la Rosa, *Laser in Research and Engineering*, Eds. w. Weidelich et al. Springer Verlag, Berlin 1996, p.308
7. Hydrogen lamp, type LGV-1.0, optimized for Lyman-α emission, All Russian Institute - Optical and Physical Measurements
8. The fluorescence decay time is taken from the 2γ-Lα signals and was found to be within 7 to 9 ns. This decay time results from the underlying two step process: collisional excitation transfer from the two-photon excited 2S levels to the 2P levels and subsequent Lα emission. The Lα Einstein coefficient A_{21} corresponds to a radiative decay time of 2.1 ns, hence the fluorescence decay time under PSI-1 plasma conditions is dominated by the collisional transfer.

RADIATIVE REDISTRIBUTION MODELING FOR HOT AND DENSE PLASMAS

C. Mossé, A. Calisti, B. Talin, R. Stamm,
Laboratoire de Physique des Interactions Ioniques et Moléculaires,
centre de Saint Jérôme, case 232,
13397 Marseille cedex 20, France.

R.W. Lee,
Department of Physics, University of California, Berkeley, CA 94720-7300, USA.

And

L. Klein.
Department of Physics, Howard University, Washington DC 20059, USA.

ABSTRACT

A model based on an extension of the Frequency Fluctuation Model (FFM) is developed to investigate the two-photon processes and particularly the radiative redistribution functions for complex emitters in a wide range of plasmas conditions. The FFM, originally, designed as a fast and reliable numerical procedure for the calculation of the spectral shape of the Stark broadened lines emitted by multi-electron ions, relies on the hypothesis that the emitter-plasma system can be well represented by a set of "Stark Dressed Transitions", SDT. These transitions connected to each others through a stochastic mixing process accounting for the local microfield random fluctuations, form the basis for the extension of the FFM to computation of non-linear response functions. The formalism of the second order radiative redistribution function is presented and examples are shown.

INTRODUCTION

The radiation scattering process or radiative redistribution plays a fundamental role in the theoretical radiation transfer studies of the high temperature laboratory plasmas or the astrophysical plasmas. Actually, for plasma lines with a large optical depth there remain difficult problems involving radiative transfer which cannot be understood simply through calculations of the one photon absorption or emission spectrum. These problems require the development of a model able to treat the scattering of near resonant radiation in hot dense plasmas. Moreover, the two-photon process study is strongly motivated by the recent progresses in the development of X-ray lasers with wavelengths appropriate for photo-pumping the ground state transitions of multi-electron ions, suggesting direct investigation of radiation scattering and further non linear spectroscopy investigations applied to laboratory plasmas[1].

One of the relevant functions for the description of a resonant radiation scattering process, is the so-called frequency redistribution function, e.g., the joint probability density for the absorption of a photon of a given frequency and the emission of a photon of a different frequency. In hot and dense plasma conditions, the calculation of such a function is a very complex problem due to the atomic structure of multi-electron emitters and the various line broadening mechanisms. The time-dependent coupling of the emitters with the plasma gives rise to Stark broadening of the resonance lines which can, therefore, be considered as partially inhomogeneous. Thus, it follows that cases where the lines are neither purely inhomogeneous nor homogeneous, will be considered. This situation results from the so-called ion dynamics effect which has been, successfully, interpreted, by the mean of the Frequency Fluctuation Model (FFM)[2,3], as a mixing effect of the inhomogeneous components. In the following, an approach of the two photon plasma spectral properties in presence of a combination of homogeneous and inhomogeneous broadening processes will be developed by extending the usual linear response formulation of the spectral line shape.

After a brief review of line shape theory in the linear response approximation and a summary of the main formal aspects of a line shape calculation involving an emitter perturbed by a stochastic electric micro field using FFM, an extension of the FFM for investigating the two-photon processes and, particularly, the radiative redistribution functions for complex emitters, will be developed. Examples of Radiative redistribution functions for simple systems are presented in order to clarify all the involved concepts. Finally, examples taken in the X-ray laser field are shown, illustrating the capability of the model to provide a sensitive method for the study of radiative transfer under conditions where the mixing of inhomogeneous components of spectral lines is not rapid enough to result in complete redistribution.

THE ONE PHOTON SPECTRAL LINE SHAPE.

The expression for the linear response of a plasma-emitter system to an unpolarized monochromatic electromagnetic wave of angular frequency, ω, is determined by the response function, $G(\omega)$, which is the one sided Fourier transform of $U(t)$, the bath averaged evolution operator of the emitter,

$$G(z) = -\frac{i}{\pi} \int_0^\infty e^{izt} U(t) dt = (z - L_o)^{-1} . \qquad (1)$$

L_o is the Liouville operator for the emitter evolution only. The line shape function in the radiative dipole approximation is related to the imaginary part of the Fourier transformed dipole autocorrelation function. This can be written as a normalized Liouville space matrix element of the response function,

$$I(\omega) = \text{Im}\langle\langle d^+ | G(\omega) | d\rho_o \rangle\rangle , \qquad (2)$$

where d and ρ_0 are, respectively, the dipole and the equilibrium density matrix operator for the active quantum system. Although not necessary, a sum over the polarization directions is assumed for simplicity. As it has been noted in the introduction, the plasma-emitter coupling is time-dependent. If we consider the interaction fluctuations or collisions to be random, this means that a stochastic Liouville equation (SLE) must be solved to obtain $G(\omega)$.

Instead of attempting to exactly solve the SLE, the FFM relies on the idea that the quantum system in the electric micro field behaves like a set of dressed two level transitions, involved in a collision-type mixing process induced by the field fluctuations. Before considering the field fluctuations, the quantum mechanics for the pseudo system is built for static ion perturbers. The most appropriate basis when the emitter is imbedded into a space and time constant electric field is the eigen basis, or the Stark basis, in which the Liouville evolution operator becomes diagonal. For each radiative transition, this procedure yields a spectral line shape function that can be written as a sum of rational fractions or generalized Lorentzian spectral components of the line. These are the static Stark components, each of which is characterized by a generalized frequency and complex intensity. These contributions resulting from the whole micro field and generally from several atomic states, cannot be distinguished from each others. In order to use objects having more physical meaning, the Stark Dressed Transitions (SDT) have been defined. Each SDT is the effective two level radiative transition leading to an equivalent contribution than a set of Stark components located in a small neighborhood in the frequency-width plan. These objects are built numerically with the constraint to preserve the inhomogeneous structure of each Stark broadened radiative transition and to retain the inhomogeneous structure due to the various radiative transitions simultaneously considered in the spectral domain of interest. As the primitive Stark components, the SDT can be interpreted like two level (e, g) elementary radiators characterized by two complex numbers, respectively, a generalized intensity a_k+ic_k and a complex frequency $f_k+i\gamma_k$. The static line profile is positive definite and can be written as an average on the initial states and a sum on the final states of the SDT contributions. In an extended Liouville basis $\{|e,g;i\rangle\rangle\}$, the intensity distribution takes the simple following form;

$$I(\omega) = \mathrm{Im} \sum_{i,f} p_i \left(<< \mathbf{D}|\mathbf{G}_0(\omega)|\mathbf{D}\rho_0 >>\right)_{i,f}, \quad (3)$$

where

$$\mathbf{G}_0(\omega) = (\omega.\mathbf{1} - \mathbf{L}_0)^{-1}, \quad (4)$$

with **1** the unit operator. L_0 is a diagonal operator which contains the complex frequencies of the SDT,

$$(\mathbf{L}_0)_{kl} = (f_k + i\gamma_k)\delta_{kl}. \quad (5)$$

The p_i represent the weights of the SDT, $\sum_i p_i = 1$, while the effective dipole **D** is defined for convenience by,

$$(\mathbf{D}_i)_{eg} = \sqrt{(1+ic_i/a_i)\sum a_k}, \quad (6)$$

with $\sum_k a_k = r^2$, the square of the reduced matrix element of the transition involved in the emission of the ith SDT.

At this stage, the ion dynamics is included through the hypothesis that the time dependence of the fluctuating ion microfield causes a mixing of the SDT. When only one Stark broadened radiative transition is considered at a time, the slowly varying ion Stark effect is assumed to transfer population between different SDTs. This transfer is observed in the absorption or emission spectrum as an exchange mechanism which mixes the formerly separate radiative channels. The mixing of the levels and radiative channels through the exchange mechanism is assumed to be a stationary Markov process parametrized by a fluctuation rate, ν, characterizing the ion field fluctuations. This process is completely determined by two sets of quantities: the instantaneous radiative channel probability vector, **p**, such that $p_i = a_i / \sum a_k$, and a transition rate operator, **W**. Concerning the one-photon spectrum, only the parameters required are the elements of **p** describing the probability of occurrence of a particular radiative channel transition, $(p_i)_{eg}$, and the coherence mixing rate matrices, $(\mathbf{W})_{eg}$ or $(\mathbf{W})_{ge}$. To ensure detailed balance, the probability and rate elements are then connected by,

$$(\mathbf{W}_{ij})_{eg} = -\nu (p_i)_{eg} \ (i \neq j), \ (\mathbf{W}_{ii})_{eg} = \nu [1 - (p_i)_{eg}]. \tag{7}$$

The diagonal and off-diagonal elements, $(\mathbf{W}_{ii})_{eg}$ and $(\mathbf{W}_{ij})_{eg}$, are, respectively, the inverse lifetime and the mixing rate of the radiative channels. The linear response line shape is, then, written in the same form as eq.(3) but with $\mathbf{G}_0(\omega)$ replaced by,

$$\mathbf{G}_W(\omega) = (\omega.\mathbf{1} - \mathbf{L}_0 - i\mathbf{W})^{-1}. \tag{8}$$

Before going further, a discussion on the time dependent characteristics of the microfield is necessary. For low coupling plasma parameter conditions, the random ion motion tends to supersede the collective motion and the relevant characteristic time is no longer the oscillation plasma frequency, ω_p. Due to the long range of the Coulomb interaction, even if it is used with a Debye screening, the field seen on the emitter results generally from a number of ions at the same time. The ion configuration that creates this field at a given time, changes when the ions move. It will be assumed that the correlation is lost when the configuration is different, that is, when the average displacement is about the mean distance between the plasma ions. The corresponding correlation time is defined as, $\frac{r_0}{v}$, where v is the ionic thermal velocity and r_0 is the mean distance between the plasma ions.

Calculations of the spectral line shape of ion emitters in hot, dense plasmas have been performed with the FFM, including the ion dynamics effect, and comparisons with experiments have verified the accuracy of the theory[4,5].

The following section is devoted to the development of the theoretical tools relevant for the case of two photon pump-probe processes.

REDISTRIBUTION.

The present approach refers to a theory developed earlier[6] with the purpose of evaluating the radiative response of an atomic system simultaneously submitted first to a monochromatic external radiation, e.g., a weak harmonic electric field not strong enough to produce a significant ac Stark effect, and second to the perturbation induced by a plasma-like thermal bath, e.g., a strong Markovian stochastic electric field with some additional damping due, for instance, to electron impact. The field interaction operators, V_S or V_L, standing respectively for the coupling of the spontaneous emission or the incoming pump radiation, respectively, with the generalized dipole operator of the dressed two-level FFM emitter, are time-dependent and have the form,

$$V_R(t) = V_R^+ e^{-i\omega_R t} + V_R^- e^{i\omega_R t}. \tag{9}$$

$V_R^\pm(t) = -\mathbf{D}.\mathbf{E}_R^\pm$, is the emitter-radiation field dipole interaction with field amplitude, E_R, and frequency, ω_R, describing the pump field for R=L and the spontaneous emission field for R=S. A pump-probe theory developed in the framework of a Model Microfield Method accounting for atomic systems embedded in strong perturbing fields, is used as a basis and modified to be used with the field dressed systems[7,8].

The two-photon redistribution is characterized by continuing the expansion of the response function to higher order than linear. The power spectrum of the radiation emitted at frequency ω_S, by a system pumped at frequency ω_L, is written

$$I(\omega_S, \omega_L) = \lim_{\eta \to 0} \mathrm{Tr}\{[V_S \tilde{G}_W(i\eta) V_L]_{av} \rho_0\}. \tag{10}$$

The average over the Markovian micro field states noted $[...]_{av}$, is transformed into a sum over the SDT weighted with their probability, p_i. Thus, equation (10) can be written,

$$I(\omega_S, \omega_L) \propto \mathrm{Im} \sum_{i,f} p_i \left(<< V_S | \tilde{G}_W(0) | V_L \rho_0 >>\right)_{i,f}, \tag{11}$$

where $\tilde{G}_W(z)$, the one-sided Fourier transform of the evolution operator for the SDT, now contains the interaction operators, V_S and V_L.

To remove the explicit time dependence in equation (11), the Liouville space basis is augmented to include Floquet numbers, n_L, that count the number of harmonics of the pump frequency present in each order of the response calculation and to include the photon numbers[9], n_e and n_g. The states of the extended Liouville space are, thus, written as $|e,g;i;n_L,n_e,n_g>>$ and the resolvent operator is given by,

$$\tilde{G}_W(z) = (z-L)^{-1} = [z - L_0 + \Omega + iW - (V_S + V_L)]^{-1}, \tag{12}$$

with $\Omega |e,g;i;n_L,n_e,n_g>> = (n_L \omega_L + (n_e - n_g)\omega_S) |e,g;i;n_L,n_e,n_g>>$.

The above propagator satisfies a Dyson equation, in terms of $\tilde{G}_W^0$, the field free linear response operator in the extended Liouville-Floquet space,

$$\tilde{G}_W(z) = \tilde{G}_W^0(z) + \tilde{G}_W^0(z)(V_s + V_L)\tilde{G}_W(z). \tag{13}$$

In the following, the studied cases are such that the pump field can be considered to be weak and the iteration of the Dyson equation can be completed at the lowest non-vanishing order. After some algebra, one gets using equation (11),

$$I(\omega_S,\omega_L) \propto \mathrm{Im}\sum_{i,f} p_i (<< V_S | \tilde{G}^0_W(\omega_S)[V_L \tilde{G}^0_W(\omega_S-\omega_L)V_S \tilde{G}^0_W(-\omega_L)$$
$$+ V_S \tilde{G}^0_W(0) V_L \{\tilde{G}^0_W(\omega_L) + \tilde{G}^0_W(-\omega_L)\}] | V_L \rho_0 >>)_{i,f}. \tag{14}$$

The two terms in this expression correspond, respectively, to the Rayleigh scattering term giving rise to a spectral line centered on the incident frequency, ω_L, and the redistribution term that depends strongly on the characteristics of the exchange mechanisms fixed in the transition rate matrix **W**.

At this stage, it is necessary to define some relevant quantities required for the discussion of radiation redistribution:

- To describe the photo pumped fluorescence, the redistribution function, $R(\omega_L, \omega_S)$, which is the joint probability density for the absorption of a photon of frequency ω_L and the emission of a photon of frequency ω_S, is used,

$$R(\omega_L,\omega_S) = \frac{I(\omega_L,\omega_S)}{\iint I(\omega_L,\omega_S)d\omega_L d\omega_S}. \tag{15}$$

- A conditional function, $P(\omega_L, \omega_S)$, is defined from the redistribution function as,

$$R(\omega_L,\omega_S) = f(\omega_L)P(\omega_L,\omega_S). \tag{16}$$

where $f(\omega_L)$ is the normalized absorption line shape describing the probability of the absorption of a photon at ω_L.

- In the simplest case of an isolated resonance, $P(\omega_L, \omega_S)$, is the conditional probability of observing a photon scattered at ω_S with the absorption of a photon at ω_L and we have,

$$\int P(\omega_L,\omega_S)d\omega_S = g(\omega_L) = 1. \tag{17}$$

- In case of complete redistribution, an additional relation holds,

$$P(\omega_L,\omega_S) = f(\omega_S), \tag{18}$$

implying that,

$$R(\omega_L,\omega_S) = f(\omega_S)f(\omega_L). \tag{19}$$

- When several transitions must be considered at the same time in the scattering process, we have,

$$f(\omega_L)g(\omega_L) = \int P(\omega_L,\omega_S)d\omega_S, \tag{20}$$

and

$$\int f(\omega_L)g(\omega_L)d\omega_L = 1. \tag{21}$$

In the following section, simple examples are presented in order to clarify the main intuitive concepts and to illustrate the contribution to redistribution of the various line broadening mechanisms, e.g., the homogeneous broadening of inhomogeneous components and component mixing. Finally a last example is presented illustrating the use of the model to investigate a real and complex case.

EXAMPLES AND DISCUSSION.

Two fundamentally different cases must be addressed. The first one concerns the more common case of broadened, but isolated lines, and the second, a more complicated situation where several radiative transitions must be considered at the same time. The latter case occurs mainly when at least two lines share the same upper level, or whenever a collision process couples two or more radiative transitions with a rate high enough to perturb the radiation scattering. For both these situations the radiative transitions cannot be considered separately because, for the two lines sharing the same excited level and for the transitions coupled by collisions, radiation absorbed by one transition can be emitted by the other.

We consider the redistribution function for a pattern composed of several transitions coupled by a mixing process by considering the two transitions systems in Fig.1 below.

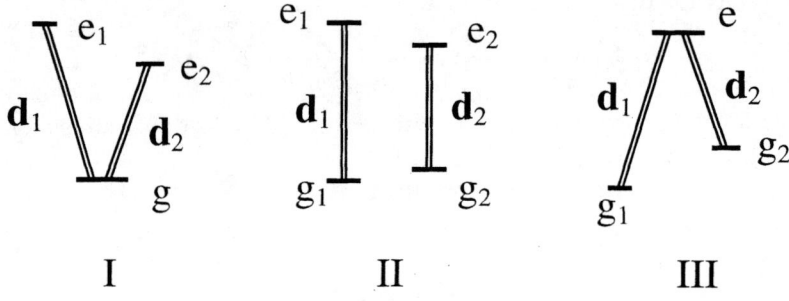

Figure 1.

The letters "e" and "g" denote, respectively, the ground and the excited level, d_1 and d_2 stand for arbitrary dipole moments.

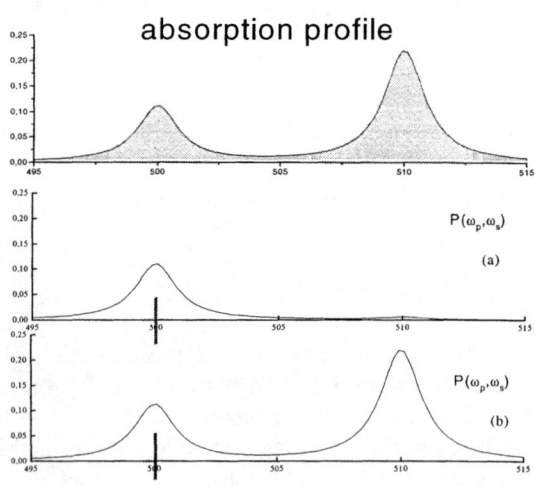

Figure 2.

The redistribution calculations have been limited to second order not accounting for saturation effects. Thus, both configurations I and II, shown in Fig.1, are equivalent and yield a non complete redistribution. The second case of interest is that of lines that share the same upper level, a situation resulting in complete redistribution.

These two types of situations are resumed in Fig.2 where the one photon profile is compared to the scattered spectra corresponding to the both cases, e.g., complete and partial redistribution. The spontaneous emission rate is supposed small enough to make the Rayleigh peak negligible. Case (a) corresponds to configurations I and II in which the redistribution is non-complete and strongly depends on the mixing rate. Case (b) corresponds to configuration III in which a photon pumping the system is scattered with the same frequency distribution as the one photon absorption probability distribution independently of the pump frequency. This situation is modeled by equalizing the populations of the levels in the upper manifolds of Eq.14.

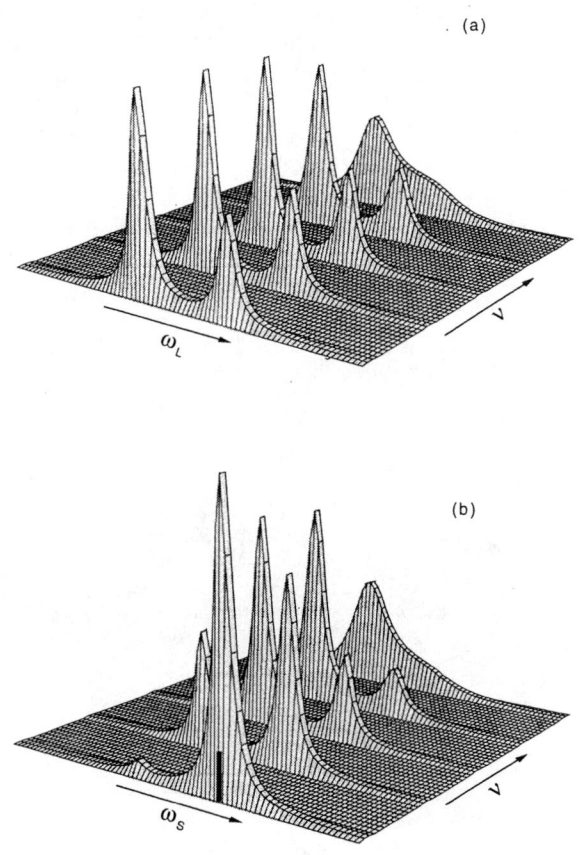

Figure 3.

To better illustrate the effect of component mixing on the spectra, the behavior of the one photon and the scattered spectra for a series of increasing fluctuation rates, ν, has been plotted on Fig.3 (a) and (b) respectively, in the case of configurations I and II. For the one photon case the effect of component mixing becomes important when the fluctuation rate is of the same order of magnitude as the component separation. The behavior is completely different for the two photon case, where, for a null or vanishing fluctuation rate, the scattered profile is unlike the corresponding one photon profile. This is a consequence of incomplete redistribution. When the fluctuation rate is greater than the life time of the upper levels, redistribution becomes complete as can be seen comparing the rear curves of Fig.3 (a) and (b).

The next example will illustrate the use of the model for the investigation of a real and complex case, the 3d-2p Balmer alpha transition of hydrogen like Carbon. Because this transition does not involve the ground state, it is more appropriate for studies related to radiation transfer than for a straightforward scattering measurement. However, this case is apropos because it is both simple and involves all the important elements relevant to radiation redistribution. For the considered plasma conditions, $N_e=5.10^{19} \text{cm}^{-3}$ and $T_e=10$ eV, the in-homogeneity of the radiative pattern is due to the fine structure and the Stark effect. The two dimensional redistribution function, $R(\omega_L,\omega_S)$, is given in Fig.4. This surface represents the elementary redistribution required for the radiation transfer equation.

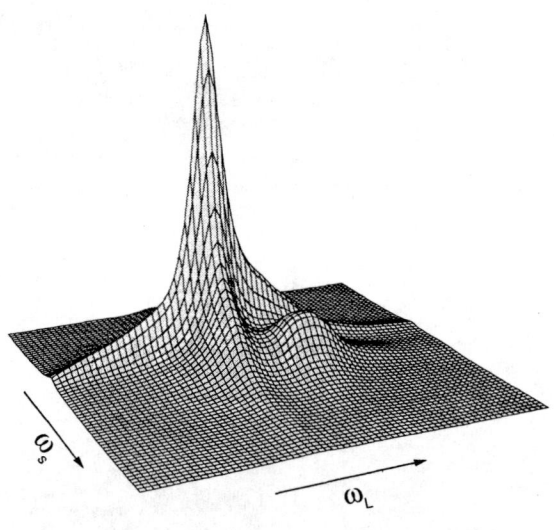

Figure 4.

As an application of the theory, the feasibility of a photo pumping experiment has been studied[10] targeting a resonance between MgIV plasma and an X-ray Zr laser at 146.515 Å. The plasma conditions used in the calculations were limited by the restriction to ground state transitions of a weakly ionized emitter and by available X-ray laser wavelengths. For these conditions, only a small ion Stark broadening effect is present in the line profile and therefore the presented studies had no redistribution associated with the ion dynamics effect. Nevertheless, these ionic transitions are composed of a number of merged and overlapping fine structure transitions and Stark components, which yield a resonance fluorescence spectrum sensitive to the photo pumping of particular inhomogeneous components.

CONCLUSION.

The FFM approach, which has been used successfully in the past to provide complex ionic absorption or emission spectral line shapes for emitters in hot, dense plasmas, has been extended to enable calculations of radiative redistribution. The procedure constitutes a powerful new method for the investigation of radiative transport in plasmas. Since the use of the FFM provides novel diagnostic capabilities that permit the study of cases where ion dynamics effects might be important, the extension to redistribution also provides a more sensitive method for the study of radiative transfer under conditions where the mixing of in-homogeneous components of spectral lines is not rapid enough to result in complete redistribution. In particular, line transitions in plasmas with conditions which involve substantial ion Stark broadening are of interest in the study of fluctuating ion micro fields. The micro field fluctuation, or ion dynamics effect, has not been definitely observed at the present time, although there are numerous examples of its suspected presence[11] affecting spectral data. The associated redistribution is a sensitive probe for the study of the ion dynamics effect

Since this investigation is based on an extension of the FFM interpretation of emitted line shapes to higher order radiative properties, the system of radiative channels or dressed two level emitters defined in this model, permits a straightforward extension to different plasma conditions, laser pumps and ionic emitters. In this manner, photo pumping calculations of other complex multi electron ionic systems can be performed in a relatively uncomplicated manner as described above. Inclusion of saturation effects for large Rabi frequencies such as those related to the ac Stark effect splitting of lines can also be handled within the framework of the present calculations.

The FFM is particularly capable of taking into account the consequences of ion dynamics on the redistribution of the incident radiation and it is planned to extend the present studies to plasmas for which this is an important effect.

[1] R.Elton, X-Ray Lasers (Academic Press, San Diego, 1990).
[2] B.Talin, A.Calisti, L.Godbert, R.Stamm, R.W.Lee, and L.Klein, Phys. Rev. A **51**, 1918 (1995).
[3] A.Calisti, L.Godbert, R.Stamm and B.Talin, JQSRT **51**, 59 (1994).
[4] L.Godbert, A.Calisti, R.Stamm, B.Talin, R.W.Lee, and L.Klein, Phys. Rev. E **49**, 5644 (1994).
[5] L.Godbert, A.Calisti, R.Stamm, B.Talin, J.Nash, R.W.Lee, L.Klein, S.Glenzer and H.J.Kunze, Phys. Rev. E **49**, 5889 (1994).
[6] B.Talin and L.Klein, Phys. Rev. A **26**, 2717 (1982).
[7] B.Talin, R.Stamm, V.Kaftandjian and L.Klein, Ap. J. **322**, 804 (1987).
[8] B.Talin, Y.Botzanowsky, C.Calmes and L.Klein, J. Phys. B **16**, 2313 (1983).
[9] A.Ben Reuven, Phys. Rev. A **22**, 2585 (1980).
[10] C.Mossé, A.Calisti, R.Stamm, B.Talin, R.W.Lee, J.A.Koch, A.Asfaw, J.Seely, J.Wark and L.Klein, JQSRT **58**, **4-6**, 803 (1997).
[11] N.C.Woolsey, B.A.Hammel, C.J.Keane, A.Asfaw, C.A.Back, J.C.Moreno, J.K.Nash, A.Calisti, C.Mossé, R.Stamm, B.Talin, L.Klein and R.W.Lee, Phys. Rev. E **56**, 2314 (1997).

Exact Analytical Solution for the Ion Impact Broadening of Hydrogen Lines in Plasmas at High Densities or High Principal Quantum Numbers

E. Oks

Physics Department, Auburn University, Auburn, AL 36849

Abstract. It was usually assumed that with the increase of the density N and/or the principal quantum number n, the Ionic Contribution to the Impact Width (ICIW) vanishes. We show here that this is not true and that even at high N and/or n, the ICIW could be comparable to the electron impact width.

1. INTRODUCTION

Experimental profiles of high Balmer and Paschen lines of hydrogen are used nowadays for diagnostics of tokamak plasmas [1,2]. Up to now, the theory of Stark broadening, employed in the diagnostics for deducing the electron density N_e from the experimental profiles, was based primarily on the quasistatic treatment of the broadening by ions (while the electron broadening was calculated in the impact approximation) [1-3].

This is not to say that the role of the ion dynamics was not recognized. It is well-known by now that the ion dynamics, generally speaking, is very important for Stark broadening of spectral lines in high-T plasmas. However, it was usually assumed that with the increase of the plasma density N and/or of the principal quantum number n of the upper level of the radiator, the ion dynamics becomes less and less important and the Ionic Contribution to the Impact

Width (ICIW) tends to zero. Therefore, even the application of codes based on simulation models for ion dynamics, such as the Frequency Fluctuation Model (FFM), for calculating profiles of high Balmer lines emitted by tokamak plasmas, resulted only in minor corrections to the results of the standard theory, where the ion broadening was assumed quasistatic [4].

In distinction to that paradigm, by finding an analytical result for the ion dynamical broadening of hydrogen lines, we show here that with the increase of N and/or n, the ICIW does not decrease. Moreover, for practically important ranges of T, N, and n, this "residual" ICIW, being virtually independent of n, can be comparable to the standard electron impact width.

2. OUTLINE OF THE DERIVATION

The derivation proceeds from our Generalized Theory (GT) of Stark broadening of hydrogen lines in plasmas [5]. The GT is based on the non-perturbative treatment of one component of the field caused by charged particles passing by the radiator. Therefore the GT is intrinsically more accurate than the fully-perturbative standard theories [6,7] - see Appendix A.

Analytical results of the GT depend on the ratio of two characteristic impact parameters:

Weisskopf radius $W = Z_i n^2 \hbar / (m_e v)$ and

mean interperturber distance $D = [3/(4\pi N)]^{1/3}$.

Here v is the perturber velocity, N is the perturber density.

Typically, for the electron broadening the ratio $D/W \gg 1$. Physically this means that most of the electrons are "impact" (i.e., cause the impact broadening of the particular hydrogen line).

For the ion broadening, one finds usually $D/W \ll 1$, what means physically that most of the ions are not impact. However, there is a (tiny) minority of ions that cause some impact broadening of the line.

By applying the GT formalism to the ion broadening in the limit $D/W \ll 1$, we have found that the ICIW γ_i is described by the following, relatively simple expression:

$$\gamma_i = (3\pi N_e/Z_i)^{1/3}[8T/(\pi\mu)]^{1/2}$$

$$\approx 3.2990 \times 10^6 [N_e(\text{cm}^{-3})/Z_i]^{1/3}[T(\text{eV})M_p/\mu]^{1/2}. \tag{1}$$

Here Z_i is the charge of the perturbing ions, μ is the reduced mass of the perturber-radiator pair, M_p is the proton mass.

The "limiting" value of the ICIW given by Eq. (1) is independent of the principal quantum number n. Figure 1 shows the reduced (dimensionless) ICIW versus the ratio of the effective principal quantum number $n_{eff} \equiv (n^2+n-n'^2-n')^{1/2}$ to its critical value $n_{eff}{}^{cr} \equiv (m_e/\hbar)^{1/2}(T/\mu)^{1/4}(\pi N_e Z_i^2/9)^{-1/6}$ (we note that $n_{eff} \sim n$ at $n \gg n'$). In the standard theories, with the increase of n_{eff} the ICIW reaches the maximum and then rapidly diminishes to zero (dashed line). In distinction to that, in the GT, with the increase of n_{eff}, the ICIW saturates (solid line).

3. PHYSICAL INTERPRETATION AND JUSTIFICATION

These analytical results can be easily understood in physical terms. It is well-known that the impact width can be estimated as

$$\gamma_i \sim Nv_i\sigma, \tag{2}$$

where σ is the "optical cross-section". It is the effective cross-section for virtual transitions (between the states of the same multiplet) that lead to the broadening of the spectral line. In the conventional case of $W \ll D$, the cross-section is determined by the Weisskopf radius

$$\sigma \sim W^2 \propto v_i^{-2}, \tag{3}$$

so that

$$\gamma_i \propto N/v_i \propto N/T^{1/2}. \tag{4}$$

However, for sufficiently high densities, the mean interperturber distance becomes smaller than the Weisskopf radius. In this case the cross-section is controlled by the mean interperturber distance:

$$\sigma \sim D^2 \propto N^{-2/3}. \qquad (5)$$

Therefore, the impact width becomes as follows:

$$\gamma_i \sim N v_i D^2 \propto T^{1/2} N^{1/3}. \qquad (6)$$

Let us now discuss some conceptual issues to justify the above results. The impact width in the GT generally consists of the contributions from adiabatic and nonadiabatic terms, as defined in the GT [5]. In the limit D << W, the ICIW yielded by the GT originates predominantly from the adiabatic term. The adiabatic impact term in the GT was calculated in the spirit of the "old adiabatic theory" (see, e.g., [12]) but with the vector summation of the contributions from the individual perturbers (in distinction to the scalar summation in the impact approximation of the old adiabatic theory).

It should be emphasized that the adiabatic broadening theory reproduces in a unified fashion both the impact and the quasistatic parts of the correlation function $C(\tau)$. In our case of D << W, the impact part of $C(\tau)$ corresponds to $\tau \leq D/v_i$, while the quasistatic part of the $C(\tau)$ corresponds to $D/v_i \leq \tau \leq W/v_i$. So the impact and quasistatic regimes are characterized by significantly different interaction volumes V'_{imp} and V'_{qs} (the interaction volume, a.k.a. "collision volume", is one of the central concepts of the adiabatic broadening theory). Indeed, in the case of D << W, we have: $V'_{imp} \sim D^3 << V'_{qs} \sim W^3$.

Consequently, first, quasistatic ions outnumber impact ions by the factor of $W^3/D^3 >> 1$, so that the impact ions constitute indeed a tiny minority. Second, the number of impact ions in the corresponding interaction volume is $N_i V'_{imp} \sim 1$, so that the applicability of the binary impact approximation is not violated.

Finally, to employ the impact approximation, one should be able to introduce a coarse-grain time scale dealing with time intervals Δt such that $\gamma_i < (\Delta t)^{-1} < \Omega$, where γ_i os the resultant ion impact width, Ω is the rms value of the frequency of any component $E_{comp}(t)$ of the ion-produced electric field. For an individual passage of the perturber by the radiator, by calculating analytically both the

Fourier transform $\Phi(\omega)$ of the time dependence $E_{comp}(t)$ and the subsequent average of ω^2 over $\Phi(\omega)$, we have found the frequency $\Omega = \omega_{rms} = 3^{1/2}v/\rho$. Then the cutoff ρ_{max}, involved in the calculation of $\gamma_i(\rho_{max})$, is defined by the requirement: $\gamma_i(\rho_{max}) < \Omega(\rho)$ for $\rho < \rho_{max}$. Thus the result presented by Eq. (1) is derived in compliance with the physical basis of the impact approximation.

4. DISCUSSION OF DIAGNOSTIC CONSEQUENCES

We have derived an analytical expression for the ion dynamical contribution to the width of hydrogen lines in the limit of high densities N and/or high principal quantum numbers n. It should be emphasized that we have obtained the first ever **non-perturbative** analytical result for the ion dynamical broadening.

Five years ago a benchmark experiment conducted by the Kunze's group in the gas liner pinch at the electron densities $10^{18}cm^{-3}$-$10^{19}cm^{-3}$[10], had puzzled the lineshape community by revealing dramatic discrepancies (up to a factor of 2) between the experimental FWHM of the H_α line and the FWHM predicted by the Kepple-Griem's code [6] (see Fig. 3). Since then, efforts have been made to improve the theory, especially by allowing for the ion dynamics - one way or another.

Up to now, one of the most successful were model numerical calculations by Stamm's group based on the Frequency Fluctuation Model (FFM) - see, e.g., [13]. The FFM results for the conditions of the experiment [10] are also shown in Fig. 3 [14]. It is seen that, on one hand, the FFM noticeably improves the overall agreement between the experimental and theoretical FWHM. On the other hand, the "slope" of the dashed FFM-line in Fig. 3 is significantly greater than the "slope" of a line that could be drawn through the experimental bar-points. In other words, the FFM overestimates the broadening at the upper end of the experimental density range and underestimates the broadening at the lower end of this density range.

Our theoretical results, based on the GT for both electrons and ions (that is, the ICIW is included), are presented in Fig. 3 by the solid line. It is seen that the "slope" of our solid line is noticeably closer to the experimental "slope" than the dashed FFM-line.

Thus, the inclusion of the ICIW, derived analytically from the first principles without model assumptions, yields the best agreement with the experiment [10]. Minor differences remaining at the lower end of the experimental density range might be due to some optical depth of the H_α line at these densities. Indeed, the lower end points in the experiment [10] had been obtained at the expansion stage of the pinch, so that for the lower densities the number of radiators along the line of sight could be significantly greater than for the higher densities. For this reason, Kunze is considering a diagnostically-improved re-make of the experiment at the lower end of the density range [15].

Let us now focus at the consequences for the diagnostics of tokamak plasmas employing high Balmer and Paschen lines. The standard theory yields the FWHM of a highly excited hydrogen lines in the form

$$\Delta\lambda_{1/2}^{high} \approx A_e(n,n')N_e/T_e^{1/2} + B_i(n,n')N_e^{2/3}, \qquad (7)$$

where the term $A_e(n,n')N_e/T_e^{1/2} \equiv \Delta\lambda_e$ is due to electrons and the term $B_i(n,n')N_e^{2/3} \equiv \Delta\lambda_i$ is due to the ions. Here the coefficient B does not depend on the temperature or on the density, but strongly depends on the principal quantum number; the coefficient A has only a very weak, logarithmic dependence on N_e and T_e.

Our theory, being intrinsically more accurate than the standard theory, yields significantly different results. First, for high Balmer lines, the ICIW does not depend on the principal quantum number n, but strongly depends on the temperature ($\propto T_i^{1/2}$). Second, for highly excited hydrogen lines near the limit of the spectral series, the total (electronic plus ionic) impact width significantly exceeds the quasistatic splitting (produced by the majority of ions). Therefore, the line profiles correspond to the regime of merged (collapsed) Stark components [8] and the quasistatic contribution from ions has practically no effect on the FWHM:

$$\Delta\lambda_{1/2}^{high} \approx A_e(n,n')N_e/T_e^{1/2} + C_iT_i^{1/2}N_i^{1/3}, \qquad (8)$$

where the coefficient C_i does not depend on the principal quantum number n.

It is interesting to note that for low Balmer and Paschen lines, the majority of perturbing ions produce the impact broadening. In this case, the ionic contribution to the FWHM predominates:

$$\Delta\lambda_{1/2}^{low} \approx A_e(n,n')N_e/T_e^{1/2} + A_i(n,n')N_i/T_i^{1/2} \approx A_i(n,n')N_i/T_i^{1/2}, \quad (9)$$

since $A_i/A_e \sim (m_i/m_e)^{1/2} \gg 1$. Therefore, in the reality, there is virtually *no practical range of principal quantum numbers where the ion quasistatic term $B_i(n,n')N_e^{2/3}$ could enter the FWHM*: with the increase of n, **the FWHM gradually transforms from a linear dependence on the density (see Eq. (9)) to a quasilinear dependence on the density (see Eq. (8))**.

An interpolative formula for the ionic contribution to the FWHM, that incorporates both limits presented in Eqs. (8), (9) and is therefore valid for a broad range of densities, can be expressed as follows

$$\Delta\lambda_i = \lambda_{nn'}^2 (\pi c)^{-1} \{1 - \exp[-(C_i T_i)^{-1} A_i(n,n')(N_e/N_{cr})^{2/3}]\} \gamma_i. \quad (10)$$

Here γ_i is given by Eq. (1) and the critical density N_{cr} has the form:

$$N_{cr} = (9/\pi) Z_i^{-2} n_{eff}^{-6} (m_e/\hbar)^3 (T_i/\mu)^{3/2}$$
$$\approx 1.7309 \times 10^{18} cm^{-3} Z_i^{-2} n_{eff}^{-6} [T(eV) M_p/\mu]^{3/2}, \quad (11)$$

where the effective principal quantum number n_{eff} is defined as

$$n_{eff} \equiv (n^2 + n - n'^2 - n')^{1/2}. \quad (12)$$

In Eq. (10), $\lambda_{nn'}$ is the unperturbed wavelength of a hydrogen line, corresponding to the radiative transition from the upper level n to the lower level n'.

The coefficient A_e in the electronic contribution to the FWHM $\Delta\lambda_e$ can be reliably obtained as follows. First, our GT [5] applied to the electron broadening, produces much more accurate results than the standard theories. Second, it was shown in [9] that in the regime of merged (collapsed) Stark components, the electron broadened profiles of all hydrogen lines are Lorentzian. Thus, combining the results of the GT [5] and of Ref. 9, the coefficient $A_e(n,n')$ can be represented in the form

$$A_e(n,n') = \lambda_{nn'}^2 (\pi c)^{-1} 3(2\pi)^{1/2} g_{nn'} \hbar^2 m_e^{-3/2} \{1.1$$

(13)
$$+ \ln[T_e g_{nn'}^{-1/2} \hbar^{-1}/(4\pi e^2 N_e/m_e + (6.3 n \hbar Z_i^{1/3} N_e^{2/3}/m_e)^2)^{1/2}],$$

where

$$g_{nn'} \equiv (n^2 - n'^2)^2 - n^2 - n'^2. \qquad (14)$$

All formulas are given in the CGS units, except the second lines of Eqs. (1), (11), and the Eqs. (15)-(19) below.

To enhance the accuracy of our results, we also analytically calculated the ion quasistatic correction to the FWHM, as follows. First, we calculated the correction at a fixed value of the ion microfield F. Second, we averaged the correction over the Holtsmark distribution [11] of the field F. The resulting final formula for the FWHM has the form:

$$\text{FWHM}(\text{Å}) = \Delta\lambda_L(\text{Å})$$

$$+ [0.94761 \times 10^{-19} Z_i^{3/2} N_e n^6 n'^6 /(n^2-n'^2)^{3/2}]/[\Delta\lambda_L(\text{Å})]^{1/2}. \quad (15)$$

Here the Lorentzian contribution $\Delta\lambda_L$ to the FWHM is given by the following practical formula

$$\Delta\lambda_L = \Delta\lambda_e + \Delta\lambda_i, \qquad (16)$$

$$\Delta\lambda_e(\text{Å}) = 2.1200 \times 10^{-20} n^4 n'^4 (n^2-n'^2)^{-2} g_{nn'} N_e T^{-1/2} \{1$$

$$+ \ln[1.5192 \times 10^{15} T g_{nn'}^{-1/2} (3.1826 \times 10^9 N_e + 53.2 n^2 Z_i^{2/3} N_e^{4/3})^{-1/2}]\}, \quad (17)$$

$$\Delta\lambda_i(\text{Å}) = f(T)(N_e/N_{cr})^{1/3}\{1 - \exp\{-9\pi Z_i g_{nn'}[2f(T) n_{eff}^2]^{-1}[1$$

$$+ \ln[1+6.2847 \times 10^8 (T/Z_i)(N_e g_{nn'} \mu/M_p)^{-1/2}]](N_e/N_{cr})^{2/3}\}\}, \quad (18)$$

$$f(T) = 0.34941[n^4 n'^4/(n^2-n'^2)^2] T M_p/[\mu Z_i n_{eff}^2], \quad (19)$$

where N_{cr}, n_{eff}, and $g_{nn'}$ are given by Eqs. (11), (12), (14). *In all of the above practical formulas, the electron density N_e is in cm^{-3}, the temperature T is in eV.*

As an illustration, we apply our analytical results to

the interpretation of the widths of the high-n Balmer lines measured at the tokamak Alcator C-Mod [1,2]. In ref. 1, experimental profiles of the lines H_8, H_9, H_{10}, and H_{11} were shown in Fig. 2 and their Stark widths (FWHM) were deconvoluted to be 4.9, 5.9, 7.5, and 9.6 Å, respectively. Then Griem, the leading theoretical co-author of Ref. 1, by using the Kepple-Griem's code [6], found that these Stark widths correspond to the electron densities 5.4, 5.6, 5.3, and 5.6×10^{14} cm^{-3}.

We note that the Kepple-Griem's code does not allow for the ion dynamics. Consequently, the widths produced by their code, have a relatively weak temperature dependence. Therefore, Griem was not able to **deduce** the temperature from the experimental widths based on that code. Instead, he **assumed** the temperature to be T=4 eV.

In distinction to this, our analytical results for the widths accurately incorporate the ion dynamics and, therefore, are much more sensitive to the temperature than the Kepple-Griem's code. Thus, while analyzing the same experimental widths, we were able to **derive** both the density $N_e = 6.5 \times 10^{14}$ cm^{-3} and the temperature T=7.7 eV. It should be emphasized, that at T=7.7 eV we obtained the **same** value of $N_e = 6.6 \times 10^{14}$ cm^{-3} from **each** Balmer line individually.

To further demonstrate the superior quality of our theory, we calculated the ratios of Stark widths $\Delta\lambda_{n+1}/\Delta\lambda_n$ of the adjacent Balmer lines and compared them with both the experimental results [1,2] and the Griem's theoretical results (presented in [1,2]). The idea behind calculating these ratios is that they should be less sensitive to the electron density than the widths of the individual lines. Therefore, the ratios are conventionally considered as a tool for evaluating the intrinsic self-consistency of a particular theory. It turns out that *the rms deviation of our ratios from the experimental ratios is almost 4 times smaller than for the Griem's ratios.*

In summary, the novel results presented in this paper should lead to:
A) a substantial revision of the past diagnostic conclusions for a variety of plasma experiments, especially those dealing with highly excited hydrogen lines from the edge and divertor regions of tokamaks;
B) a much better possibility to deduce from experimental line widths not only the plasma density but also the plasma

temperature;
C) a significant enhancement of the accuracy of the density and the temperature obtained from the experimental line profiles.

ACKNOWLEDGMENTS

I would like to thank R.W. Lee for helpful theoretical discussions, R. Stamm for making available to me his unpublished FFM calculations, H.-J. Kunze for valuable discussions of the gas liner experiments, and S. Alexiou for the help in computations.

REFERENCES

1. B.L. Welch, H.R. Griem, J. Terry, C. Kurz, B. LaBombard, B. Lipschultz, E. Marmar, and G. McCracken, Phys. Plasmas **2**, 4246 (1995).
2. B.L. Welch, H.R. Griem, J.L Weaver, J.L. Terry, R.L. Boivin, B. Lipschultz, D. Lumma, E.S. Marmar, G. McCracken, and J.C. Rost, 10th Conf. "Atomic Processes in Plasmas", AIP Conf. Proceedings 381, AIP Press, New York, 1996, p. 159.
3. R.D. Bengtson, J.D. Tannich, and P. Kepple, Phys. Rev. A **1**, 532 (1970).
4. B.L. Welch, H.R. Griem, J.L Weaver, J.U. Brill, J.L. Terry, B. Lipschultz, D. Lumma, G. McCracken, S. Ferri, A. Calisti, R. Stamm, B. Talin, and R.W. Lee, 13th Int. Conf. "Spectral Line Shapes", AIP Conf. Proceedings 386, AIP Press, New York, 1997, p.113.
5. Ya. Ispolatov and E. Oks, J. Quant. Spectr. Rad. Transfer 51, 129 (1994).
6. P. Kepple and H.R. Griem, Phys. Rev. **173**, 317 (1968).
7. G.V. Sholin, A.V. Demura, and V.S. Lisitsa, Sov. Phys. JETP, **37**, 1057 (1973).
8. M.L. Strekalov and A.I. Burshtein, Sov. Phys. JETP 34, 53 (1972).
9. C. Stehle, Astron. Astrophys. **305**, 677 (1996).
10. St. Böddeker, S. Günter, A. Könies, L. Hitzschke, and H.-J. Kunze, Phys. Rev. E **47**, 2785 (1993).
11. J. Holtsmark, Ann. Phys. **58**, 577 (1919).
12. V.S. Lisitsa, Sov. Phys. Usp. **20**, 603 (1977).
13. B. Talin, A. Calisti, L. Godbert, R. Stamm, R.W. Lee, and L. Klein, Phys. Rev. A **51**, 1918 (1995).
14. R. Stamm, private communication.
15. H.-J. Kunze, private communication.

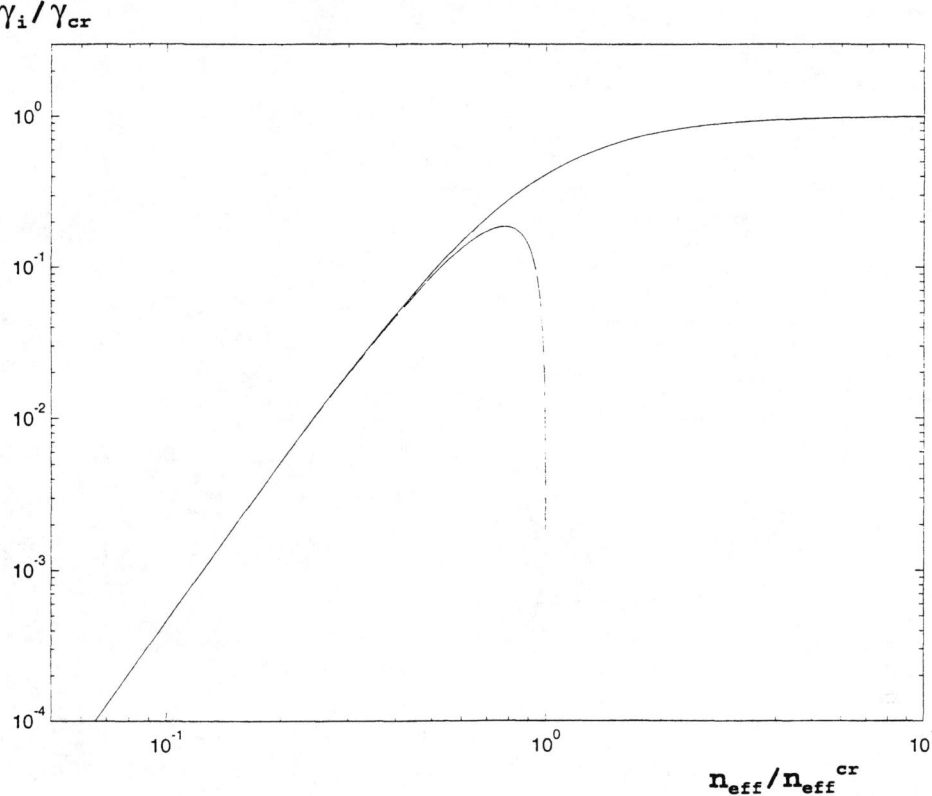

Fig. 1. The reduced ionic contribution to the impact width γ_i/γ_{max} versus the ratio of the effective principal quantum number $n_{eff} \equiv (n^2+n-n'^2-n')^{1/2}$ to its critical value $n_{eff}^{cr} \equiv (m_e/\hbar)^{1/2}(T/\mu)^{1/4}(\pi N_e Z_i^2/9)^{-1/6}$. Here $\gamma_{max} \equiv (3\pi N_e/Z_i)^{1/3}[8T/(\pi\mu)]^{1/2}$.

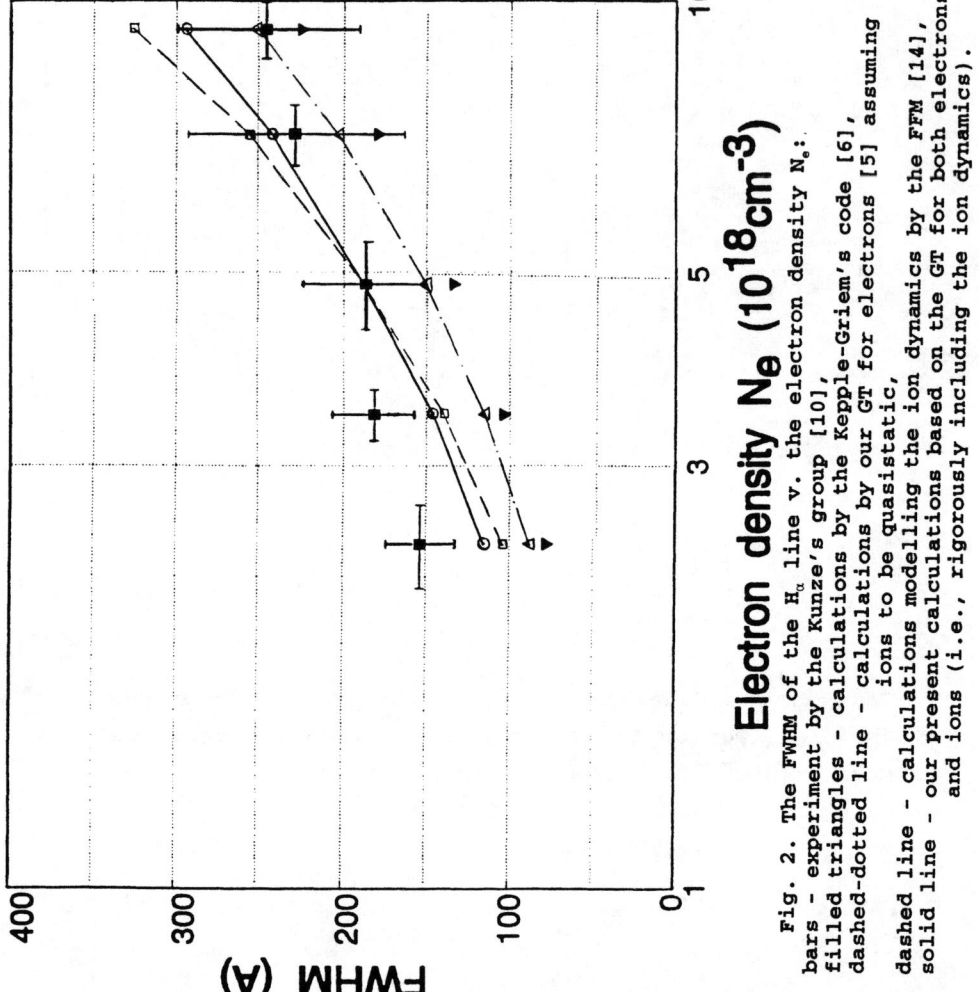

Fig. 2. The FWHM of the H_α line v. the electron density N_e:
bars - experiment by the Kunze's group [10],
filled triangles - calculations by the Kepple-Griem's code [6],
dashed-dotted line - calculations by our GT for electrons [5] assuming ions to be quasistatic,
dashed line - calculations modelling the ion dynamics by the FFM [14],
solid line - our present calculations based on the GT for both electrons and ions (i.e., rigorously including the ion dynamics).

Polarization Capture of Electrons by Ions as a New Channel of Recombination: Shapes and Total Recombination Rates

L. Bureyeva* and V. Lisitsa**

*Scientific Council on Spectroscopy of the RAS,
Leninski Pr. 53, Moscow, 117924, Russia E-mail: bureyeva@sci.lebedev.ru
** RRC "Kurchatov Institute", Kurchatov Sq., 46, Moscow, 123182, Russia
E-mail: lisitsa@qq.nfi.kiae.su

Abstract A new channel for recombination of free electrons on multicharged ions with complex core is under investigation. The channel is connected with a dynamic polarization of an ion's core by the colliding electron which results in radiation of the core and capture of the electron. This channel (called polarization recombination-PlR) is estimated in the frame of statistical (Thomas-Fermi) model of the complex ion. It is shown that the contribution of PIR to the total recombination rates may be compared or exceed the standard contribution of radiative recombination.

The essence of the effect is as shown in Fig. 1.

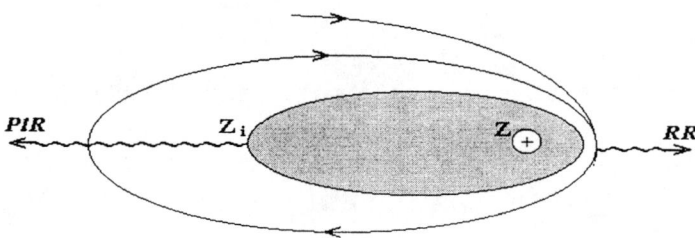

Figure 1: A scheme of radiation (RR) and polarization (PLR) recombinations

The estimation of polarization recombination (PlR) rates is based on the comparison of spectral radiation intensities determined by Fourier components of the dipole moment d_ω of the incident electron $I^{RR}(\omega)$ and by a dipole moment $D_\omega = \alpha(\omega)F_\omega$ induced in the ion's core by the electric field $\mathbf{F}(t)$ of the colliding electron $I^{PlR}(\omega)$, $\alpha(\omega)$ being a dynamical polarizability of the core, F_ω is a Fourier component of the electric field:

$$\frac{I^{PlR}}{I^{RR}} = \frac{|D_\omega|^2}{|d_\omega|^2} = \left|\frac{\omega^2 \alpha(\omega)}{Z_{eff}}\right|^2 \equiv R(\omega). \qquad (1)$$

The relationship (1) is a shape's ratio for two channels of recombination (RR and PLR) in terms of the core polarizability $\alpha(\omega)$ and effective charge Z_{eff}.

We will follow the statistical atomic model developed in [1]. According statistical models the core polarization is expressed in terms of functions depending on a percentage ionization $q = Z_i/Z$, a dimensionless ion radius $X(q)$ and a reduced frequency $\beta = \omega/Z$.

The ratio $R(\omega)$ takes the form

$$R(\omega) = (0.885X)^6 \beta^4 F_X(\beta)(Z/Z_{eff})^2, \quad F_X(\beta) = (Xa_{TF})^{-3} \mid \alpha(\beta) \mid^2. \quad (2)$$

The function $F_5(\beta)$ for q=0.3, X=5 obtained from numerical data [1] is shown on Fig. 2 where the parameter β is present in eV. The values of $Z_{eff}(\omega)$

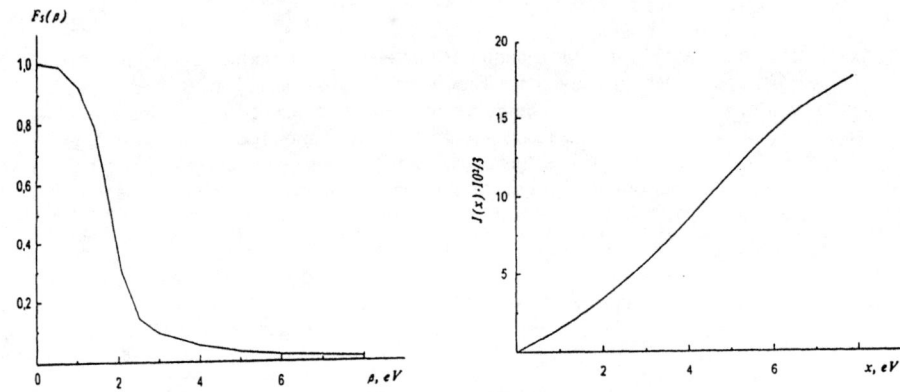

Figure 2: The function $F_5(\beta)$ describing the spectral distribution of the polarization recombination.
Figure 3: The function $J(x)$ describing the total polarization recombination rate.

and β are determined by the initial electron energy E and an energy E_n of the captured electron with upper (β_{max}) and lower (β_{min}) values of the parameter β

$$\beta \equiv \omega/Z = (E + |E_n|)/Z, \quad \beta_{max} = (E + |I_i|)/Z, \quad \beta_{min} = E/Z, \quad (3)$$

where I_i is the ionization potential of the ion under consideration.

The factor $R(\omega)$ is expressed in simple analytical forms in limiting cases $\omega \to 0$ and $\omega \to \infty$ using static and high frequency polarizabilities of the ion.

The boundary value β_X^* is determined by equating expressions for $\alpha(\omega)$ for limiting cases $\omega \to 0$ and $\omega \to \infty$. Putting $\alpha(0) = (X \cdot a_{TF})^3$ for $\omega = 0$ and $|\alpha(\omega)| = N/\omega^2 = Z(1-q)/\omega^2$ for $\omega \to \infty$ we obtain for the boundary value of

β_X^* (a.u.):

$$\beta_X^* = \frac{\sqrt{(1-q)}}{(0.885X)^{3/2}} \simeq \frac{\sqrt{1.5(1-q)}}{X^{3/2}} \qquad (4)$$

The estimation of the factor $R(\omega)$ in eq. (2) for uranium ion U^{+28} in the case of intermediate domain of frequencies gives it's value near the factors from 3 to 4, that is polarization recombination may spectrally dominate over radiative one.

The **total recombination rate** are determined by an integration of corresponding differential recombination rate over frequencies from β_{min} to β_{max}.

Using the simplest Kramers formulae for radiative recombination rate (see, for example [1]) we obtain the following ratio R^{tot} for total recombination rates

$$R^{tot} = Z^2(0.885X)^6 \frac{\int_{\beta_{min}}^{\beta_{max}} d\beta \beta^3 F_X(\beta)}{\int_{\beta_{min}}^{\beta_{max}} (d\beta/\beta) Z_{eff}^2(\beta)} \qquad (5)$$

Slow electron case In this case $Z_{eff}(\beta)$ in eq. (5) is close to the ion charge Z_i and eq. (5) reduced to

$$R^{tot} = (Z/Z_i)^2 (0.885X)^6 \frac{\int_0^{\beta_{max}} d\beta \beta^3 F_X(\beta)}{\ln(I_i/E)}; \quad \beta_{max}(E \ll I_i) = \frac{qZ^{1/3}}{0.885X} \text{ (a.u.)} \quad (6)$$

Functions $F_X(\beta)$ of two variables X and β are approximately described by a universe function $F(\beta/\beta_X^*)$ of one variable $\tilde\beta = \beta/\beta_X^*$, where β_X^* is given by eq. (4):

$$F(\tilde\beta) = F_X(\tilde\beta \cdot \beta_X^*) = F_X(\beta). \qquad (7)$$

The same scalling follows from local plasma frequency approximation for $q \geq 0.2$ [2].

The function $F(\tilde\beta)$ is just the one presented on Fig. 2 with corresponding scaling $\beta \to \tilde\beta \cdot \beta_5^*$. Changing variables in the integrand in eq. (6) in accordance with the scaling (7) we obtain

$$R^{tot}(q, Z) = \left(\frac{1-q}{q}\right)^2 \frac{J(\tilde\beta_{max})}{\Lambda}. \qquad (8)$$

Here $\tilde\beta_{max} = \beta_{max}/\beta_X^*$ (β_X^* is given by eq. (4)), $\Lambda = \ln(I_i/E)$ and

$$J(\tilde\beta_{max}) = (\beta_5^*)^{-4} \int_0^{y_{max}, \, a.u.} dy y^3 F_5(y, \text{ a.u.}), \qquad (9)$$

were $y_{max} = \tilde{\beta}_{max} \cdot \beta_5^* \approx 0.1 \, \tilde{\beta}_{max}$ (a.u.) in accordance with eqs. (6), (7) and the function $F_5(y)$ is presented on Fig. 2. In limiting cases of small and large values of $\tilde{\beta}$ the integral $J(\tilde{\beta})$ (9) takes the form

$$J(\tilde{\beta}) \approx \begin{cases} \ln \tilde{\beta}; & \tilde{\beta} \gg 1 \\ \tilde{\beta}^4/4; & \tilde{\beta} \ll 1 \end{cases} \quad (10)$$

The values of the function $J(x, eV)$ are presented on Fig. 3. It is seen that the typical values of the integral in eq. (8) are near 10-20. Let's estimate the parameter R^{tot} for uranium ions ($Z = 92$) in two cases: 1) $q = 0.3$ ($Z_i = 30$), $X = 5$ and 2) $q = 0.1$ ($Z_i = 9$), $X = 10$. In the first case we obtain $R^{tot}(0.3, 92) \simeq 1$, whereas in the second case we have $R^{tot}(0.1, 92) \simeq 10$.

The consideration above demonstrates the essential contribution of the new recombination channel (polarization recombination - PlR) to recombination rates of free electrons on complex ions. It is the most essential for heavy ions ($Z \gg 1$), low ionization stages ($q = Z_i/Z \ll 1$) and low ($E \ll I_i$) and intermediate ($E = I_i$) electron energies.

Our consideration bases on the simplest statistical (Thomas-Fermi) theory of atoms and ions. From one hand it makes possible to obtain universal estimations for the effect under consideration but from the other hand it doesn't take into account a contribution of resonance transitions (of dielectronic types) connected with discrete atomic levels as well as the shell structure of atomic systems. The effects mentioned can change the estimations presented above. The more refined statistical theories can also lead to modifications of the estimations as it has beam already pointed out in connection with large discrepancies between present statistical theories.

The estimations above don't take into account the interference effects between standard radiative (RR) and polarization (PlR) recombinations. Really the radiation of a quantum $\hbar\omega$ is provided by the total dipolar momentum of the system "colliding electron + heavy ion" which is determined by the sum $d_\omega + D_\omega$ of both dipolar momenta of the electron and of the core. It seems however that the contribution of interference term into the total recombination rate will be small.

This work was performed with the support of the Russian Mimistry of Science (Project "Spectroscopy of Rydberg States") and the RFBR (project 98-02-16763).

References

[1] Vinogradov A V and Tolstikhin O I 1989 *Zh. Eksp. Teor. Fiz.* **96** 1204

[2] Brandt W and Lundqvist S 1965 *Phys. Rev.* **139** A612

IMPROVED CALCULATIONS ON HIGH-N BALMER AND PASCHEN SERIES PROFILES

S. Ferri, A. Calisti, R. Stamm and B. Talin
*UMR 6633, Université de Provence, Centre St Jérôme
13397 Marseille, cedex 20, France*

R.W. Lee
*Department of Physics, University of California
Berkeley, CA 94720-7300*

ABSTRACT

A study of the Balmer and Paschen series from n=8 to 13 for plasma regimes relevant to the magnetic fusion devices is carried out. For such conditions the main improvement that can be inferred from standard electron broadening theories is that the line profiles change from the impact to the static line shape. A method using a frequency dependent electron broadening operator is implemented in the line shape code based on the Frequency Fluctuation Model. The line shape code thus modified is used to synthesize the Balmer et Paschen series with the aim to improve diagnostics of electron density based on spectroscopy.

INTRODUCTION

High-n atomic hydrogen or isotope lines in low density plasmas relevant to the magnetic fusion devices are mostly Stark broadened. The comparison of the entire synthetic Balmer and Paschen spectra from n=8 up to n=13 with the experimental data can therefore be used as a diagnostic method. The Balmer and Paschen series, presented in a previous paper[1], are calculated by a code, based on the Frequency Fluctuation Model[2,3]. These calculations are updated by using a frequency dependent electronic collision operator to reproduce more accurately electronic effects.

THE SPECTRAL LINE SHAPE

In the Frequency Fluctuation Model, the quantum system perturbed by the electric microfield can be considered as a set of Stark Dressed Transitions (SDT), which are characterized by a set of two complex numbers, respectively, a generalized intensity $a_k + i c_k$ and a complex frequency $f_k + i g_k$. This procedure allows to include many more levels than is customary. The field fluctuations induced by the ion motion are taken into account through a mixing of the SDT obeying a stochastic Markovian process. The line shape, written in the Liouville space extended to the STD, takes the simple following form:

$$I(\omega) = \frac{1}{\pi} \Im m \sum_{i,j} p_i \langle\langle d_i | (\omega \mathbf{1} - L_0 + iW)^{-1} | d_j \rangle\rangle \qquad (1)$$

Here p_i is the weight of the SDT labeled i, L_0 is a diagonal operator containing their complex characteristic frequencies and W is the Markov transition rate matrix, while the effective dipole d_i is given by :

$$d_i = \sqrt{(1+i\frac{c_i}{a_i})\sum_k a_k} \qquad (2)$$

The electron density domain investigated for this paper goes from 3×10^{19} to $10^{22} m^{-3}$ for a temperature range from 0.1 to 10eV. For these conditions, the ion dynamics effect has been found unimportant. It only slightly smoothes the central regions of the lines and has been ignored in the following. The effects of the $\Delta n \neq 0$ interactions have also been investigated. Again the effect of such couplings on the lines has been found unimportant for the studied domain. A small increase of the line widths occurs when Stark couplings involving $\Delta n = \pm 1$ are included.

At the highest density of the range of interest, due to the high principal quantum number, the impact approximation for electrons breaks down. Then, it is necessary to account for a more quasi-static description of the electronic effects. Theoretical approaches such as the unified theory[4,5] or semi-empirical procedures[6,7] have been developed in the past years. They provide non Lorentzian line shapes with three well identified regions: the line center and the static wings connected by a transition region. For this study, a semi-empirical collision operator has been used. This operator involves a new frequency dependent cutoff which is adjusted using an accurate numerical simulation protocol.[8] In some cases, the implementation of this new collision operator in the code results in a 50% increase of the electron density. Fig.1 shows the Balmer series from n=8 up to 13, calculated assuming the impact approximation for the electrons (dotted line), and using the frequency dependent electronic collision operator (solid line).

DIAGNOSTICS OF ELECTRON DENSITY

The most common way for diagnosing the electron density relies onto the use of the Stark widths, corrected for Doppler and instrumental broadening. Following Bengston et al,[9] in order to design a practical method, the calculated Stark widths have been fitted with a simple law involving both a quasi-static $N_e^{2/3}$ scaling and a linear impact theory scaling.

$$\Delta\lambda(\overset{0}{A}) = 2.5 \times 10^{-13}(\alpha N_e^{2/3} + \beta N_e) \quad (N_e \text{ in } m^{-3}) \qquad (3)$$

The next table presents the parameters α and β derived from the fit of the Balmer and Paschen series line for an electronic temperature of 4 eV.

	Balmer		Paschen	
n	α	$\beta(10^{-9})$	α	$\beta(10^{-8})$
8	0,216	5,65	1,158	3,57
9	0,288	6,37	1,691	3,76
10	0,372	8,95	2,042	5,26
11	0,453	12,3	2,655	5,52
12	0,597	11,9	2,786	9,52
13	0,736	12,3	4,016	6,28

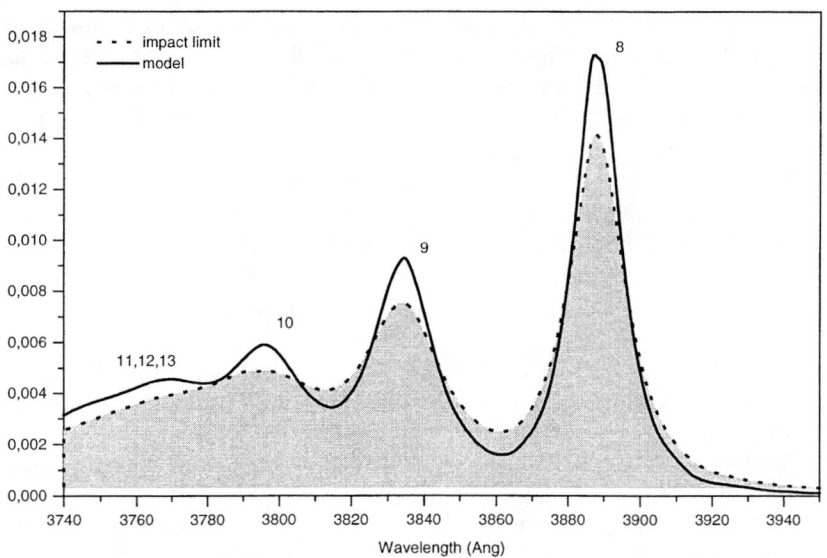

Figure 1 : Comparisons of the Balmer series calculated with the impact approximation for the electron (dotted line) and using the frequency dependent electronic collision operator (solid line)

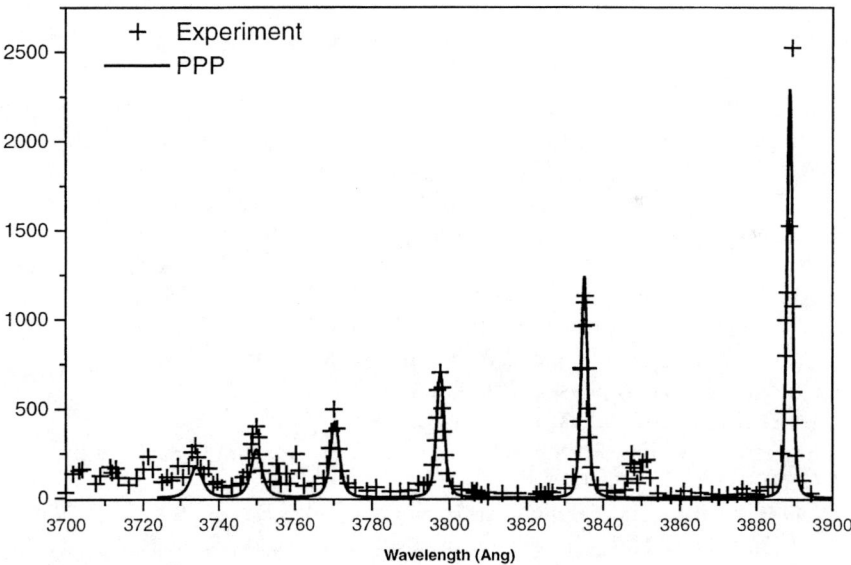

Figure 2 : Balmer series from n=8 up to 13 : experimental data (crosses), PPP calculation (solid line)

The first column lists the upper-level principal quantum numbers of the transitions. The columns 2 and 3 contain the parameters corresponding to the Balmer series lines, while the columns 4 and 5 contain the parameters corresponding to the Paschen series lines.
We present on Fig.2 the Balmer series from n=8 up to n=13, for a temperature of 4eV. From the experimental data, presented by B. L. Welch[10], we have measured the width of the Balmer n=10 to 2 transition. LTE populations have been used. The electron density is found to be equal to $8.5 \times 10^{19} m^{-3}$. Note that at short wavelengths, the spectrum neither accounts for the higher series members nor for the continuum edge.

CONCLUSION

The code, based on the Frequency Fluctuation Model, that benefits from an efficiently optimized algorithms allows to calculate the line shapes from very large quantum systems. As a result, this study allows to better understand the features connected to non-binary electronic collisions. On the other hand, a practical diagnostic tool for the electron density for plasmas occurring in magnetic fusion devices is proposed. A more sophisticated method could be obtained with the association of the improved line shape code together with a population kinetic code. Such a method should allow to synthesize and compare to experiments large portions of the spectra involving several lines. This should result in an increased accuracy and the possibility to infer the electron temperature.

ACKNOWLEDGMENTS

The authors are grateful to Professor Hans Griem for helpful discussions and comments on this work.

REFERENCES

1. B. L. Welch, H. R. Griem, J. L. Weaver and J.U. Brill, J. Terry, B. Lipschultz, D. Lumma and G. McCracken, S. Ferri, A. Calisti, R. Stamm and B. Talin, and R.W. Lee, AIP Conf. Proc. **386** (1), 113-6 (1997)
2. B. Talin, A. Calisti, L. Godbert, R. Stamm, R. W. Lee and L. Klein, Phys. Rev.A **51**, 1918 (1995)
3. B. Talin, A. Calisti, S. Ferri, M. Koubiti, T. Meftah, C. Mossé, L. Mouret, R. Stamm, S. Alexiou, R. W. Lee and L. Klein, J. Q. S. R. T. **58**, 4-6, 953 (1997)
4. E. W. Smith, J. Cooper and C.R. Vidal, Phys. Rev. **185**, 140 (1969)
5. C. R. Vidal, J. Cooper and E.W. Smith, J.Q.R.S.T **10**, 1011 (1970)
6. H. R. Griem, Astrophys. J. **136**, 422 (1962)
7. H. R. Griem, Astrophys. J. **147**, 1092 (1967)
8. S. Ferri, A. Calisti, R. Stamm, B. Talin, R.W. Lee and L. Klein, Phys. Rev. E **58** (6), (Dec. 1998)
9 R.D Bengston, J.D. Tannich and P. Kepple, Phys. Rev. A **1**, 532 (1970)
10. B. L. Welch, H.R. Griem, J.L. Weaver, J. L. Terry, R. L. Boivin, B. Lipschultz, D. Lumma, E. S. Marmar, G. McCracken and J. C. Rost, AIP Conf Proc **381**, 159-166 (1997)

HIGH DENSITY EFFECTS ON HYDROGEN STARK PROFILES

A. Calisti, S. Ferri, M. Koubiti, L. Mouret, R. Stamm and B. Talin
Laboratoire de Physique des Interactions Ioniques et Moléculaires, UMR 6633
Université de Provence, centre Saint Jérôme, 13397 Marseille, France

M. Gigosos and V. Cardenoso
Universidad de Valladolid, 47071 Valladolid, Spain

Abstract. Hydrogen lines of low principal quantum number have been recently measured for high densities and low temperatures. For such conditions, the use of a binary collision electronic impact operator is shown to overestimate the width of the line.

Plasma diagnostic achieved by a study of the Stark profile is a commonly used technique in plasma physics and the physics of stellar atmospheres. For the case of line shapes dominated by Stark broadening a common and widely used model consists in using an impact approximation for the electronic broadening, and a static approximation for the ionic component [1]. Although this standard model has been very successful for many types of plasmas, observational conditions exist today for which either one or the other of these two approximations is no longer valid. For these cases, the main difficulty is due to the dynamic many body problem of the emitters interaction with a large number of ions or electrons. In the two last decades, several models have been developed which provide an approximate solution to this problem [2,3,4]. Among those, an accurate simulation technique [2] has been recently applied to extensive hydrogen line shape calculations. The technique used is specialized for hydrogen, and is based on the decoupling of the Schrödinger equation obtained by the use of the Runge-Lenz vector. This formulation provides an efficient algorithm for solving the emitters evolution equation, and leads to a solution having a number of advantages : It is simple to numerically evaluate, and gives a very accurate result. It also has a property of numerical stability which is useful for a numerical integration over long time configurations. In this paper we compare the results of the Frequency Fluctuation Model (FFM) [4] with line shape simulation calculations [2] and experimental profiles. For high density ($N_e = 10^{17} - 10^{18}$ cm^{-3}) and low temperature plasma conditions (T≅1 eV), these comparisons allow to analyze the validity of the electronic binary approximation for low lying hydrogen lines.

By plotting the width of the hydrogen Paschen-α line versus electronic density at a temperature T=10000 K, it can be seen on fig. 1 that there is a good agreement between the FFM and the simulation for densities around 10^{16} cm^{-3}, but that for higher densities, our model predicts width which are larger than the simulation calculations. For such conditions the ratio of the electronic Weisskopf radius to the average distance $N_e^{-1/3}$ is no longer negligible compared to one, favoring the occurrence of simultaneous strong collisions, and casting a doubt on the validity of a binary approach.

By using devices such as capillary discharges, or wall stabilised discharge plasmas, rather high densities can be reached, while keeping the temperature close to one eV.

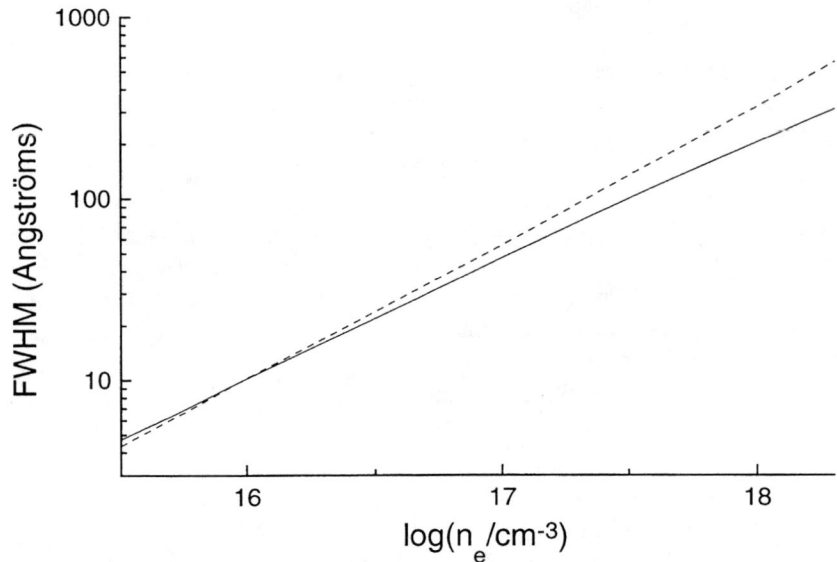

Fig. 1 Logarithmic plot of the width (FWHM) of the hydrogen P_α at T=10000 K versus the electron density. The results of our model (dashed line) are compared to the simulation (solid line).

It is thus possible to compare the high density and low temperature experimental results to the simulation [2], and to the FFM profiles. We can note that for the conditions considered, the FFM almost reduces to the standard model for Stark broadening, i.e. an impact electron broadening, and quasistatic ion approximation. We have plotted on fig.2 a comparison of the experimental and calculated Full Widths at Half Maximum of the hydrogen H_α in an argon plasma, as a function of density for the conditions of the experiment (triangles) of Vitel [5], which have been done for temperatures between 16000 K and 18700 K. The simulation calculation (crosses) is an a rather good agreement (within 10%) with the experiment, while the standard model (solid line), which is obtained here by the FFM in the static ion limit, and using an electronic binary impact approximation, predicts widths which are 40% larger.

Similar conditions are found in the recent measure of the hydrogen P_α width by Döhrn et al [6] done for temperatures between 11720 K and 13900 K. A difference

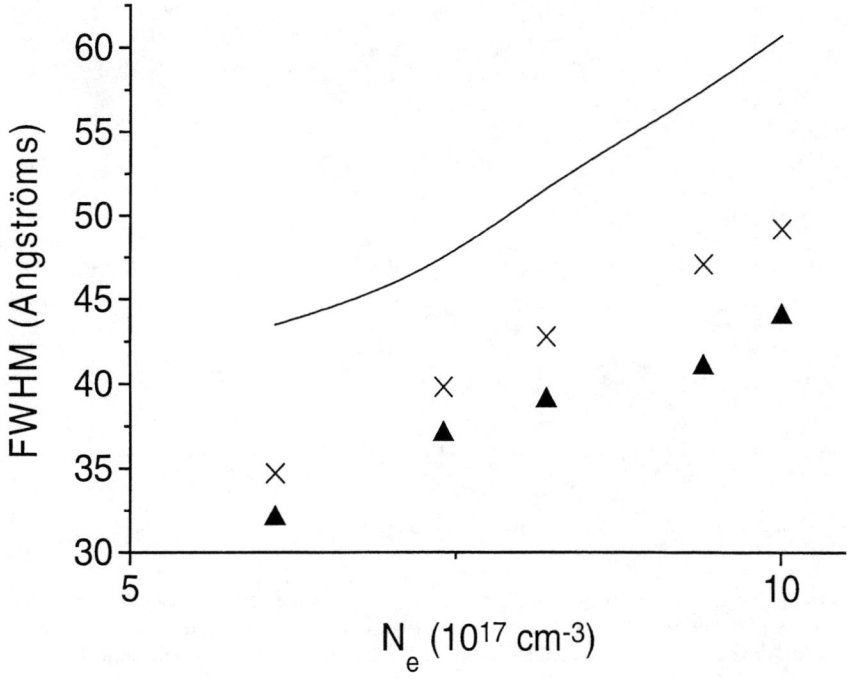

Fig 2. Full width of H_α at T≈1.5 eV in an argon plasma, as a function of electron density in the conditions of the experiment of Vitel[5]. The experiment (triangles) is compared to the simulation of Gigosos et al[2] (crosses), and to the calculations of the FFM with a binary electronic collision operator (solid line).

increasing with density, and reaching 30% at the highest density measured, is seen on fig.3 between the standard model (solid line) and both the experiment (triangles) and the simulation (crosses). For the conditions of figs.2-3, the use of a binary collision operator can severely overestimate the electronic broadening. A possible improvement for treating the electronic component for such case, is to generalize the FFM to the two components of the plasma.

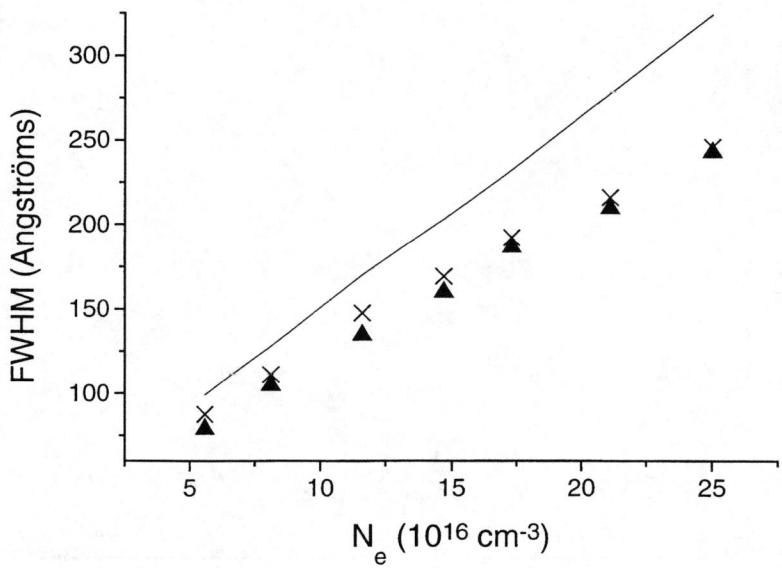

Fig 3. Full width of P_α at $T \approx 1$ eV in an argon plasma, as a function of electron density in the conditions of the experiment of Döhrn et al [6]. The experiment (triangles) is compared to the simulation of Gigosos et al[2] (crosses), and to the calculations of the FFM with a binary electronic collision operator (solid line).

In summary comparisons of our model to simulations and experiments in high density and low temperature cases indicate that a binary collision operator for the electrons may seriously overestimate the broadening. For these conditions, we are actually for the electrons in the same situation as for the ion dynamic problem, namely with simultaneous effects of strong collisions on the emitter, and any solution of the many body ion dynamic problem should also be applicable to the electrons.

[1] H. R. Griem, Spectral Line Broadening by Plasmas, Acad. Press (New York 1974).
[2] M. A. Gigosos and V. Cardenoso, J. Phys. B: At. Mol. Opt. Phys. **29**, 4795 (1996).
[3] C. Stehle, Astron. Astrophys. **292**, 699 (1994).
[4] B. Talin, A. Calisti, L. Godbert, R. Stamm and R.W. Lee, Phys. Rev. A **51**, 1918 (1995).
[5] Y. Vitel, J. Phys. B.: At. Mol. Phys., **20**, 2327(1987).
[6] A. Döhrn, P. Nowack, A. Könies, S. Günter and V. Helbig, Phys. Rev.E **53**, 6389(1996).

A New Spectroscopic Effect Depressing the Electron Impact Broadening in Dense Plasmas

E. Oks

Physics Department, Auburn University, Auburn, AL 36849

Abstract. We show that an acceleration of perturbing electrons by the nearest ion translates into a decrease of the electron broadening.

1. We consider a hydrogen atom at the presence of the nearest perturbing ion of a nuclear charge Z_p at a distance R from the atom. A perturbing electron passes by the atom at a distance of the closest approach $\rho \ll R$. Due to the presence of the perturbing ion, the electron follows a hyperbolic (rather than straight) path (see Fig. 1).

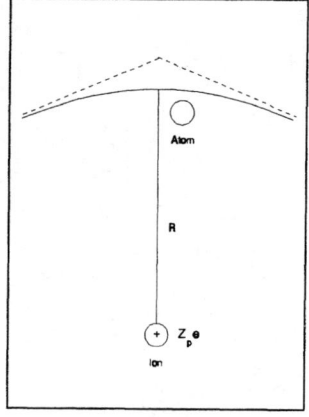

Due to the attraction to the perturbing ion, the electron flies by the atom with an increased velocity V compared to its asymptotic velocity v at distances $r \gg R$. From the energy balance

$$mV^2/2 \approx mv^2/2 + Z_p e^2/R \qquad (1)$$

we find:

$$V^2(R) \approx v^2 + g^2(R), \quad g(R) \equiv [2Z_p e^2/(mR)]^{1/2}. \qquad (2)$$

Let us consider a segment of the electron path within a distance $r_0 \sim (\rho R)^{1/2}$ from the atom. First, since $r_0 \gg \rho$, this segment yields practically the same contribution to the impact broadening of spectral lines of the atom as the entire path of the electron. Second, since $r_0 \ll R$, this segment is very close to a straight line and the electron travels over this segment with a practically constant velocity equal to the value V(R) given by Eq. (2).

Therefore, the electron impact broadening by electrons having the asymptotic velocity v can be described by the standard formulas with the substitution of v by the increased value of V(R) given by Eq. (2). In other words, at the stage preceding the averaging over velocities, instead of the standard expression for the electron impact broadening operator in the dipole approximation[6,7]

$$\Phi_{ST}(v) = K/v \qquad (3)$$

we should use

$$\Phi(v) = \Phi_{ST}(V) = K/[v^2+g^2(R)]^{1/2} = \Phi_{ST}(v)v/[v^2+g^2(R)]^{1/2}, \quad (4)$$

where K is a well-known operator that practically does not depend on velocity.

The next step is the usual averaging over the Maxwell distribution of velocities M(v). The analytical integration yields

$$\Phi(R) = \Phi_{ST} f(\eta(R)),$$

$$f(\eta) \equiv (\eta/2)[\exp(\eta/2)][K_1(\eta/2) - K_0(\eta/2)], \qquad \eta(R) \equiv Z_p e^2 R^{-1}/T, \quad (5)$$

where

$$\Phi_{ST} = \int_0^\infty dv M(v) \Phi_{ST}(v), \qquad (6)$$

$K_0(z)$, $K_1(z)$ are the modified Bessel functions. Physically, the argument of the function f is simply the ratio of the maximum absolute value of the potential energy of electrons in the field of the perturbing ion to their temperature.

2. We performed numerical calculations of the H_α profiles for the conditions of the underwater laser plasma experiment[1,2], where it was observed up to 20% reduction of the linewidth compared to the standard theory[18]. In addition to incorporating the narrowing due to the acceleration of electrons by the ion field, our code has other advanced features as well. Namely, it uses the advanced generalized theory describing a strong coupling between the electron and ion broadenings[9]. Further, the code includes the ion dynamics, based on our analytical solution for the ion-dynamical broadening of hydrogen lines[10].

From Table 1 it is seen that the allowance for the narrowing phenomenon dramatically decreases the discrepancy between the theoretical and experimental results. Moreover, the standard theory[18] did not include the ion dynamics. If we would include only the ion dynamics and use the generalized theory for the electron broadening, but would not incorporate the narrowing phenomenon, this would actually bring up the overestimation of the width to about 30-35% (at the two highest density points). So the narrowing phenomenon is significantly stronger than might appear from Table 1: it reduces the width by 30-35% (at the two highest data points), thus bringing it in agreement with the experimental values of the width.

Table 1.
Comparison of the experimental[1,2] FWHM W_{exp} of the H_α line with the theoretical FWHM W_G, interpolated from the Griem's tables[6] by the authors of Ref. 2, and with the theoretical FWHM W_O of the present paper; δW_G and δW_O are the discrepancies of W_G and W_O, respectively, benchmarked at W_{exp}.

$N_e, 10^{18} cm^{-3}$	$T, °K$	$W_{exp}, Å$	$W_G, Å$	$\delta W_G, \%$	$W_O, Å$	$\delta W_O, \%$
1.9	7900	85.2	82.4	-3.3	78.2	-8.2
2.3	8500	92.0	93.6	1.7	88.8	-3.5
2.55	8500	93.5	100.3	7.3	92.9	-0.6
2.7	8500	95.4	104.2	9.2	95.3	-0.1
3.75	8900	111.8	129.7	16.0	113.1	1.2
6.1	10000	147.3	179.4	21.8	153.3	4.1

3. Now let us apply our analytical description of the narrowing effect to the phenomenon of dips in the spectral line profiles. This phenomenon, being observed in a broad range of densities at different plasma devices by a variety of experimental groups, had opened up a new area of plasma spectroscopy (the Intra-Stark Spectroscopy) and had become a powerful diagnostic method[3]. However, it had not been understood yet, why even at high electron densities, such as $10^{18} cm^{-3}$, the dips have been observed not only in the L_α profiles but also in the H_α profiles[4,5]. By conventional estimates for these densities, possible dips in the H_α profiles should have been washed out by a relatively large electron impact broadening.

The dips result from the resonance between the frequency of Langmuir oscillations $\omega \approx \omega_{pe}$ and the ion quasistatic splitting of the upper or lower multiplets of the principal quantum numbers n or n', respectively[3]. In other words, the dips can correspond to only two particular values of the distance between the atom and the perturbing ion. At these values of the distance, the argument η of the function f can be calculated by this practical formula:

$$\eta = 2.60 \times 10^{-5} n_*^{-1/2} [N_e (cm^{-3})]^{1/4} / T(eV), \qquad n_* = n \text{ or } n_* = n'. \qquad (7)$$

As an example, we now calculate the narrowing for the conditions of the z-pinch experiment by Finken et al[4], where the dips in the H_α profiles were observed: $N_e \approx 2 \times 10^{18} cm^{-3}$, $T \approx 4$ eV, n=3. For these conditions we find: $\Phi/\Phi_{ST} \approx 0.75$. Thus in this case we deal with the 25% suppression of the electron broadening at the locations of the dips.

Using the standard theory, Griem[11] estimated for these conditions the FWHM due to electron collisions to be ≈40 Å. However, our above result for the same conditions shows that the FWHM due to electron collisions is only ≈30 Å at the locations of the dips.

Finally let us translate this suppression in the relative depth δ of the dip defined as

$$\delta \equiv 2(I_{bump} - I_{dip})/(I_{bump} + I_{dip}). \qquad (8)$$

Here I_{dip} is the intensity at the center of the dip, I_{bump} is the halfsum of the intensities of the two bumps surrounding the dip. Given that the experimental separation between the bumps in the H_α profile is ≈20 Å, we have calculated the theoretically expected relative depth ε of the dip in two versions: A) by using the Griem's value of 40 Å for the electronic FWHM; B) by using the more consistent value of 30 Å derived above.

The result of the calculation is the following. The ion-caused 25% suppression of the electron broadening translates into the increase of the relative depth of the dips by an order of magnitude: from ~1% to ~10%. The latter theoretical value of the relative depth of the dips agrees with the experiment[4].

Finally, we would like to emphasize that the narrowing due to acceleration of electrons by the ion field is a universal phenomenon that applies to spectral lines of all atoms and ions. For non-hydrogen atomic spectral lines, the above formulas can be used as the first approximation.

Thanks are due to A. Escarguel for providing the experimental data in parallel with its submission for publication and to S. Alexiou for computations.

REFERENCES

1. A. Escarguel, A. Lesage, and J. Richou, 14th Int. Conf. "Spectral Line Shapes", PennState, USA, June 1998.
2. A. Escarguel, B. Ferhat, A. Lesage, and J. Richou, submitted to JQSRT.
3. E. Oks, "Plasma Spectroscopy: The Influence of Microwave and Laser Fields", Springer Series on Atoms and Plasmas, Vol. 9, Springer, New York (1995).
4. K.H. Finken, R. Buchwald, G. Bertschinger, and H.-J. Kunze, Phys. Rev., 1980, **A21**, 200.
5. E. Oks, S. Böddeker, and H.-J. Kunze, Phys. Rev., 1991, **A44**, 8338.
6. H.R. Griem, "Spectral Line Broadening by Plasmas", New York, Acad. Press, 1974.
7. V.S. Lisitsa, Sov. Phys. Usp., 1977, **20**, 603.
8. Ya. Ispolatov and E. Oks, J. Quant. Spectr. Rad. Transfer, 1994, **51**, 129.
9. E. Oks and J. Touma, this issue.
10. E. Oks, INVITED PAPER, this issue.
11. H.R. Griem, "Spectral Line Shapes", v. 7, Nova Science Publishers, Commack, NY, 1993, p.3.

New advances in redistribution of resonance radiation in dense plasmas

A.V. Demura*, N. Feautrier**, I.N. Kosarev***, V.S. Lisitsa*,
C. Stehlé****

*Russian Research Centre "Kurtachov Institute", 123182, Moscow, Russia
**DAMAP et URA812 du CNRS, Observatoire de Paris, 5 Place Jules Janssen, F-92195 Meudon Cedex, France
***Institute of Physics & Power Engineering, 249020, Obninsk, Russia
****DARC et UPR176 du CNRS, Observatoire de Paris, 5 Place Jules Janssen, F-92195 Meudon Cedex, France

Abstract. We present here a new method for the calculation of the redistribution function in dense plasmas based on the atomic density matrix formalism. We are able, for the first time, to include both the ion dynamics effects and the consistent coupling between the atomic populations and coherences and the radiative polarizations. This allows us to go beyond the usual approximation of Complete Redistribution. The method is applied to a three-level model relevant for transitions in helium-like ions with a forbidden component.

In recent years, the problems of emission and absorption profiles and redistribution of radiation of multiply charged ions in plasmas were reconsidered on the basis of kinetic theory in the frame of the atomic density matrix formalism in the context of the study of X-ray lasers [1-4]. From this initial point, it occurs that even the starting formula for radiation from such a medium defined through a traditional correlation function should be substituted by a more complicated expression for the emitting power, which contains the time evolution of populations of radiating levels. Moreover, the coherences of radiating sublevels of the multiplet, which determine the radiation dynamics, and their populations are coupled together by the interaction with the slowly varying in time ion plasma microfield, formed by many body interactions. This important mixing exhibits the essence of the quantum interference effects, which were called nonlinear interference effects (NIEF) similar to the analogous phenomena in laser physics.

Earlier the NIEF influence on the key chracteristics of the radiation transfer - the radiation redistribution functions was investigated in terms of the atomic density matrix formalism for the three-level model system in the quasistatic approximation [1-2] and by Molecular Dynamics (MD) simulations for Ly-α [3]. It was found that these effects are very important quantitatively but also qualitatively. Namely, due to NIEF, the Complete Redistribution Approximation (CR) fails even at very large densities and the physically meaningful definition of the redistribution function is impossible without the inclusion of NIEF. During Molecular Dynamics simulations it was also found [3] that ion dynamics considerably influence the behaviour of redistribution functions.

However, MD is too time consuming for this type of study and one could not get sufficiently large information on the various features of redistribution functions. On the other hand it is known that a good candidate for incorporation of the effects of the ion motion is the Model Microfield Method (MMM). That is why in the present work the next significant advance was achieved - the above formalism of the atomic density matrix was developed for this problem within the MMM. Firstly, it was done for line profiles [4] and now for the radiation scattering [5], where more details may be found.

The particular calculations of the incoherent part of the redistribution function $R(\omega, \omega')$ have been performed for the spectral doublet $(1s2s)^1S - (1s4p)^1P$, $(1s2s)^1S - (1s4d)^1D$ of helium-like ion Al^{+11} in hot dense aluminium plasmas. Here data are reported for $T = 350$ eV, $N_e = 10^{20}$ cm^{-3} with the frequency of the incident photons fixed and equal to $\omega' - \omega_{21} = -5 \times 10^{14}$ rd.s^{-1} versus the scattered frequency ω (see Figs.1-4). This function has the following peculiarities:

1. The account of NIEF is essential in the case where the absorbed radiation frequency ω' is close the frequency of the forbidden transition ω_{31}, and increases when the plasma density decreases (see fig.1, 2). For $\omega' \simeq \omega_{21}$ the influence of NIEF is not essential.

2. The most principal point in the calculation of the redistribution function in this range of plasma parameters is the account of the ion dynamics due to a huge value of the ion microfield time variation characteristic frequency in comparison with other characteristic frequencies of the problem. Therefore, in comparison with the results in the static microfield, the thermal ion motion leads to an important decrease of $R(\omega, \omega')$ in the centres of the allowed and forbidden components, a drastic increase of $R(\omega, \omega')$ in the nearest wings and an essential shift of the redistribution functions versus ω at fixed frequencies of the absorbed radiation (see fig. 3). The latter effect is increased as the frequency of the absorbed radiation goes from the line centre to the wing of the line.

3. The complete redistribution approximation (CR) fails in the present conditions although the electron relaxation rate is large compared with the radiative decay rate of the radiating state (see fig.4), the coherent part of the line profile being small.

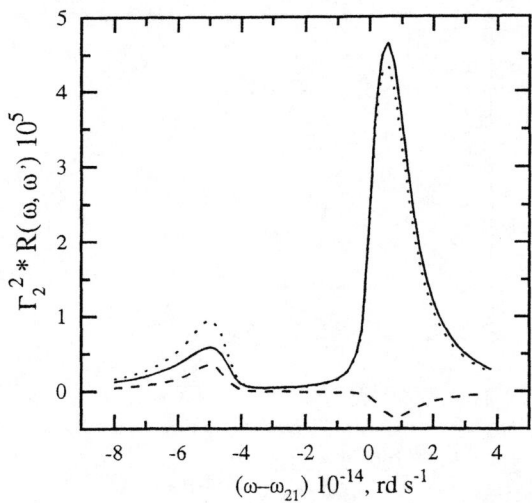

FIGURE 1. The incoherent redistribution function (see for details the text above): without NIEF (full curve), with NIEF (dotted curve), NIEF contribution (dashed curve).

REFERENCES

1. Anufrienko A.V., Godunov A.L., Demura A.V., Zemtzov Yu.K., Lisitsa V.S., Starostin A.N., Taran M.D., Schipakov V.A., 1990, *Sov. Phys.-JETP* **71**, 728.
2. Demura A.V., Anufrienko A.V., Godunov A.L., Zemtzov Yu.K., Lisitsa V.S., Starostin A.N., Taran M.D., Schipakov V.A., 1990, *SLS* **v.6**, Eds. L. Frommhold, J.W. Keto (New York: AIP) pp.227-254
3. Bulyshev A.E., Demura A.V., Lisitsa V.S., Starostin A.N., Suvorov A.E., Yakunin I.I., 1995, *Sov. Phys.-JETP* **81**, 113.
4. Kosarev I.N., Stehlé C., Feautrier N., Demura A.V. and Lisitsa V.S., 1997, *J. Phys. B: At. Mol. Opt. Phys.* **30**, 215.
5. Demura A.V, Feautrier N., Kosarev I.N., Lisitsa V.S., Stehlé C., 1998, *J. Phys. B: At. Mol. Opt. Phys.* **31**, 4283.

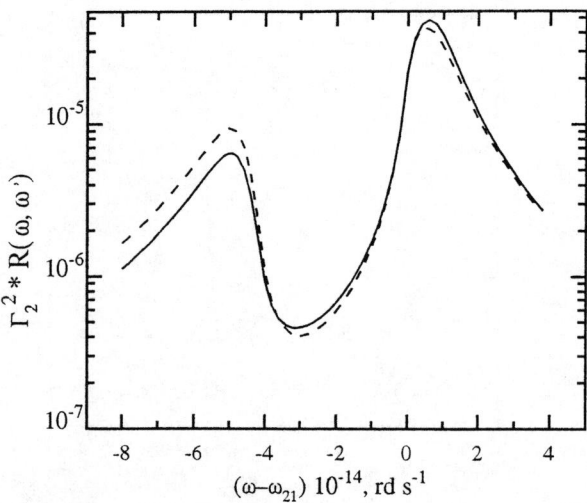

FIGURE 2. The same as fig.1: within the radiative-collisional model for the kinetics of the radiating states (full curve), including the ion field in the kinetics of radiating states (dashed curve).

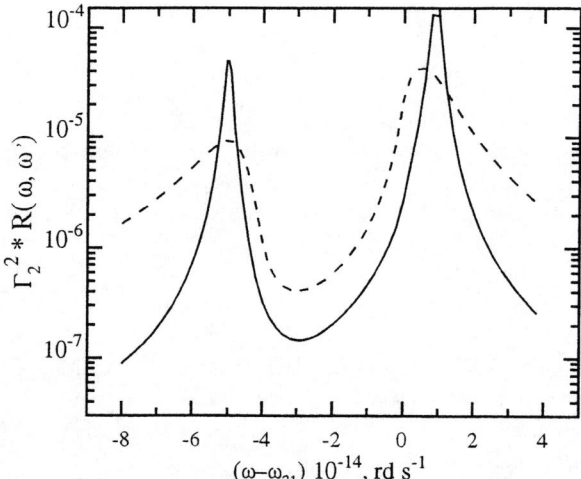

FIGURE 3. The same as fig.1: within the static ion field approximation (full curve), including ion dynamics effects (dashed curve).

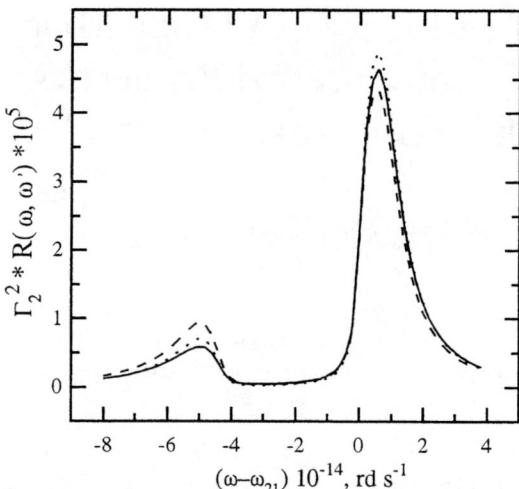

FIGURE 4. The same as fig.1: without NIEF (full curve), with NIEF (dashed curve), within the complete redistribution approximation (dotted curve).

The Influence of Plasma Microfields on Spectral Properties of Lithium-like Ions in Non-ideal Plasma

A. N. Starostin, I. I. Yakunin

State Research Centre
Troitsk Institute for Innovation & Thermonuclear Research
Troitsk 142092, Russia

For spectral features description of plasma containing lithium-like (or hydrogen-like) ions a model was developed which is based on steady state kinetic equations for density matrix. A solution of such equations is obtained from numerical code TRINITY. This code permits calculations of excited level populations as well as spectral gain and line profiles for emission and absorption for transitions between levels with principal quantum numbers up to 5. The model takes into consideration the influence of plasma microfields which leads to mixing of sublevels and Stark broadening of line shapes including non-linear interference effects due to coherent excitation of sublevels.

In a number of experiments population inversion or lasing of transitions of Li-like ions has been observed where three-body recombination is predicted to play a dominant role in achieving the population inversion. For this type of inversion scheme to be effective high electron densities are needed in a rapidly cooling plasma to obtain a high three-body recombination rate. These conditions imply that Stark broadening will dominate the line profiles of lasing transitions and seriously affect the achievable gain. Also, as far as population relaxation processes (diagonal elements of the density matrix) and polarizations (nondiagonal elements) are not separated in this case the description of the atomic system is necessary to carry out on the basis of the density matrix approach. This approach permits to take into account the influence on the gain of the non-linear interference effects arising from coherent excitation of degenerate sublevels by plasma microfields in stimulated transitions.

The description lithium-like (and hydrogen-like [1]) ions in a plasma was conducted by the solving in numerical code TRINITY kinetic equations for the atomic density matrix by taking into consideration the processes of ionisation, recombination, radiating and collisional transitions between levels as well as account of interaction of ions with the plasma microfield and availability of pumping on a certain level. The code TRINITY allows to calculate in steady-state approximation the level populations, the absorption and emission line profiles and the gain of a weak electromagnetic signal. Solutions of the equations are averaged over Holtsmark microfield distribution or over one calculated by the molecular dynamics method.

To some extent results of calculation over kinetics of Li-like ions of aluminium were compared with reported ones [2]. They concerned simulations of the experiment, performed by Carillon et al [3], when it was observed an amplification at the wavelength of several transitions belonging to Li-like Al. It was demonstrated that taking into account the influence of the plasma microfield significally change the results of gain calculations. In Fig. 1 are shown results of simulation of gain profiles on transitions 4F — 3D (electron temperature $T_e = 25$ eV, ion temperature $T_i = 21$ eV, electron concentration $N_e = 3.1 \, 10^{19}$ cm^{-3}, plasma density $\rho = 1.3 \, 10^{-4}$ g/cm^3 (Fig.1a)) and 5F — 3D ($T_e = 19$ eV, $T_i = 15$ eV, $N_e = 1.0 \, 10^{19}$ cm^{-3}, $\rho = 4.4 \, 10^{-5}$ g/cm^3 (Fig. 1b)). The dashed line corresponds to calculation without taking microfield into consideration, solid line - to case with its account.

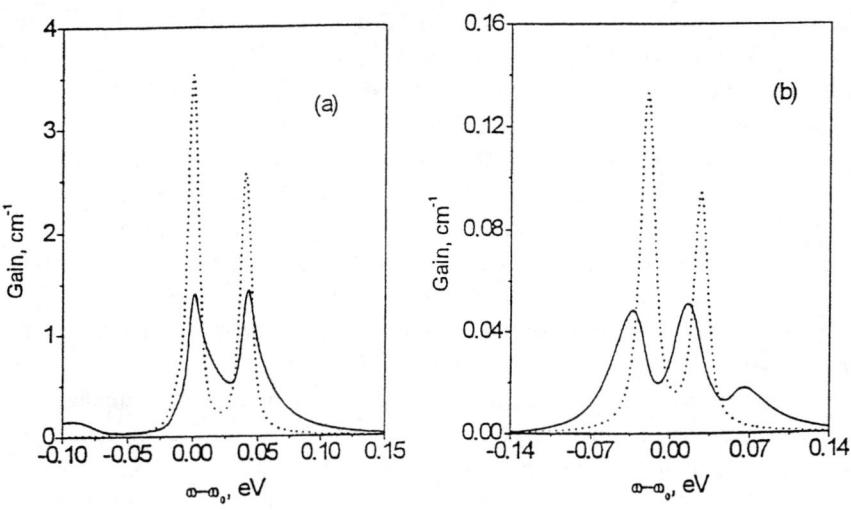

Fig. 1 Spectral gains near 4f-3d (a) and 5f-3d (b) transitions of Li-like Al for conditions of experiments [3]. Dotted curves use Voigt Line shapes, while the solid ones include ion Stark broadening.

REFERENCES

1. A. E. Bulyshev, A. V. Demura, V. S. Lisitsa et al., Zh. Eksp. Teor. Fiz. **108**, 212 (1995) [Sov. Phys. JETP **81**, 113 (1995)].
2. V. Makhrov, V. Roerich, A. Starostin et al., J. Phys. B, **27**, 1899 (1994).
3. A. Carillon et al., J. Phys. B, **23**, 47 (1990).

Measurement of the D/T Fuel Mixture in MCF Plasmas by Doppler-Free Two-Photon Spectroscopy

J. Seidel[*] and D. Voslamber[#]

[*] *Physikalisch-Technische Bundesanstalt, Abbestr. 12, D–10587 Berlin, Germany*
[#] *Ass. EURATOM-CEA, DRFC, CEA Cadarache, F–13108 St. Paul-lez-Durance, France*

Doppler-free two-photon excitation of ground-state atoms of the hydrogen isotopes, possibly created by charge exchange with a neutral beam, allows to utilize the isotopic shift of the hydrogen resonance frequencies for isotope-selective excitation of fluorescence radiation and may provide a diagnostic of the D/T fuel density ratio in magnetically confined fusion (MCF) plasmas (1,2). So far only the atomic transition scheme ("scheme I") where both excitation and observation occur in Lyman-α has been studied in detail (1) and tested experimentally (3). We have now investigated the diagnostic potential of another scheme ("scheme II"): two-photon excitation in Lyman-β and observation of the Balmer-α fluorescence. Even if powerful laser radiation at 205 nm, twice the Lβ wavelength, will be more difficult to obtain than radiation at 243 nm, twice the Lα wavelength, the Lβ/Hα scheme has the great advantage of fluorescence detection in the visible.

In a magnetic field $\boldsymbol{B}$, the excited levels ($n = 2$ or 3 for scheme I or II) of an atom with velocity $\boldsymbol{v}$ are split by the Zeeman effect and the motional Stark effect in the electric field $\boldsymbol{v} \times \boldsymbol{B}$. Only the central, nearly unperturbed components of the multiplets can be excited efficiently. The line shapes of the corresponding narrow peaks in the two-photon absorption profiles are determined by quadratic Doppler and motional Stark broadening, by plasma (ion impact) Stark broadening and by distorting effects of the fine and hyperfine splitting. Examples of these line shapes for scheme I are given in Refs. 1–3, examples for scheme II are shown in Figs. 1 and 2 here. A marked difference between the two schemes is the more involved dependence on laser polarizations (4) for scheme II, where two-photon transitions occur to 3s and 3d, while they occur only to 2s in scheme I.

In order to assess the relative merits of the two schemes I and II, we compared their signal-to-noise ratios SNR = $N/(N+N_b)^{1/2} \approx N/N_b^{1/2}$ (for numbers N and N_b of detected fluorescence and background photons). If laser irradiances are approximately optimized with respect to the main losses (three-photon ionization for scheme I, two-photon stimulated emission for scheme II) and w, E, ε, and R denote the atomic two-photon transition rate per laser intensity squared, the laser pulse energy, the efficiency of fluorescence detection, and the background radiance, respectively, the ratio of the SNRs is found to be

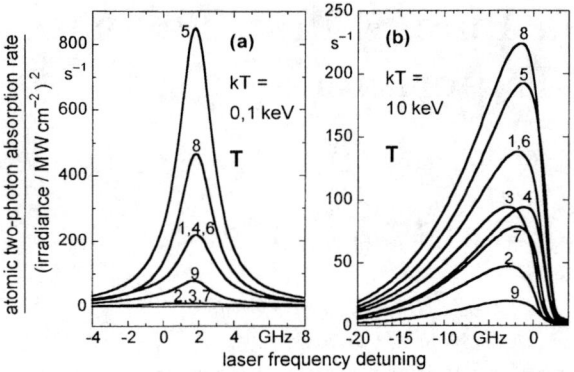
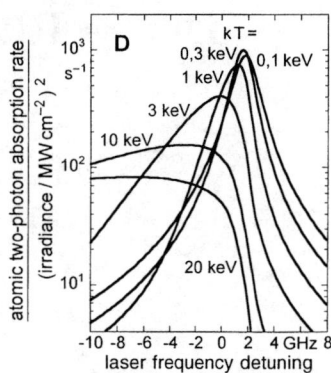

FIGURE 1. Two-photon absorption profiles for the tritium Lyman-β transition. Plasma parameters are $n_e = 2\, n_D = 2\, n_T = 10^{20}$ m^{-3}, $B = 4$ T, $kT = 0.1$ keV (a) and 10 keV (b). Numbers 1 to 9 denote different configurations of laser beam directions and polarizations. Absorption is most efficient for laser beams perpendicular to $\boldsymbol{B}$ with identical linear (no. 5) or circular (no. 8) polarizations.

FIGURE 2. Two-photon absorption profiles for the deuterium Lyman-β transition for temperatures from 0.1 keV to 20 keV. Other plasma parameters are the same as for Fig. 1. Laser beam configurations: no. 5 for $kT \leq 1$ keV, no. 8 for $kT \geq 3$ keV.

SNR$_{\text{II}}$/SNR$_{\text{I}}$ = 1.4 $(w_{\text{II}}/w_{\text{I}})\,(E_{\text{II}}/E_{\text{I}})^{3/4}\,(\varepsilon_{\text{II}}/\varepsilon_{\text{I}})^{1/2}\,(R_{\text{I}}/R_{\text{II}})^{1/2}$. Figure 2 shows SNR$_{\text{II}}$/SNR$_{\text{I}}$ as a function of temperature for three tentative parameter sets which appear realistic or at least achievable, for instance by selective spectral filtering of the background radiation, which should be possible with higher efficiency in the visible (Hα) than in the vacuum uv (Lα):

(a) $E_{\text{II}}/E_{\text{I}} = 0.1$, $\varepsilon_{\text{II}}/\varepsilon_{\text{I}} = 10$, $R_{\text{II}}/R_{\text{I}} = 0.1$;
(b) $E_{\text{II}}/E_{\text{I}} = 0.1$, $\varepsilon_{\text{II}}/\varepsilon_{\text{I}} = 10$, $R_{\text{II}}/R_{\text{I}} = 0.01$;
(c) $E_{\text{II}}/E_{\text{I}} = 0.2$, $\varepsilon_{\text{II}}/\varepsilon_{\text{I}} = 10$, $R_{\text{II}}/R_{\text{I}} = 0.01$.

Obviously, scheme II might be preferable to scheme I for D/T measurements in the hot plasma core, provided that powerful laser radiation at 205 nm becomes available and sufficiently high neutral densities can be generated by neutral beam injection.

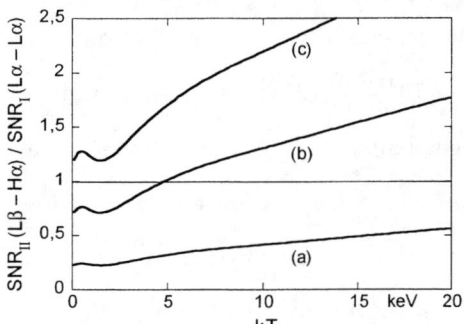

FIGURE 3. Ratio of the signal-to-noise ratios SNR of the atomic transition schemes I and II for deuterium as a function of temperature. For the definition of the parameter sets (a), (b), (c), see text.

REFERENCES

1. D. Voslamber, in *Spectral Line Shapes*, Vol. 8 (AIP Conf. Proc. 328), Eds. A.D. May, J.R. Drummond, E. Oks, New York: AIP Press, 1995, pp. 3–27
2. D. Voslamber and W. Mandl, Rev. Sci. Instrum. **69**, 1702–1715 (1998)
3. A. Steiger, K. Grützmacher, Ch. Seiser, M.I. de la Rosa, and U. Johannsen, this volume
4. J. Seidel, in *Spectral Line Shapes*, Vol. 2, Ed. K. Burnett, Berlin: de Gruyter, 1983, pp. 381–393

Dense Plasma Line Shifts: Theory and Experiment

G.C. Junkel[1], M.A. Gunderson[1], D.A. Haynes, Jr.[1],
C.F. Hooper, Jr.[1]
D.K. Bradley[2], J.A. Delettrez[2], P.A. Jaanimagi[2], S. Regan[2]

[1] *Department of Physics, University of Florida*
Gainesville, Florida 32611-8440
[2] *Laboratory for Laser Energetics, University of Rochester*
Rochester, New York 14623-1290

A full-Coulomb, second-order relaxation model of line broadening was used to produce lineshapes for hydrogen- and helium-like transitions in high-Z radiators. The theoretical profiles were used in the analysis of K-shell Ar spectra from recent high-density implosions performed at the Laboratory for Laser Energetics.

Data taken from these experiments using the sixty-beam Omega laser revealed high electron densities ($n_e = 5 \times 10^{23} - 2 \times 10^{24}$ cm^{-3}) and temperatures of ≈ 1 keV [1]. Furthermore, detailed analysis of spectra show features substantially shifted from their unperturbed locations. Figure 1 shows a time resolved spectral history of a recent implosion experiment conducted at LLE, in which a microballoon filled with D$_2$ and trace amounts of Ar and Ne were imploded using the Omega laser system. As the density increases, both β and γ lines of K-shell Ar are observed to shift to the red relative to isolated ion energy positions. Note also that the γ lines have greater shifts than the β lines. These new implosions motivated a reinvestigation of plasma-induced line-shift calculations.

Although there have been attempts at *ad hoc* shift calculations [2], a line shift does follow naturally from standard line broadening models when employing the full-Coulomb radiator-perturber interaction for the calculation of electron broadening operator. Since the monopole term of the radiator-perturber interaction provides the dominant portion of the shift for charged radiators, a full-Coulomb treatment of this interaction, which allows for orbital penetration, is necessary. In order to analyze the observed spectroscopic data, theoretical line-shapes, which included the line-shift as well as the line-width contributions, were calculated using the full-Coulomb radiator-electron interaction in a quantum mechanical, second-order relaxation theory using multi-electron, relativistic atomic physics extracted from Cowan's atomic physics code [3-5]. With increasing electron density the line width and shift both increase. One can also observe increasing distortion of the line-shape. The distortion arises from the fact that radiator states with differing angular momentum shift differently. As a result, the shift of the peak differs from the shift of the center of mass of the profile and neither correspond precisely with any particular level shift. Temperature and radiator angular momentum dependences are illustrated in Figure 2, where first order level shifts calculated with the current theory are compared with observed shifts of the resulting lineshapes.

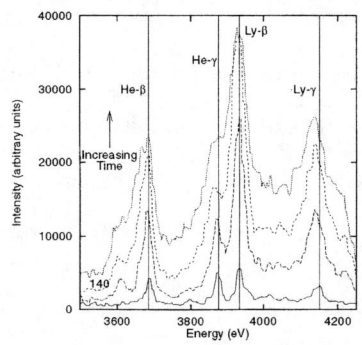 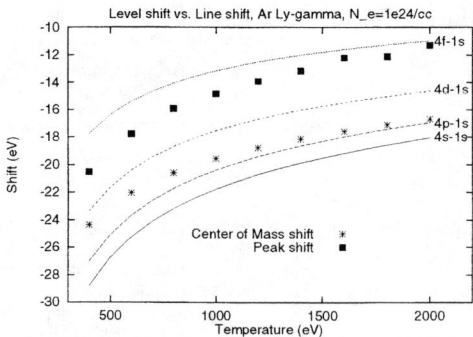

FIGURE 1. Time-resolved lineouts from LLE shot 11708 (1% Ar/2% Ne in D_2). Vertical lines represent isolated ion line positions of, going from left to right, Ar He-β, He-γ, Ly-β, and Ly-γ. Note how the γ lines shift more than the β lines.

FIGURE 2. Comparison of level shifts, first order in the interaction potential, of H-like Ar, with observed theoretical shifts of the line profile center of mass and line peak; electron density, 1×10^{24} cm^{-3}.

1. D.K. Bradley, *et al.*, Physics of Plasmas **5**, No. 5, 1870 (1998).
2. S. Skupsky, Phys. Rev. A **21**, 1316 (1980). J. Davis and M. Blaha, J. Quant. Spectrosc. Radiat. Transfer **27**, 307 (1982). X. Yan and S. Ichimaru, Phys. Rev. A **34**, 2173 (1986).
3. L.A. Woltz and C.F. Hooper, Jr., Phys. Rev. A **38**, 4766 (1988).
4. D.B. Boerker and C.A. Iglesias, Phys. Rev. A **30**, 468 (1984).
5. R.D. Cowan, *The Theory of Atomic Structure and Spectra.* (University of California Press, Berkeley and Los Angeles, 1981).

Charge exchange between two nearest neighbour ions immersed in a dense plasma

P.Sauvan, P.Angelo, H.Derfoul, E.Leboucher-Dalimier, A.Devdariani*, A. Calisti**
and B. Talin**

Physique Atomique dans les Plasmas Denses, LULI, CNRS
Université Paris VI, 4 Place Jussieu, 75252 Paris Cedex 05, France and Ecole Polytechnique, 91128, Palaiseau, France

* *Department of Optics and Spectroscopy, University of St Petersburg, St Petersburg, 198904, Russia*

** *Physique des Interactions Ioniques et Moléculaires, Université de Provence, Centre de S' Jérome, 13397, Marseille, France*

Abstract - In dense plasmas the quasimolecular model is relevant to describe the radiative properties : two nearest neighbour ions remain close to each other during a time scale of the order of the emission time. Within the frame of a quasistatic approach it has been shown that hydrogen-like spectral line shapes can exhibit satellite-like features. In this work we present the effect on the line shapes of the dynamical collision between the two ions exchanging transiently their bound electron. This model is suitable for the description of the core, the wings and the red satellite-like features. It is post-processed to the self consistent code (IDEFIX) giving the adiabatic transition energies and the oscillator strengths for the transient molecule immersed in a dense free electron bath. It is shown that the positions of the satellites are insensitive to the dynamics of the ion-ion collision. Results for fluorine Lyβ are presented.

The dynamics approach for the Two-Centre Model had been previously tested analytically with simplifying assumptions for dense plasma (1). In the present work the ion dynamics is treated selfconsistently with the plasma. The molecule formation is basically transient and the resulting dynamics correction is not negligible in the vicinity of profile singularities ω_S (satellite-like features) giving a new reference for the time of interest $(\omega - \omega_S)^{-1}$.

Let us consider an ion moving on a classical trajectory depending on the potential energy $E_p(R,R_c)$ due to the NN fixed ion and a non uniform free electron gas. This trajectory is truncated because of the cut off R_c corresponding to the interionic distance of the molecule at the limit of the confined quasineutral volume. The line profile within this approach is given by :

$$I(\omega) \propto \sum_{if} \int d\vec{R}_c d\vec{v}_{\vec{R}_c} \, P(\vec{R}_c, \vec{v}_c) \left| \int_{R_{tp}}^{R_c} \frac{dR}{v} D_{if}(R) \cos\left(\int_{R_{tp}}^{R} (\omega - \Delta E_{if}(R')) \frac{dR'}{v} \right) \right|^2$$

It depends on the history of the collision controlled through the interionic distance R and the radial velocity v. R_{tp} corresponds to the turning point and v_c to the radial velocity for the cut off. The ensemble average involves the conditional probability $P(\vec{R}_c, \vec{v}_c)$ computed by Molecular Dynamics. The transition energies $\Delta E_{if}(R)$, the dipole moments $D_{if}(R)$ and the potential energy $E_p(R,R_c)$ are computed by the code

IDEFIX (2). Electron collisional broadening, not expressed in the formula, has also been included in our calculations.

The enhancement with the dynamics correction of satellite-like features due to extrema in $\Delta E_{if}(R)$ is a highlight result. In figure 1 a comparison between the two versions (quasistatic / dynamics) of the Two-Centre model (QMSPECTRA) is given for FLyβ at high density conditions ($N_e = 4.10^{23}$ cm^{-3}, $T_e = 300$eV). The satellites involve interionic distances shorter than 3 a.u. which can be achieved for all trajectories such as $R_{tp} < 3$ a.u. In the quasistatic approach the observability of these satellites was conditionned by the more restrictive NN probability window P_{NN} ($R_c \approx 3$a.u., $v_c = 0$). The profiles computed with the dynamics version of QMSPECTRA have been compared to standard profiles issued from the ion dynamics version of Pim Pam Poum. As seen in figure 2 for FLyβ ($N_e = 2.10^{23}$ cm^{-3}, $T_e = 300$eV), our model gives the same near wings shape as Pim Pam Poum and the molecular satellites in addition.

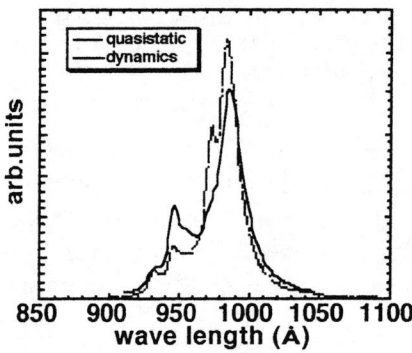

Figure 1. Comparison between quasistatic (the dotted line) and dynamics (the solid line) FLyβ profiles computed with the code QMSPECTRA.

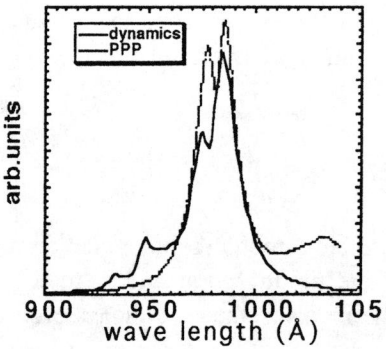

Figure 2 Comparison between FLyβ profiles computed with Pim Pam Poum (the dotted line) and the dynamics version of QMSPECTRA (the solid line). For the comparison the PPS shift (not taken into account in Pim Pam Poum) has been eliminated in our computation.

REFERENCES

1. Devdariani, A. et al., Sauvan, P. et al., *LULI Reports*, NTIS PB97- 170963 (1997)
2. Gauthier, P. et al., J.Quant.Spectrosc.Radiat.Transfer **58**, 597 (1997) ; Phys. Rev. E **58**, n°1 (1998)

Time resolved FLγ ß spectroscopic measurements in colliding foil experiments

T. Ceccotti, H. Derfoul, P. Sauvan, P. Angelo, P. Gauthier,
A. Poquérusse, C. Chenais-Popovics, E. Leboucher-Dalimier,
M. Vollbrecht*, E. Förster*

*Physique Atomique dans les Plasmas Denses,
LULI CNRS-Ecole Polytechnique-CEA-Université Paris VI
Université Paris VI, 4 Place Jussieu, 75252 Paris Cedex 05, France
and Ecole Polytechnique, 91128, Palaiseau Cedex, France.*

* *X-Ray Optik Group, Friedrich Schiller Universität
Max-Wien Platz 1, D. 07743 Jena, Germany*

Abstract - The experimental results we present here concern the time evolution of the main plasma parameters in the region between two thin laser accelerated Teflon foils as deduced from fluorine H-like and recombination continuum emission, as well as the time evolution of Fluorine Lyß line profile. Two X-ray streak cameras with a time resolution of 10 ps have then been used, recording the emission of a 25 µm sized plasma slice between the foils, to be compared to the initial distance (≈100 µm). To obtain a well resolved spectrum despite of the low streak camera photocathode spatial resolution, a high dispersive TlAP spectrometer has been used to record the Lyß line profile. In this way we can associate any particular line profile we record to the plasma conditions producing it. In conclusion this experiment allows quantitative comparison with hydrodynamics code results, post-processed by atomic physics and spectra codes.

Recently the colliding foil system known for its capability to generate dense and hot plasma has been implemented with a time-resolved spectroscopic diagnostic (experiment made at LULI). The intensity in the laser beams working at 4ω on focal spots 60µm is approximately $4 \cdot 10^{14}$ Wcm^{-2}. Two thin Teflon foils (5µm) initially separated by 100µm collide approximately at the pulse maximum. We give here three results among the different diagnostics implanted. An X-Ray 2D imaging (presented in figure 1) allowed an accurate determination of the geometric characteristics of the collision when successful. A time-resolved FLyβ spectra evolution has been recorded for the 25 µm central plasma. The resolution 50 ps was efficient for the exhibition of a red polarisation shift and a varying broadening due to strong density gradients (see figure 2). At the peak density the spectrum exhibits a Plasma Polarisation Shift (PPS) induced asymmetry on the red wing and a forbidden transition 3d-1s on the blue side (see figure 3). The codes IDEFIX and QMSPECTRA postprocessed to a 1D1/2 hydrocode (FILM-LULI) are efficient for the interpretation of the results.

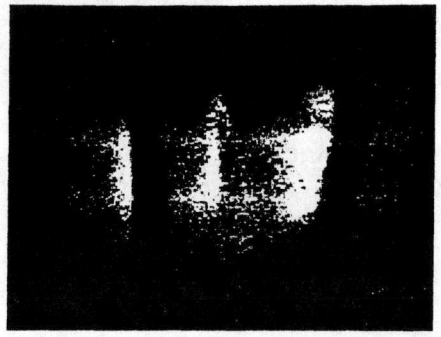

Figure 1. The colliding foil technique. X-ray 2D imaging of two 5μm CF_2 foils intially separated by 100μm. The laser beam intensities are $4.10^{14} Wcm^{-2}$

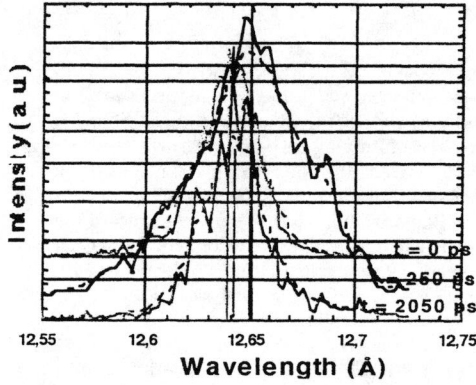

Figure 2. Time-resolved FLyβ spectra evolution. A red shift is detected at time t = 250ps

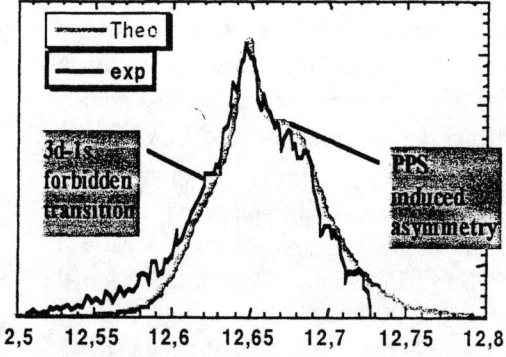

Figure 3. Satellite-like features on FLyβ spectra at the peak density

This work was supported by the HCM (contrats n° CHGET 930046) and TMR programs, (contrats n° FMGECT 950044 and ERBFMBI CT 95 0081) and by the HCM European network (contrat n° CHRXCT 930377).

Space-Resolved XUV Spectra of C VI and B V Lines from a 10 ps KrF Laser-Produced Plasma[†]

E. J. Iglesias[1*], H. R. Griem[1], R. C. Elton[1] and H. Scott[2]

[1] *Institute for Plasma Research, University of Maryland, College Park, MD 20742-3511*
[2] *Lawrence Livermore National Laboratory, P.O. Box 808, L-18, Livermore, CA 94551.*

Abstract. We produced a plasma using highly focused $\approx$ 50 mJ, 10 ps pulses from a KrF laser on graphite and boron-carbide targets. We measured space-resolved (along the plasma axis) line profiles of Hydrogen-like and Helium-like Carbon and Boron resonance lines, using a crossed-slit, 1 m grazing-incidence spectrometer, with a spatial resolution $\approx$ 50 μm. Synthetic spectra generated with the atomic postprocessor CRETIN provided preliminary estimates of the plasma electron temperature and density.

We produced a plasma using highly focused $\approx$ 50 mJ, 10 ps pulses from a KrF laser [1] on graphite and boron-carbide targets. We measured space-resolved (along the plasma axis) line profiles of Hydrogen-like and Helium-like Carbon and Boron resonance lines, using a crossed slit (5 μm $\times$ 150 μm), 1 m grazing-incidence spectrometer. The entrance slit was placed approximately 2 inches from the plasma. At the wavelengths of interest, in second order, the magnification of the instrument was about a factor of 3, along the direction of expansion of the plasma (e.g., perpendicular to the dispersion plane).

We used 101-105 film to record the data, and each run required $\approx$ 1500 shots of accumulated exposure. To register the spectral traces from the film we used either a micro-densitometer with a 10 $\times$ 100 μm slit, or a scanner with $\approx$ 15 μm pixel size. Either method provided a spatial resolution of $\approx$ 50 μm on the plasma.

¿From a preliminary study of the spectrum of both Carbon and Boron, we have observed a distinct evolution of the plasma from a mostly Hydrogen-like spectrum that extends up to 150 μm from the target surface, to only Helium-like at distances as far as 400 μm. We modeled our plasma assuming a concentric-shell, cylindrical system, in which temperature and densities peak on the axis of the cylinder. Using the atomic postprocessor CRETIN, developed at the Lawrence Livermore National Laboratory, we successfully simulated a spectrum of Carbon (in first order) at $\approx$ 100 μm from the target using a two-shell model (see figure 1). An electron density in the core of about 4×10^{21} cm^3 required in order to fit the overall spectrum,

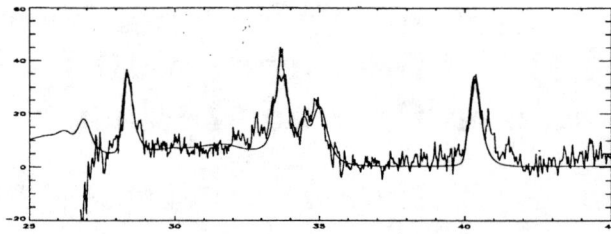

FIGURE 1. Experimental data recorded at $\approx 100\mu$m from the target surface. **Smooth line:** simulated spectrum with CRETIN using a two-shell model: **Core**(20 μm diam)$N_e = 4 \times 10^{20}$ cm^{-3}, T_e=100 ev; **Outer Shell** (10 μm thick) $N_e = 4 \times 10^{20}$ cm^{-3}, T_e=60 ev.

is consistent with the FWHM of Lyβ. For the line width CRETIN uses the results of reference 2; however, for this comparison CRETIN was integrated with TOTAL, a line broadening code developed by R. Lee at Lawrence Livermore National Laboratory. Both methods provided the same results within the resolution of the experiment. A cooler outer shell T_e=60 ev, at the same density, was required to provide the absorption of Lyα and the Helium-like contribution to the spectrum. We have also analyzed the second order spectrum of C VI, to look in more detail at Lyγ of C VI. Here we simplified our model to a homogeneous 30 μm cylindrical core, whose electron density is set as required by the width of Lyβ at different distances from the target; the results are shown in figure 2.

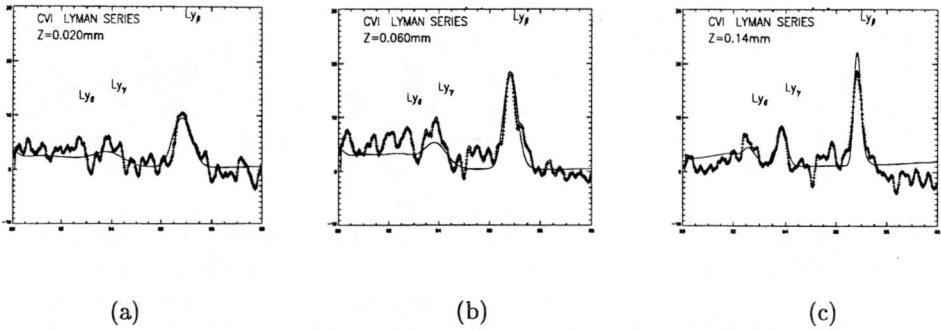

(a)   (b)   (c)

FIGURE 2. Carbon series in second order (Graphite target). Best fits using a solid 30 μm diameter cylinder. (a) Ne=10^{22}; (b) 6×10^{21}; and (c) 10^{21} cm^{-3}, $T_e \approx 100$ ev. Z is the distance to the target. Wavelength in Å.

The spectrum produced by the Boron-Carbide target shows evidence of the complexities caused by the lack of time resolution. This observation is inferred from the simultaneous presence, at one given distance from the target surface, of the Hydrogen-like and Helium-like spectrum of both Carbon and Boron. This fact is easily explained invoking the cooler plasma produced at later times. We have used again a simple model, as in the case of the carbon-only plasma, and simulated the end of the Lyman series of Boron, as shown in figure 3. After calibrating the wavelength against Lyα of Boron in both orders, Lyγ presents a shift to the red that will require further study in the future.

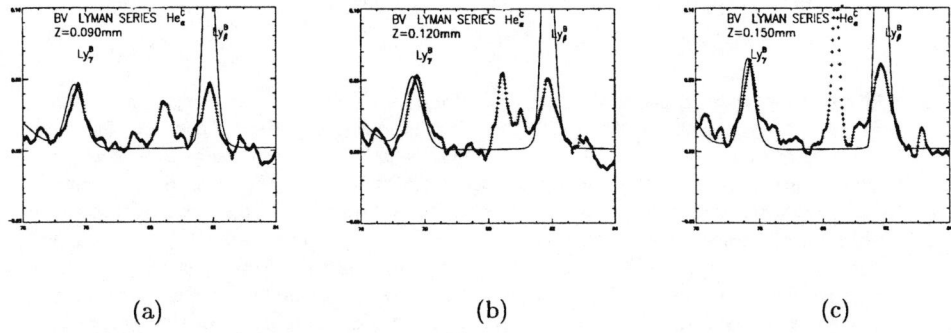

(a)   (b)   (c)

FIGURE 3. Boron series in second order (Boron-Carbide target). Best fits using a solid 30 μm diameter cylinder. (a) Ne = 5×10^{20}; (b) 5×10^{20}; and (c) 2.5×10^{20} cm^{-3}, T$_e$=50 ev. Z is the distance to the target. Wavelength in Å˙

[1] Y. Leng, V. E. Yun, J. Goldhar, and H. R. Griem, Rev. Sci. Instrum. **66**, 4045 (1995).
[2] H. R. Griem, M. Blaha and P. C. Kepple, Phys. Rev. A **19**, 2421 (1979).

†Partially supported by the National Science Foundation.
*On leave from Universidad Simón Bolívar, Caracas.

Shapes of Polarization-Induced Resonances in Stimulated Bremsstrahlung Spectra

Astapenko V.A.

Moscow Institute for Physics and Technology, Dolgoprudnyi, Moscow region 141700, Russia

In the frame of fixed classical electron current approximation the cross-sections of the stimulated Bremsstrahlung (SBr) on highly charged ions (HCI) with the core are studied. The polarization of the HCI core in the laser field is taken into consideration as well as the interference between polarization and standard (static) SBr channels. Spectral and intensity (with respect to the laser field strength) analysis of the SBr cross-sections is made for the near-resonant transitions in the alkali-like HCI core with (i) $\Delta n=0$ and (ii) $\Delta n=1$. In the first case the influence of the fine-splitting of the near-resonant transition on the SBr line shape is investigated. In the second case the nondipole interaction between incident electron and HCI core is analyzed in Coulomb approximation for projectile trajectory.

Interaction between radiation in UV and soft X-ray frequency range with HCI is of importance from the fundamental and applied points of view. SBr is an essential channel for the energy exchange between radiation and plasma. As it is known [1] the amplitude of this process in the case of target with electronic core is a sum of two terms, namely, standard (static) radiation amplitude and polarization radiation amplitude. For the quasiclassical electron radiative scattering when the radius of the spatial region responsible for the radiation is greater than the ion core radius the interference between static and polarization channels plays the significant role. Particularly, this interference produces the line shape asymmetry and dips in the cross-section spectra [2].

Previous investigations of these effects were done in the dipole approximation for the incident particle - HCI core interaction and neglecting the fine splitting of the near resonant transition in alkali-like HCI. However estimations show that the first approximation is valid for the relatively low frequency SBr corresponding to the transition in alkali-like HCI core with $\Delta n=0$ while the second one is appropriate to the transitions with $\Delta n=1$. In the present work the restrictions mentioned above are removed. Namely, the line shape of SBr is studied with account for fine-splitting (if $\Delta n=0$) and for the nondipole effects (if $\Delta n=1$).

The results of the SBr cross-section calculations for Δn=0 and Δn=1 are presented in Fig. 1 and Fig.2 correspondingly.

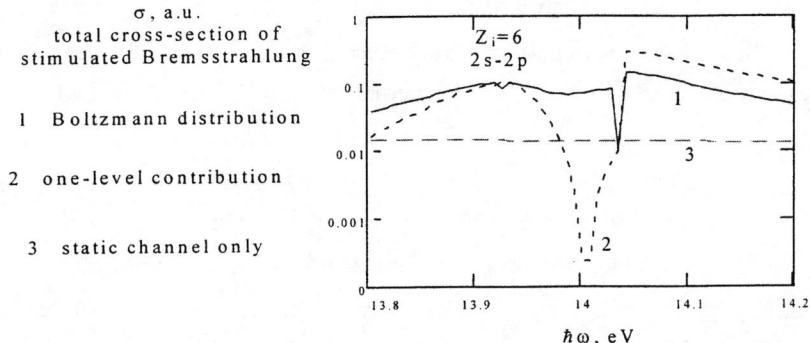

Fig.1 Spectral dependence of the near resonant SBr for 2s-2p transition in FVII ion with account for the fine-splitting of upper resonant level and for Boltzmann population distribution among levels. The contribution of the static channel only is shown as well.

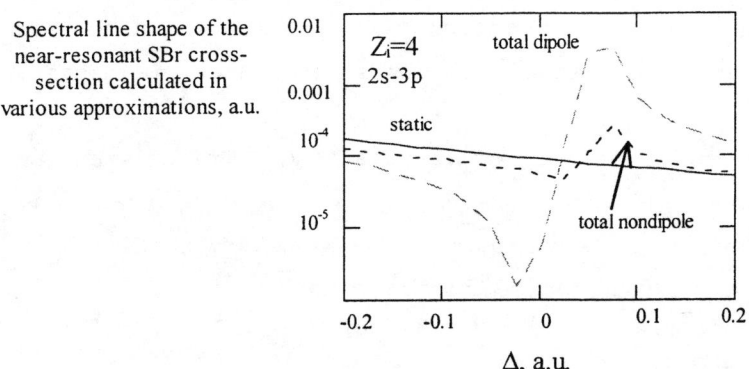

Fig. 2 Spectral dependence of the SBr cross-section for H-like ion (Z_i=4, 2s-3p near resonant transition) with account for the core polarization, interference effects and nondipole "incident particle-ion core" interaction.

REFERENCES

1. Polarization Bremsstrahlung, Edited by Tsytovich V.N. and Oiringel I.M., Plenum, 1992.
2. Astapenko V.A., Kukushkin A.B. Laser Phys. **8** (1998) 552.

Exact Solution for the Impact Broadening of the Hydrogen Lines Lyman-beta and Lyman-gamma

A. Derevianko* and E. Oks**

*Physics Dept., University of Notre Dame, Notre Dame, IN 46556, USA
**Physics Department, Auburn University, Auburn, AL 36849, USA

We have previously reported obtaining the first exact analytical solution for a multi-particle, purely dynamic Stark profile of the L_α line at the absence of a quasistatic field. In the present paper we bring up the first exact analytical solution for multi-particle, purely dynamic Stark profiles of the L_β and L_γ lines at the absence of a quasistatic field. A more complicated structure of the latter two lines compared the L_α line has made the present task much more challenging than it was to find the solution for the L_α line. However, the results obtained are of a greater practical importance than our previous result, since the lines L_β and L_γ provide more sensitive tools for plasma diagnostics than the L_α line.

For the brief history of efforts to obtain the exact solution of this problem and for the steps we followed to obtain the results, we refer to our previous paper [1]. Our exact result for the impact operator can be expressed in the form:

$$<nlm|\Phi|nlm> = -9(2\pi)^{1/2} n^2 \hbar^2 m_e^{-3/2} T_e^{-1/2} g_{nl}(D), \qquad (1)$$

where

$$g_{nl}(D) = \int_0^D [1 - <nlm|S(z)|nlm>] z \, dz. \qquad (2)$$

Here $D = \rho_D/\rho_w$ is the ratio of the Debye radius ρ_D to the Weisskopf radius $\rho_w = 3n\hbar/[2(m_e T_e)^{1/2}]$.

We have managed to <u>analytically</u> integrate over impact parameters in (2). The results are expressed through elementary functions and the integral sine [Si(x)] and cosine [Ci(x)] functions. Due to the limited space, we give here only the result for the L_β line:

$$g_{31}(D) = [3 - \pi Si(s\pi+\pi)s^2 + \pi Si(-\pi+s\pi)s^2 + 2\pi \sin(2s\pi)s^3 - 2\pi \sin(2s\pi)s$$
$$-4\pi^2 Ci(2s\pi)s^4 + 4\pi^2 Ci(2s\pi)s^2 + \cos(2s\pi) - 4\ln(s)s^2 + 4\pi^2 Ci(s\pi)s^4 + 4\ln(s)s^4$$
$$-4\pi^2 Ci(s\pi)s^2 + \pi Si(s\pi+\pi)s^4 + 2Ci(2s\pi+2\pi)s^2 + 4\ln(s\pi)s^4 + 4Ci(2s\pi)s^4$$
$$-2Ci(2s\pi+2\pi)s^4 + 12Ci(s\pi)s^2 + 2\ln(s+1)s^2 - 2Ci(2s\pi-2\pi)s^4 + 2\ln(s-1)s^2$$
$$-6Ci(s\pi+\pi)s^4 + 6Ci(s\pi+\pi)s^2 - 4Ci(2s\pi)s^2 - 4\ln(s\pi)s^2 - 6Ci(s\pi-\pi)s^4 \quad (3)$$
$$+2Ci(2s\pi-2\pi)s^2 + 2\ln(s\pi+\pi)s^2 + 2\ln(s\pi-\pi)s^2 - 2\ln(s\pi+\pi)s^4 - 2\ln(s\pi-\pi)s^4$$
$$+4\pi \sin(s\pi)s - 12Ci(s\pi)s^4 - 2\ln(s-1)s^4 - 2\ln(s+1)s^4 - 4\cos(s\pi) + 6\cos(s\pi)s^2$$
$$-4\pi \sin(s\pi)s^3 + 6Ci(s\pi-\pi)s^2 - s^2 - \cos(2s\pi)s^2 - \pi Si(s\pi-\pi)s^4] / [6s^2(s^2-1)],$$

where $s \equiv (1+D^2)^{1/2}$.

In the practically important limit of $\rho_D \gg \rho_W$, the results <u>for all Lyman lines</u> can be simplified to the form:

$$g_{nl}(D) = (2/3)(n^2 - l^2 - l - 1)[K_{nl} + \ln(D)]. \quad (4)$$

This representation of the exact solution allows to trace the correspondence with the perturbative standard theories. Indeed, in the latter, the second term in brackets would be considered as originating from "weak" collisions, while the first term (K_{nl}) would be considered as originating from "strong" collisions. However, the standard theories cannot yield K_{nl} better than by the order of magnitude. In distinction to that, we have obtained <u>exact</u> results for the constants K_{nl} corresponding to the L_α, L_β and L_γ lines as follows:

$$K_{21} = 0.82542, \quad K_{31} = -0.424548, \quad K_{41} = -0.747093. \quad (5)$$

References

1. A. Derevianko and E. Oks, in "Physics of Strongly Coupled Plasmas", World Scientific, New Jersey, 1996, p.286-291, p.292-296.

Distinctive Features of the Advanced Generalized Theory of Stark Broadening of Hydrogen Lines in Plasmas

E. Oks and J. Touma

Physics Department, Auburn University, Auburn, AL 36849, USA

The Generalized Theory (GT) of Stark broadening of hydrogen lines in plasmas, published by Ispolatov and Oks in JQSRT, v.51, p.129, 1994, is based on non-perturbative treatment of one component of the electron field. Therefore the GT is intrinsically more accurate than the fully-perturbative, standard theories, such as the theory by Kepple-Griem (KG) published in Phys. Rev., v.173, p.317, 1968, and the theory by Sholin-Demura-Lisitsa (SDL) published in Sov. Phys. JETP, v.37, p.1057, 1973. The GT embraces the standard theories as its limiting case of the zero-coupling between the ion and electron fields. In the first version of the GT by Ispolatov-Oks (1994), the theory was developed analytically only to the stage where three integrations still remained to be done (two time-integrations plus the integration over impact parameters).

The present paper introduces an Advanced Generalized Theory (AGT). The AGT was derived from the GT as follows. First, we have managed to perform analytically both the remaining time-integrations. Second, for the final integration over impact parameters we have used analytical approximations for the integrand. As a result, we have obtained closed-form expressions for the width, shift, and coupling of Stark states. Thus, from the mathematical point of view, the AGT is much simpler and more user-friendly than the GT.

This advance has been achieved by using a formal expansion of the GT results in terms of a parameter $1/u$, $u=(nq-n'q')/n$, where n and q are the principal and the electric quantum numbers of the upper state, n' and q' are the corresponding numbers for the lower state. The results turned out to be remarkably accurate (about 5%) even for $|u|=1$, as has been tested by the comparison with exact numerical GT-calculations. Thus just these results cover practically all lateral Stark components of all hydrogen lines, with the exception of only few lateral components characterized by $|u| \ll 1$. However, for the latter few components, the imposition of the unitarity restrictions effectively reduces the GT to the standard theories. So the AGT formulas can be actually used for all lateral components, yielding enhanced accuracy (compared to the standard theories) for the overwhelming majority of the lateral components. For the central Stark components, a separate set of formulas has been obtained by combining the GT-asymptotics at large impact parameters with the unitarity limit at small impact parameters.

The mathematical simplicity of the AGT results has made it possible to gain a much deeper physical insight into important intimate features of the generalized theories. The features that distinguish the AGT/GT from its predecessors and ensure its superior accuracy are as follows.

1. <u>Empirical</u> choices of important characteristic impact parameters, made by our predecessors, turned out to be, generally speaking, <u>erroneous</u>.

A) In the AGT, the **effective Weisskopf radius** R_W, being rigorously derived from the first principles, turns out to be proportional to n^2, while SDL had empirically chosen R_W proportional to n.

B) In the AGT, the effective R_W turns out to be intrinsically **individual** for each Stark component (i.e., dependent on q), while KG had empirically chosen a component-independent R_W.

C) In addition, KG had empirically chosen a wrong expression for the ion-field-dependent **upper cutoff** R_F as being proportional to $1/n^2$, while from the selection rules for the parabolic quantization it follows that R_F should be proportional to $1/n$, as it is used both in the AGT and by SDL

2. The AGT proves rigorously that the <u>coupling</u> between the ion and electron broadenings (which is effective in the high field/density range defined by the inequality $\omega_{pe} < 3n\hbar F/(2m_e e)$, where ω_{pe} is the plasma electron frequency, is significantly stronger than it was empirically introduced by both the KG- and SDL- theories, where the dependence on the ion field F entered only the argument of logarithm.

3. Even in the low field/density range ($\omega_{pe} < 3n\hbar F/(2m_e e)$), where the coupling between the ion and electron broadenings is negligible, the results of the AGT are more accurate than the results of the standard theories (due to a more accurate treatment of one component of the electric field in the AGT). First of all, the AGT still yields the rigorously derived effective Weisskopf radius, <u>individual</u> for each Stark component (as noted above) in distinction to the KG theory. Second, the AGT brings up the <u>exact</u> value of the so-called "<u>strong collision constant</u>" - in distinction to both the KG- and SDL-theories, where the choice of this constant was empirical and ambiguous.

While the difference between the AGT and the standard theories affects any hydrogen line, the consequences are especially important for highly excited (high-n) lines. Diagnostics based on measuring profiles of high-n hydrogen lines are utilized primarily in magnetic fusion and in astrophysics. In magnetic fusion, Balmer and Paschen lines up to n=15 are employed for tokamak plasmas of densities (10^{14} - 10^{15}) cm^{-3}. In solar physics, the diagnostics of the chromospheric plasma relies on Balmer lines up to n=20 for solar flares (densities of the order of 10^{13} cm^{-3}) and on Balmer lines up to n=30 for the quiet Sun (densities of the order of 10^{11} cm^{-3}). For both tokamak and solar plasmas, the AGT should result into a significant revision of diagnostic conclusions derived previously using the standard theories.

LOWER TEMPERATURE/DENSITY PLASMAS

Stark Broadening of the 2S-Level of Hydrogen at Low Electron Densities Measured by Doppler-Free Two-Photon Polarization Spectroscopy

M. Schmidt, K. Grützmacher, A. Steiger

Physikalisch-Technische Bundesanstalt, Abbestr. 2-12, D-10587 Berlin, Germany

Abstract. Doppler-free two-photon polarization spectroscopy was used for measuring the Stark-broadening of the 2S-level (Lyman-α transition) of hydrogen in an arc plasma at electron densities around 10^{21} m^{-3} in order to examine corresponding theories. This electron density range, not accessible so far, is of special interest because Stark-broadening is comparable with the fine-structure splitting. Previous investigations in arc plasmas with somehow higher electron densities already demonstrated, that atomic fine-structure is causing a distinct asymmetry in the central part of the Stark-profiles. Precise profile measurements were possible using a recently developed pulsed UV-laser spectrometer with a bandwidth of 300 MHz tunable around 243 nm. In addition, the electron densities were determined with highly sensitive dispersion interferometry.

INTRODUCTION

Stark broadening of spectral lines provides a valuable tool for spectroscopic diagnostics of plasmas. For the basic understanding and for application as well, atomic hydrogen is especially suited because it can be treated easily and precisely by quantum mechanics. At larger electric field strengths, its Stark splitting shows the linear Stark effect, hence the line shapes react very sensitively to the properties of the electric microfield in plasmas. Of particular importance for proving Stark broadening calculations is the principle resonance line Lyman-α. The corresponding two-photon transition allows for Doppler-free measurements as required for electron densities below 10^{23} m^{-3}. Such measurements are only possible so far by Doppler-free two-photon polarization spectroscopy [1]. This technique gives direct access to the 1S-2S transition which also exhibits the Stark broadening of the first excited level only. However, in contrast to the Lyman-α line, it splits at larger electric field strengths exclusively into one red and one blue shifted Stark component.

Two-photon polarization spectroscopy combines some special characteristics, already verified by previous experiments [2, 3], which indeed allow for precise Stark broadening measurements and a very sensitive proof of Stark broadening calculations:
- The first order Doppler effect can be eliminated and the measured profiles have no Doppler broadened background at all.

- Elastic and inelastic collisions do not perturb or limit the measurements.
- The measured profiles are a superposition of the squared absorption profile and the squared dispersion profile. They allow therefore for a very sensitive test of the central part of the Stark profiles which depends sensibly on ion-dynamical effects etc.

First systematic experimental investigations of Doppler-free Stark-broadened line profiles of hydrogen and deuterium were carried out in arc plasmas at electron densities from $3 \cdot 10^{21}$ m^{-3} up to $1.2 \cdot 10^{22}$ m^{-3} [1,2]. Although these measurements confirmed the predictions that ion dynamical effects dominate the Stark broadening, the measured line profiles exhibited in contradiction to calculated ones a pronounced asymmetry, increasing towards lower electron densities. This discrepancy was explained as a deviation from the linear Stark effect approximation due to the atomic fine structure [4], which was ignored so far by Stark broadening calculations. In contradiction to previous common understanding, this asymmetry is mainly attributed to different strengths of the blue and red shifted Stark component, while the deviation from linear splitting itself is negligible. It should be mentioned, that the asymmetry shows up in the line profile already at an electron density as high as 10^{22} m^{-3}, where the corresponding Holtsmark normal field strength leads to a splitting of the Stark components about one order of magnitude larger than the fine-structure splitting of the unperturbed atom. Improved calculations [5] which accounted additionally for the correct Stark splitting achieved finally a remarkably improved agreement with calculated profiles, although the predicted half widths exhibited the tendency to be smaller than the measured ones at lower electron densities, i.e. 20 % less at $n_e = 3 \cdot 10^{21}$ m^{-3}.

Consequently, these previous investigations call for measurements at even lower electron densities. This is furthermore of special interest because at these low densities Stark-broadening is comparable with the fine-structure splitting and the linear Stark splitting approximation is not valid at all. However, the decreasing line width of the Stark profiles requires high power pulsed laser radiation tunable around 243 nm with better spectral resolution than commercially available.

For the Stark broadening measurements down to 10^{20} m^{-3} reported in this paper, a recently developed UV-laser spectrometer with exceptional spectral quality and pulse power was employed. Additionally, highly sensitive dispersion interferometry was used to determine the electron density via the electronic part of the refractive index of the plasma. The hydrogen plasma source is a wall stabilized cascaded arc of extraordinary stability and reproducibility, which was previously established as a standard of atomic hydrogen densities [3].

EXPERIMENTAL SETUP AND MEASUREMENT PROCEDURE

Two-photon polarization spectroscopy was used for the Doppler-free measurements of the 1S-2S Stark-broadened profiles in a wall-stabilized arc plasma, 100 mm in length and 8 mm in diameter, operated in pure hydrogen at a pressure of 10 kPa and dc-discharge currents varying from 18 A to 63 A. This plasma source and the laser-spectroscopic technique as well are described in detail in literature [3]. However, in

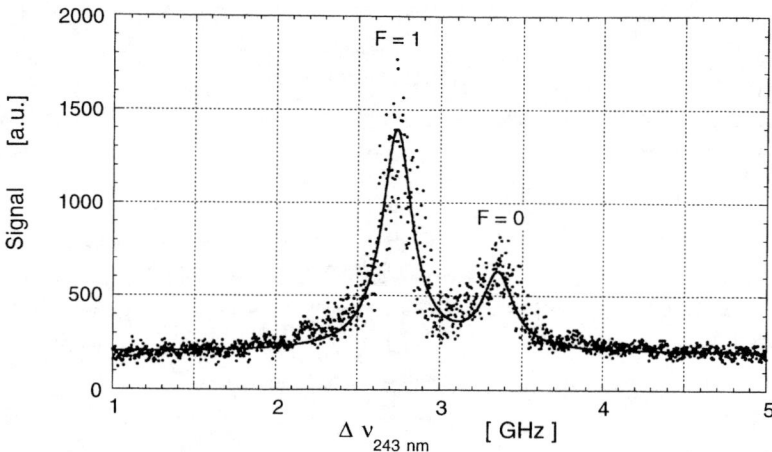

FIGURE 1. The optogalvanic spectrum with the recorded signal of individual pulses while slowly scanning the laser demonstrates the high resolution of the UV-laser spectrometer. On top of a small Doppler background, the two hyperfine components of 1S-2S transition of atomic hydrogen can be described by two Lorentz functions (solid line) with the correct splitting and the intensity ratio of 3:1.

contrast to this previous experiment, a new type of UV-laser spectrometer has been used. Its development is based on a novel concept to generate tunable pulsed UV-radiation with unprecedented efficiency, i.e. conversion of the harmonics of a Nd:YAG laser by frequency splitting in optical-parametric processes and sum-frequency generation [6].

The laser spectrometer is based on a commercial laser system consisting of an injection seeded Nd:YAG laser, a single longitudinal mode optical-parametric KTP oscillator and a BBO amplifier. It provides tunable radiation around 772 nm and this output is finally converted into UV-radiation by sum-frequency generation with the residual third harmonic. Within a bandwidth of about 300 MHz, 10 mJ pulse energy in 2.5 ns are obtained from moderate input powers of 10 mJ for the second and 200 mJ for the third harmonic. The excellent spectral resolution can been seen in figure 1 where the hyperfine components $F = 0$ and $F = 1$ of the 1S-2S transition are clearly resolved. This two-photon absorption profile was measured by an optogalvanic signal which is created when a third laser photon ionizes the 2S excited hydrogen atoms. The atomic hydrogen is provided by thermal dissociation at an electrically heated tantalum wire inside a small vacuum cell filled with hydrogen at a pressure of a about 400 Pa. The optogalvanic detection is very sensitive, less than 1 % of the total laser output energy is needed.

The experimental setup for two-photon polarization spectroscopy is shown in figure 2. Doppler-free measurements require counter-propagating pump and probe beam of equal wavelength. Approximately 90 % of the 243 nm radiation is directed into the pump beam. A half-wave plate and a linear polarizer serve as a variable attenuator and

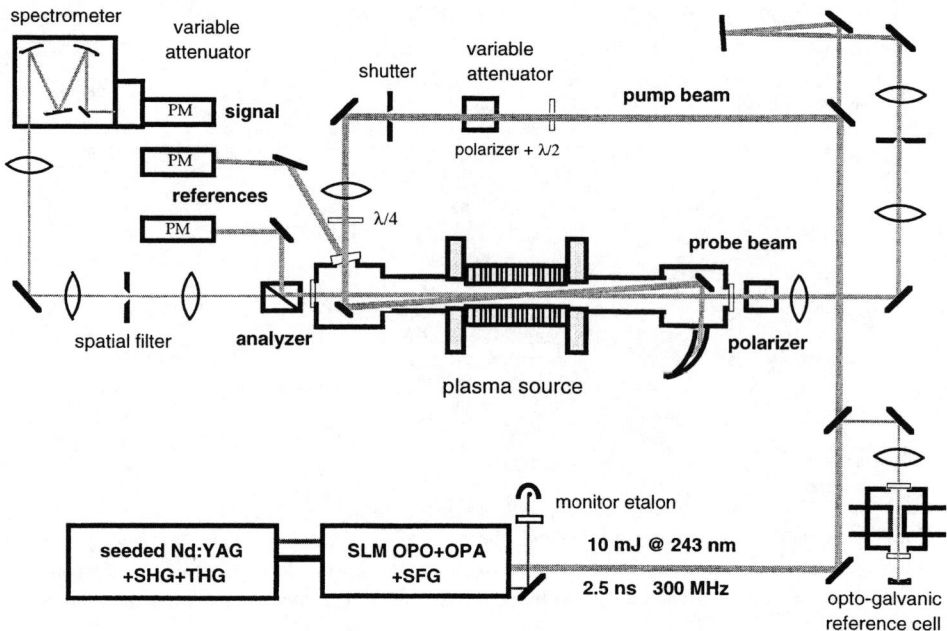

FIGURE 2. The experimental setup for Doppler-free two-photon polarization spectroscopy is shown together with the pulsed UV-laser spectrometer and the optogalvanic reference cell. The overlap of pump and probe beam defines the measurement volume (ø = 200 µm, 30 mm long) on axis in the center of the plasma column. After spatial and spectral filtering, the signal is detected by a photo multiplier (PM). Polarizer and polarization analyser provide a residual transmittance as small as 10^{-8} while measuring the Stark profiles.

finally the pump beam becomes circularly polarized by a quarter-wave plate. Less than 10 % of the pulse energy are used for the probe beam. After passing a spatial filter, the central part is sent through a high quality polarizer before it enters the low pressure chamber. The counter-propagating beams are focused by lenses of 1 m focal length into the center of the arc plasma with 16 mrad crossing angle. The focus diameter of the probe beam is 200 µm while the pump beam has a spot size of 300 µm which results in a 30 mm long overlapping volume which is relatively insensitive to small beam displacements. The irradiance provided by the pump beam was around 1 GW/cm^2.

By two-photon absorption, the circular pump beam induces a small change of the linear polarization state of the probe beam. As signal, this polarization change is detected very sensitively by a crossed analyzer, because a very small residual transmittance of 10^{-8} could be achieved. Also, two reference signals are recorded to account for laser output variations in two different ways. Every laser pulse, the original probe beam which is reflected out by the crossed analyzer is detected for normalization of the polarization signal. As there is a nonlinear signal dependency of the effective pump irradiance, the pump beam reference is used only after being averaged over 20 laser pulses to account for the drift of the laser output power.

The profiles were recorded with two identical scans but with opposite circular polarization of the pump beam. As part of each scan, additionally 100 signals with blocked pump beam were registered at the beginning and at the end of each scan, which allowed to determine the actual residual transmittance contributing to the signal as background. The correct polarization profile is obtained finally after subtracting the residual transmittance and by averaging the profiles of the two scans obtained with left and right circularly polarized pump beam. This procedure cancels out polarization imperfections of the probe beam caused by a small deviation from perfect crossing of the polarizers or by the residual birefringence of windows of the plasma chamber, which otherwise would result in small deviations from the correct line shapes of the line wings.

For measuring the Stark broadened profiles of the 1S-2S transition of atomic hydrogen the laser was continuously scanned over a frequency interval of typically 350 GHz, which is more than 1000 times its bandwidth. The linearity of tuning the laser frequency was checked by recording the transmission of a plan parallel solid Fabry-Perot etalon with a free spectral range of about 7.5 GHz. Furthermore, the optogalvanic signal of the reference cell was recorded in order to obtain the absolute resonance frequency of the unperturbed 1S-2S transition.

MEASURED 1S-2S STARK PROFILES

The Stark broadened profiles were measured at various electron densities in the range from about $3 \cdot 10^{21}$ m^{-3} down to $5 \cdot 10^{20}$ m^{-3}. As an example which demonstrates the quality achieved for the measurements by Doppler-free two-photon polarization spectroscopy, a Stark profile at an intermediate electron density of $1.8 \cdot 10^{21}$ m^{-3} (36 A discharge current) is plotted in figure 3 on a logarithmic scale. Each data point is the average over at least 40 laser pulses, while the laser frequency is detuned continuously. Additionally shown is the optogalvanic reference signal and the transmittance curve of the monitor etalon on a linear scale. The sharp central peak of the optogalvanic signal corresponds to the Doppler-free 1S-2S resonance in the reference cell, indicated by v_{ref}.

The line wings exhibit the typical asymptotic behavior, which is a general characteristic of all line profiles measured by polarization spectroscopy [7]. However, the interesting central part of the profile shows ion dynamics being so dominant, that the individual contributions of the underlying red and blue shifted Stark components can not be distinguished anymore, i.e. the Stark broadening leads to a single maximum. On the other side, the central part of the profile exhibits a pronounced asymmetry. This is caused by the deviation from linear Stark effect approximation, which causes the strength of blue shifted Stark component being remarkably smaller compared to the red shifted component. As all these effects show up in the central part of the profile, a linear scale is chosen for the following discussion.

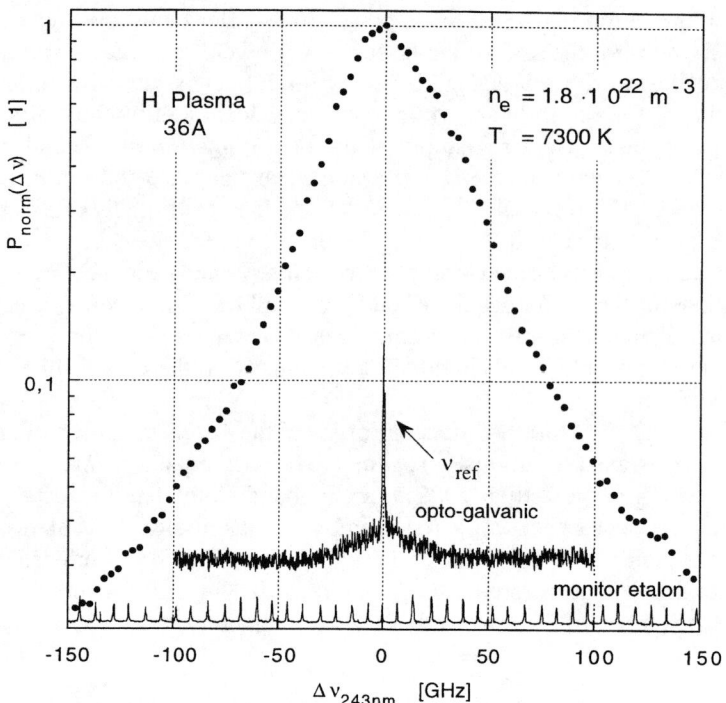

FIGURE 3. The Doppler-free Stark profile of atomic hydrogen measured at a discharge current of 36 A is shown on a logarithmic scale versus frequency. At the bottom the two reference signals are plotted on a linear scale, i.e. the transmission curve of the monitor etalon and the optogalvanic signal of the reference cell, which shows the Doppler-free 1S-2S resonance on top of a Doppler background.

For the highest and lowest electron density, the measured Stark profiles, normalized with respect to their maximum value [8], are shown in figure 4. For $n_e = 3.1 \cdot 10^{21}$ m^{-3} (63 A discharge current) the central profile exhibits well pronounced the asymmetry with a distinct shoulder on its blue side, as discussed above. However, for $n_e = 5 \cdot 10^{20}$ m^{-3} (18 A discharge current) the asymmetry seems to disappear and the measured Stark profile can be fitted quite well by a Lorentz-profile having a half width still 80 times larger than the bandwidth of laser. Obviously, ion dynamic effects become increasingly dominant with decreasing electron density. However, a detailed comparison between the Stark profile and the fitted Lorentz profile still exhibits a minor asymmetry, which mainly shows up as a small red shift of the central part of the profile.

Finally it should be mentioned, that the Stark profile at the highest electron density of this work agrees with that measured in the previous experiment [4] within the measurement uncertainties, although completely different laser spectrometers were used and the wall stabilized hydrogen plasma sources had different arc channel diameters and were operated at different pressures and discharge currents.

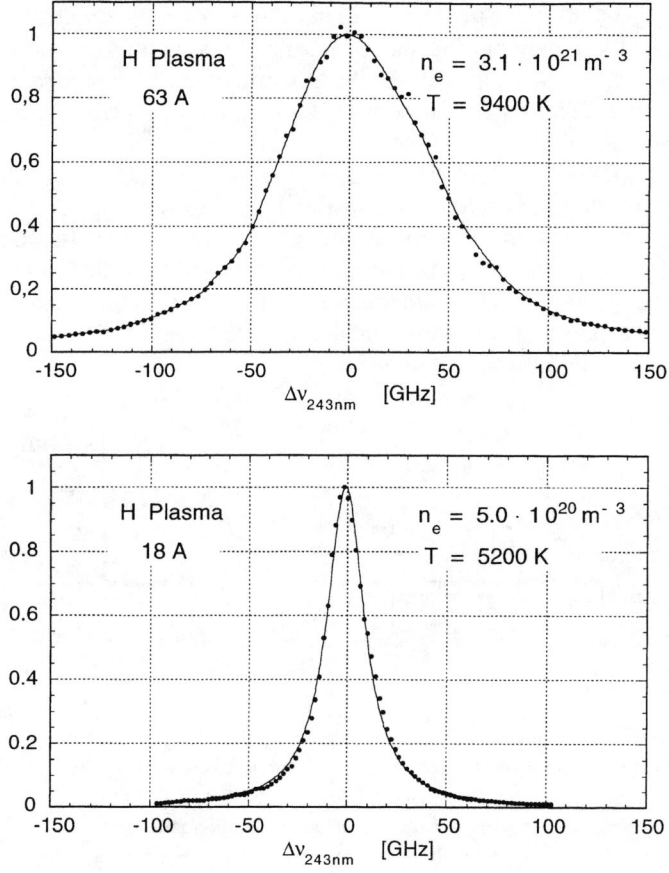

FIGURE 4. For comparison, the Stark-broadened profiles at the highest and lowest measured electron density are plotted on linear scale. The Stark profile at the smallest electron density of $5 \cdot 10^{20}$ m^{-3} can be fitted quite well by a Lorentz profile. This is not anymore a reasonable approximation at the largest electron density, because ion dynamic is less important. The upper Stark profile and its smooth curve fit clearly exhibits therefore the asymmetry in the central part, which is due to different strength of the underlying red and blue Stark component.

MEASUREMENT OF ELECTRON DENSITY

In the previous Stark broadening investigations at higher electron densities [4], n_e was fixed by emission spectroscopy, in particular via the Stark broadening of the Balmer-β and Balmer-γ lines. The aim of this work was not only to extend the Stark broadening measurements to lower electron densities but to perform also an independent precise electron density determination. A suitable new tool is nonlinear dispersion interferometry [9] which measures the dispersion part of the refractive index at the fundamental laser frequency and its second harmonic along a common beam pass. It allows therefore a direct measurement of the electronic part of the refractive index of plasmas at such low electron densities were conventional interferometry is already facing serious limitations.

Using a mode-locked Nd:YAG laser, the principle of a dispersion interferometer has been successfully demonstrated in Novosibirsk [10]. In its simplest case it consists of two frequency doublers and in-between the plasma under investigation. The main advantages of this interferometer type are compactness, insensibility to vibrations and enhanced sensitivity with respect to classical interferometers. These favorable properties result from the fact that the two optical channels have no different geometrical path but are separated in radiation frequency.

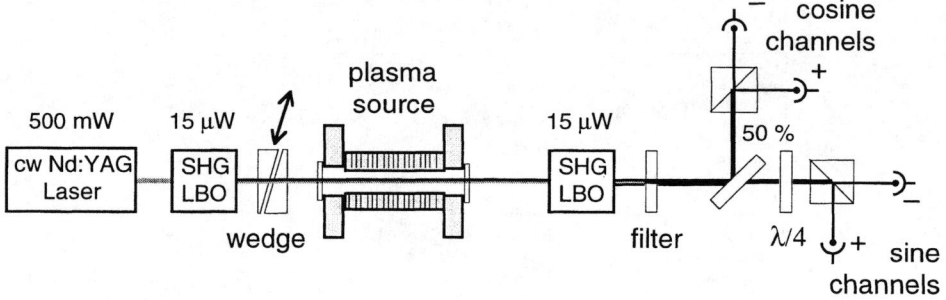

FIGURE 5. Schematic experimental setup for measuring the electron density of the plasma by dispersion interferometry. Before and behind the plasma source, a small part of the 500 mW power of a cw Nd:YAG laser is converted into its second harmonic (15 µW) in LBO crystals. The crystals are rotated by 90° with respect to each other. After the infrared radiation is filtered out, the interference of both second harmonics with orthogonal linear polarization is detected by four channels with equal intensity but multiple $\pi/2$ phase difference with respect to each other ($\pm$ sine and $\pm$ cosine channels). This is achieved by a beam splitter (50 % for both s and p polarization) in combination with a quarter wave plate and two polarizers set to 45 ° with respect to the s and p polarization orientation.

Figure 5 clarifies the principle of operation. The Nd:YAG laser radiation is partially converted into its second harmonic in the first doubler. Both waves propagate along the same optical pass through the plasma. Behind the plasma, once more a part of the Nd:YAG laser radiation is converted into the second harmonic and the remaining fundamental is suppressed by a dichroic mirror. Thus, the interference between the two second harmonics can be measured with adequate detection channels. In this work a four channel detection system was set up, which allows to compensate intensity fluctuations and to eliminate residual plasma background radiation. In addition, highest sensitivity is provided by $\pi/2$ phase separation. A laser-diode pumped Nd:YAG laser was used as stable cw light source. For frequency doubling, non-critical phase matching in LBO at 148 °C was chosen, which avoids any walk-off of the second harmonic with respect to its fundamental. Imaging of the beam waist from the first doubler into the second doubler with an additional focus in the center of plasma column is performed by two achromatic lenses. For calibration purpose, the dispersion of quartz is used by moving a wedge plate across the beam path.

As a special characteristic, the dispersion interferometer is only sensitive to changes of the difference of the refractive indices at the fundamental $n(\lambda)$ and the second harmonic $n(\lambda/2)$ of the laser, i.e. the phase difference $\Delta\varphi$ being detected is given as

$$\Delta\varphi = 2\varphi(\lambda) - \varphi(\lambda/2) = 4\pi/\lambda \int \left(n(\lambda) - n(\lambda/2)\right) dl \quad (1)$$

In figure 6, the measured phase difference is shown for a time interval of 200 μs when the plasma discharge current of 36 A is switched off at time zero. The extraordinary stability of the plasma source shows up in very small signal variations when the plasma is on (as small as 3‰ of $\pi/2$ phase difference). This corresponds to an electron density of change of 10^{19} m^{-3} only. Indeed, the plasma is providing such a stability for a long period which allows a precise calibration of the detection channels while the plasma is on. The calibration procedure takes about one minute.

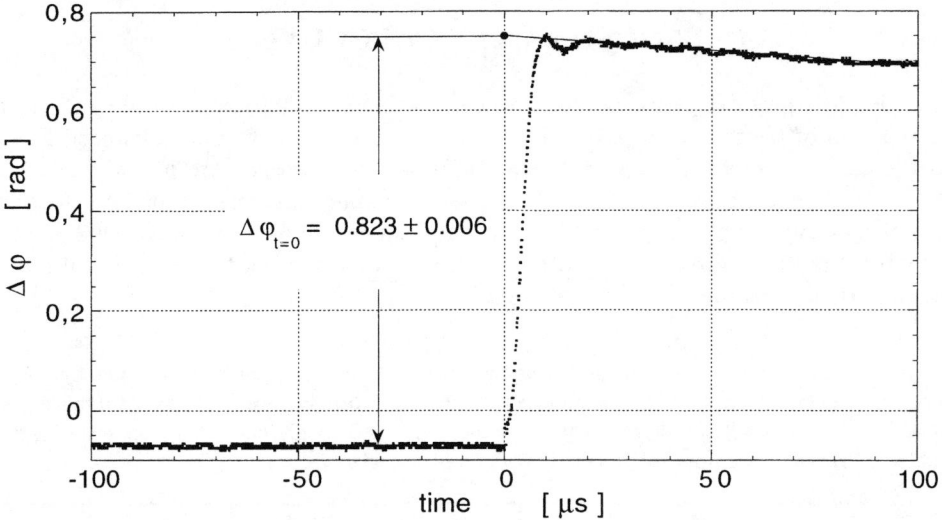

FIGURE 6. The phase-signal of the dispersion interferometer is shown for a time interval of ± 100 μs when the discharge current of 36 A of the hydrogen plasma is switched off (t = 0 s). The change of the phase difference of 0.823 corresponds to an electron density of $1.8 \cdot 10^{21}$ m^{-3}.

Switching off the plasma current results in the change of the phase difference as shown in figure 6 in the time interval from zero to 100 μs. The 5 μs slope of the signal exhibits the rise time of the detection channels only, while the electronic recombination of the hydrogen plasma occurs on a much shorter time scale [11]. After the electronic recombination, the neutral particles of the plasma are cooling down and the molecular density is increasing. However, this takes place on a much longer time scale and can be seen as variation of the phase difference in the time interval from 20 μs to 100 μs in figure 6. Therefore, the extrapolation of this slope to time zero allows to determine the change of the phase difference due to electronic recombination only, i.e. the electron density is given by

$$\Delta\varphi_{t=0} = 3 \cdot r_e \, \lambda/2 \int n_e \, dl \qquad (2)$$

re denotes the classical electron radius, l is the effective length of the plasma column and n_e the electron density of interest. The factor of three in front of the expression is a special characteristic of dispersion interferometry, which is three times more sensitive compared with conventional two wavelength interferometry at λ and $\lambda/2$.

The interferometric electron density determination was carried out several times, and the measured phase differences turned out to be reproducible within 1 % for the higher electron densities and 5 % for the lowest electron density. Finally, the uncertainty for the absolute number density of electrons is conservatively estimated to be ± 5 % and ± 7 % respectively. It originates mainly from the limited knowledge of the effective length of the plasma column.

COMPARISON WITH CALCULATIONS

A detailed proof of Stark broadening calculation by experiment would require a comparison of the entire Stark profile. However, recently performed advanced Stark profile calculations for the 1S-2S transition of atomic hydrogen are providing so far mostly values for the FWHM (full width at half maximum). At least at the low electron densities of this work, the FWHM is a reasonable measure for a first test. This can be seen from the measured Stark profiles in figure 4, which show a smooth, slightly asymmetric line profile with one maximum only.

TABLE 1. In the upper part, discharge conditions and corresponding parameters of atomic hydrogen are given. The lower part shows the electron density (relative uncertainty varies from 5% for the largest value to 7% for the smallest value), the FWHM of the measured Stark profiles (relative uncertainty 5%), and the FWHM obtained from two advanced Stark profile calculations.

discharge current / A	63	36	27	18
kinetic temperature / K	9400	7300	6400	5200
atomic density n_H / 10^{22} m^{-3}	7.0	9.0	9.8	9.3
electron density n_e / 10^{20} m^{-3}	**31**	**18**	**12**	**5.0**
1S-2S measured FWHM / GHz	**93**	**62**	**48**	**23**
FWHM by Seidel [12] / GHz	75	50	35	16
FWHM by González [13] / GHz	84	53	38	19

The results are given in table 1 for four different plasma conditions covering the whole electron density range of the present investigation. To allow for an estimation of e.g. additional resonance broadening, number densities (5 % relative uncertainty) and kinetic temperatures (4 % relative uncertainty) of atomic hydrogen are listed as well. These values are known from a previous experiment [3], where exactly the same hydrogen arc plasma source as used for this Stark broadening investigation was established as a standard of atomic hydrogen.

In general, all values for the FWHM of calculated profiles are smaller than measured ones. However, for the largest n_e the calculation of González [13] are missing 11% of broadening only and obtain the best agreement with the experiment, while the calculations of Seidel [12] lead to 25 % of missing broadening. Obviously, details in the calculations can lead to different results, although both calculations are taking into account the correct Stark splitting, the measured kinetic temperatures, and both are treating ion dynamic effects by computer simulations. Towards smaller n_e value, deviations are increasing, i.e. at the lowest measured plasma parameter, the calculated half width would have to be increased by 25 % for González and nearly 50 % for Seidel to reach the measured value.

CONCLUSION

This investigations of the Stark broadening of the 12-2S transition of hydrogen revealed excellent agreement with previous ones [4] and fairly good agreement with advanced Stark profile calculations [12,13] at an electron density $3 \cdot 10^{21}$ m^{-3}. Nevertheless, with the electron densities decreasing down to $5 \cdot 10^{20}$ m^{-3}, remarkable discrepancies between theory and experiment show up. Actually, there is no explanation for this so far, neither neutral broadening nor laser power broadening seems to close the gap. The discrepancies ask therefore for theoretical explanation and improvement. From a simple point oft view, i.e. from the fact, that Stark broadening due to ion dynamic effects is dominating the underlying Stark splitting, it can be concluded, that the broadening mechanisms, which are acting on a single Stark component, are still underestimated by the theories.

In order to clarify this situation, the present investigation will be extended to measurements at about a factor of two lower electron densities. This will be possible by lowering the pressure of the cascaded arc discharge. Furthermore, we will also change from hydrogen to deuterium. At equal plasma parameters the line shape of deuterium will be the same as hydrogen unlike the ion dynamic contributions of the Stark broadening due to the factor of two in reduced mass.

REFERENCES

1. K. Grützmacher, A. Steiger, *Spectral Line Shapes, Vol. 5*, Ed. J. Szudy, Ossolineum Publ. House, Wroclaw Poland 1989, p. 35
2. A. Steiger, K. Grützmacher, *Spectral Line Shapes, Vol. 7*, Eds. R. Stamm and B. Talin, Nova Science Publ., New York 1993, p. 141
3. R. Dux, K. Grützmacher, M.I. de la Rosa, B. Wende, Phys. Rev. E **51**, 1416 (1995)
4. A. Steiger, *Ph.D. thesis*, Technische Universität Berlin, 1993
5. J. Seidel, A. Steiger, K. Grützmacher, *Spectral Line Shapes, Vol. 8* (AIP Conf. Proc. 328), Eds. A.D. May, J.R. Drummond, E. Oks, AIP Press, New York 1995, p. 32
6. A. Steiger, K. Grützmacher, M.I. de la Rosa, *Laser in Research and Engineering*, Eds. W. Waidelich et al, Springer-Verlag, Berlin 1996, p.308
7. A. Steiger, K. Grützmacher, "Line shapes obtained by polarization spectroscopy", contribution to this volume
8. It has to be mentioned, that the usual normalization of absorption profiles to unit area is not valid for profiles measured by polarization spectroscopy.
9. G.N. Alferov, S.A. Babin, and V.P. Drachev, Opt. Spectrosc. **63**, 348 (1987)
10. V.P. Drachev, Yu.I. Krasnikov and P.A. Bagryansky, Rev. Sci. Instrum **64**, 1010 (1993)
11. H.W.P. van der Heijden, "Time resolved LIF of atomic hydrogen in an expending cascaded arc plasma", report VDF/NT 97-31, Eindhoven University of Technology, Netherland, August 1997
12. J. Seidel, private communication
13. M.A. González, *Ph.D. thesis*, Universidad de Valladolid, Spain, 1998

Radiative Collisions and the Sonoluminescence Continua

Lothar Frommhold

Physics Department, University of Texas, Austin TX 78712

Abstract. Current views concerning the light emitting processes of sonoluminescence (SL) vary widely, even where certain spectroscopic facts seem to be well established. In search of the actual light emitting processes of sonoluminescence, we consider a variety of neutral-neutral and electron-neutral collision processes that take place under SL conditions. Such spectra are computed for comparison with the observed SL spectra. Calculated electron-atom bremsstrahlung spectra and measured profiles of SL of the rare gases are remarkably consistent with regard to both, spectral shape and absolute intensity — more so than the spectra of alternative collisional systems, e.g., pairs of neutral atoms or molecules. If indeed electron-atom bremsstrahlung were the principal emission process of SL of the rare gases, a number of puzzling observations could find rather natural explanations.

I INTRODUCTION

Large-amplitude sound waves in water generate light (sonoluminescence, SL) [1,2]. It has long been known that in the tensile phase of sound cavitation occurs [2,3]. During the subsequent compression phase, the cavities ("bubbles") collapse so that their gas content is compressed. As a consequence, the temperature of the gas increases, either adiabatically or by an ensuing shockwave mechanism. Under certain conditions, the cavities collapse at speeds amounting to roughly four times the speed of sound so that a spherical, convergent shockwave is driven [4,5]. The shock front is reflected at the bubble center, creating momentarily ($\approx 10^{-11} \cdots 10^{-10}$ s) a region of high gas density ($\approx 10^{22}$ particles per cm^3) and high temperature which may emit a flash of light. Estimates of the peak 'temperatures' vary widely; they range from several thousand to millions of kelvin. Compelling temperature estimates at the low end of this range have been given for SL in organic liquids [6,7], while the highest temperatures were claimed for SL in water [5]. Therefore water has been considered by some a very special liquid. It is, however, unclear which specific properties of water would make that liquid so special.

The specific nature of the light emitting processes of SL is unknown, but a number of spectroscopic facts seem well established and provide certain clues. SL spectra in water, in which various gases or gas mixtures were dissolved in controlled ways, have been recorded [8–13]. According to such studies,

- SL of rare gas bubbles is much brighter than that of the molecular gases;
- SL intensities increase as we go from neon to helium, argon, krypton, xenon;
- admixtures of molecular gases to rare gases have a striking dimming effect;
- if the water temperature is lowered from room temperature to near 0 ^{0}C, stronger SL emission of the rare gases results [5].

We note that SL spectra in water are continuous, without any discernible line or band structures in the spectral "window" of water (i.e., at wavelengths between 200 and 700 nm). Apart from the striking intensity variations between the various gases, all observed spectral profiles are of a similar shape: intensities fall off nearly exponentially with increasing wavelengths ($\lambda \to 700$ nm), from a broad maximum at short wavelengths [8–12]. In some cases such a maximum seems to fall just outside of that window ($\lambda < 200$ nm) where however it cannot be observed.

In a recent review article entitled 'Defining the unknowns of SL' [5] the various attempts to understand the nature of the light emitting processes of SL are briefly discussed and then discarded — there are problems with all of them. For the purpose of discussion, however, electron-ion bremsstrahlung from a hypothetical, optically dense plasma was mentioned as the "most complete candidate emission model of SL" — despite a number of inconsistencies with known facts. For example, the emission from a dense plasma should probably not much depend on whether the plasma was generated from a rare gas or from a molecular gas, or what the precise water temperature was — but it certainly does. Moreover, no trace of an afterglow (or recombining plasma) was yet discovered, nor has the application of strong magnetic fields (20 T) affected SL behavior in any discernible way [14] so that one may think that a dense plasma actually does not exist under SL conditions.

Existing SL temperature estimates are often based on Planck's blackbody formula — or rather on a small part of it, with wavelengths between 200 and 700 nm. Planck's law is applicable if and when thermal equilibrium is established. However, the formation of thermal equilibrium takes time, perhaps more time than is available in the picosecond SL event. Moreover, Planck's law applies to optically thick sources. Below we will show some theoretical evidence that the SL source may actually be optically thin.

In search of a better understanding of the origin of the SL continua we consider a variety of emission processes that are known from neutral and weakly ionized, hot environments. Specifically, we focus on radiative collision processes of the type

$$A^{(*)} + B^{(*)} + \Delta E \to A + B + h\nu , \qquad (1)$$

where A and B may be neutral atoms or molecules, possibly in an excited rotovibrational or electronic state. Alternatively, in weakly ionized environments, A and

B may be an electron or ion interacting with a neutral particle; ΔE is the change of translational, rotovibrational and/or electronic energy in the collision; and $h\nu$ is the photon emitted in the process.

II COLLISIONS OF NEUTRAL PARTICLES

If A and B, Eq. (1), are both neutral atoms or molecules in their electronic groundstates, the radiative process is called collision-induced emission (CIE). Four mechanisms are known that induce dipoles — and thus emission — in intermolecular interactions: induction by exchange forces, dispersion forces, molecular multipole induction, and collisional frame distortion [15]. The *absorption* spectra of such induced dipoles are well known both from experiment and theory. For the simpler systems, e.g., He–Ar, H_2–He, H_2–H_2, *ab initio* calculations of the spectra exist which are in very close agreement with the measurements [15]. From the knowledge of the interaction-induced dipole and the potential surfaces *emission* spectra can also be computed. Such calculations may shed some light on the nature of the SL emission if the calculated spectra resemble the observations.

In binary encounters (1), at angular frequencies $\omega = 2\pi f = 2\pi c \nu$ between ω and $\omega + d\omega$, at the temperature T, the power [15]

$$I(\omega, T)\, d\omega = \frac{4\omega^4}{3c^3} n_A n_B V g_e(\omega, T)\, d\omega \qquad (2)$$

is emitted per unit volume. Here, n_A and n_B are the number densities of the species A and B; V is the volume and g_e is the so-called spectral function for emission, which is given by

$$g_e(\omega, T) = \sum_{ss'tt'} P_s(T) P_t(T) |\langle t\, |\boldsymbol{\mu}_{ss'}|\, t'\rangle|^2\, \delta\left(\omega_{tt'} + \omega_{ss'} - \omega\right). \qquad (3)$$

In this expression the s and t designate the sets of molecular and translational quantum numbers, respectively; the $\hbar\omega_{ss'}$ and $\hbar\omega_{tt'}$ are the energy differences of the molecular and translational transitions, respectively; $\langle t\, |\boldsymbol{\mu}_{ss'}|\, t'\rangle$ is the translational matrix element of the collision-induced transition dipole moment $\boldsymbol{\mu}_{ss'}(\boldsymbol{R}) = \langle s|\boldsymbol{\mu}|s'\rangle$; $\boldsymbol{\mu} = \boldsymbol{\mu}(\boldsymbol{R}, \boldsymbol{r})$ is the collision-induced dipole surface, a function of the separation $\boldsymbol{R}$ of the collision partners and the vibrational coordinate $\boldsymbol{r}$ of the diatom; P_s and P_t are the normalized population factors of rotovibrational and translational states, respectively, which we will assume to be given by Boltzmann factors, for lack of better information. Further details may be found elsewhere [15].

We mention that the absorption coefficient α can similarly be computed from the knowledge of the induced dipole and potential surface, according to

$$\alpha(\omega, T) = \frac{4\pi^2}{3\hbar c} n_A n_B\, \omega \left[1 - \exp\left(-\frac{\hbar\omega}{kT}\right)\right] V g_a(\omega, T). \qquad (4)$$

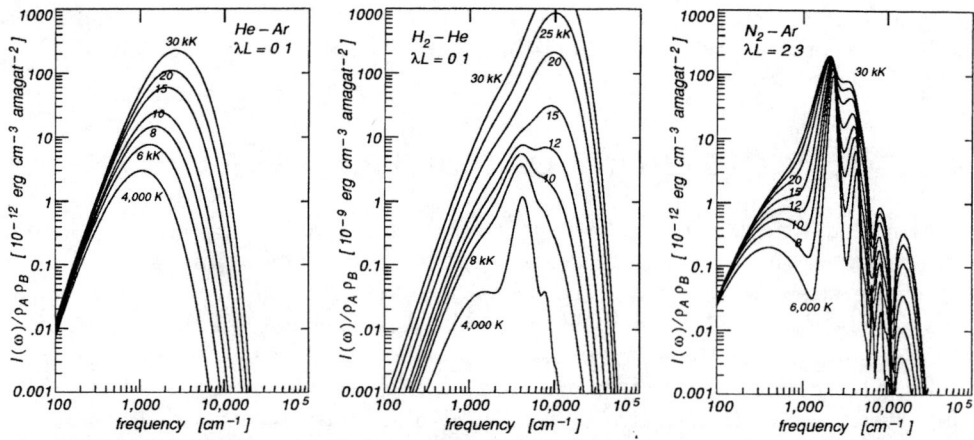

FIGURE 1. (left panel:) Emission of He-Ar pairs at temperatures from 4 to 30 kK.
FIGURE 2. (center panel:) Emission of H_2-He pairs at temperatures from 4 to 30 kK.
FIGURE 3. (right panel:) Emission of N_2-Ar pairs at temperatures from 6 to 30 kK.

The spectral function g_a for absorption differs from g_e, Eq. (3), merely by the different sign of the sum $\omega_{ss'} + \omega_{tt'}$; the energy of the initial state is lower than the final state energy in the case of absorption.

Densities ϱ are given in units of Loschmidt's number, $N_L = 2.68675 \cdot 10^{19}$ cm^{-3}. Number densities n are related to ϱ according to $n = N_L \varrho$; subscripts A or B will be attached. Units of density are thus practically identical with the amagat.

He–Ar. For the He–Ar pair an accurate *ab initio* induced dipole surface exists [16]. With a refined model of the interaction potential [17], the collision-induced emission spectra can be computed with confidence. Figure 1 shows the results for seven temperatures, from 4,000 to 30,000 K. The spectra are relatively weak (if compared with other spectra shown below); they are unstructured and extend barely into the visible and ultraviolet ($\nu \approx 14,000 \cdots 50,000$ cm^{-1}) unless temperatures are well above 30,000 K. In other words: the He–Ar collisional pair in the electronic groundstates does not generate emission comparable to that of SL of the rare gases. (More about intensity below.)

Empirical collision-induced dipoles of other rare gas pairs are known [15], but we refrain from showing more of the spectra of this type. We merely state that virtually all of the common pairs of dissimilar atoms in their electronic groundstates are emitters too weak to shed much light on the SL of the rare gases, even where the induced dipole strengths of the more massive atomic pairs are an order of magnitude greater than that of He–Ar pairs.

H_2–He. At comparable temperatures hot molecular, neutral gases emit more strongly at optical frequencies than monatomic gases. This is so because in the case of atom pairs the photon energy must come from the translational energies

($\sim \frac{3}{2}kT$). On the other hand, if molecules are involved, rotovibrational transitions occur as well during the radiative interactions so that the average photon energy is increased by one or more kT. This may be illustrated for the system H_2–He for which an *ab initio* dipole surface and reliable potentials exist [18–22].

The H_2–He spectra, Fig. 2, differ from He–Ar, Fig. 1, by the collision-induced rotovibrational bands at 4,200 cm^{-1} (the fundamental band, $v = 1 \to 0$, and 'hot' bands with $v' = v - 1 > 0$) and the higher overtone and other 'hot' bands ($v - v' > 1$) that are somewhat discernible at the lower temperatures. These bands are of course quasi-continuous, owing to the short duration Δt of the average collisions, $\Delta \omega \, \Delta t \approx \frac{1}{2}$, with $\Delta t \approx 10^{-13}$ s. At the higher temperatures, the H_2–He spectra extend into the visible and the near ultraviolet. Intensities are, however, still much lower than, for example, the observed SL intensities of the rare gases.

N_2–Ar. Other systems, such as N_2–Ar and N_2–N_2, were previously thought to be of interest in SL studies, especially in the case of SL of air [23]. However, more recently it has been argued that the emission of air bubbles may actually be due to the argon content of air [13], because the nitrogen concentrations fall off to virtually zero after forcing strong SL activity for several minutes or so. Nevertheless, it is interesting to look at the quadrupole-induced spectra of N_2–Ar pairs briefly, Fig. 3. That system is much like H_2–He, but differs from the latter by its greater dissociation energy of N_2 (10.2 *versus* 4.5 eV for H_2), the greater quadrupole moment of N_2; the vibrational spacings (which are roughly only one-half of those of H_2) and the smaller rotational constants ($B \approx 2$ *versus* ≈ 60 cm^{-1}). The differences between these spectra, Figs. 2 and 3, are striking.

We mention that in Fig. 3 only the quadrupole-induced dipole component is considered; there are also fairly strong overlap-induced and a hexadecapole-induced dipole components which however are not very well known. A crude estimate of the former was previously attempted [23,24]; we note that inadvertently in that work the labeling of the intensity axes was in error. The intensities of these other components of N_2–Ar are probably not much stronger than the intensities reported in Fig. 3; more accurate data for these other components are desirable but do at present not exist.

III ELECTRON-ATOM COLLISIONS

Since collisions of neutral atoms or molecules do not produce the amount of light seen in SL of the rare gases, we consider alternatives. The most obvious extension of such work is to focus on weakly ionized environments. We think weakly ionized environments are likely to exist under SL conditions because of the picosecond time scales: starting with a strictly neutral gas bubble ($\to$ cavitation), in the absence of strong external electric fields, ionization in adiabatically or shockwave heated fluids proceeds by neutral-neutral collisions. Ionization cross sections of atom-atom collisions are generally extremely small at energies near the threshold of ionization [25] so that typically very many collisions are required, even for generating but

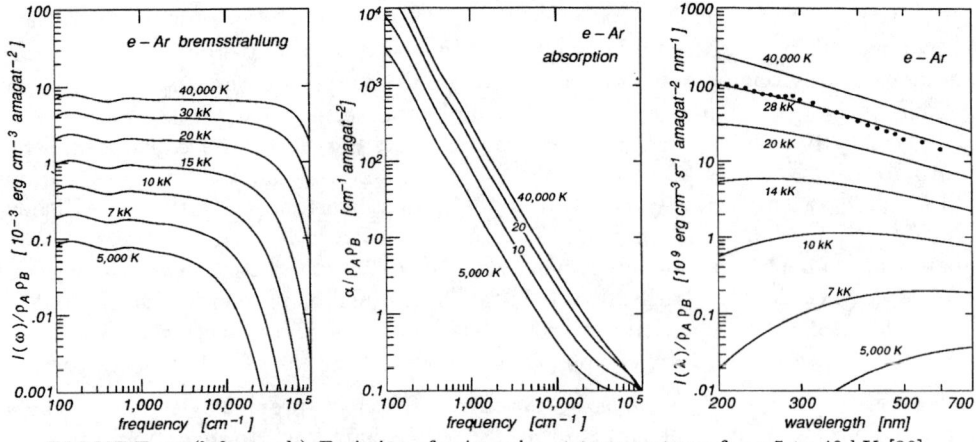

FIGURE 4. (left panel:) Emission of e-Ar pairs at temperatures from 5 to 40 kK [38].
FIGURE 5. (center panel:) Absorption of e-Ar pairs at temperatures from 5 to 40 kK [38].
FIGURE 6. (right panel:) The e-Ar emission, from Fig. 4, in the spectral window of water [38]. A measured SL spectral profile of argon bubbles is also shown (dots; from Ref. [5]).

small concentrations of electrons. Typical SL time scales are $\approx 10^{-11}$ s and electron concentrations are likely to be much smaller than those given by the equilibrium condition (Saha equation) at comparable electron 'temperatures.'

When electrons collide with neutral atoms bremsstrahlung is emitted. We want to distinguish electron-*atom* bremsstrahlung from the electron-*ion* emission, also called bremsstrahlung. The former is important at lower temperatures; both bremsstrahlung processes are similarly efficient in generating light over a wide band of frequencies. The spectra of electron-rare gas atom collisions are well known both from theory [26–30] and experiment [31–36]; a comprehensive bibliography is given elsewhere [37]. If a weakly ionized environment may be assumed, reliable predictions of such spectra to be expected under SL conditions can be made, using the modified Hartree-Fock-Slater potential that was used previously for such calculations [26,38].

Figure 4 shows the calculated bremsstrahlung spectra of electrons colliding with argon atoms. At the lower frequencies a 'white' continuum is observed, with intensities nearly independent of frequency, up to a point where the product of optical frequency and the duration of a collision are of the order of unity; this is the case in the visible or near ultraviolet region of the spectrum. At such high frequencies, the bremsstrahlung spectrum falls off rapidly. Intensities in the visible and near ultraviolet of the spectrum increase sharply with increasing temperature.

Figure 5 shows the absorption coefficient α as function of frequency, normalized by the densities of the electrons and argon atoms. Knowledge of the absorption coefficient permits an estimate of the optical density of the bremsstrahlung light source. The mean free path of a photon is given by the reciprocal absorption

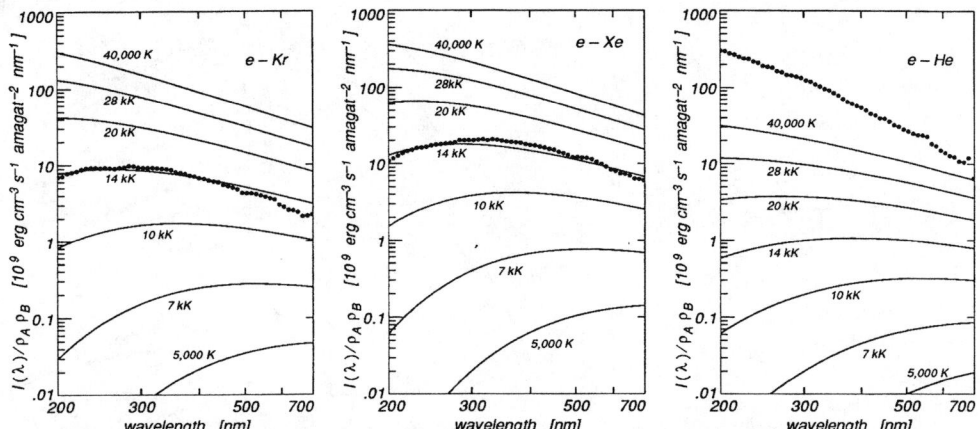

FIGURE 7. (left panel:) The e-Kr emission spectra in the spectral window of water [38]. A measured SL spectral profile of krypton bubbles is also shown (dots; from Ref. [5]).

FIGURE 8. (center panel:) The e-Xe emission spectra in the spectral window of liquid water [38]. A measured SL spectral profile of xenon bubbles is also shown (dots; from Ref. [5]).

FIGURE 9. (right panel:) The e-He emission spectra in the spectral window of liquid water [38]. A measured SL spectral profile of helium bubbles is also shown (dots; from Ref. [5]).

coefficient, $\lambda_{\rm ph} = 1/\alpha$. In the log-log scale, we note a nearly linear fall-off over several orders of magnitude with increasing frequency, Fig. 5, roughly as $\alpha \sim \omega^{-2}$.

In order to compare the calculations, Fig. 4, with a measured SL lineshape, we replot in Fig. 6 that part of the data with wavelengths in the spectral window of water. The intensity $I(\lambda)$ is related to $I(\omega)$, Eq. (2), according to $I(\omega) \, d\omega = I(\lambda) \, d\lambda$. Also shown is a measured SL line shape (dots) [9] of argon bubbles, which is similar to the computed profiles at a temperature of roughly 20,000 K. For this comparison we have subjected the measured profile to arbitrary vertical shifts, without regard to absolute intensity. We note, however, that we will show below (Fig. 11) that the calculated and measured intensities are actually in harmony at such an electron 'temperature.'

The e-Ar spectra, Figs. 4 and 5, are characteristic of all electron-neutral atom spectra of the rare gases. The most striking differences among the rare gases are the higher intensities arising from e-Kr and e-Xe collisions, and the lower intensities of e-He and e-Ne interactions relative to e-Ar.

Figures 7 through 10 compare measured profiles of krypton, xenon, helium and neon bubbles [5] with our calculations. For e-Kr and e-Xe pairs, agreement is observed at temperatures around 14,000 K. For e-He and e-Ne pairs, on the other hand, the logarithmic slopes of the measured profiles seem to be somewhat steeper than our calculations suggest. We note that the measured spectra shown in Figs. 7 through 10 are raw spectra, uncorrected for the transmission of the monochromator, etc., whereas the e-Ar spectrum of Fig. 6 was corrected. Such corrections will affect

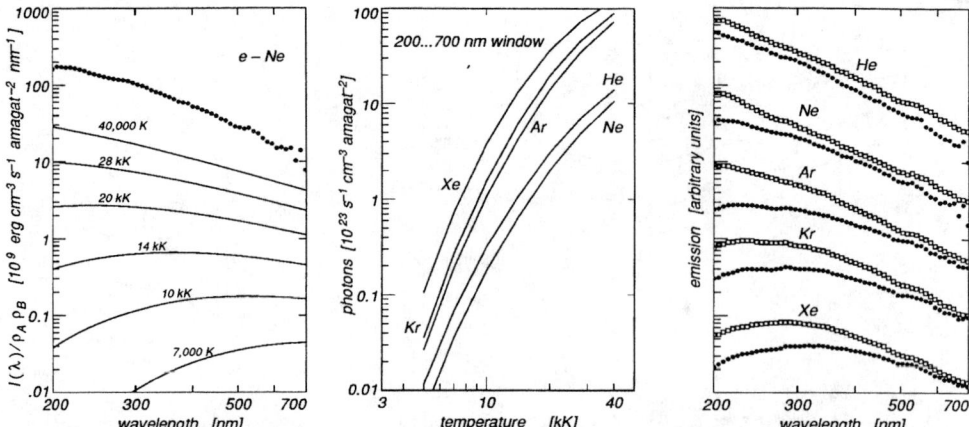

FIGURE 10. (left panel:) The e-Ne emission spectra in the spectral window of water [38]. A measured SL spectral profile of neon bubbles is also shown (dots; from Ref. [5]).

FIGURE 11. (center panel:) Number of photons emitted per second, per cm^3, per amagat^2, by collisional electron-rare gas atom pairs, in the spectral window of water [38].

FIGURE 12. (right panel:) Comparison of measured spectral profiles of SL rare gas bubbles, taken at near freezing temperatures (squares) and at room temperature (dots) [38].

the line shapes; one can hardly expect perfect agreement with calculations unless the corrections are made. Moreover, for the line shape calculations, a Boltzmann distribution of the electron energies was assumed (Eq. 3). More realistic choices (if they could be made) would affect the line shapes. In any case, at least for the heavier rare gases, the agreement of calculated and observed spectral profiles is remarkable.

Next we turn our attention to the absolute intensities. Each argon SL light flash emits roughly $10^5 \cdots 10^6$ photons [5]. Figure 11 shows the number of photons emitted per second, per unit volume, under stationary conditions, for the rare gases, that is the integral of Figs. 6 through 10, after division by $\hbar\omega$. For the comparison of calculated and observed photon yields, these integrals must be multiplied by four factors, namely **i.** the duration τ of an SL flash; **ii.** the volume V of the source; **iii.** the density ϱ_{Ar} of argon, etc.; and **iv.** the density ϱ_e of the electrons. The duration τ is not well known. Previously, it has been argued that τ must be 'much smaller than 50 ps,' but recently values around 100 ps have been demonstrated [39]. The volume V must be estimated from the smallest bubble radius R_{min} of roughly 0.5 μm. The density of argon at maximum compression is roughly 800 amagat, see Ref. [5] for details. The exact value of the product of τ, V, ϱ_{Ar} and ϱ_e for electron densities of roughly 1% of the neutral density, amounts to something like

$$\tau V \varrho_1 \varrho_2 \approx 10^{-10}\,\text{s} \cdot 10^{-12}\,\text{cm}^3 \cdot 6400\,\text{amagat}^2 = 0.64 \cdot 10^{-18}\,\text{s cm}^3\,\text{amagat}^2.$$

Applying such a factor to the coordinate values, Fig. 11, we get the right photon numbers per SL flash. For example, for argon bubbles, for temperatures around

20,000 K, we thus expect roughly 10^6 photons per flash. (Uncertainties of the above factor may be an order of magnitude or more in either direction.)

Figure 11 also shows that theory suggests a 50% greater intensity for helium than for neon. This is in (quantitative!) agreement with the measurements [5] if we may assume that SL in helium and neon generates comparable temperatures. SL of all the other rare gases glows brighter than that of helium, again in reasonable quantitative agreement with the computations shown in Fig. 11.

IV OTHER EMISSION MODELS

Specifically, we have also looked at the spectra of ion-neutral collisions, e.g., Ar^+–Ar. Ions when accelerated in the field of the atom emit like other charged particles, only at much lower frequencies so that ion bremsstrahlung emission may safely be ignored here. Ions also polarize atoms during interaction, which gives rise to a quasi-molecular, 'collision-induced' emission, again generally at frequencies in the infrared, unless temperatures are higher than assumed here [15]. Moreover, one can hardly ignore certain impurities, specifically H_2O and perhaps the OH radical, which are strongly polar and also give rise to collision-induced emission (e.g., by polarizing the neutral atom in the electric dipole field of H_2O, and via exchange forces [15]). In the case of molecular impurities, collision-induced bands appear that may very well extend significantly to frequencies in the spectral window of water. At present, we have looked at a number of such spectra, but so far those spectra seem to be either much weaker or strong only at lower frequencies. Supramolecular spectra may however be important for SL studies of molecular gases [23,24].

Under conditions of weak ionization, electronic excitations of the rare gas atoms are likely to be rare. If, however, electronic excitations are significant, a number of further emission processes should be considered that generate quasi-continuous spectra. Some of these are also collision-induced [40], others have to do with pressure- or Stark-broadened, shifted atomic lines, line merging, etc. [41]. At present, we have not considered in any detail the quasi-continua involving electronic excitations but intend to do so in the future.

V DISCUSSION AND CONCLUSIONS

Relative intensities. We have seen that reliable computations of the spectra of electrons colliding with neutral atoms at 'temperatures' around 10,000 or 20,000 K resemble the observed SL profiles of the rare gases — certainly for the heavier rare gases if not for all. Moreover, at such temperatures, the computed intensities are consistent with the observed emission of roughly 10^6 photons per flash. We conclude that the free-free emission mechanism involving electrons and neutral atoms is most likely significant for SL of the rare gases. The theory of bremsstrahlung from collisions of electrons with neutral atoms suggests that SL bubbles of helium glow $\approx 50\%$ brighter than those of neon, while all the other, more massive gases glow brighter than helium at comparable 'temperatures', in remarkable agreement with the experimental facts.

Degree of ionization. Essential for the above considerations is the assumption of a weakly ionized SL environment. Under stationary (non-SL) conditions, at temperatures of 10,000 or 20,000 K and near liquid state densities, a significant fraction (given by the Saha equation) of the atoms would be ionized and the emission spectra significantly modified, for a number of reasons (e.g., pressure- and Stark-broadening; line merging; opacity of source, etc.). These facts need, however, not necessarily invalidate the assumption of a weakly ionized SL environment, because the duration of the SL process is measured in picoseconds, that is just a few times the duration of an average neutral-neutral collision. The neutral-neutral collisional cross sections for ionization are typically much smaller than, for example, the gas-kinetic cross sections so that many collisions must occur before even small numbers of electrons are generated. The SL environment may not exist long enough for establishing the ionization equilibrium that is typically seen in hot environments other than SL.

Opacity. The reciprocal absorption coefficient α, Fig. 5, is the mean free path for absorption of a photon, $\lambda_{ph} = 1/\alpha$. Multiplying the data shown in Fig. 5 by the gas density (800 amagat) and electron density (assume 8 amagat, or 1% ionization), in the spectral window of water, we obtain a value of $\alpha \leq 640$ cm^{-1}, or a mean free path of 16 μm or greater. That is much greater than the size of the SL source (≈ 0.5 μm). In other words: at wavelengths from 200 to 700 nm the electron-Ar source is optically thin — just like all the other electron-rare gas atom source shown above. Under such conditions, Planck's blackbody law, which has often been used to obtain temperature estimates of SL, is not applicable to the analysis of measured SL spectra. If there was an effect due to significant optical densities, it would first show up at the longest wavelengths, reducing emission in the red.

Molecular gases. If indeed electron-atom bremsstrahlung is an important SL emission process as we think, one might wonder why the SL of argon bubbles is orders of magnitude brighter than, say SL of nitrogen bubbles. Doubtlessly, the bremsstrahlung hypothesis would suggest that electrons interacting with molecules such as N_2 emit about the same spectrum as e-Ar pairs *if the electron energies are the same* and if the same number of atoms is considered. But it is well known that pure nitrogen SL bubbles glow much dimmer compared to argon bubbles. We think that an explanation of this unshakable fact must focus on the differences of the electron temperatures in monatomic and diatomic gases: In molecular gases electrons undergo inelastic collisions, transferring translational energy to the rotovibrational levels and reducing their mean energies in the process [42,43]. In monatomic gases, on the other hand, such "cooling" cannot occur. In molecular gases the energies of the electrons generated in neutral-neutral collisions will therefore rapidly decrease by such inelastic collisions, and with it the emission (which decreases strongly with decreasing electron energies, Fig. 11). As a matter of fact: other types of radiative collisions (e.g.; Refs. [23,24]) may then contribute to the weak emission (if any) as much or more than the above bremsstrahlung process.

Water temperature. In the same way, another observation that has been rather puzzling may perhaps be understood: the increase of the SL emission intensity with decreasing water temperature. The vapor pressure of water increases by an order of magnitude when temperatures increase from 0° to 35° C, and the intensity of SL bubbles must decrease as the inelastic collisions of the electrons with H_2O molecules increase in frequency.

Figure 12 compares the SL line shapes of the rare gases, taken near 0° C (squares) and at room temperature (dots) [5]. (Intensity scales are here arbitrary.) We notice that for all rare gases (with the possible exception of helium) the room temperature profiles seem to peak consistently at longer wavelengths than their freezing point equivalents. According to what we have seen in the figures 6 through 10, this again would suggest lower electron temperatures when the water vapor pressure is increased (assuming the shock or heating mechanism itself is not much affected — just the emission).

Concluding remark. We note that the electron-atom bremsstrahlung model of SL, as presented above, is of course a simplified one. We have considered the radiation from binary collisions of electrons with atoms, but at near liquid state densities many-body effects may contribute to the observable radiation. Moreover, we consider largely electron–atom collisions, but the gases of the SL bubbles are not really very pure. Contaminations, such as water vapor, residual gas impurities, ions, and possibly the effects that the dense SL environment may have on the observable emission are more or less ignored. Furthermore, at densities approaching liquid state densities, the spectroscopy of the rare gases is affected by pressure broadening, Stark broadening, line merging, and other more or less familiar processes of the dense fluid states that need to be considered eventually — an enormous task which may take years to evolve. In spite of all these uncertainties that must eventually be considered we feel that the concept of electron-atom bremsstrahlung is significant for the SL environment of the rare gas bubbles.

Acknowledgments: The support of the R. A. Welch Foundation, grant 1346, is gratefully acknowledged.

REFERENCES

1. H. Frenzel and H. Schultes, Z. Phys. Chem. **B 27**, 421 (1934).
2. T. G. Leighton, *The Acoustic Bubble* (Academic Press, London, 1994).
3. L. Rayleigh, Phil. Mag. **34**, 94 (1917).
4. C. C. Wu and P. H. Roberts, Proc. Roy. Soc. (London) **A 445**, 323 (1994).
5. B. P. Barber *et al.*, Phys. Reports **281**, 65 (1997).
6. K. S. Suslick, Science **247**, 1439 (1990).
7. E. B. Flint and K. S. Suslick, Science **253**, 1397 (1991).

8. R. Hiller, S. J. Putterman, and B. P. Barber, Phys. Rev. Letters **69**, 1182 (1992).
9. R. Hiller, K. Weninger, S. J. Putterman, and B. P. Barber, Science **266**, 248 (1994).
10. R. A. Hiller and B. P. Barber, Sci. American **272**, 96 (1995).
11. D. F. Gaitan et al., Phys. Rev. E **54**, 525 (1996).
12. K. Weninger et al., J. Phys. Chem. **99**, 14195 (1995).
13. D. Lohse et al., Phys. Rev. Letters **78**, 1359 (1997).
14. J. B. Young, T. Schmiedel and W. Kang, Phys. Rev. Letters **77**, 4816 (1996).
15. L. Frommhold, *Collision-induced Absorption in Gases* (Cambridge University Press, Cambridge, New York, 1993).
16. W. Meyer and L. Frommhold, Phys. Review A **33**, 3807 (1986).
17. G. C. Maitland, M. Rigby, E. B. Smith, and W. A. Wakeham, *Intermolecular Forces* (Clarendon Press, Oxford, 1981).
18. W. Meyer and L. Frommhold, Phys. Review A **34**, 2771 (1986).
19. A. Borysow, L. Frommhold, and M. Moraldi, Astrophys. J. **336**, 495 (1989).
20. A. Borysow and L. Frommhold, Astrophys. J. **341**, 549 (1989).
21. A. Borysow, L. Frommhold, and W. Meyer, Phys. Review A **41**, 264 (1990).
22. D. Hammer, L. Frommhold, and W. Meyer, *Ab initio* dipole surface of H_2–He and D_2–He, in prep. See also the report by Hammer and Frommhold in this volume.
23. L. Frommhold and A. A. Atchley, Phys. Rev. Letters **73**, 2883 (1994).
24. L. Frommhold and W. Meyer, in *Spectral Line Shapes*, edited by M. Zoppi and L. Ulivi (Am. Inst. Physics, N.Y., 1996), Vol. 9, pp. 471 – 484.
25. D. R. Bates and H. S. W. Massey, Phil. Mag. **45**, 111 (1954).
26. S. Geltman, J. Quant. Spectroscopy and Rad. Transfer **13**, 601 (1973).
27. A. Rutscher and S. Pfau, Beitr. Plasmaphys. **8**, 315 (1968).
28. R. H. Pratt and I. J. Feng, in *Applied Atomic Collision Physics, vol. 2: Plasmas*, C. F. Barnett and M. F. A. Harrison, eds. Acad. Press, Inc., New York, 1984, p. 307.
29. M. Ashkin, Phys. Review **141**, 41 (1966).
30. R. R. Johnston, J. Quant. Spectroscopy and Rad. Transfer **7**, 815 (1967).
31. C. Yamabe, S. J. Buckman, and A. V. Phelps, Phys. Review A **27**, 1345 (1983).
32. R. L. Taylor and G. Caledonia, J.Q.S.R.T. **9**, 657 (1969).
33. R. T. V. Kung and C. H. Chang, J.Q.S.R.T. **16**, 579 (1976).
34. S. Pfau and A. Rutscher, Zeitschrift fur Naturforschung **8**, 2129 (1967).
35. S. Pfau and A. Rutscher, Beitr. Plasmaphys. **8**, 73 (1968).
36. A. Rutscher and S. Pfau, Physica (Utrecht) **81C**, 395 (1976).
37. J. W. Gallagher, Technical Report No. 16, Joint Institute for Laboratory Astrophysics, University of Colorado, Boulder, U.S.A. (unpublished).
38. L. Frommhold, Phys. Rev. E, 1998, to appear.
39. B. Gompf et al., Phys. Rev. Letters **79**, 1405 (1998).
40. P. S. Julienne, in G. Birnbaum, ed., *Phenomena Induced by Intermolecular Interactions*. Plenum Press 1985, p.479.
41. H. Griem, *Principles of Plasma Spectroscopy*. Cambridge Univ. Press, N.Y., 1997.
42. L. G. H. Huxley and R. W. Crompton, *The Diffusion and Drift of Electrons in Gases*. Wiley, New York, 1974.
43. H. S. W. Massey, *Electronic and Ionic Impact Phenomena*, vol. 2. Clarendon, Oxford, 1969.

Estimation of the spectral line profiles in the high-frequency discharge

G. Revalde, A. Skudra

*Institute of Atomic Physics and Spectroscopy, University of Latvia,
Raina blvd. 19, LV 1586, Riga, Latvia*

Abstract. The paper is devoted to the estimation of the true profiles of spectral lines and equipment influence basing on experimental complex line shapes in high-frequency discharge by means of non-linear multiparameter chi-square fit. The self-absorption and energy transfer effects are taken into the consideration. A good agreement with experimental interferometric data are achieved.

INTRODUCTION

High-frequency electrodeless light sources (HFELS) are known to be extremely bright radiators of line spectrum, characterised by high intensity of spectral lines, high discharge stability and long life time[1]. Exploitation of the HFELS in modern equipment demands specific conditions with respect to the source radiation, working time and other characteristics. In order to optimize lamp design, filling and working regimes, we have to understand the physical processes taking place in HFELS and to estimate parameters characterizing the high-frequency discharge. One criterion what we use for the lamp quality control is measurements and modeling of spectral line profiles.

EXPERIMENTAL

For the measurements of emission spectral line profiles a new performance Hg and Hg-Cd HFELS were used. These lamps was made as quartz sphere or cylinder of 2 cm diameter with a short side branch [2]. HFELS samples were filled with Hg 202-isotope of 99.8 % abundance and/or Cd-114 isotope of 96.7% abundance. As buffer gas different noble gases of various pressure were used. The inductive coupled discharge was effected by means of the electromagnetic field, created by HF-oscillator working on 100 MHz frequency. For the line profile measurements, a pressure scanned Fabry-Perrot interferometer was used. The line shapes were registered in dependence on the current in the coil of HF-oscillator for resonance and visible triplet spectral lines of Hg and Cd.

MATHEMATICAL MODELING

The spectral line profile, registered by means of Fabry-Perrot interferometer, differs from the true one. The observed distribution is given by a convolution of the true profile of a spectral line and instrumental function. To determinate the true spectral line profile it is necessary to solve the inverse problem [3]. We used another method: the mathematical modeling and fitting the model to an experimental interferogram by means of minimizing of the chi square function. Our model [4] include the basic factors causing the spectral line broadening in the high-frequency electrodeless discharge: Doppler, natural, resonance and collision broadening. Commonly these effects are accounted by using a Voigt profile. In the case of HF discharge the line profiles, in particular resonance line profiles, are distorted by self-absorption. To describe the influence of self-absorption correctly we have to know the excitation function of the source. In our case very little is known about exact form of the excitation function. We used the approximation of uniformly excited source and source with spatially separated emitting and absorbing atoms. We found that by taking the excitation function as suggested in [5] it is possible to determine the self-absorption in HF-discharge by changing the degree of the homogenity of the source. In this case, the profile of spectral line could be expressed as follows:

$$I(\nu-\nu_0) = I_0 P(\nu-\nu_0) e^{-\mu} \sum_{j=0}^{\infty} \frac{n! \mu^{2j}}{(2j+n)!}$$

$$\mu = k_0 l * \frac{P(\nu-\nu_0)}{P(0)}$$, where (1)

and $P(\nu-\nu_0)$ is the contour function of radiation in unity volume, $k_0 l$ is the optical density in the center of the line, n is an integer characterizing the homogenity of the radiation source.

Hyperfine splitting and isotope shifts are considered basing on multiple overlapping lines. We take into account the energy transfer from the atoms of one isotope to another. In the case of mercury HFELS, this process can be described as follows:

$$^{j}Hg(6^3P_1) + {}^{i}Hg(6^1S_0) \rightarrow {}^{j}Hg(6^1S_0) + {}^{i}Hg(6^3P_1), \quad (2)$$

where i, j are the numbers of the isotopes. In this case, we must include in the group of fitting parameters the intensities of components. The changes of spectral line profile, due to the registration process is considered at the end. The non-linear multiparameter chi-square fit allows to determine such characteristics of HF discharge plasma as the temperature of radiating atoms, the resulting Lorentzian width (FWHM), the optical density, homogenity of the source and equipment influence. In figure 1 are shown the examples of the model and experimental line shapes of Hg resonance line 253.7 nm from HFELS filled with Hg 202 isotope of 99.8 % abundance at two values of mercury vapor pressure in a) case $p_{Hg}=1.95*10^{-1}$ Tor and in b) case $p_{Hg}= 5.5*10^{-1}$ Tor. The experimental lines was registered by the HF-oscillator current of 160 mA. In the case a) the estimated source type is nearly to uniform: n=2. In this case, the density of Hg atoms in the ground state is low and the discharge current of I=160 mA is high enough to create a more uniform excitation of atoms as in the b) case. In the b) case the line profile is more self-reversed due to higher density of the ground state Hg atoms and

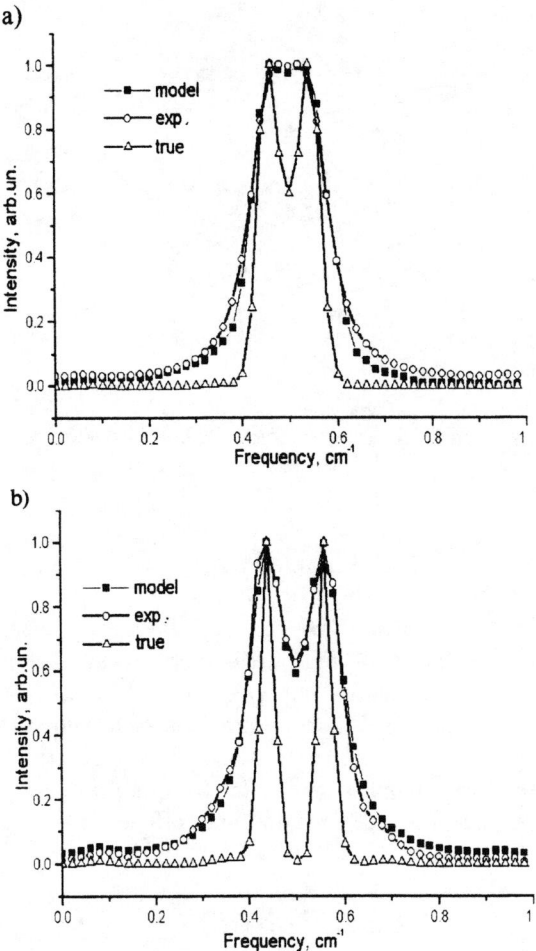

FIGURE 1. Examples of modeling of spectral line profile of Hg 202 isotope of 99.8 % abundance HFELS. a) $p_{Hg}=1.95*10^{-1}$ Tor, estimated parameters: FWHM of Gaussian $\Delta\nu_D=0.082$ cm^{-1}, FWHM of Lorentzian $\Delta\nu_L=0.0002$ cm^{-1}, $k_ol=3.5$, n=2, effective refraction coefficient of mirrors $R_{eff}=0.8$, b) $p_{Hg}=5.5*10^{-1}$ Tor, $\Delta\nu_D=0.084$ cm^{-1}, $\Delta\nu_L=0.0002$ cm^{-1}, $k_ol=8$, $R_{eff}=0.8$.

the best agreement is by the approximation of strongly non-homogenous source (n=13). The obtained temperature of radiating atoms changes from 500°C to 1500°C as a function of HF-oscillator working regime. The confidence regions, standard deviations of parameters are obtained using the statistical modeling. Figure 2 shows an example of 80% confidence region for the model allowing to obtain three parameters: FWHM of the Gaussian ($\Delta\nu_D$), the FWHM of the Lorentzian ($\Delta\nu_L$) and optical density (k_ol), at effective reflection coefficient of mirrors (R_{eff}) of 70%.

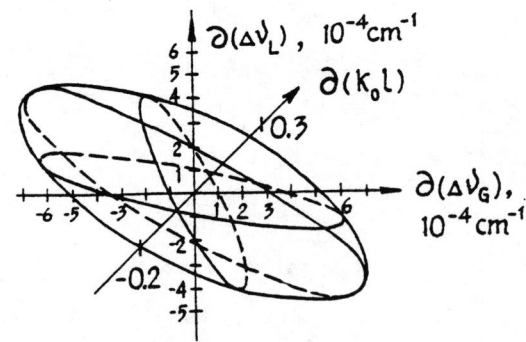

FIGURE 2 80% confidence region for the three parameter model at effective reflection coefficient of mirrors of 70%.

CONCLUSIONS

We conclude that the homogenity of the high-frequency electrodeless source changes dramatically in dependence on the source type, working and cooling regime.

Although the parameters, obtained by means of mathematical modeling of spectral line profiles, are numerous and they are correlated we have shown that this method allows to diagnose important characteristics of the low temperature HF discharge plasma. This knowledge is very important in order to optimise the HFELS and to use them in every concrete application with a maximum of efficiency.

REFERENCES

[1] Wharmby D. O., IEE Proceedings "Science measurements and Technology", Vol.140, Iss.6, 1993, pp. 465-473.
[2] Berzina D., Putnina S., Skudra A., Patent SU, No: 1702454 (1989).
[3] Preobrazhenskii N. G., *Spectroscopy of Optical Thick Plasma*, Nauka, Novosibirsk, 1971, pp. 177.
[4] Revalde G., Skudra A., *JQSRT* (in press).
[5] Cowan R. D., Dieke G. H., *Rev. Mod. Phys.* **2**, 418 (1948).

Profiles of neutral lines emitted from weakly non-ideal helium plasmas produced in flashlamps

Yves Vitel, Mohammed El Bezzari, Lev G. D'yachkov*
and Yuri K. Kurilenkov*

*Laboratoire des Plasmas Denses, Université P et M Curie, Tour 12 E5, 4 place Jussieu,
F-75252 Paris Cedex 05, France*
**Institute for High Temperature, Russian Academy of Sciences, Izhorskaya 13/19,
Moscow 127412, Russia*

Abstract. High pressure helium arcs are created in linear flashlamps. Plasma diagnostics taking into account non-ideality effects, give on axis electron densities in the range $2\,10^{17}$-$1.7\,10^{18}$ cm^{-3} and temperatures included between 20000 and 30000 K. In these conditions of dense plasmas, profiles of emitted neutral lines are recorded and compared with other experimental values and theoretical calculations.

INTRODUCTION

High pressure pulsed arcs produced in flashlamps filled with pure helium are applied for generation of weakly non-ideal plasmas characterized by on axis electron densities in the range $3\,10^{17}$ cm^{-3} - $1.7\,10^{18}$ cm^{-3} and temperatures around $(2\text{-}3)\times10^4$ K. giving a plasma parameter $\Gamma = E_p/E_k$ closed to 0.1. Those pulsed arcs have a good cylindrical symmetry, are reproducible and in local thermodynamic equilibrium. Therefore, flashlamp as plasma source is a very convenient and efficient tool to study spectra emitted in dense plasmas. Radiative emission in the spectral range 300-850 nm is recorded. The profiles of neutral lines are studied and the evolutions of their width and shift versus electron density, are presented and compared with other experimental and theoretical results.

EXPERIMENT SETUP, DIAGNOSTICS AND RESULTS

We use fused quartz linear flashlamps whose inner diameter is 5 mm and distance between electrodes 100 mm. They are filled with pure helium at initial pressures from 50 to 500 Torr. The main discharge is applied 12 ms after ignition and setting up of the simmer mode. It has a maximum current intensity in the range 0.5–1.5 kA and its pulse duration at half height is around 100 µs. Two spectrographs with different dispersions and coupled with intensified photodiode array or CCD, are used to record spectra. All spectral measurements are performed during a short period of time (less than 1 µs) at the maximum of the current, when the best filling of the plasma is obtained inside the flashlamp. Spectral lamps and calibrated deuterium and tungsten ribbon lamps are used for standardisation purposes. Different methods of diagnostics are applied to determine the radial profiles of particles and temperature. They are based on measurements of continuum intensities, neutral line intensities and opacities, and infrared laser interferometry (3.39 µm). The on axis temperatures deduced from continuum and line intensities on the basis of classical plasma calculations, are in disagreement with those deduced from opacity measurements. Their values are by far too high compared to the ones deduced from optical thickness, and would have to lead to observe ionic helium lines on spectra but we have never seen anyone. On the other hand, if we take into account the effect of the statistical ionic microfields on the atomic levels in the calculation of the continuum and the line intensities [1,2], a good agreement is obtained in the evaluation of the plasma parameters by those different methods. Radial profiles of temperature and particle densities so deduced are relatively flat and show a good filling of the plasma inside the flashlamp.

The profiles of some neutral lines are recorded and fitted to lorentzian profiles calculated from those radial profiles. The width and shift of these lines versus electron density are presented and compared with other experimental and theoretical results [3,4,5]. In the figure 1, examples are shown for the line at 587.5 nm ($2p^3 P_2^0 - 3d^3 D_3$).

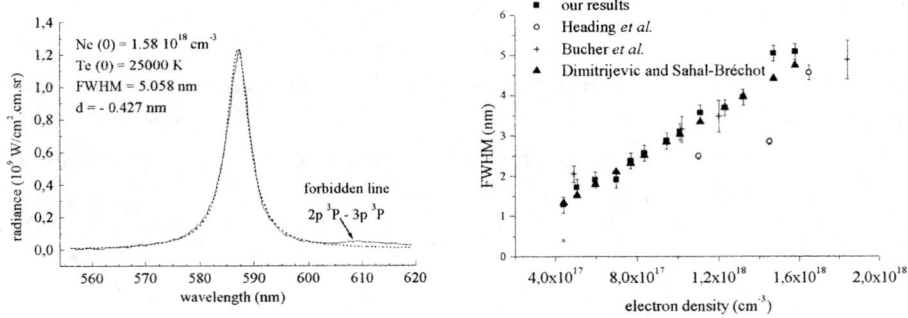

FIGURE 1. HeI line at 587.5 nm: a- Experimental and fitted profiles, b- width versus electron density.

REFERENCES

1. Vitel Y., El Bezzari M., D'yachkov, L. G., Kurilenkov, Y. K.," Radiation of dense helium plasma produced in flashlamps", proceeding 23th ICPIG, Toulouse, France, pp 190-191 (1997)
2. D'yachkov, L.G., Pankratov, P.M., *J. Phys. B* **24,** 2267- (1991); **27,** pp 461- 472 (1994).
3. Heading, D.J., Marangos, J.P., Burgess, D.D.,*J. Phys. B* **25,** pp 4745- 4753 (1992).
4. Büscher S., Glenzer, S., Wrubel, TH., Kunze, H.-J., *JQSRT* **54,** pp 73- 80 (1995)
5. Dimitrijevic, M.S., Sahal-Brechot, S., Bull.Obs. Astro. Belgrade **141,** pp 57- 86 (1989)

The Hydrogen Balmer α emission line in underwater laser plasmas

Alexandre ESCARGUEL*, Alain LESAGE**, Jacques RICHOU*

*Laboratoire d'Optoélectronique, Université de Toulon et du Var, BP 132, 83957, LA GARDE, FRANCE

**Observatoire de Paris-Meudon, DASGAL,URA 335, 92195 MEUDON CEDEX, FRANCE

High density plasmas are found in fields as various as laser fusion and stellar atmospheres. Such plasma behaves as strongly interacting system, when the temperature is low enough, so that a large potential energy-kinetic energy ratio is established for the interaction between plasma particles [1, 2, 3]. Due to the initial high density medium, underwater laser plasmas correspond to these conditions. It is therefore a convenient source to study high density, low temperature plasmas. However, to our knowledge, no hydrogen emission lines have been observed, up to now, in plasma created with single pulse laser in water.

For that a Nd:YAG laser of 10 ns duration with an energy ranging from 50 mJ to 115 mJ is focused into the liquid by a 50 mm focal length lens. Study samples are composed of distilled water with Ca trace to allow plasmas parameters N_e and T measurements independently from hydrogen lines. The spark light is observed perpendicular to the laser path. Two different spectroscopic systems are used to analyse Balmer Hα profiles and Calcium emission lines. A 1024 pixels time-gated intensified photodiode array coupled with a Jobin-Yvon spectrometer provides simultaneous detection of the whole Balmer Hα profile. Calcium emission lines used to determine plasmas parameters are observed with another gated intensified photodiode array coupled with a Chromex spectrometer to have a better spectral resolution.

Due to the high hydrogen concentration, Balmer Hα opacity effects have to be taken into account. To correct Hα from reabsorption, a concave mirror placed behind the plasma is used. Figure 1 shows a typical reabsorption corrected profile compared to the initial one.

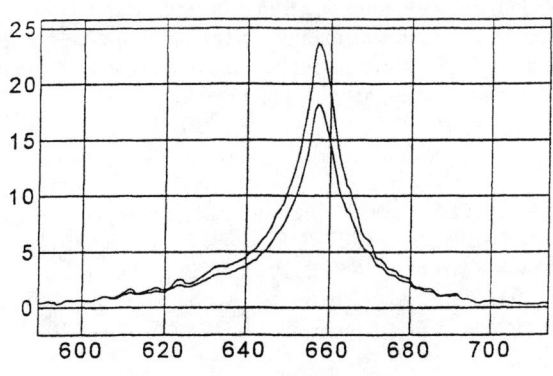

Figure 1. Hα profile in underwater laser plasma before(lower curve), and after (upper curve) reabsorption correction.

Hα profiles are observed between 200 and 300 ns after laser pulse, with an integration time of 100 ns. The strong background continuum is substracted from line emission spectrum. Local thermodynamic equilibrium is checked for the treatment of the observed spectra using Griem's theoretical relations [4]. The Stark broadened CaII (3968.5 Å) emission line is used to measure electronic density whereas CaI (4226.7 Å) and CaII (3968.5 Å) spectral integrated intensities ratio permits temperature measurements. The unfolded CaII (3968.5 Å) line is corrected from self-absorption if optical depth is non negligeable using a "doublet intensity ratio method" with CaII (3968.5 Å) and CaII (3933.4 Å) lines. The temperature and electronic density results are in good agreement with those of D. A. Cremers and al. [5]. After deconvolution, experimental Hα $FWHM_{exp}$ were measured and compared to theoretical Hα $FWHM_{th}$ obtained with Griem theory using experimental N_e and T measured with Ca traces. To characterize plasma non-ideality, the electronic Coulomb coupling constant Γ_{ee}, which is the ratio of the electrostatic energy of two neighbouring electrons to their thermal energy is inferred from plasma parameters. Table I resumes these results.

N_e (10^{18} cm^{-3})	T (K)	$FWHM_{th}$ (Å)	$FWHM_{exp}$ (Å)	Δ_{FWHM} (%)	Γ_{ee}
1.8	8400	78.8	65.3	17.1	0.39
3.9	8900	132.9	111.8	15.9	0.47
6.2	10,000	181.5	147.3	18.8	0.49

Table I : Experimental and theoretical Hα FWHM results in underwater laser spark. Δ_{FWHM} is the difference between **$FWHM_{th}$** and **$FWHM_{exp}$** in percents.

As can be seen, $FWHM_{th}$ are larger than experimental ones. This disagreement may be due to the statical electric field involved in Griem Stark broadening calculations which could be modified because of non ideality effects. Indeed, as can be seen on Table I, coupling constants Γ_{ee} found in underwater laser plasma are greater than 1/3, which corresponds to less than one particule in the Debye sphere. Therefore, the Debye shielding which allows to replace a fluctuating microfield by a stationary distribution of screening charge, may be inapplicable.

References
[1] V. Sevastianenko, 1985, Beitr. Plama-phys., 25, 151-197.
[2] Y. Vitel and M. Skowronek, 1987, J. Phys. B, 20, 6477-6491.
[3] Y. Vitel and M. Skowronek, 1987, J. Phys. B, 20, 6493.
[4] H. R. Griem, 1974, *Plasma spectroscopy*, Academic Press, New York.
[5] D.A. Cremers and al., 1984, App. Spectro., 38, n°5, 721-729.

Interpretation of Measured Stark Broadened Profiles of Highly Excited Hydrogen Lines Emitted from a Low-Density, Low-Temperature Plasma

R.D. Bengtson
Fusion Research Center, University of Texas at Austin, Austin, TX 78712, USA

E. Oks, J. Touma
Physics Department, Auburn University, Auburn, AL 36849, USA

Almost 30 years ago, Bengston, Tannich, and Kepple (Phys. Rev. A, v.1, p.532, 1970) had presented experimental profiles of hydrogen Balmer lines with the principal quantum number of the upper level in the range from n=6 to n=12, emitted from a radio-frequency discharge. The independently-measured temperature and electron density had been 1850 K and 1.2×10^{13} cm^{-3}, respectively. In that paper, a comparison had been also made with corresponding profiles calculated by the Kepple-Griem (KG) code, based on their theory published in Phys. Rev., v.173, p.317, 1968. The comparison had demonstrated that the experimental FWHM was systematically lower (by up to 20%) than the FWHM calculated by the KG code. Bengtson had also measured simultaneously a profile of the Balmer line from n=15. However, this line Balmer-15 had not been mentioned in the above paper of 1970, since the KG code had not been able to generate a meaningful theoretical FWHM for this line (for the plasma parameters under consideration, the line Balmer-15 is beyond the validity limit of the KG theory by their own estimates).

In the current paper we apply our newly-developed Advanced Generalized Theory (AGT) of Stark broadening of hydrogen lines for the analysis and interpretation of all of the above experimental profiles. The main features distinguishing the AGT from the KG theory are discussed in another paper by Oks and Touma presented at this conference. For highly excited hydrogen lines the cutoff for electron impact parameters, that KG introduced empirically (and rather arbitrarily), is much smaller than the actual cutoff following from the first principles, as shown in the AGT. In other words, if KG would have used the proper cutoff (without any other changes in their theory), they would have ended up with even larger discrepancy between the calculated and experimental FWHM. Due to this reason, we knew in advance that our AGT code would not demonstrate a significantly better agreement with the experiment than that claimed on the basis of the KG-code with the wrong cutoff.

At that point we have recalled that all theories and codes developed over the last 40 years used so-called "statistical" intensities of Stark components. However, at low densities one should use the "dynamical" (rather than statistical) intensities, as had been experimentally demonstrated by Mark and Wierl in 1929 and described by Hans Bethe in his famous books of 1933 and 1957.The assumption of the dynamic intensities results into a much better agreement with the experiment. Indeed, while the KG code yielded the rms discrepancy of 14%, our code in its "dynamical" version resulted into the rms discrepancy of 5% (see Table 1). Moreover, while the KG code totally fails for the line Balmer-15, our code reproduced the experimental width of this line with an amazing accuracy of 2%. Thus, we have arrived to the conclusion that the results produced by our AGT code in its dynamic-intensity- version are in a very good agreement with the experimental profiles of highly excited Balmer lines emitted from this low-density, low-temperature plasma.

Table 1

n: Upper principal quantum number (the lower one is 2)
W_{Ex}: Experimental HWHM
W_{KG}: HWHM by the Kepple-Griem theory
D_{KG}: Discrepancy between the Kepple-Griem theory and the experiment
$$(W_{KG}-W_{Ex})/W_{Ex} *100$$
W^S_{AGT} : HWHM by the Advanced Generalized Theory with statistical intensities

D^S_{AGT}: Discrepancy between the Advanced Generalized Theory with statistical intensities and the experiment
$$(W^S_{AGT} -W_{Ex})/W_{Ex} *100$$

W^D_{AGT}: HWHM by the Advanced Generalized Theory with dynamic intensities
D^D_{AGT}: Discrepancy between the Advanced Generalized Theory with dynamic intensities and the experiment
$$(W^D_{AGT} - W_{Ex})/W_{Ex} *100$$

n	W_{Ex} (Å)	W_{KG} (Å)	D_{KG}	W^S_{AGT} (Å)	D^S_{AGT}	W^D_{AGT} (Å)	D^D_{AGT}
6	.115	.137	19%	.133	16%	.122	6.0%
7	.140	.160	14%	.166	18%	.149	6.4%
8	.185	.207	12%	.197	6.6%	.173	-6.7%
9	.223	.249	12%	.256	15%	.226	1.3%
10	.284	.316	11%	.297	4.7%	.257	-9.4%
11	.335	.372	11%	.375	12%	.331	-1.2%
12	.397	.451	14%	.436	9.8%	.379	-4.4%
15	.673	Fails/Invalid	100%	.734	9.0%	.659	-2.1%
			RMS 38%		RMS 12%		RMS 5%

Systematic Experimental Study of the Stark Broadening of C II, C III, N II, N III, O II and O III Spectral Lines

B.Blagojević, M.V.Popović and N. Konjević

Institute of Physics
11080 Belgrade, P.O.Box 68, Yugoslavia
e-mail: blagojevic@atom.phy.bg.ac.yu

Abstract. We report the experimental Stark widths of plasma broadened lines belonging to 3s-3p and 3p-3d transitions of singly and doubly ionized C, N and O emitters. The light source was a low pressure pulsed arc. The plasma electron densities were determined from the width of the HeII P_α line while the electron temperatures were measured from the relative line intensities of five N II spectral lines.

EXPERIMENT

The light source was a low pressure pulsed arc with a quartz discharge tubes 10 and 24 mm internal diameter. The distance between aluminum electrodes was 162 mm and 3 mm diameter holes were located at the center of both electrodes to allow end-on plasma observations. A 30 mm diaphragm placed in front of the focusing mirror ensures that light comes from the narrow cone about the arc axis. All plasma observations are performed with 1-m monochromator with inverse linear dispersion 0.833 nm/mm in the first order of the diffraction grating, equipped with the photomultiplier tube and stepping motor. The discharge was driven by: 15.2 µF low inductance capacitor charged to 2.2 kV and 3.0 kV, pressure of the gas mixtures p = 400 Pa, compositions: 4.8% C_2H_2 in He, 2% N_2 in He and 0.6% O_2 in He. Greatest care was taken to find the optimum conditions with the negligible line self absorption.

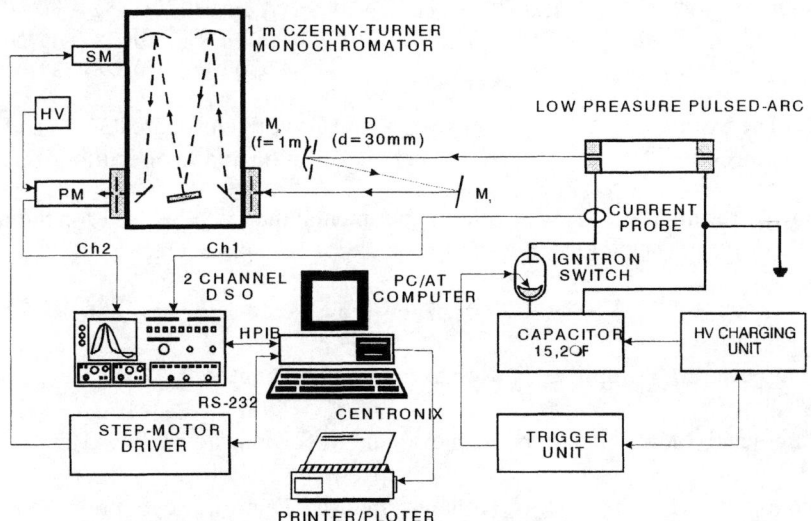

Figure 1. The experimental setup.

The spectral line profiles were recorded with instrumental half widths of 0.0084 nm. The stepping motor and oscilloscope are controlled by a personal computer which was also used for data acquisition. Recordings of spectral line shapes were performed shot-by-shot. At each wavelength position of the monochromator the oscilloscope recorded time evolution and decay of the plasma radiation. Eight such signals were averaged at each wavelength. To determine the Stark width from the measured profile, a standard deconvolution procedure for the Lorentzian (Stark) and Gaussian (instrumental+Doppler) profiles was used. The experimental apparatus and procedure are briefly described in Refs.[1-3].

Emitter	Transition	$N_e(10^{17} cm^{-3})$	$T_e(K)$	w_m (Å)	Transition	$N_e(10^{17} cm^{-3})$	$T_e(K)$	w_m (Å)
C II	$3s^2S_{1/2}-3p^2P^0_{3/2}$	0.233	17500	0.200	$3p^2P^0_{1/2}-3d^2D_{3/2}$	0.307	17100	0.350
	$3s^2S_{1/2}-3p^2P^0_{1/2}$	0.233	17500	0.206	$3p^2P^0_{3/2}-3d^2D_{5/2}$	0.307	17100	0.350
C III	$3s^3S_1-3p^3P^0_2$	0.389	19500	0.188				
	$3s^3S_1-3p^3P^0_1$	0.389	19500	0.190				
	$3s^3S_1-3p^3P^0_0$	0.389	19500	0.194				
N II	$3s^3P^0_0-3p^3D_1$	0.233	17500	0.094	$3p^3D_1-3d^3F^0_2$	0.233	17500	0.068
	$3s^3P^0_2-3p^3D_3$	0.233	17500	0.099	$3p^3D_2-3d^3F^0_3$	0.233	17500	0.066
					$3p^3D_3-3d^3F^0_4$	0.233	17500	0.064
N III	$3s^2S_{1/2}-3p^2P^0_{3/2}$	0.308	18800	0.100	$3p^2P^0_{1/2}-3d^2D_{3/2}$	0.308	18800	0.131
	$3s^2S_{1/2}-3p^2P^0_{1/2}$	0.308	18800	0.105	$3p^2P^0_{3/2}-3d^2D_{5/2}$	0.308	18800	0.130
O II	$3s^2P_{3/2}-3p^2D^0_{5/2}$	0.308	18800	0.087	$3p^2D^0_{5/2}-3d^2F_{7/2}$	0.308	18800	0.115
	$3s^2P_{1/2}-3p^2D^0_{3/2}$	0.308	18800	0.086				
O III	$3s^3P^0_2-3p^3D_3$	0.464	19400	0.097	$3p^3D_1-3d^3F^0_2$	0.828	39700	0.103
	$3s^3P^0_1-3p^3D_2$	0.464	19400	0.104	$3p^3D_2-3d^3F^0_3$	0.828	39700	0.100
					$3p^3D_3-3d^3F^0_4$	0.828	39700	0.100

Table 1. The experimental Stark widths w_m (FWHM) of investigated 3s-3p and 3p-3d transitions. N_e and T_e are electron concentration and temperature of plasma.

The reported results together with other experimental data will be used for the testing of various theoretical calculations. Further investigations are in progress.

References

1. B.Blagojević, M.V.Popović, N. Konjević and M.S.Dimitrijević, Phys.Rev.E **50**, 2986 (1994).
2. B.Blagojević, M.V.Popović, N. Konjević and M.S.Dimitrijević, Phys.Rev.E **54**, 743 (1996).
3. B.Blagojević, M.V.Popović, N. Konjević and M.S.Dimitrijević, in press JQSRT (1998).

Plasma Broadened 419.07 nm and 419.10 nm Neutral Argon Lines

D. Nikolić, S. Djurović, Z. Mijatović, R. Kobilarov and N. Konjević*

Institute of Physics, Trg Dositeja Obradovića 4, 21000 Novi Sad, Yugoslavia
*Institute of Physics, P.O. Box 68, 11080 Belgrade, Yugoslavia

1. Introduction

Large number of papers are devoted to study of shapes and shifts of plasma broadened neutral argon lines (see Ref. 1 and references therein). Some lines have not been studied experimentally although theoretical data are available. For neutral atom lines in many cases this is due to the vicinity of neighbouring overlapping lines which makes separation of the line profile difficult.

Here, experimental results of Stark broadening parameters for two close neutral argon lines 419.07 nm and 419.10 nm will be given.

2. Experimental

The atmospheric pressure wall stabilized argon arc was used as a plasma source. Working gas was the mixture of 99 % Ar and 1 % H_2 - introduced for diagnostics purposes. An averaging technique, for spectral intensity recordings from the arc plasma and reference source of radiation used for shift measurements, is described in [2]. The spectral intensities from the arc are recorded side-on at twelve different positions along the plasma column radius. Numerical procedure for separation and deconvolution [3] was applied to Abel inverted spectra. After that the contributions of resonance and Van der Waals broadening were subtracted.

Plasma electron densities in the range $(0.74 - 2.90) \, 10^{16}$ cm^{-3} were determined from Stark width of hydrogen H_β line [4], while the temperatures, ranging 9300 K - 10800 K, were determined from plasma composition data [5].

3. Results and discussion

The dependence of widths and shifts (at the halfwidth) upon electron density for 419.07 nm and 419.10 nm lines are given in Figs. 1 and 2 respectively, together with estimated error bars.

Measured Stark broadening parameters for the 419.07 nm line are compared with the theoretical ones [6]. The results of this comparison are given in Table 1. The mean ratio of measured to theoretical values is 0.55 for widths and 0.85 for shifts, see Table 1. Since theoretical values for widths and shifts for the 419.10 nm line are not available so only measured values are presented, see Table 1.

In both cases comparison with the experimental results of other authors has not been performed; experimental data for these lines have not been found in literature.

The results of this work suggest that theory for 419.07 nm line should be improved while missing theoretical values of Stark parameters for 419.10 nm line should be calculated.

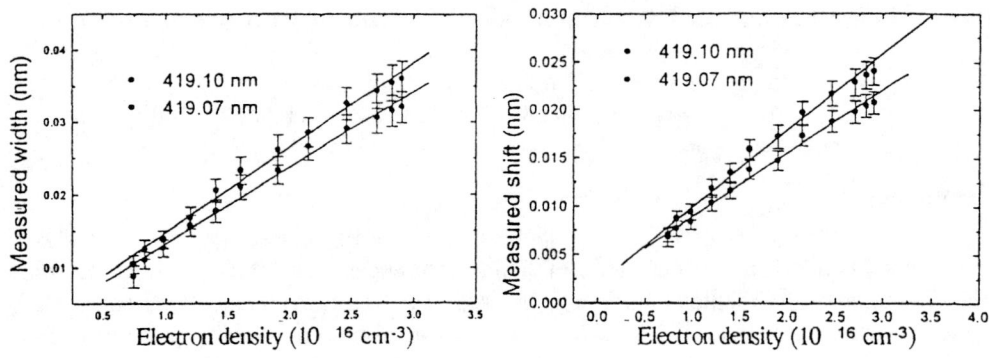

Figure 1   Figure 2

Table 1

N_e $(10^{16}$ cm$^{-3})$	T (K)	419.07 nm				419.10 nm	
		w_m (0.1nm)	w_m/w_{th}	d_m (0.1nm)	d_m/d_{th}	w_m (0.1nm)	d_m (0.1nm)
2.90	10800	0.320	0.50	0.207	0.79	0.360	0.240
2.82	10750	0.315	0.50	0.204	0.80	0.354	0.236
2.70	10700	0.305	0.51	0.198	0.80	0.343	0.229
2.46	10550	0.290	0.53	0.188	0.83	0.325	0.217
2.15	10400	0.266	0.56	0.173	0.87	0.285	0.197
1.90	10250	0.233	0.56	0.147	0.83	0.261	0.172
1.60	10050	0.210	0.61	0.138	0.92	0.233	0.159
1.40	9900	0.177	0.59	0.116	0.88	0.206	0.135
1.20	9700	0.157	0.61	0.104	0.91	0.168	0.119
0.98	9500	0.127	0.61	0.084	0.90	0.138	0.094
0.83	9400	0.111	0.63	0.077	0.96	0.125	0.087
0.74	9300	0.088	0.57	0.068	0.97	0.105	0.070

4. References

1. N. Konjević and J.R. Roberts, J. Phys. Chem. Ref. Data 5, 209 (1976); N. Konjević, M.S. Dimitrijević and W.L. Wiese, J. Phys. Chem. Ref. Data 13, 619 (1984); N. Konjević and W. L. Wiese, J. Phys. Chem. Ref. Data 19, 1307 (1990).
2. Z. Mijatović, N. Konjević, R. Kobilarov and S. Djurović, Phys. Rev. E 51, 613 (1995).
3. see the paper of the same group of authors in this book.
4. C. R. Vidal, J. Cooper and E. W. Smith, Astrophys J. Suppl. Ser. 25, 37 (1973).
5. W. B. White, S. M. Jonson and G. B. Dantzig, J. Chem. Phys. 28, 751 (1958)
6. H. R. Griem, *Spectral Line Broadening by Plasmas,* (Academic, New York, 1974)

Deconvolution procedure for plasma broadened neutral atom lines

D. Nikolić, Z. Mijatović, R. Kobilarov, S. Djurović and N. Konjević*

Institute of Physics, Trg Dositeja Obradovića 4, 21000 Novi Sad, Yugoslavia
*Institute of Physics, P.O. Box 68, 11080 Belgrade, Yugoslavia

Plasma broadened neutral atom spectral lines may be described by the convolution of Gaussian profile (Doppler + instrumental broadening) and the Stark profile which is asymmetric, due to interaction between emitters and ions and can be well described by $j_{A,R}(\lambda)$ function defined as[1,2]:

$$j_{A,R}(\lambda) = \frac{1}{\pi} \int_0^\infty \frac{H_R(\beta)\, d\beta}{1 + \left((\lambda - (\lambda_0 + d_{se}))/w_{se} - A^{4/3}\beta^2\right)^2} \tag{1}$$

where $H_R(\beta)$ represents ion field-strength distribution function[3] dependent on dimensionless parameter R that accounts for Debye shielding and ion-ion correlations[1], while A is a dimensionless parameter[1] which is the measure of the relative importance of ion broadening. In order to obtain parameters of Stark broadening (shifts and especially widths) it is necessary to apply deconvolution procedure to experimental line profiles. In the cases of asymmetric line profiles, as in this case, this procedure is very difficult. Until recently[4], the deconvolution procedure for symmetric Voigt profiles is used. In this systematic error up to 25 %[4] is introduced in determination of Stark widths. The problem of deconvolution is particularly difficult in the case of overlapping lines.

We report here new method for deconvolution of asymmetric line profiles which includes calculation of $j_{A,R}$. This method uses software package Mathematica.

The model convolution function (continuum reduced) which is fitted to experimental profiles is:

$$\hat{K}(\lambda) = C_n \cdot \Psi_{1,0,0}^R(w_{se}, d_{se}, \alpha, C_G | \lambda) \tag{2}$$

with C_n as normalising factor, and

$$\Psi_{a,b,c}^R(w_{se}, d_{se}, \alpha, C_G | \lambda) = \int_{\lambda_0 - \Delta\lambda}^{\lambda_0 + \Delta\lambda} d\lambda' (\lambda')^b \cdot e^{-C_G(\lambda' - \lambda)^2} \int_0^{10} \frac{(\beta)^c \cdot H_R(\beta)}{\left(1 + \left((\lambda' - (\lambda_0 + d_{se}))/w_{se} - \alpha\beta^2\right)^2\right)^a} d\beta \tag{3}$$

where $C_G = (4 \cdot \ln 2)/(w_D^2 + w_I^2)$ includes Doppler and instrumental broadening effects, while $\Delta\lambda$ determines integration limits (as well as accuracy and computing time) and is set to be ten half-widths of experimental profile. λ_0 is wavelength of unperturbed line, while w_I could be measured independently[5] and R is determined from plasma parameters[1]. Five parameters in (3) are fitted to experimental profiles: width of Doppler profile w_D, electron shift d_{se} and width w_{se}, ion broadening parameter A and normalising factor C_n. The values of these quantities must be initially estimated. The starting values for Stark broadening parameters w_{se}, d_{se} and A could be taken from the literature[1] when available.

Because of non-linear dependence of model function (2) on the set of unknown parameters $w_D, w_{se}, d_{se}, \alpha \equiv A^{4/3}$ and C_n, minimisation of χ^2 merit function must proceed iteratively. Levenberg-Marquardt[6] procedure is used to improve initial values for adjustable parameters during iteration. The procedure is then repeated until χ^2 stops to decrease (or effectively stops), fulfilling condition $\Delta\chi^2/\chi^2 \leq 5\cdot 10^{-3}$. The Levenberg-Marquardt method can be implemented as a model-trust region method for minimisation[7] applied to the special case of least squares function. This method works very well in practice and has become the standard of non-linear least-squares routines[8]. Described procedure is also applicable (although computing time is 80% longer) on two overlapping spectral lines when optimisation of linear combination of convoluted profiles is performed, that is model function (2) has to be taken as:

$$\hat{K}(\lambda) = C_{n1} \cdot \Psi^R_{1,0,0}(w_{se1}, d_{se1}, \alpha_1, C_G|\lambda) + C_{n2} \cdot \Psi^R_{1,0,0}(w_{se2}, d_{se2}, \alpha_2, C_G|\lambda) \quad (4)$$

with nine parameters $(w_D, w_{se1}, d_{se1}, \alpha_1, C_{n1}, w_{se2}, d_{se2}, \alpha_2, C_{n2})$ to be optimised. The estimated errors in determination of Stark broadening parameters are less then 9%, but these values are influenced by the error of the measurements of experimental points. Deconvolution results for neutral argon spectral line Ar I 430.01 nm are given in Fig. 1.

As it can be seen from presented example, described method results in fitted profiles which are in very good agreement with the experimental ones. The only drawback of this method is long computing time - few hours for two overlapping lines - on pentium based PC with 32 MB RAM.

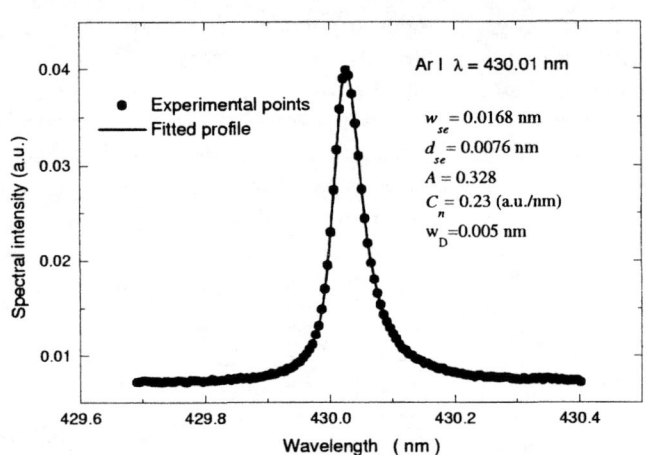

Figure 1

References

1. H. R. Griem, *Spectral Line Broadening by Plasmas*, (Academic Press, New York, 1974).
2. L. A. Woltz, JQSRT **36**, 547 (1986).
3. C. F. Hooper Jr., Phys. Rev. **165**, 215 (1968).
4. Z. Mijatović, R. Kobilarov, B.T. Vujičić, N. Konjević and D. Nikolić, JQSRT **50**, 339 (1993).
5. Z. Mijatović, N. Konjević, M. Ivković and R. Kobilarov, Phys. Rev. E **51**, 4891 (1995).
6. W. H. Press, S.A. Teukolsky, W.T. Vetterling and B.P. Flannery, *Numerical Recipes in C*, 2nd ed., (Cambridge University Press, Cambridge, 1995).
7. J. E. Dennis and R. B. Schnabel, *Numerical methods for Unconstrained Optimization and Non-linear Equations*, (Prentice-Hall, Englewood Cliffs, 1983).
8. S. Wolfram, *The Mathematica Book*, 3rd ed., (Wolfram Media/Cambridge University Press, 1996).

Anomalous Asymmetry of a Helium Spectral Line – Anomalous Electric Fields in a Current Sheet Plasma

A. G. Frank*, V. P. Gavrilenko**, N. P. Kyrie*

* General Physics Institute of the Russian Academy of Science, Vavilov St. 38, Moscow 117942, Russia

** Center for Surface and Vacuum Research of the Russian State Committee for Standards, Andreevskaya nab. 2, Moscow 117334, Russia

The problem of possible correlation between magnetic reconnection phenomena and excitation of plasma microinstabilities was advanced many years ago. We experimentally studied non-equilibrium electric fields in a curent sheet plasma by examining the profiles of two helium spectral lines: HeI 667.8 nm (transition $3^1D \to 2^1P$) and HeI 587.6 nm (transition $3^3D \to 2^3P$). Experiments were performed at the device "Current Sheet". Initial plasma ($N_e \sim 10^{16}$ cm^{-3}) was produced in the He-gas at a pressure 0.3 Torr in 2D quadrupole magnetic field with the null-line. An excitation of electric current along the null-line resulted in a formation of plane current sheet and plasma compression into the sheet. Typical plasma parameters in the sheet were the following: $N_e = (4-9) \cdot 10^{16}$ cm^{-3}, $T_e \approx T_i \approx 2$ eV.

Plasma emission was registered in the direction of the null-line. We observed that the profiles of the line HeI 667.8 nm were anomalously asymmetrical: the blue wing was considerably more intensive than the red one (see Fig. 1, solid curves). At the same time, the profiles of the line HeI 587.6 nm were approximately symmetrical with respect to the ordinate axis drawn through the maximum of these profiles. The Full Widths at Half Maximum (FWHM) of the profiles of the line HeI 667.8 nm were (0.3 – 0.4) nm, whereas the FWHM of the profiles of the line HeI 587.6 nm were (0.12 – 0.2) nm. Assuming that the broadening of the line HeI 667.8 nm was due to the Stark effect in individual ion microfields, we could obtain an asymmetry of this line, namely, with the red wing more intensive than the blue one. However, the experimental profiles of HeI 667.8 nm demonstrate the opposite type of the asymmetry, with more intensive blue wing. To explain this effect, the existence of anomalous one-dimensional quasistatic electric field of the strength $\vec{F}$ was assumed in a current sheet plasma. The direction of $\vec{F}$ should be orthogonal to the direction of the observation. The reason of the registered asymmetry of the profiles of the spectral line HeI 667.8 nm can be qualitatively understood from Fig. 2. The action of the electric field $\vec{F}$ leads to the Stark splitting of the line HeI 667.8 nm into three components (components "0", "1", and "2", see Fig. 2). The component "K" corresponds to the radiative transition $(3^1D, |M| = K) \to 2^1P$, where M is the magnetic quantum number of the state belonging to level 3^1D ($M = 0, \pm 1, \pm 2$). One of these components (the component "2") is not shifted in the field $\vec{F}$. Two other components (the components "0" and "1") are shifted in the field $\vec{F}$ towards the frequencies $\Delta\omega < 0$ (the frequency $\Delta\omega = 0$ corresponds to the unshifted position of the line HeI 667.8 nm). As a result, the envelope drawn through the tops of these three components (see the dashed line in Fig. 2) has an asymmetry of the same type as the asymmetry of the experimental profiles of the line HeI 667.8 nm.

In order to perform a quantitave analysis of experimental profiles of the line HeI 667.8 nm, we calculated a set of theoretical profiles of this line taking into account besides Stark effect in the field $\vec{F}$ other broadening mechanisms, such as instrumental broadening, Doppler effect, and Stark effect due to the individual plasma microfields. The best fitting was achieved when the field strength was $F = 100 - 120$ kV/cm. The corresponding theoretical profiles of the line HeI 667.8 nm are shown by the dashed line in Fig. 1. The absence of the appreciable asymmetry and smaller values of the FWHM of the experimental profiles of the line HeI 587.6 nm (in comparison with the line HeI 667.8 nm) are due to smaller value of the Stark constant for this line than for to the line HeI 667.8 nm.

Thus, on the basis of comparing the experimental and theoretical profiles of the spectral line HeI 667.8 nm, a hypothesis was advanced that anomalous quasistatic electric fields with the strength up to 100 kV/cm were excited in the peripheral regions of the current sheet (in the regions with strong electron density gradients).

This work was sponsored by the INTAS (project No. 96-456) and by the Russian Foundation for Basic Research (project No. 96-02-18546).

REFERENCES

1. S.Yu. Bogdanov, V.B. Burilina, A.G. Frank, JETP, V. 114, No 9 (1998)
2. S. Buscher, N.P. Kyrie, H.-J. Kunze, A.G. Frank, Plasma Phys. Reports, V. 24 (1998) (in press).

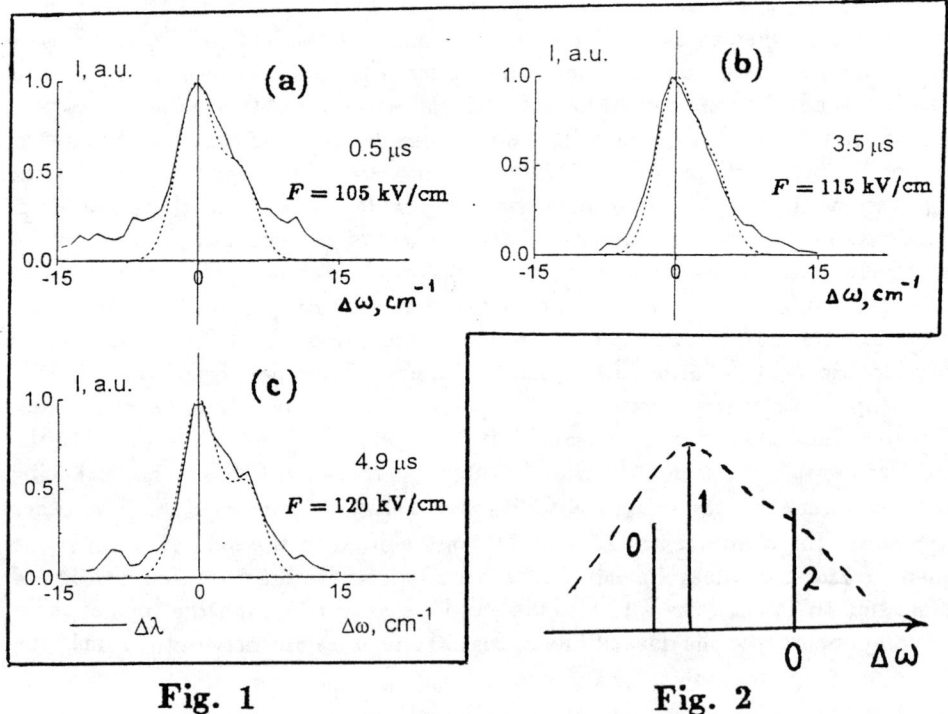

Fig. 1   Fig. 2

Modified Helium Microwave - Induced Plasma Discharge chamber
M.M. Mohamed* and Z.F Ghatass**
*Medical Research Institute and **Institute of Graduate Studies and Research, Alexandria University, Alexandria, EGYPT

Introduction

Among the various plasma excitation sources used for spectrochemical analysis, the microwave -induced plasma (MIP) is unique because of its low power and gas consumption. Performance characteristics of the MIP include good detection limits, wide range and linearity. Its high excitation temperature [1] makes it useful for the determination of most elements. Until recently, the major disadvantage of the MIP has been its low tolerance to liquid aerosol introduction. The low power (<100w) commonly coupled to the microwave discharge does not provide sufficient plasma density to both desolvate and excite elemental emission from directly nebulized solutions. In the present work to minmize problems caused by sample introduction into a helium-MIP, a new microwave discharge chamber was designed.

Instrumentation

The microwave generator used in the present work has been previously described in detail [2]. The plasma is sustained in a quartz discharge tube (1.5-mm inner diameter and 3 mm outer diameter) which is located axially in an outer tube (4-mm inner diameter and 6 mm outer diameter). The plasma torch (Fig. (1)) was placed in a 7-mm diameter discharge aperture with centre of the cavity. The outer flow (sample gas) was introduced tangentially to the axis of the torch. During emission measurements, the plasma is viewed axially. Helium gas carried sample aerosol (generated by a cross-flow nebulizer) to the plasma discharge. In this torch a portion of the plasma is kept operating by partially isolating it from the rest of the plasma within the plasma chamber. This auxiliary plasma is passed during sample or solvent injection and is therefore not affected by such event. An 8 cm focal length lens focuses an image (1:1) of the plasma on the entrance slit of the monochromator (1200 groves and 500 blaze) Czerner-Turner spectrometer, output radiation from the monochromator is detected by a photomultiplier tube (PMT) (RC AIP28) operated at 1200 volt by a suitable power supply. The PMT output signal is filtered with an R-C integrator with time constant 1 s. The output signal is recorded using x-t recorder.

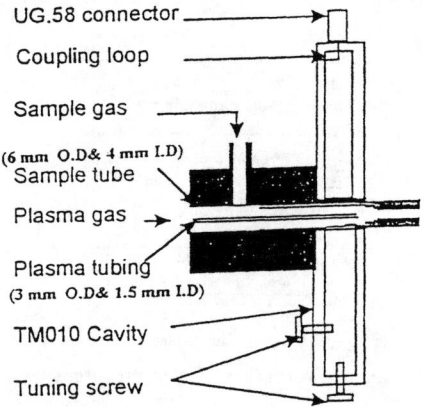

Fig. (1): Schematic diagram of the discharge tube.

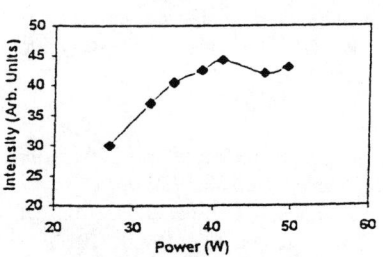

Fig.(2): Effect of the microwave power on the Ca (I) 393.37 nm intensity.

Experimental Results

Optimization of the experimental conditions to increase the sensitivity of the system for metal determination in solution, plasma microwave power, plasma and sampling gas flow rates, and discharge tube position were investigated. The results obtained are presented in Fig.(2-4). Detection limits for Ca, Cd, Cu, Fe, Mg, and Zn were determined and are listed in Table (1). Samples containing 1 ppm of each of the previous elements was selected and used to evaluation of detection limits. Detection limits were calculated with the use of the method of Boumans and Vrakking [3]. Analytical calibration curves for Ca, Cd, Cu, Fe, Mg, and Zn are presented in Fig.(5).

Table (1): Detection Limits (ppb).

Element	Wavelength (nm)	Detection Limits (ppb)
Ca (I)	393.37	1.6
Cd (I)	361.05	9
Cu (I)	324.75	0.92
Mg (I)	383.83	0.85
Fe (I)	371.99	30
Zn (I)	213.86	8

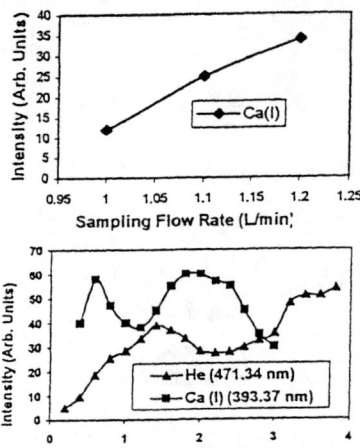

Fig.(3): Influence of He flow rate on the intensities.

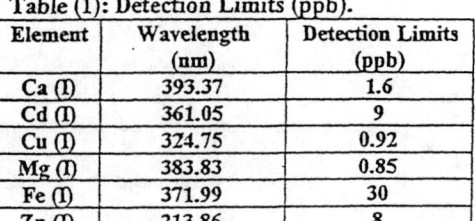

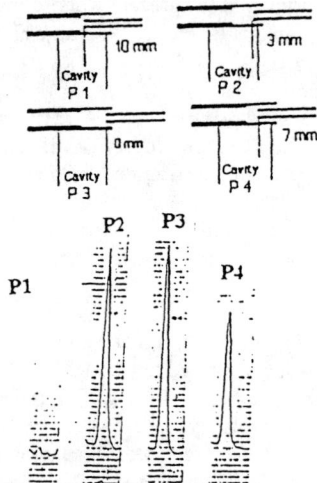

Fig. (4): Discharge tube at different position and corresponding emission intensities of Ca (I).

Each reading corresponds to an integration time of 1 sec. Analytical calibration curves obtained with the present system show linear relationships between emission intensity and concentration. Three- decades of linear dynamic ranges are achieved.

References

1. M.M. Mohamed, T.Uchida, and S. Minami, Bynko, Kenkyo 38,288,1989.
2. M.M. Mohamed, T.Uchida, and S. Minmai, Appl. Spectrosc. 43, 129 (1989).
3. P.W.J.M. Boumans and J.J.A.M. Vrakking, Spectrochim Acta 43 B, 553 (1987).

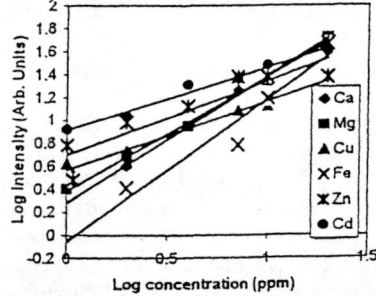

Fig. (5): Emission spectrum and calibration curve for Fe (I) 371.9 nm.

New Three Phase Double Arc Plasma for Spectrochemical Analysis of Solid and Powder Samples

M.M.Mohamed[*], Z.F.Ghatass[**], E.A.Shalaby[**], M.M.Kotb[**], and M.El-Raey[**]

* Medical Research Institute and ** Institute of Graduate Studies and Research
Alexandria University, Alexandria, EGYPT.

Introduction

The revival of interest in the field of atomic emission spectroscopy (AES) during the past decade has been prompted by the development of numerous electric discharge plasma sources. Despite of its wide use, plasma AES remains limited in routine applications by samples introduction methods that require large volume of aqueous solutions[1]. The separation of the sampling and the excitation steps in atomic emission spectroscopy have often resulted in enhanced analytical performance[2,3]. In the present work, a three-phase current is used to support double arc argon plasma for sampling and excitation of solid and powdered samples.

Materials and Methods

The power used to sustain the argon plasma arc was derived from three-phase ac line voltage. It consists of three sine waves 120° out of phase from each other. The voltage between any two of the three phase was 380 V. This voltage was reduced with a three phase wye "Y" connected transformer (7.5 kW) to 95 V. 32 Amp. circuit breakers were placed in each leg. To drop the voltage to the plasma and to regulate the plasma current to 27 Amp., a 1.55 Ohm rheostat was connected in series between each transformer phase and its corresponding electrode.

As shown in Fig. (1), aerosol is generated from the sample (first electrode) with the use of an ac arc (sampling arc) ignited between itself and the second electrode. The resultant material is swept, by using sampling argon gas through a hole in the second electrode, into the argon arc generated between the second and third electrode (excitation arc). The three electrodes were cooled with water. The sampling arc, which burns between the first and second electrodes, is surrounded with a cylindrical glass housing. With the system applied, particles can be expected to leave the interaction cylinder in the axial direction. The diameter of the exhaust and argon inlet orifices, position of the tangential gas inlet, and geometry of the housing were evaluated by varying dimensions until a stable plasma could be sustained for one hour. A tangential stream of argon gas is introduced through the inlet orifice as a

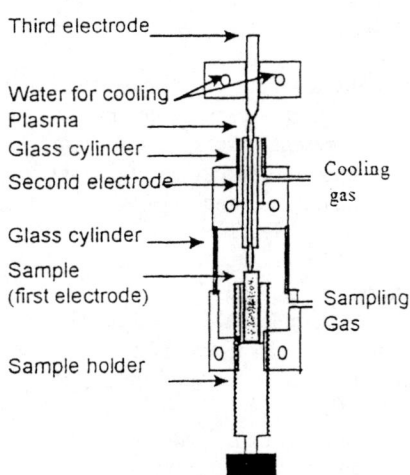

Fig. (1): Schematic diagram of the double arc.

coolant gas for the second electrode and to induce rotation of the excitation arc discharge. This increases the stability of the excitation arc plasma. Emission of the plume discharge depends on the swept material from the first arc discharge. A 12-cm focal length lens focuses an image (1:1) of the plasma into entrance slit of a one-meter spectrometer and detected by a photomultiplier tube. X-t recorder records the output signal.

Disc samples of 0.5 g of carbon powder containing different amounts of Fe was pressed at 9 tons 1 cm to form a disc with dimension of 9-mm, diameter and 2-mm thickness.

Results and Discussion

Experimentally controllable parameters investigated include sampling efficiency, effects of cooling flow rate on the signal to background ratio (S/B), plasma current, and excitation arc gap on the emission intensities of Ar (I) 425.9 nm as plasma gas emission line and Fe (I) 371.9 nm as analytical line. Results obtained are presented in Figures (2, 3 and 4). Emission spectrum and analytical curves for Fe (I) 371.9 nm, are presented in Fig. (5). The present work double arc technique shows good linearity, low detection limits, and low gas consumption.

References:
1. MM Mohamed, MM Mossad, MK Nasra, FI Nasr, and NM Fikry, Indian J. Pure & Appl. Phys. 32, 471, (1994).
2. MM Mohamed, T Uchida, and S Minomi, Appl. Spectrosc., 43, 794, (1989).
3. MM Mohamed, T Uchida, D. Coleman and S Minomi, Proceeding of the (1989) FACSS meeting, Chicago, USA, October, (1989).

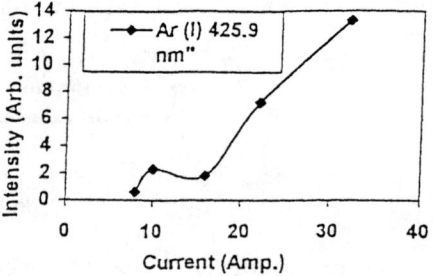

Fig. (2): Effect of the current on argon line.

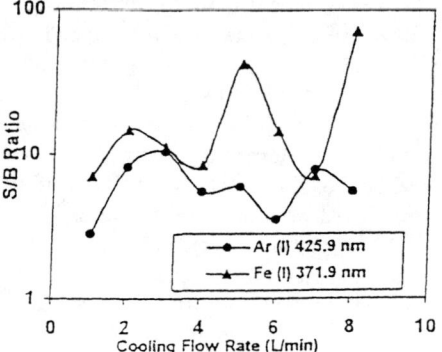

Fig. (3): Effects of cooling flow rate on S/B ratio.

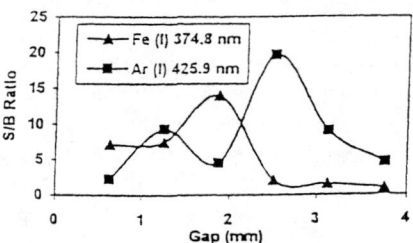

Fig. (4): Effect of excitation arc gap on the signal to background ratio.

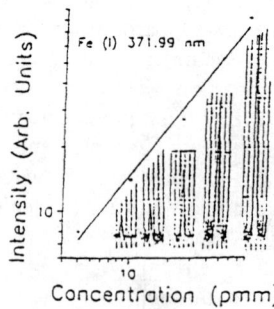

Fig. (5): Emission spectrum and calibration curve for Fe (I) 371.9 nm.

Line Shapes Obtained by Polarization Spectroscopy

A. Steiger, and K. Grützmacher

Physikalisch-Technische Bundesanstalt, Abbestr. 2-12, D-10587 Berlin, Germany

Polarization spectroscopy is a very sensitive laser technique for measuring atomic or molecular one- and two-photon transitions. Two-photon polarization spectroscopy was applied for the determination of absolute local ground-state densities of hydrogen atoms in non-L.T.E. environments [1] and two-photon absorption cross sections of xenon [2]. The method was developed for Doppler-free measurements of the Stark broadening of the 2S-level of hydrogen and deuterium in arc plasmas [3, 4, 5]. Highest detection sensitivity is achieved if the change of the linear polarization state of a signal wave induced by a circularly polarized pump wave is detected behind a crossed polarization analyzer: $\delta E_s(\Delta v) = \frac{1}{16} \Delta \alpha_s^2 \cdot P(\Delta v) + R$, where δE_s denotes the relative transmitted signal, i.e. the signal behind the analyzer relative to the suppressed one which is taken from the second output of the analyzer. $\Delta \alpha_s = \alpha_{s0} - \alpha_{s2}$ is the difference of the frequency-integrated two-photon absorption value $\alpha_s = \sigma N E_p l$ induced by the pump beam (irradiance E_p) for a signal beam with circular polarization which is either complementary ($\Delta m_L = 0$ transition) or identical ($\Delta m_L = 2$) with respect to the pump beam polarization. Acting as background, R represents the residual transmittance of the crossed polarizers measured as signal without any pump beam.

The line shape $P(\Delta v)$ of the polarization signal is remarkably different from that observed in linear emission or absorption spectroscopy. It is the sum of the squares of the absorption line shape function $L(\Delta v)$ and the corresponding dispersion function which are interconnected by a Kramers-Kronig relation:

$$P(\Delta v) = \left[L(\Delta v)\right]^2 + \left[\frac{1}{\pi}\int_{-\infty}^{\infty} \frac{L(v)}{v - \Delta v} dv\right]^2.$$

The properties of the polarization profile $P(\Delta v)$ follow from the mathematical structure of this nonlinear relation. Even if $P(\Delta v)$ can always be obtained from any area normalized absorption profile $L(\Delta v)$ at least numerically, there is still the inversion problem which occurs when measured polarization data $\delta E_s(\Delta v) - R$ are analyzed with respect to their underlying absorption profile $L(\Delta v)$.

The important characteristics which are necessary for quantitative measurements and for the interpretation of polarization profiles as well have been studied in detail [6]. In the center of a spectral line, the polarization line shape exhibits a strong gain of information. However, there is also a loss of information especially in the line wings. Regardless of the shape of $L(\Delta v)$, the line wings of $P(\Delta v)$ always approach the same asymptote $(\pi \Delta v)^{-2}$ which is symmetric to the center of mass of $L(\Delta v)$.

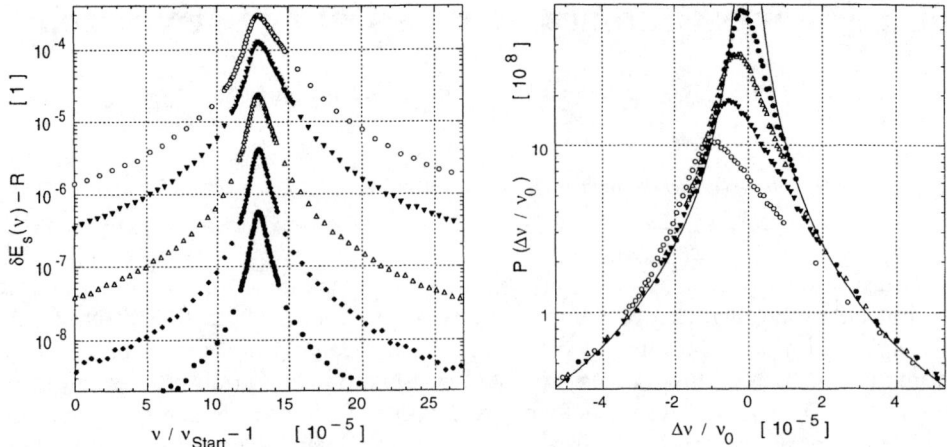

FIGURE 1. Hydrogen 1S-2S polarization spectra δE_s -R (left side) were measured with different powers of the pump beam. After signal normalization and frequency correction of the ac-Stark shift, the resulting line profiles $P(\Delta\nu/\nu_0)$ and the common asymptote (—) are plotted on the right side.

In detail, the area normalization is lost and the line integral is proportional to the inverse of the half width. Any structure of the line center is enhanced, differences of local extrema are relatively amplified by about a factor of two, the half width and the separation of local maxima are increased. The asymptotic behaviour of P takes over the role of the area normalization of L: The separation into signal size and line shape function can be achieved and the line center ν_0 can be found, even in the case of an asymmetric line.

As an example, Doppler-free 1S-2S line profiles are presented in figure 1 which were measured in a hydrogen arc plasma at a discharge current of 18 A and a pressure of 7.5 kPa. The pump beam irradiance E_p was raised from about 200 MW/cm² to more than 10 GW/cm². The signal size increases quadratically with E_p. The ac-Stark effect linearly shifts the center of mass ν_0 to the blue (visible only in the line wings of the signal) and asymmetrically broadens the line due to temporal and spatial variations. Additional broadening is caused by the photo-ionization of the 2S level which is also induced by the intense light field of the pump beam.

REFERENCES

1. R. Dux, K. Grützmacher, M.I. de la Rosa, B. Wende, Phys. Rev. E **51**, 1416 (1995)
2. R. Dux, K. Grützmacher, M.I. de la Rosa, A. Steiger, B. Wende, to be accepted by Phys. Rev. A
3. A. Steiger, K. Grützmacher, *Spectral Line Shapes, Vol. 7*, Eds. R. Stamm and B. Talin, Nova Science Publ., New York 1993, p. 141
4. J. Seidel, A. Steiger, K. Grützmacher, *Spectral Line Shapes, Vol. 8* (AIP Conf. Proc. 328), Eds. A.D. May, J.R. Drummond, E. Oks, AIP Press, New York 1995, p. 32
5. M. Schmidt, K. Grützmacher, A. Steiger, "Stark broadening of the 2S-level of hydrogen at low electron densities by Doppler-free two-photon polarization spectroscopy", contribution to this volume
6. A. Steiger, K. Grützmacher, to be published

Transition probability measurement in a NeI plasma

J. A. del Val
Departamento de Fisica Aplicada, E. U. Politécnica,
Universidad de Salamanca, 05071, Avila, Spain

J. A. Aparicio and S. Mar
Departamento de Optica y Fisica Aplicada, Universidad de Valladolid, 47071
Valladolid, Spain

This work reports a collection of 31 transition probabilities of lines in the spectral region 590 - 810 nm, corresponding to the most intensive Ne I transitions (3p-3s, 3d-3p), all of them measured in an emission experiment[1]. Relative intensity measurements have been made on a pulsed discharge lamp and the absolute A_{ki} values have been obtained by using an extensive set of data taken from the literature (usually we have taken the mean value between the available works). The electron density has been determined by one-wavelength interferometry and ranges from 0.1 to 1.8×10^{23} m^{-3} in the plasma. The NeI temperature (10000-21000 K) has been determined from the Boltzmann-plot of the lines and also estimated by the Saha law. The experimental arrangement and the diagnostic methods can be seen elsewhere [2,3].

The plasma has been demonstrated to be well described by a partial local thermodynamic equilibrium model. The absence of self-absorption and the spectral calibration have been very carefully taken in account. The high number of measurements (25) performed for each line along the plasma life, and its very controlled features, allows us to obtain a very good A_{ki} value by the mean value, and its uncertainty by the standard deviation (usually <15%).

The Table shows, for the 31 measured lines, the mean value taken from the literature like reference, $<A_{ki}>_{lit}$, the mean value resulting of this experiment, $<A_{ki}>_{exp}$, and their standard deviations σ. All the values must be multiplied by $10^7 \, s^{-1}$.

λ (nm)	$<A_{ki}>_{lit}$	$<A_{ki}>_{exp}$	σ$_{lit}$	σ$_{exp}$	λ (nm)	$<A_{ki}>_{lit}$	$<A_{ki}>_{exp}$	σ$_{lit}$	σ$_{exp}$
594.48	1.11	1.03	0.10	0.11	653.29	1.13	1.08	0.09	0.06
597.55	0.38	0.81	0.37	0.03	659.90	2.37	2.48	0.25	0.23
603.00	0.55	0.58	0.13	0.05	671.70	2.20	2.32	0.28	0.13
607.43	5.84	5.53	0.80	0.43	692.95	1.85	1.94	0.19	0.14
609.62	1.74	1.65	0.17	0.13	702.40	0.21	0.24	0.06	0.02
612.85	0.12	0.08	0.04	0.05	703.24	2.49	2.57	0.37	0.30
614.31	2.86	2.28	0.33	0.55	705.13	0.40	0.41	0.08	0.19
616.36	1.52	1.53	0.22	0.15	717.39	0.33	0.40	0.06	0.05
618.21	0.36	0.39	0.05	0.01	724.52	0.94	1.12	0.11	0.10
621.73	0.68	0.54	0.13	0.07	743.89	0.27	0.33	0.06	0.03
626.65	2.56	2.25	0.37	0.25	747.24	0.53	0.38	0.13	0.06
630.48	0.45	0.46	0.07	0.04	748.89	2.69	2.66	0.31	0.25
633.44	1.58	1.58	0.26	0.25	753.58	3.92	3.84	0.23	0.17
638.30	3.30	2.60	0.37	0.28	754.40	4.75	5.03	0.63	0.01
640.22	5.11	3.38	1.38	0.27	813.64	1.66	1.73	0.17	0.29
650.65	2.96	2.27	0.72	0.42					

Acknowledgements : Authors thank S. González for his work in the experimental device, Drs. I. de la Rosa, C. Pérez and M. Gigosos for his help, the DGICYT (MEC) of Spain for its financial support under Contract No. PB-94-0216 and the Consej. Educ. y Cult. de la Junta de Castilla y León (VA96-96).

References :

[1] J.A. del Val, Tesis doctoral, Universidad de Valladolid, 1997.
[2] M.A. Gigosos, S. Mar, C. Pérez, I. De la Rosa, Physical Review E, 49, 1575 (1994)
[3] J.A. Aparicio, M.A. Gigosos, S. Mar, J. Phys. B: At. Mol. Opt. Phys., 30, 3141 (1997)

ASTROPHYSICAL AND ATMOSPHERIC APPLICATIONS

Collision – induced absorption in dense atmospheres of cool stars

Aleksandra Borysow[1] and Uffe Gråe Jørgensen

Niels Bohr Institute, for Astronomy, Physics and Geophysics, University Observatory, Juliane Maries vej 30, DK-2100 Copenhagen, Denmark

Abstract. In the atmosphere of the Sun the major interaction between the matter and the radiation is through light absorption by ions (predominantly the negative ion of hydrogen atoms), neutral atoms and a small amount of polar molecules. The majority of stars in the universe are, however, cooler and denser than our Sun, and for a large fraction of these, the above absorption processes are very weak. Here, collision-induced absorption (CIA) becomes the dominant opacity source. The radiation is absorbed during very short mutual passages ("collisions") of two non-polar molecules (and/or atoms), while their electric charge distributions are temporarily distorted which gives rise to a transient dipole moment. We present here a review of the present-day knowledge about the impact of collision-induced absorption processes on the structure and the spectrum of such stars.

INTRODUCTION

Collision induced absorption (CIA) is known to be the major opacity source in dense atmospheres composed of neutral and nonpolar molecules, such as the atmospheres of the giant planets. In recent years interest increased concerning the possible impact of CIA on stellar atmospheres of cool stars as well.

We have computed a grid of self–consistent atmospheres representing oxygen-rich stars (Borysow et al., [1]) and carbon-rich stars (Jørgensen et al., [2]), where we varied the effective temperature (T_{eff}), metallicity (Z) and gravity (g), i.e. the three fundamental parameters characterising real stars. We examined the effect of including CIA as one of the opacity sources on the total flux and on the atmospheric structure.

We review here the outcome of our work and that of others, and conclude for *real* stars (characterised by the three fundamental stellar parameters), which atmospheres are most affected (and in what way) by collision-induced absorption.

---

[1] On professional leave of absence from Physics Department, Michigan Technological University, Houghton, MI 49931, USA

Our investigation is the first *systematic* study of the effect of CIA relative to other opacity sources for a wide range of real stars. Our results show that CIA is of considerable importance for a large part of the stars in the HR diagram, principally for those of low metallicity, high gravity, and low temperatures ($\leq 4{,}000$ K).

In 1993 an international conference, IAU Colloquium 146, on "Molecules in the Stellar Environment" took place in Copenhagen at the Niels Bohr Institute. It assembled a selection of astronomers together with spectroscopists whose work was related to applications of their work towards stellar atmospheric research. Up to 1993 the interest in CIA in stellar models has been sporadic, though definitely crucial for future development. The first review paper on the importance of CIA in stellar environments was presented at this conference [3], and resulted in a significantly increased awareness of the role of CIA in stellar atmospheres. As a result several astrophysicists attempted to test the impact of the CIA on specific tasks they were currently working on.

Two main reasons motivate the present work. One purpose is to make a grid of stars which will answer the question: for which stars is CIA really important? Besides reviewing this question and updating the most recent work in the field, we also attempt to make results understandable to spectroscopists and not merely to astrophysicists. To this end we include a small glossary of astrophysical terms commonly used in the field.

HISTORICAL REVIEW

Early CIA models

Linsky (1969, [4]) and Tsuji (1969, [5]) were the first to indicate the necessity of accounting for collision-induced absorption in the opacities of cool stars. Linsky [4] made the first high temperature predictions of the rototranslational (RT), the fundamental, and the first overtone CIA spectral bands. His analytical models, however, covered only the RT and the fundamental band. intended to be used at temperatures from 600 K to 3,000-4,000 K, and at wavenumbers below $\sim 8{,}000$ cm^{-1}. Astrophysicists, however, sometimes used these models freely at temperatures and frequencies seriously exceeding the intended limits. Linsky pointed out the importance of weak, continuous opacity sources (such as CIA) for cool, late type stellar atmospheres, where temperatures are less than 4,000 K. In his opinion such opacities, even though intrinsically weak, might affect the radiation flux in an even more pronounced way than the strong, narrow, molecular bands. Naturally, CIA does become the dominant continuous opacity in the absence of other sources, like those due to H$^-$. Linsky determined that CIA could be *the* most important opacity source in regions of stellar atmosphere with $T \leq 2{,}500$ K. He estimated that at these temperatures little besides neutral molecular H$_2$ exists. Under such conditions CIA due to molecular hydrogen plays an important role in the atmospheric (infrared) opacity.

Unfortunately, although analytical models were given for RT and the fundamental bands, Linsky concluded that the first overtone intensity is too weak to be of importance and he therefore didn't present an analytical model for the first overtone band. Instead, he only gave figures of his estimates of the spectra of that band at a few selected temperatures. Due to the lack of an analytical model for the first overtone, this crucial input was missing (or neglected) until just recently.

At around the same time Tsuji (1969, [5]) studied various molecular opacities in cool stellar atmospheres. Similarly to Linsky, Tsuji independently concluded that the opacity due to CIA may be very important for cool stars. Using a semi-empirical technique he was able to construct rotovibrational (RV) CIA band spectra of hydrogen pairs at 2,500 K. Tsuji's study gave convincing evidence that CIA is indeed a crucial source of opacity at high gas densities, and that future model atmospheres must account for it in order to lead to reliable results for the emerging flux. Unfortunately, Tsuji's work remained relatively unnoticed for many years, which might have been an effect of the fact that it was not published in one of the main journals, but in a conference proceedings only. In addition, it was not easy for other users to use the spectra at temperatures other than the one (T=2,500 K) shown in a figure in the paper.

In 1971 Patch [6] performed *ab initio* computations of RV CIA in the fundamental band of H_2–H_2 pairs. His work covered temperatures up to 7,000 K and frequencies from 100 to 40,000 cm^{-1}. Together with his results, he has given an analytical form which could model this band in the abovementioned region. Since his model was considered more reliable, being based on the first principles (correctly – for comparison see [3,15]), it often replaced Linsky's original data of that band by

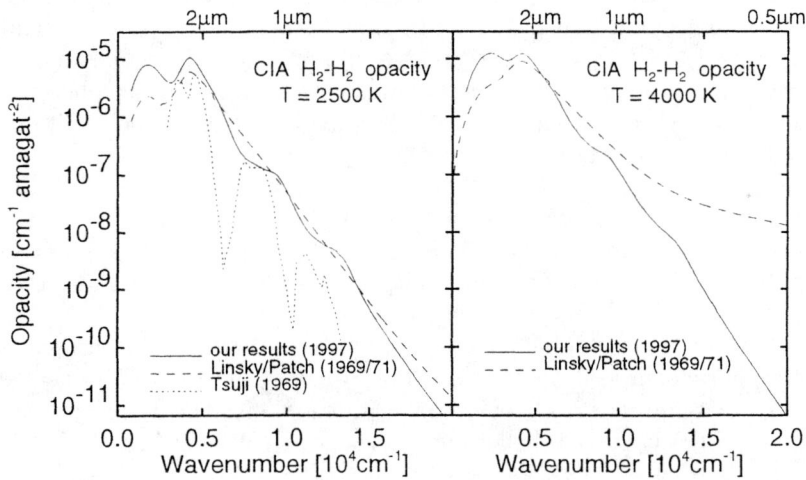

FIGURE 1. Comparison of the pioneering results (at T=2,500 K) by Linsky(1969)/Patch(1971) and by Tsuji (1969), with our recent computations [1]. Left panel is for T=2,500 K and right panel is for T=4,000 K (Tsuji's results are given only for T=2,500K).

various model atmosphere producers.

In Fig. 1 we compare these two early predictions with our recent computations [1]. It is seen that Linsky&Patch's models for T=2,500 K agree surprising well with our detailed results far into the high wavenumbers. Such an agreement must, however, be considered rather accidental. The presented models were not designed to be used beyond modelling the fundamental band, which might be relevant up to ~8,000 cm^{-1}. Even though Patch's data extended up to 40,000 cm^{-1} they neglected the first and the second overtone bands. Therefore, combined Linsky/Patch's results should be considered reliable only up to ~8,000 cm^{-1}. To confirm this, we have compared their model also at 4,000 K (which is the upper limit of Linsky's model), with our CIA estimates. Whereas at low wavenumbers agreement is found similar to the one at 2,500 K, the extrapolated model exceeds the realistic absorption spectra by as much as three orders of magnitude at 20,000 cm^{-1} (see Fig. 1). This should serve as a warning to modellers not to use models *beyond* limits for which they were designed.

Early applications of CIA to stellar atmospheres

Linsky's pioneering paper triggered some interest in other astronomers. A few papers emerged, trying to apply Linsky's new CIA models. A more detailed review of these early attempts was given in [3]. Here, we only refer to these efforts briefly.

As early as in 1971, Tsuji [7] computed Rosseland mean opacity of various low temperature non zero metallicity stars. For the first time, he included also Linsky's CIA due to H_2 as opacity source. When comparing the difference between accounting for continuum alone, and continuum with CIA, Tsuji demonstrated a very strong impact of CIA on computed Rosseland mean opacity values, χ_R, at low temperatures, high pressures, and especially in metal-deficient atmospheres.

Shipman (1977) [8] computed the first model atmospheres of cool white dwarfs that accounted for CIA. He considered pure hydrogen atmospheres at temperatures between 4,000 and 8,000 K. He found that at a temperature of 4,000 K CIA contributes essentially all the opacity at the wavelengths where the flux is emitted! At 5,000 K CIA provided only 10% of the total opacity, on account of the larger abundances of electrons and ions which give rise to relatively much stronger opacities.

Mould & Liebert (1978) [9] included CIA in the studies presented in their paper on the atmospheric composition of cool white dwarfs. They computed new atmospheric models at temperatures between 4,000 and 7,000 K. The overall importance of CIA is not explicitly stated, but a question was raised regarding the reliability of Linsky's predictions.

Palla (1985) [10], while working on primordial stars of zero-metallicity, postulated for the first time for this kind of stars that the more conventional opacity sources (e.g., Rayleigh scattering by H, He and H_2, e^- scattering, bound-to-free (b-f) and free-to-free (f-f) absorption by H, H^-, and H_2^+) are not adequate when dealing with such stars at low temperatures (1,000 K and up). When he included CIA in his computations of the Rosseland mean opacity, the magnitude of the effect of including CIA was striking at all temperatures below 3,500 K. Depending upon temperature, he found the differences in the resulting value of χ_R, when he did/did not account for CIA, to be between up to one order of magnitude (at gas density ρ of 10^{-6} amagat), up to three orders for $\rho=10^{-3}$ amagat, and up to five orders of magnitude for $\rho=0.1$ amagat.

One year later, Stahler et al. (1986) [11] confirmed Palla's findings. The conclusion was now made even stronger: at *all* densities between 10^{-8} and 0.1 amagat, and temperatures below 2,000 K, CIA was found to be the dominant source of all continuous opacity. At higher temperatures, up to 3,000 K, CIA remains dominant, but at higher densities.

At around that time L. Frommhold focused attention at the possibility of computing all H_2–H_2 and H_2–He CIA bands straight from first principles, also at high temperatures. Most of his pioneering work in this field, as well as many of his low-temperature quantum mechanical computations, have been collected in his recent book [12]. The first high-temperature quantum mechanical computations were performed in collaboration between Frommhold and one of the present authors [13–15], with application to model atmospheres in mind.

In 1991, Lenzuni et al. [16] presented a detailed analysis of zero-metallicity stars. Temperatures between 1,000 and 7,000 K, and densities from 10^{-8} to 10^3 amagat were considered. Monochromatic opacities were computed, which included collision-induced absorption of H_2–H_2 and H_2–He pairs. For the first time available models based on quantum mechanical computations, [15] for H_2–H_2 in the fundamental band, and [13,14] for all rotovibrational bands of H_2–He, were used in the analysis. The authors demonstrated how the importance of CIA depends upon *both* temperature and density. For a fixed temperature at high enough density, CIA became the major opacity source. At fixed density and low enough temperatures, more or less the same effect was achieved. The need for reliable models of the first and the second overtone band of H_2–H_2 CIA at high temperatures was strongly emphasised. A shortcoming of the authors' approach is, however, that comparisons were made for 'virtual' atmospheric layers, and not ones predicted to exist in *real* stars. Furthermore, zero-metallicity stars have never been observed, and it is not obvious from a theoretical point of view that such stars exist.

Summarising, early work focused on "the CIA effect" by studying various aspects of it; complete model atmospheres were never developed, however. Considerable effort has been put in on figuring out at which temperatures and pressures the opacities due to CIA will dominate over other opacity sources without paying strict attention to whether such conditions as specified by the authors indeed exist. Little effort has been expended on inferring whether CIA will have *any* effect on the radiated flux. However, the examples presented above had a profound effect in pointing out the importance of CIA, and served as inspiration for carrying out further studies in this direction.

Selected Work After 1993

Once the importance of CIA became apparent, applications to stellar atmospheres began to emerge. In this period, stellar research has been performed using quantum mechanical CIA data listed in [3]. Allard et al. [17,18] showed the effect CIA has on the radiated flux of certain low T_{eff} M dwarfs, M subdwarfs, brown dwarfs and Population III stars. The authors concluded that CIA has the strongest effect at lowest T_{eff} and for the most metal-poor stars ($Z/Z_\odot = 10^{-4}$), see Fig. 7 and Fig. 8 of [17]. Fig. 3 of [18] demonstrates a strong effect of metallicity on the emerging flux when CIA is accounted for.

Burrows et al. [19,20] and Saumon et al. [21,22] showed the effect CIA has on the radiated flux of zero–metallicity, low T_{eff} hypothetical brown dwarfs. The impact of CIA, was obviously powerful: in the absence of other opacity sources (i.e., at low T_{eff} and $Z/Z_\odot = 0$), it was almost *only* CIA which was of importance in radiative transfer calculations. The work shows the impressive difference between the black body flux radiated from the bottom of the atmosphere, and the emitted flux, strongly affected by all RV CIA bands. A shift in the peak intensity of the spectrum moving towards shorter wavelengths (towards "blue"), occurred when CIA was included.

The work by Tsuji and Ohnaka [23] appeared also at that time, and included the quantum mechanical (QM) CIA available at that time, see [3]. We will refer to their work near the end of this paper.

The newly emerging work did not use the "Linsky & Patch" CIA opacity models anymore. Instead, the authors used the available CIA QM models, computed by Borysow and Frommhold (for reference see [3,24]). It appeared, however, that the database was highly incomplete. It did not cover sufficiently the high temperatures necessary for modelling the deep atmospheric layers that affect the atmospheric structure. In addition, the frequency range of the QM data was limited. It covered neither the high temperature rototranslational (RT) band nor the even more crucial first and second overtone bands centred around 1.2 and 0.83 μm. Additionally, hot bands, important at higher temperatures, were also missing in the H_2–H_2 data. Since the bands are extremely broad at high temperatures, the region where the QM CIA input was missing included all wavenumbers beyond 8,000 cm^{-1}. It is difficult to pinpoint exactly which frequencies are most important when modelling stellar atmospheres of cool (i.e., low T_{eff}) stars. In the most simple picture, the radiation from a star can be approximated by a black body radiation of a temperature T_{eff}, which is characterised by the position of its maximum, $\lambda_{max}(T_{eff})$. In such a simplified picture, for stars with T_{eff} between 1,000 K and 4,000 K, the peaks would lie between 2.9 and 0.72 μm, respectively. This range includes all the fundamental and the first and second overtone bands. In reality, it appears that *all* frequencies matter up to $\sim$20,000 cm^{-1}, and also at temperatures exceeding T_{eff}, since the real spectrum is affected by radiation from many different depths (and hence temperatures) throughout the atmosphere, which makes the emergent spectrum deviate considerably from that of a black body.

Thus it is necessary to model the missing bands at high temperatures. This is done in Borysow *et al.* [1]. (see Figures 1 and 2 therein, and Fig. 9 near the end of this review).

WHY IS COLLISION INDUCED ABSORPTION IMPORTANT IN STELLAR ATMOSPHERES ?

It is by now well known that CIA plays the major role in the opacity of the outer planets (H_2-H_2, H_2-He and H_2-CH_4) [25–27], of Titan (N_2-N_2, N_2-CH_4, N_2-H_2, CH_4-CH_4, H_2-CH_4) [28], and of Venus (CO_2-CO_2) [29]. In general, CIA is important for environments composed of neutral molecules. Usually this happens in *cool* environments, which ensures the absence of free electrons and ions. However, since collision-induced dipoles are extremely weak, high gas densities are needed in order for CIA to be appreciable. This can be accomplished in the laboratory, and it happens to be well satisfied in planetary atmospheres as well. The idea of CIA being important also in stellar atmospheres was initially surprising. Even after the paper by Linsky [4] very few astrophysicists took his message seriously. Most atmospheric modellers seemed to overlook its existence at all.

However, when considering why CIA would be important in stellar atmospheres, we may point to the existence of *cool* stars (i.e., with low T_{eff}), where electrons and ions (especially H^-, an ion whose opacity is very strong) are rare, and where neutral, molecular hydrogen (and helium) are abundant. The additional requirement is, that in order for CIA to matter, the density of hydrogen molecules needs to be significant (though the importance of CIA strongly depends upon other factors like the absence of other opacity sources).

Do we find stars satisfying such conditions? It turns out that actually the majority of the stars do satisfy these conditions. Most of the stars in our Galaxy are cool and have considerably high gravity (i.e., are cool dwarfs). The most abundant candidates for CIA to be significant are M dwarfs, cool white dwarfs, brown dwarfs and carbon dwarfs. In addition, we shall see that low metallicity also favours CIA. Reasons for it are two-fold. First, in the absence of molecules or atoms other than H_2 and He, there is only a very limited number of "electron-donors". The low metallicity produces a feedback for CIA: wherever CIA is most likely to be the only opacity source, a very low absorption in general results, making the atmosphere more compact. This fact in turn favours CIA, which is dependent on the square of the density. Not surprisingly, early works, which studied zero-metallicity stars only, concluded CIA is dominant. At low T_{eff} there is nothing else to be expected, since only an increased temperature may give rise to the existence of H^- and other ions, which compete with CIA as opacity sources. One finds most low metallicity stars in the halo of our Galaxy and in globular clusters, which contain stars older than those in the disk of our Galaxy. Having stated that, we find there is good reason for a serious quantitative investigation of how CIA affects the spectra of real stars as functions of T_{eff}, $\log(g)$, and metallicity (Z). The task is important, as modifications in the radiated flux of a star will also affect the spectral classification of stars in the HR diagram and hence our understanding of how they evolve.

In order to estimate which kind of stars could be affected by CIA, we start with a known star where we expect CIA to be of importance, and then vary the fundamental stellar parameters [2] toward less and less favourable conditions for CIA in order to quantify the limiting values where CIA is relevant. The opacity from CIA is proportional to ρ^2 (as opposed to other opacity sources which are proportional to ρ), and we therefore search for a dwarf star[2] as a starting point of our investigation. Another condition is that the star has to be cool enough that hydrogen exists primarily in molecular form. This leaves us in the lower right region of the HR diagram[2]. Furthermore, low metallicity[2] (metallicity is defined as the mass density of all elements heavier than helium divided by the mass density of all elements in the gas) favours the effect of CIA, because the competing opacity sources in the stellar atmosphere are (1) polar molecules including atoms of carbon, nitrogen and oxygen and other "metals", and (2) continuum opacities from negative ions (dominated by the H^- ion) where the free electron to the ion mainly comes from Ca, K, Na, and a few other elements with low ionisation energy. To understand which are the most extreme stars in this parameter space, but which are still abundant and normal in our Galaxy, we need to consider the following:

When the Universe was created in the Big Bang, basically only hydrogen and helium was formed. None of the elements astronomers call "metals" (i.e., all elements heavier than helium) were created in the Big Bang, and the first objects in the Universe must therefore have had zero metallicity. However, these objects are not likely to have been stars in our common sense, but rather a few enormous gas contractions of maybe a million times the mass of our Sun. Regular stars are not expected to have formed before these first mega objects had produced small amounts of metals and through explosions had spread them all over the collapsing Galaxy. The oldest stars we observe today are found in the globular clusters which are spherical systems of typically 100,000 stars each, and which, together with a more diffuse halo of the stars, are distributed in a large sphere around our present-day Galaxy. The metallicities of the most metal poor stars known are around $10^{-3} - 10^{-4}$ times the solar value, but clearly larger than zero (models of stellar atmospheres with even this small amount of metals are qualitatively different from stellar models with metallicity strictly zero). Stars of approximately 0.8 times the mass of the Sun (0.8 $M_\odot$) take a time equal to the present age of the Universe ($\approx$ 15 billion years) to live through their life as main sequence[2] stars (=dwarfs, corresponding to $\log(g) = 5$), after which time they become giant stars of $T_{eff} \approx 2,800$ K. Higher mass stars become warmer giants. Generally, the coolest giants we find in the universe have effective temperatures $T_{eff} \approx 2,800$ K. We have therefore now defined the lowest right corner in the HR diagram from where it will be meaningful to explore variations of fundamental stellar parameters one by one, in order to investigate where CIA is of importance for the stellar atmospheric structure and

---

[2] See glossary of astrophysical terms at the end of this paper.

spectrum. This corner must have $T_{\text{eff}} \approx 2{,}800\,\text{K}$, $\log(g) \approx 5$, and $Z/Z_\odot \approx 10^{-4}$.

Due to the large distance of the globular clusters and the stars in the Galactic halo, no detailed spectra exist yet of dwarf stars in these systems, but with the present generation of new telescopes under installation it will very soon be possible to obtain such spectra, and we predict that the infrared spectrum of dwarf stars in the halo and in the globular clusters will be completely dominated by CIA (see Figs. 2, 3, 4).

It should be mentioned that even though, in general, we will not find low metallicity giant stars with $T_{\text{eff}} < 2{,}800\,\text{K}$ (such that the effect of variation of $\log(g)$ in our models can have a realistic meaning and be compared with real stars for these low temperatures), there will be dwarf stars with masses below $0.8 M_\odot$ at cooler temperatures than $T_{\text{eff}} = 2{,}800\,\text{K}$, and there will be white dwarfs with $\log(g)$ larger than 5 (although maybe not with T_{eff} as low as $2{,}800\,\text{K}$). These specific regions of the HR diagram have been studied very recently by Tsuji [23] for $M < 0.8 M_\odot$ and by Saumon&Jacobson [31] for $\log(g) = 8$ (but $Z = 0$), based on our CIA opacities from [3] and from [1], respectively. These studies therefore supplement our recent investigation very well and will be described in detail below.

OUR RESULTS

Before we could begin a thorough testing, we needed to supplement missing CIA data. We have done so, and in a semi-empirical way (for details, see [1]). We collected a set of high temperature CIA spectra of H_2–H_2 and H_2–He at temperatures from 1,000 K to 7,000 K (see Figs 1 and 2 of [1] and Fig. 9 near the end of the present paper). We included rototranslational (RT), fundamental, and first and second overtone bands. We also accounted for hot bands present at temperatures above 5,000 K. We have based our CIA database on existing work, but many intensities were missing and needed to be extended semi-empirically. Having CIA spectra covering complete temperature and wavenumber ranges $(0-20{,}000\,\text{cm}^{-1})$, we were the first in a position to test the impact of CIA on all kinds of stars.

The work we present here is done for all oxygen-rich stars with $T_{\text{eff}} \geq 2{,}800\,\text{K}$, $\log(g) \leq 5$, and $Z/Z_\odot \geq 10^{-4}$. Besides CIA opacities, we include molecular (b-b) opacities due to H_2O, TiO, CO, CH, CN from the SCAN molecular data base (Jørgensen 1997, [32]) and due to SiO from Langhoff & Bauschlicher (1993 [33]). We assume no presence of free atoms because they are not important at the temperatures of interest here. As continuous sources of opacity we included b–f and f–f absorption by H^-, f–f absorption by H, H_2, H_2^-, H_2^+, He, He^-, Rayleigh scattering by H_2, H, He, and Thompson scattering by e^-. We attempt to present what impact CIA has on both radiated flux and atmospheric structure. We also show the opacities of various atmospheric layers.

In Fig. 2 we show the emitted (normalised) flux of stars with various values of $\log(g)$; i.e., $\log(g) = 1$ (giant), 2, and 5 (dwarf). T_{eff} and $Z/Z_\odot$ are fixed in this figure at 2,800 K and 10^{-3}, respectively. We clearly observe a strong influence of

CIA on the emitted spectrum. The effect of CIA is to move the peak of the radiated intensity towards shorter wavelengths, and to smooth out the infrared part of the spectra considerably, making them almost featureless. Energy radiated most in the far infrared (longer λ) is all absorbed, and, because the total flux must be preserved (for a given T_{eff}), we note the excess of energy at higher frequencies. We can also observe the limiting cases for gravity. For the case presented, $\log(g)=1$ shows no effect of CIA at all (the gas in the atmosphere is too rarefied), whereas for $\log(g)=5$ the effect is strong.

Fig. 3 presents the effect of T_{eff} on the radiated flux. Here we fixed $\log(g)$ and $Z/Z_\odot$ to be equal to 5 and 10^{-3}, respectively. Panels show the emitted spectra for $T_{\text{eff}} = 3{,}800$, $3{,}400$ and $2{,}800$ K. Again, whereas the right panel (corresponding to the lowest T_{eff}) shows the most dramatic effect, the left one is barely affected by the inclusion of CIA. T_{eff} close to 3,800 K appears to be an upper limit above which CIA needs not be considered in order to reproduce an accurate emitted spectrum of the star at the low spectral resolution shown. When we compare Figs 2 and 3, we clearly observe that the effect CIA has on the observed flux is quantitatively similar. Where it is most important, the effect of CIA is to smooth out the infrared part of the spectra and to shift the emission peak towards shorter wavelengths.

Fig. 4 is the last figure of the series (completing the "grid") and shows the $Z/Z_\odot$ dependence. This time, we keep $T_{\text{eff}} = 2{,}800$ K, and $\log(g) = 5$, the two most favourable values where CIA is significant. We vary $Z/Z_\odot$ from 0.0001 to 0.1. Whereas for $Z/Z_\odot = 10^{-3}$ and $Z/Z_\odot = 10^{-4}$ the effect due to CIA is pronounced, it decreases with increasing Z, until it is unimportant at $Z/Z_\odot$ equal to 0.1.

It is actually very interesting to study various opacities (we select "continuum" opacities (all but CIA), molecular b–b opacities, and CIA opacities). We plot the opacities for changing $Z/Z_\odot$ ($T_{\text{eff}} = 2{,}800$ K and $\log(g)=5$) for two important atmo-

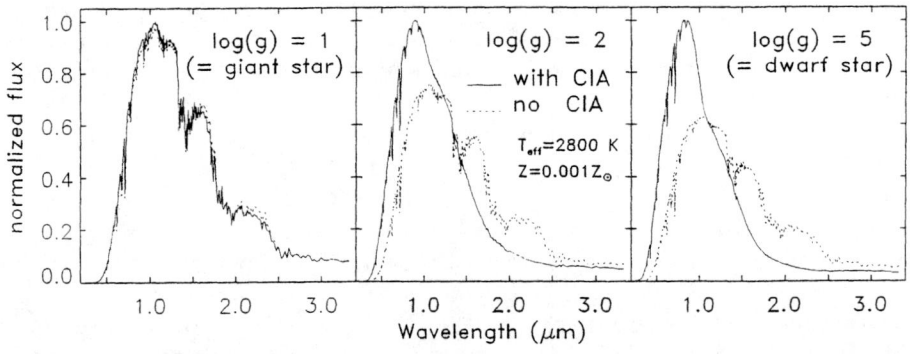

FIGURE 2. The effect of varying gravity g for models with $T_{\text{eff}} = 2{,}800$ K and $Z/Z_\odot = 10^{-3}$. Compared to models with no CIA (dotted curve), the emerging flux from models with CIA (full line) is more suppressed and smooth at infrared wavelengths beyond $\sim 1.5\mu$m, and with a more narrow and pronounced flux peak in the 1μm region.

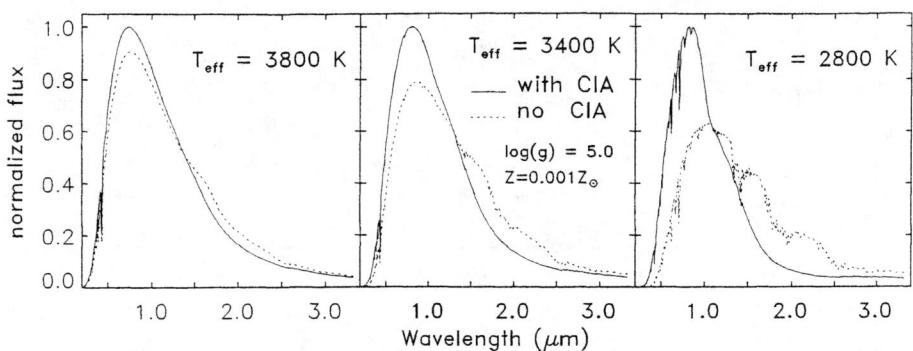

FIGURE 3. The effect of changing the effective temperature, T_{eff}. The other two fundamental stellar parameters are kept fixed at $\log(g) = 5.0$ and $Z/Z_\odot = 10^{-3}$, respectively, in this figure.

spheric "layers" between which most of the continuum is formed (they correspond roughly to middle layers with $\tau_R = 0.01$ and $\tau_R = 1.0$). We have made similar plots for various values of g and T_{eff} (refer to [1]) but we find the one with variable $Z/Z_\odot$ most educational and informative. We present our opacities in Fig. 5. We can see that CIA (dotted curve) seriously competes with the continuum sources (dashed curve) even at the highest $Z/Z_\odot$ considered. We can also see that whereas the molecular opacities (solid curve) are very strong for $Z/Z_\odot = 0.1$ and 0.01, they become less intense at $Z/Z_\odot = 0.001$. This relationship holds particularly at $\tau_R = 0.01$, i.e. high in the atmosphere. In deeper layers ($\tau_R = 1.0$), densities are higher. At the same time, the temperature becomes higher. There is, therefore, competition be-

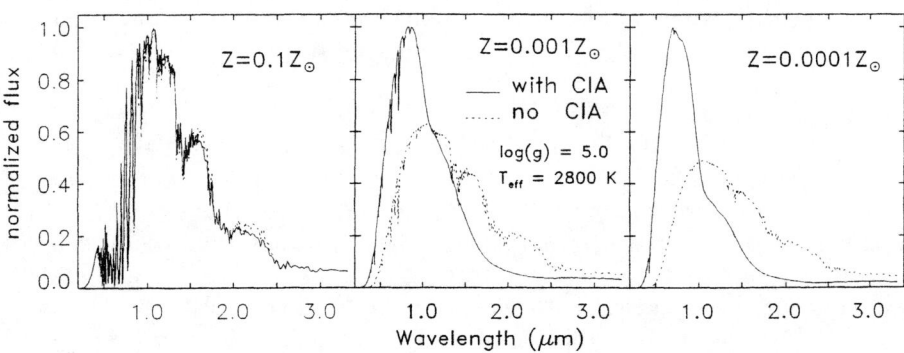

FIGURE 4. The effect of changing the metallicity, $Z/Z_\odot$. T_{eff}=2,800 K and $\log(g)$=5 (cgs units).

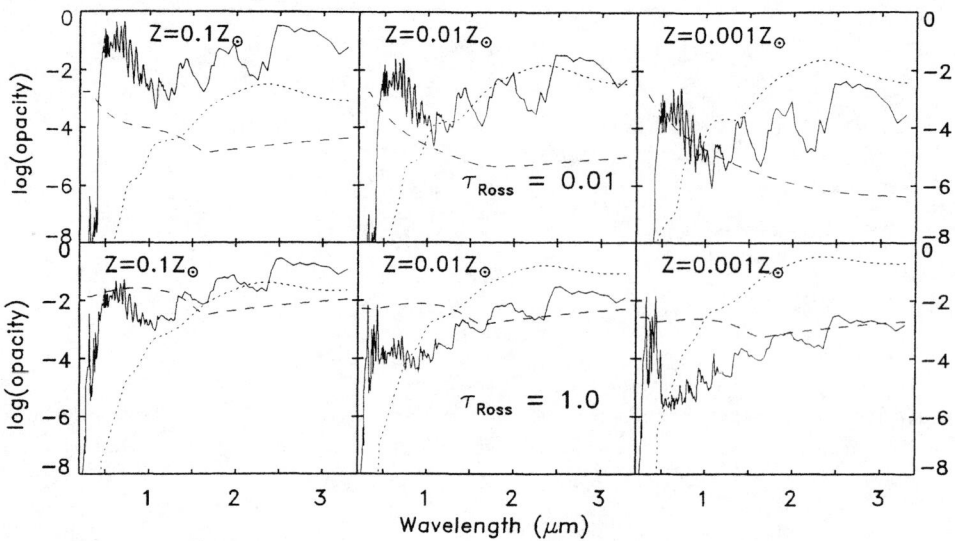

FIGURE 5. Opacities (continuum, molecular and CIA) at two atmospheric layers, with τ_R=0.01 and 1.0.
$Z/Z_\odot$ varies from 0.1 to 0.001. T_{eff} =2,800 K and $\log(g)$=5. Full drawn lines correspond to molecular b-b opacity, dashed lines correspond to continuum (all but CIA) opacity, and dotted lines correspond to the CIA opacity.

tween the increasing amount of CIA, which is proportional to ρ^2, and an increasing number of electrons, heavily contributing to the continuum opacity. In this more dense environment CIA dominates over other opacities even for $Z/Z_\odot$ as great as 0.01.

An alternative way to show the effect of CIA on the emitted spectra is presented in Fig. 6. In this figure the *innermost* curve shows the flux resulting from a self-consistent model computation, which accounts for all opacity sources, including CIA (solid line). Now, *keeping* the atmospheric structure of that specific model, we can study how much each opacity source contributes to the final spectrum. The outermost curve shows the flux resulting from continuum alone (dashed line), the next line results from accounting for continuum and molecular lines (solid "wiggly" curve). Therefore, the difference between the outermost and the middle curves shows, explicitly, the effect of "molecules only". The next line (dots) shows the effect of CIA + continuum on the radiated flux. Therefore, we can make the statement that the difference between the dashed and the dotted lines shows the effect of "CIA only", or the "missing flux due to the CIA". It is apparent that the effect of the CIA is to smooth out the infrared part of the spectra. It is also obvious that when accounting for CIA, the details of the structure, due to various bound-bound molecular transitions, may disappear entirely (depending on T_{eff}, $\log(g)$, and $Z/Z_\odot$). The effect of CIA is clearly to reduce the spectrum at longer wavelengths,

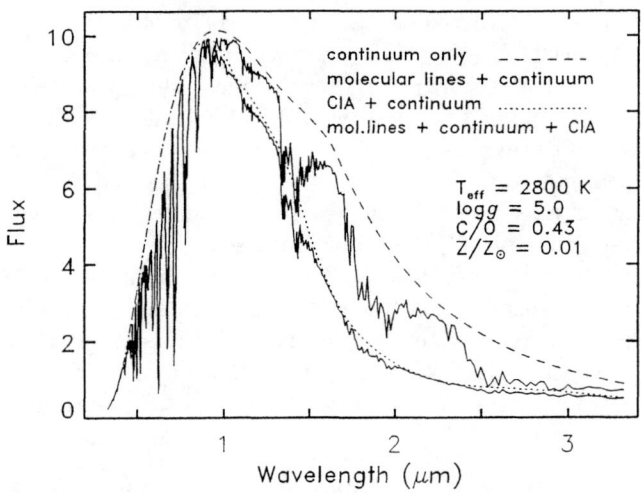

FIGURE 6. The contribution of various opacity sources to the total spectrum of a star with $T_{\text{eff}} = 2{,}800$ K, $\log(g) = 5$, and $Z/Z_\odot = 0.01$. Explanations given in the text.

and to push the peak of the flux towards shorter λ. Of course, the integrated flux is always preserved. It is essential to understand that we have shown the effect of "continuum only" for a model with all opacity sources taken into account. We must be aware that if we took continuum as the only opacity source, we would get a completely *different* atmospheric structure (compare Figures 2, 3 and 4 which show different fluxes resulting from assuming different opacity sources.)

It is perhaps interesting to compare the densities of the stellar atmospheres where our results show CIA to be significant. For the "most dense atmospheric environments" considered in our work, i.e. $\log(g) = 5$, $T_{\text{eff}} = 2{,}800$ K and $Z/Z_\odot = 10^{-4}$ (high gravity, low T_{eff}, and low $Z/Z_\odot$), we find that the density of molecular hydrogen varies from 0.01 amagat at the surface to 35 amagat at the bottom of the atmosphere. For a star with lesser gravity ($\log(g) = 2$) the density varies from 0.001 amagat (surface) to only 4 amagat at the bottom. It is also quite interesting to observe how the stellar atmosphere is stratified, i.e. what kind of temperatures, densities and pressures it acquires for a given T_{eff}. As an example, we take again one of our densest stars, an M dwarf with $T_{\text{eff}} = 2{,}800$ K and $Z/Z_\odot = 0.001$. Near to the surface of this star (where $\tau_R = 10^{-4}$), T=1,600 K, ρ=0.01 amagat, and pressure is equal to 0.1 atmospheres. At the middle layer ($\tau_R = 1$) the temperature is equal to 3,200 K (close to T_{eff}), ρ=5 amagat and the pressure is 50 atm. At the bottom of the atmosphere ($\tau_R = 100$), T=4,100 K, ρ=10 amagat, and the pressure is 200 atm.

It may be noted that typical stellar densities are *low*, lower than those common in the planetary atmospheres, and typically lower than those used in laboratory measurements of CIA. It should also be noted that our results are significantly different from those of other authors. Early works indicated that CIA may be important at densities as low as 10^{-6} amagat, depending on temperature. The authors of the early works, when studying the effect of CIA, did not work with real model atmospheres. Instead, they often studied χ_R alone, which is very sensitive to *low* opacities, and therefore prone to large numerical errors in case of uncertain values of assumed opacities. In addition, they often assumed "virtual" atmospheric layers that may, or may not exist in real stars. Our results depend on internally consistent stellar atmospheric structure models and new CIA data. In addition, most of the previous work assumed $Z/Z_\odot=0$. This condition, naturally, allows for higher densities, because in such case (at low T_{eff}) CIA becomes the only (and extremely weak!) opacity source. Summarising, it is apparent that for low metallicity, low temperature dwarf stars (i.e., with high $\log(g)$), CIA must be included in their model atmospheres. To be more precise, we formulate our findings in the following way:

1. There exist *real* stars with T_{eff} up to $\approx$ 4,000 K where CIA is still important (but for $\log(g) \geq 5$ and $Z/Z_\odot \leq 10^{-3}$).

2. There exist *real* stars with $\log(g)$ as low as 2 (but with $Z/Z_\odot \leq 10^{-3}$ and $T_{eff} \leq 2800$ K) where CIA is still important.

3. There exist *real* stars even with $Z = Z_\odot$, but with $T_{eff} \leq 1,500$ K and $\log(g)=5$, where CIA is still important[3].

Qualitatively, we confirm the statements made by previous authors about the importance of the CIA under certain physical conditions. We have also determined quantitative limits for the three fundamental stellar parameters ($\log(g)$, T_{eff}, $Z/Z_\odot$) to be satisfied by the real stars, above or below which CIA is significant.

OTHER RECENT WORK

In this section we refer to the two other recent works mentioned earlier which complement our work and extend the limits of our grid.

Tsuji & Ohnaka [23] studied very low mass M dwarf stars, subdwarfs and brown dwarfs. They calculated atmospheric models including CIA from [13-15,30] as opacity sources. They demonstrated clearly the importance of CIA especially for brown dwarfs. In their figures (Figs. 3, 4, 5; our Fig. 7 corresponds to Fig. 5 of [23]) the authors present the total flux, the black body corresponding to $T_{eff} = $ 1,000 K and 1,500 K, and separately, the opacity due to continuum only. Their models do account for molecules, i.e. cover non-zero metallicity stars. One can observe several interesting features: For the model at $T_{eff}=$ 1,000 K, $\log(g)=5$ (substellar brown dwarf), the authors consider two different metallicities, $Z/Z_\odot=1.0$ and 0.01. For *both* metallicities, when the opacity due to the continuum (which

---

[3] We refer our readers to Ref. [23], and to description of that work later in this paper.

does include CIA) is investigated alone, one can observe CIA's importance (spectral features specifically due to CIA are marked on the figures). Whereas it is natural that the overall flux is more affected at $Z/Z_\odot=0.01$, and the dips due to various RV CIA bands are clearly visible also in the total flux, the case with $Z/Z_\odot=1$ is very interesting. In our grid, which did not include effective temperatures as low as 1,000 K, we concluded that a metallicity of the order of $Z/Z_\odot = 0.1$ is the limiting case for stars of $T_{eff} \geq 2,800$ K and $\log(g) \leq 5$. Tsuji & Ohnaka's work clearly shows that CIA, for very low temperature stars, is also important even at *solar* metallicities! It is also worth noticing that while CIA is important, its spectral features are completely obscured by molecular spectra. In other words, while one definitely needs to include CIA in the list of opacity sources, one would not be able to tell, a priori, that CIA is present (neither from the model computations, nor from the observed spectra).

The other work which we'd like to mention here, extends our work to higher gravities. Even though we could have expected that at increasing $\log(g)$, CIA will gain in importance, we would like to present the quantification of this statement by the very recent results by Saumon and Jacobson [31]. These authors consider atmospheres of very cool (low T_{eff}) white dwarf stars (i.e., $\log(g)=8$) with pure hydrogen composition (i.e., $Z/Z_\odot=0$). White dwarfs are remnants of stars which have used up all their nuclear energy resources. They are the hot ashes after the nuclear burning which made the star shine during its lifetime. Usually, the

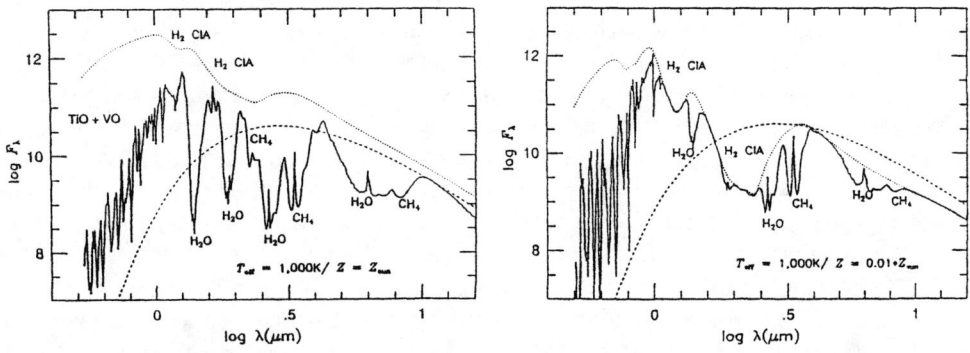

FIGURE 7. Predicted spectral flux distribution, for models with $Z/Z_\odot =1.0$ at temperature 1,000 K (substellar brown dwarf). Metallicities shown: $Z/Z_\odot =1$, and $Z/Z_\odot =0.01$, gravity $\log(g) =5$.
The solid, dotted and dashed lines represent the total flux, continuum, and black body radiation, respectively. Major opacity sources are indicated. Reproduced from Ref. [23] with kind permission from the authors.

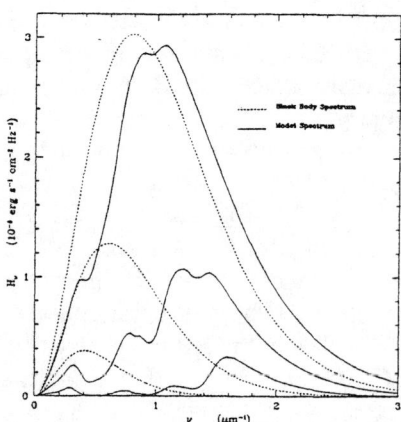

FIGURE 8. Synthetic spectra of very cool white dwarf stars with atmospheres composed of pure H_2 (i.e., $Z/Z_\odot = 0$), $\log(g)=8$, and T_{eff} vary (from top to the bottom), from 4,000 K, through 3,000 K to 2,000 K. Dotted lines are black body spectra, while full lines represent the emitted flux when CIA is included. From [31], reproduced prior to publication, with kind permission from the authors.

material a white dwarf is made of is very compact, degenerate, oxygen or helium. Slowly this stellar remnant will cool down while it radiates its energy into space at a decreasing rate throughout billions of years. The cooling rate of a degenerate object is a fairly simple and basic computation, which can be performed with very small uncertainty. It is therefore possible to relate the effective temperature of a white dwarf to its age with quite high accuracy. The coolest white dwarfs are the oldest white dwarfs, and basically all stars end their lives as white dwarfs. If one could identify the coolest white dwarfs in the sky and estimate their effective temperatures one would immediately have a very accurate estimate of the age of our Galaxy. However, it is only possible to relate the observed quantities (colors or spectra) of a star to its effective temperature by use of a model atmosphere. The authors [31] demonstrated how this transformation is crucially dependent on the CIA input data, for the coolest white dwarfs ($Z/Z_\odot=0$), as is shown in our Fig. 8.

The authors assume $\log(g)= 8$, and T_{eff} between 2,000 and 4,000 K. At temperatures below 3,000 K molecular H_2 is abundant in their atmospheres. Since in the $Z/Z_\odot=0$ case CIA is almost the only opacity source at very low temperatures, the authors compare the modelled spectrum with the black body spectrum (dotted line) which would be radiated in the absence of the atmosphere (i.e. absorbers). One can clearly see the impact of CIA opacity on the emitted flux (solid line). Whereas at $T_{eff}=4,000$ K the effect is mainly limited to shifting the peak of the emitted spectrum towards 'blue', at temperatures $\leq 3,000$ K the effect is so pronounced that it changes the shape of the flux entirely (see Fig. 8). The work is similar to [21,22], which concerned cool, zero metallicity brown dwarfs ($\log(g)=5$). The assumption

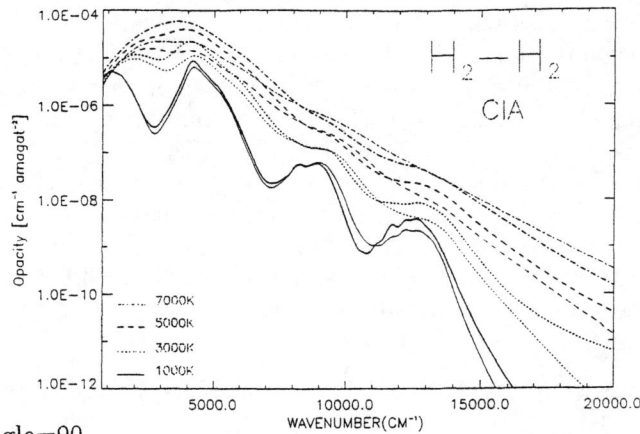

FIGURE 9. Comparison between our high temperature CIA data for H_2–H_2, at temperatures between 1,000 K and 7,000 K. Thick lines: new (unpublished) data. Corresponding thin lines: same as used in [1].

of $Z/Z_\odot=0$ may, however, in the present work have more resemblance to the stars investigated in that work than the ones investigated in their earlier works. This is because the authors in the new work selectively search for stars with extreme low metallicities (i.e., $Z/Z_\odot$ close to zero). The oldest white dwarfs are remnants of the first stars in our Galaxy, when it was still not contaminated with "metals" produced in later generations of stars, just as the case for the globular clusters. Furthermore, this new work uses our high temperature H_2–H_2 CIA opacities [1].

FUTURE OUTLOOK

Finally, we would like to mention that recently we were able to compute new high temperature CIA of H_2–H_2, based on newly developed *ab initio* collision-induced dipoles. These quantum mechanical computations account for RV bands up to the second overtone and take into account all hot bands important at temperatures up to 7,000 K. At lower temperatures (T $\sim$ 1,000 K) the results are relatively similar to those deduced from available data [1]. At higher temperatures the integrated absorption coefficients of the new data become up to a factor 2 larger than those of [1], and monochromatic absorption coefficients around 2000–5000 cm^{-1} (where the difference is biggest) are up to factor of 3 larger than the old data.

The real effects on stellar atmospheric structures and stellar spectra are therefore expected to be somewhat larger than described in this review. Detailed information on how we obtained the dipoles, and accompanying tables, can be obtained from www site, at http://www.astro.ku.dk/~aborysow/. The work is a joint effort of C. Zheng, Y. Fu and A. Borysow (1994–1998), and is part of Zheng's and Fu's Ph. D. Thesis. The new dipoles have been thoroughly tested against newly performed low temperature laboratory measurements of the second hydrogen overtone, in collaboration with C. Brodbeck, J. P. Bouanich, and Nguyen-van-Thanh (1998, J. Chem. Phys., submitted). In Fig. 9 we present the comparison between our "old" data [1] and our new (1998, to be published) ones. In our future work we will use these newly obtained data. The CIA H_2-H_2 and H_2-He absorption coefficients used for producing the atmospheric structures and the spectra presented in Fig. 2−6 and in [1] and [2] can be obtained by anonymous ftp to stella.nbi.dk (with your email as password, and cd to pub/scan), where also other opacity data for stellar astrophysics are available and where our new CIA data will be obtainable too after being prepared for public access.

SMALL GLOSSARY OF ASTROPHYSICAL TERMS

1. *The stellar atmosphere* is the outer region of the star where the spectrum is formed. Iterative numerical solution of a set of coupled differential equations allows us to compute the structure of the stellar atmosphere when the fundamental stellar parameters (see below) are specified, and the absorption coefficients of the relevant constituents in the atmosphere are known at the relevant pressures and temperatures. The result of this computation is called *the stellar model atmosphere*, and it gives us the values of all relevant physical quantities in all layers of the atmosphere: for example the temperature, the gas pressure, the partial pressures of the individual molecules, the total opacity, etc. The atmosphere of a dwarf star (see below) is typically a few hundred km from bottom to top, whereas the atmosphere of a giant can be hundreds of million km thick (and its gas will therefore have very low density). Peculiar types of stars can have atmospheres that are only few meters or even mm thick.

2. *The three fundamental stellar parameters* which describe the stellar atmosphere (of the kinds we are interested in here), are the effective temperature, the gravity, and the chemical composition. *The effective temperature* (T_{eff}) of a star is the temperature a black body would have in order to emit the same integrated flux per cm^2 per second as emitted by the star. Notice, however, that the star can have (and in general will have) a flux distribution which is completely different from a black-body flux distribution. *The gravity, g*, at the surface of the star is usually expressed in cm/s^2 (usually astronomers refer to $\log(g)$, rather than g). Most stars (those that burn hydrogen to helium in their center) have $\log(g)$ around 5 (the Earth has $\log(g)$=3). Giant stars (like many of those we see with the naked eye in the sky) typically have $\log(g)$ around zero, and very compact stellar remnants which we see as white dwarfs have $\log(g)$ around 8. Finally, *the "chemical" composition* actually means the relative abundance of the elements (and therefore has absolutely nothing to do with chemistry). Since it is usually possible to know the relative abundance of only a handful of the elements (even for those stars for which we have very

good spectra) it is assumed that the relative abundances of all elements heavier than helium (usually called the "metals") are the same as in the Sun, and that stars therefore only differ by their total amount of these elements relative to hydrogen and helium (and that the number of helium atoms per cm^3 is one tenth that of hydrogen atoms, as in the Sun). In the Sun the mass fraction of the "metals" is 2% of the total mass of hydrogen, helium and "metals". We refer to this mass fraction as the solar *metallicity*, $Z_\odot$, and say that a star has, for example, a metallicity $Z/Z_\odot = 10^{-4}$, meaning that the mass fraction of the "metals" is 10^4 times smaller than in the Sun.

3. *The Hertzsprung-Russell (HR) diagram* is the most common way of showing the classification of stars. The x-axis of the diagram has T_{eff} increasing from right to left (for historical reasons), and the y-axis shows stellar luminosity (usually the logarithm of) increasing upwards. The bulk of all stars fall along the so called *main sequence* in the HR diagram, a curve stretching from the upper left to the lower right and representing stars burning hydrogen to helium in their center, with the high-mass stars in the upper left of the sequence, and stellar mass decreasing toward the lower right. *Main sequence stars* are also called *dwarf stars*, as opposed to the fewer stars more luminous that the main sequence stars, which are called *giant stars*. The sequence of dwarf stars of low metallicity fall below the corresponding sequence of solar metallicity stars in the HR diagram, and these stars are therefore called *subdwarfs*. Continuing beyond the lower right end of the main sequence is the sequence of *sub-stellar objects* or *brown dwarfs*, which are gas-objects that are too small for possible nuclear burning to stop their contraction.

4. The Rosseland mean opacity, χ_R, is defined as follows:

$$1/\chi_R \sim \left(\int_0^\infty 1/\chi_\nu \times \frac{\partial B_\nu(T) d\nu}{\partial T} \right) \bigg/ \left(\int_0^\infty \frac{\partial B_\nu(T)}{\partial T} d\nu \right).$$

where χ_ν is the total absorption coefficient at wavenumber ν. $B_\nu(T)$ is the black body (Planck) flux distribution at temperature T.

5. The optical depth, $\tau = \int_{-\infty}^h \chi \, dh$, is equivalent to the quantity which in physics or chemistry is usually known as $\alpha \, l$ in Beer's law. Note, however, that while Beer's law is determined under the assumption that the temperature and pressure are constant over the cell-length l, both α ($\sim \chi$) and l ($\sim \int_{-\infty}^h dh$) are functions of h in the definition of the optical depth. In particular, the meaning of the Rosseland optical depth, $\tau_R = \int_{-\infty}^h \chi_R \, dh$, is to measure, approximately, the probability $P(\tau_R)$ that an "average photon" emitted at a certain atmospheric layer h will escape the star. For example, for $\tau_R=1$ (or x), $P(\tau_R)$ equals $1/e$ (or $\exp(-x)$). The atmosphere does not literally have a surface and a bottom, but boundary conditions in the computations of the model atmosphere are typically $\tau_R \approx 10^{-6}$ and $\tau_R \approx 100$, and one would refer

to atmospheric regions with τ_R close to these values as the "surface" and the "bottom", respectively. Where $\tau_R \sim 1$, the physical temperature in the gas is usually approximately equal to the effective temperature of the star.

6. *Globular clusters, the halo* and *the Galactic disk*: It is believed that our Galaxy formed from a rotating, spherical gas cloud which slowly collapsed. Early in the collapse, condensations in the cloud formed spherical gas systems, each containing about 10^5 stars. These systems of stars are called globular clusters, and today they consist of only very old stars (formed during the early days of our Galaxy). Since the gas in the universe had not yet been enriched in "metals" at that time, stars in globular clusters have low $Z/Z_\odot$ and CIA will therefore play a particularly large role in the atmospheres of globular cluster stars. Today the gas cloud has flattened to a disk, and new stars are only formed in the disk. The metallicity in the gas in the disk is approximately that of the Sun, and young stars therefore have solar metallicity. That is why CIA is of smaller importance in young stars, except for the very coolest and most dense ones. Around the disk is a halo of stars formed within a few billion years after the globular clusters. The metallicity of some of the halo stars can be even lower than the metallicity of stars in typical low-metallicity globular clusters.

Acknowledgements We acknowledge grant # NAG5-3689 from NASA, Astrophysics Theory Program, and support from the Danish Natural Science Research Council. We thank Dr. Robert E. Samuelson for the critical reading of the manuscript and for his valuable comments.

REFERENCES

1. Borysow, A., Jørgensen, U. G., and Zheng, C., *Astronomy and Astrophysics*. **324**, 185–195 (1997).
2. Jørgensen U. G., Borysow A., and Höfner S., in Chan, K. L., Cheng, K., S., and Singh, H. P., (eds.), *1997 Pacific Rim Conference on Stellar Astrophysics*, ASP Conference Series **138**, 157–160 (1998).
3. Borysow, A. *Molecules in the Stellar Environment*, ed. U. G. Jørgensen, Lecture Notes in Physics, Berlin: Springer–Verlag, 1994, pp.209–222.
4. Linsky, J. L., *Astrophys. J.* **156**, 989–1005 (1969).
5. Tsuji, T., *Low Luminosity stars*, New York, London, Paris: Gordon and Breach Science Publishers, 1969, pp. 457–482.
6. Patch, R. W., *J. Quant. Spectroscopy and Rad. Transfer.* **11**, 1331 – 1353 (1971).
7. Tsuji, T., *Publ. Astr. Soc. Japan* **23**, 553–565 (1971).
8. Shipman, H. L., *Astrophys. J.* **213**, 138–144 (1977).
9. Mould, J., and Liebert, L., *Astrophys. J.* **226**, L29–L33 (1978).
10. Palla, F., *Molecular Astrophysics*, D. Reidl Publishing Company, 1985, pp. 687–693.
11. Stahler, S. W., Palla, F., and Salpeter, E. E., *Astrophys. J.* **302**, 590–605 (1986).

12. Frommhold, L., *Collision-Induced Absorption in Gases*, New York: Cambridge University Press, 1994.
13. Borysow, A., Frommhold, L., and Moraldi, M., *Astrophys. J.* **336**, 495 – 503 (1989).
14. Borysow, A., and Frommhold, L., *Astrophys. J.* **341**, 549–555 (1989).
15. Borysow, A., and Frommhold, L., *Astrophys. J. Lett.* **348**, L41–L43 (1990).
16. Lenzuni, P., Chernoff, D. F., and Salpeter, E. E., *Astrophys. J. Supplement Series.* **76**, 759–801 (1991).
17. Allard, F., and Hauschildt, P. H., *Astrophys. J.* **445**, 433–450 1995).
18. Allard, F., Hauschildt, P. H., Alexander, D. R., and Starrfield, S., *Annu. Rev. Astron. Astrophys.* **35**, 137–177 (1997).
19. Burrows, A., Hubbard, W. B., Lunine, J. I., and Saumon, D., *Arizona Theoretical Astrophysics Preprint* 94-16 (1994).
20. Burrows, A., Hubbard, W. B., Saumon, D., and Lunine, J. I., *Astrophys. J.* **406**, (1993).
21. Saumon, D., Bergeron, P., Lunine, J. I., Hubbard, W. B., and Burrows, A., *Astrophys. J.* **424**, 333–344 (1994).
22. Saumon, D., Bergeron, P., Lunine, J. I., and Burrows, A., *Astrophys. J.* **424** (1994).
23. Tsuji, T., and Ohnaka, K., *Elementary Processes in Dense Plasmas*, S. Ichimaru and S. Ogata (eds), Addison Wesley, 1995, pp. 193-200.
24. Borysow, A., *Collision–induced absorption in the infrared: A data base for modeling planetary and stellar atmospheres.* A detailed report updated annually, distributed on request.
25. Conrath, B. J., Hanel, R. A., and Samuelson, R. E., *Origin and Evolution of Planetary and Satellite Atmospheres*, Tucson:University of Arizona Press, 1989, pp. 513–538.
26. Hanel, R. A., Conrath, B. J., Jennings, D. E., and Samuelson R. E., *Exploration of the Solar System by Infrared Remote Sensing*, Cambridge, New York: University Press, 1992.
27. Trafton, L., *Induced spectra in planetary atmospheres*, NATO Advanced Research Workshops, Dordrecht: Kluver, 1995, pp. 517-528.
28. Samuelson, R. E., Nath, N., and Borysow, A., *Planetary & Space Sciences.* **45/8**, 959–980 (1997).
29. Pollack, J. B., Toon, O. B., and Boese, R., *J. Geophys. Research.* **85**, 8223 – 8231 (1980).
30. Zheng, C., and Borysow, A., *Astrophys. J.* **441**, 960–965 (1995).
31. Saumon, D., and Jacobson, S. B., in preparation, (1998).
32. Jørgensen U. G., in E. F. vanDishoeck (ed.), *Molecules in Astrophysics: Probes and Processes*, (IAU Symp. 178), Kluwer, 1997, pp. 441–456.
33. Langhoff S. R., Bauschlicher C. W., *Chem. Phys. Lett.*} **211**, *305 (1993).*

LYMAN SERIES PROFILES: FROM LASER-PLASMAS TO WHITE DWARF STARS

J.F. Kielkopf

University of Louisville, Louisville, Kentucky 40292, USA

N.F. Allard

Observatoire de Paris-Meudon, France and Institut d' Astrophysique, Paris, France

Abstract.
The low energy interactions of neutral and ionized hydrogen atoms are fundamental processes which also have important applications to the diagnostics of laboratory and astrophysical plasmas. Satellites in the far wings of Lyman α and Lyman β have been identified as ultraviolet absorption features in the spectra of white dwarf and λ Bootis stars, and they are seen in the emission spectra of plasmas produced when a pulsed laser excites a target H_2 gas. The observed Lyman series profiles agree with unified line shape theory which includes variation of the dipole transition moment during the radiative collision.

I INTRODUCTION

The Lyman series of atomic hydrogen is a challenge to spectral line shape theory and experiment. It is fundamentally interesting because, while we understand the binary interactions of hydrogen atoms and ions, we are only beginning to be able to predict the spectrum of a dense gas in which radiative collisions are likely. The far ultraviolet from astronomical sources is absorbed by the Earth's atmosphere. First with the International Ultraviolet Explorer (IUE), then with the Hubble Space Telescope (HST), the Hopkins Ultraviolet Telescope (HUT) and the Orbiting Retrievable Far and Extreme Ultraviolet Spectrometers (ORFEUS), and now with the Far Ultraviolet Spectroscopic Explorer (FUSE), this window has been opened for spectroscopic investigation. Recently in the laboratory we have found that laser produced plasmas generate conditions that are similar to those in white dwarf stellar atmospheres, and permit repetitive well-diagnosed experiments in support of the theory and astronomical observation. The conditions in this source, modeled with a shock wave theory which includes dissociation and ionization of the gas, are found to be similar to the outer atmospheres of hydrogen rich white dwarf stars [1–4]. In this paper we will review and compare the recent results, and look at future directions.

II LABORATORY SPECTRA

Self-focusing of a 1.064 μm, 6 ns, 600 mJ laser causes most of its energy to be delivered suddenly to a cylindrical volume only a few μm in diameter and a few mm long. As a consequence, a shock wave propagates outward and leaves in its wake a cooling mixture of neutral H, H_2, ions and electrons. The shock front and the post-shock gas provide a source for studying radiative collisions of atomic H experimentally with time-resolved emission spectroscopy and laser-induced fluorescence. The prompt atomic emission from the plasma arises from a thin expanding shell outside of which the H_2 is not dissociated, and inside of which the H is fully ionized [5–7].

Figure 1 shows two typical direct images of the laser-produced plasmas. The laser is incident from the right, and in the upper one, recorded at the laser wavelength of 10,640 Å, the scattering due to self-focusing and beam breakup is evident. The plasma responds dynamically while the laser is on, and the beam is confined by a channel of its own making. In the lower image, recorded in the bremsstrahlung continuum at 5500 Å, the channel down the axis is visible. This is an integrated exposure, showing emission from a shock front moving initially at speeds of about 20 km/s.

A cylindrical blastwave model of the shock is shown in Fig. 2. Near the axis the high temperature results in a totally dissociated and ionized plasma with very low density. A dense ridge of neutral gas follows the shock outward. The motion of the gas and the presence of this shell are confirmed by diagnostic shadowgraph images. Excited atoms are also in a narrowly confined shell. The unexcited atomic hydrogen outside, and the ions inside, overlap the region emitting Lyman α. As Fig. 3 shows, this region is 0.04 cm off-axis 20 ns after the pulse, and about 0.01 cm thick. Excited atoms in this volume are perturbed by collisions with about 10^{20} neutral atoms/cm^3 and 10^{19} protons/cm^3. As the shell moves outward the neutral and proton perturber densities decrease. Experiments with laser produced plasmas may use temporal and spatial resolution to select densities of interest.

III LINE SHAPE THEORY

New theoretical unified theory line shape calculations have been made which include the effects of the variation of the electric dipole transition moment with internuclear separation during atom-atom and atom-ion collisions. The details of the theory are described in recent papers [3,4,8], and in two papers at this conference [9,10]. In contrast to the often made assumption that the dipole moment is independent of R, for H_2 and H_2^+ there is a strong radial dependence and $D(R)$

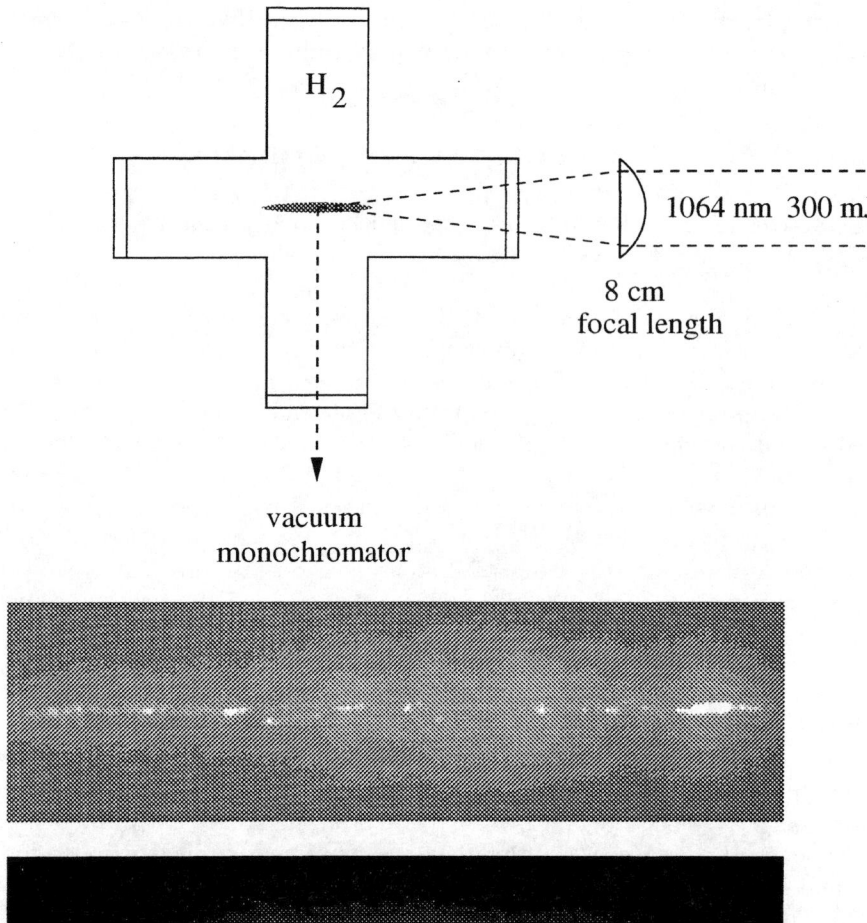

FIGURE 1. A laser is focused from the right to produce a plasma in H_2. The upper image shows scattered incident laser light for 602 Torr H_2 and the lower one shows integrated emission in the bremsstrahlung continuum around 5500 Å for 400 Torr. The images are 2.3 mm wide.

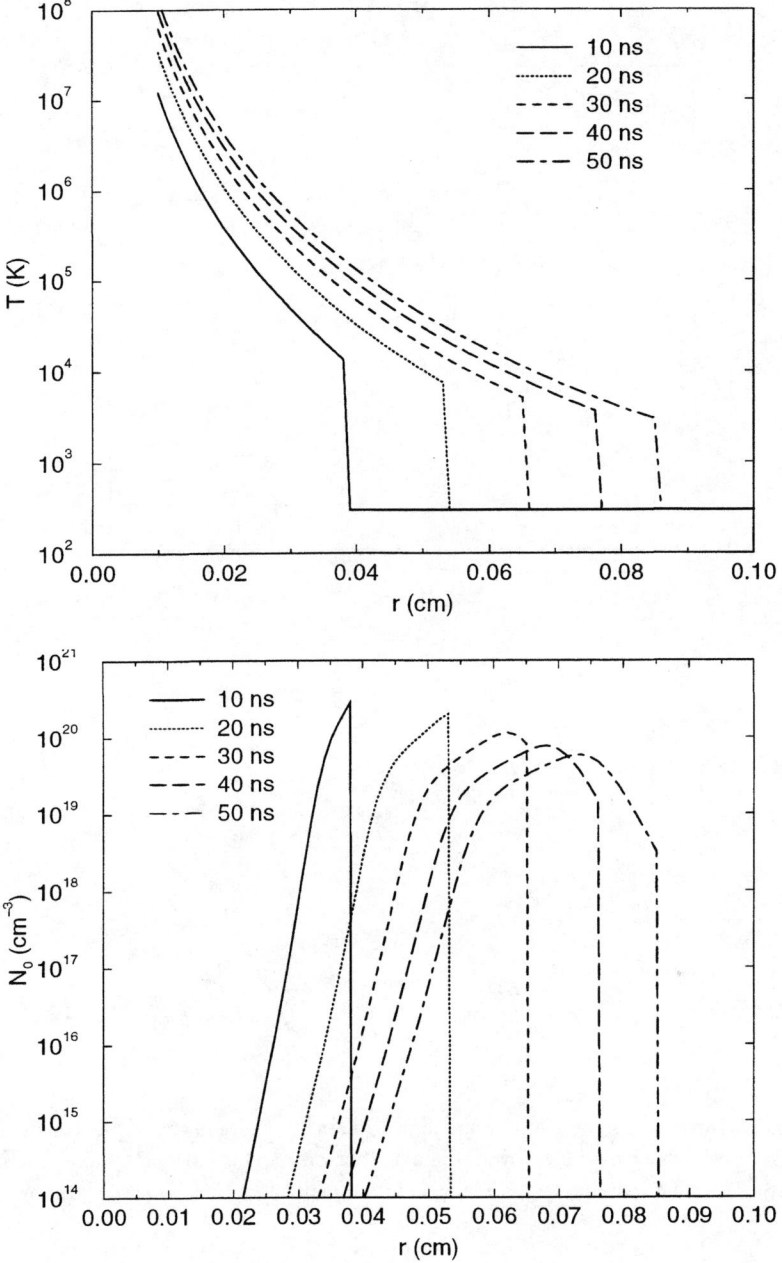

FIGURE 2. Shock models of plasma temperature and neutral atomic H density for a 600 mJ pulse into 800 Torr H_2.

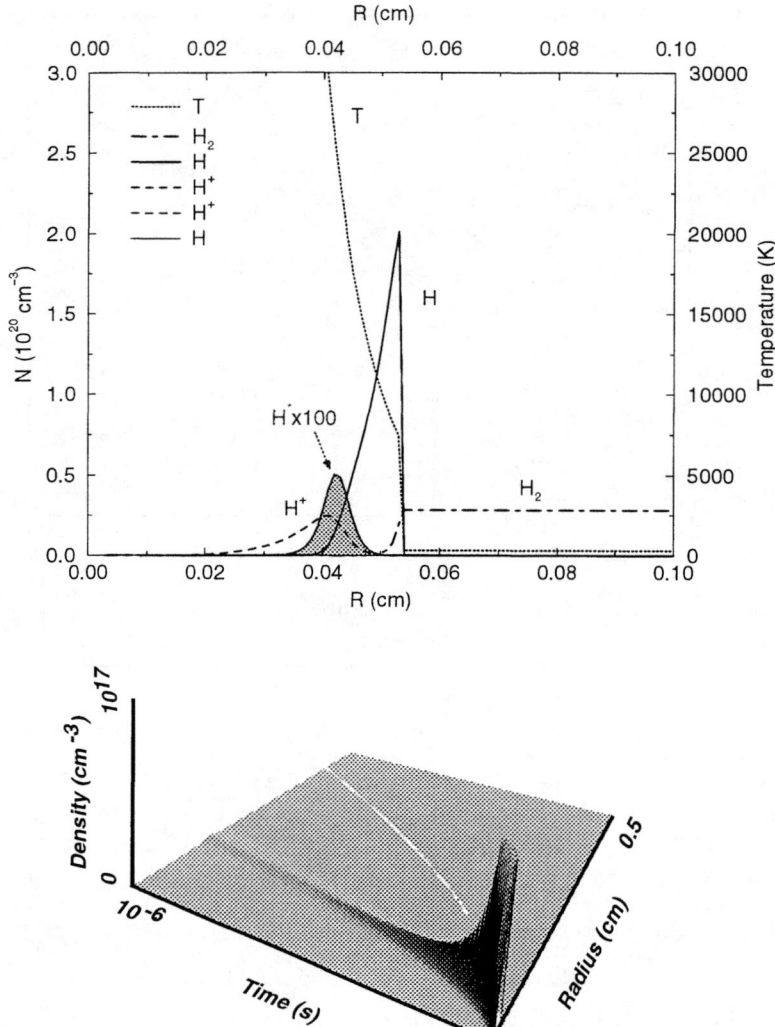

FIGURE 3. Dissociation, excitation, and ionization in the shocked gas. Atoms in the n=2 state of H are in a shell that moves away from the axis (bottom) and is 0.04 cm off axis 20 ns after the pulse (top). The shell makes selected regions of high neutral and ion densities visible.

increases significantly in the regions of the potentials where the satellites originate. Unified theory calculations shown in Fig. 4 take this into account. It results in a significant enhancement of the satellite at 1600 Å.

Lyman α from the laser-produced plasma exhibits the satellites which are predicted by the unified line shape theories. Typical experimental profiles are shown in Fig. 5. The satellites due to ions were predicted by Stewart, Peek, and Cooper [11], and those due neutrals were predicted by Sando, Doyle, and Dalgarno [12], in their early theoretical work on the Lyman α wing. This experimental confirmation of the theory adds support to the identification of these features in the Lyman α spectra of white dwarf and λ Bootis stars [9].

Figure 5 also shows two satellites far from the line center, one at 1400 Å assigned to H$^+$ perturbers, and the other at 1600 Å assigned to neutral H. When the electron density is high it contributes to the observed spectrum through a bremsstrahlung continuum that increases to longer wavelengths, as well as through electron broadening of Lyman α. The 1230 Å region is compared with a theoretical profile for $N_{H^+} = 10^{19}$ ions/cm^3 $N_H = 2 \times 10^{20}$ atoms/cm^3 in Fig. 6. Satellites due to neutrals at ≈ 1180 Å and to ions at ≈ 1230 Å appear at the observed strengths, and satellites at 1400 Å and 1600 Å appear in the far wing. The experimental data are shown with the electron-ion bremsstrahlung contribution removed.

In Fig. 7 a spectrum of the 1600 Å region is compared with theoretical models, with and without variable $D(R)$, and scaled for a neutral H density of 10^{19} atoms/cm^3. The line wing is optically thin in this region. The observed shape of the satellite is in good agreement with the variable $D(R)$ theory; the constant D theory underestimates the strength of the satellite. Both theories predict oscillatory structure between the satellite and the line, and the experiment appears to confirm that it is present. Oscillatory structure near the 1600 Å satellite was predicted by Sando, et al. [12], and recent velocity averaged calculations [4] lead us to conclude that it should be observable.

IV ASTRONOMICAL SPECTRA

The line shape theory has proven useful in diagnostics of ion and neutral densities, and thereby temperature, for stars in which the far ultraviolet continuum from atomic hydrogen is not obscured by other contributions [2,9,10]. Experimental confirmation of the theory in laboratory spectra supports the identification of these features in the Lyman α spectra of white dwarf and λ Bootis stars as seen with the International Ultraviolet Explorer (IUE) and Hubble Space Telscope (HST) [2]. Recent theoretical work on both Lyman α and Lyman β is presented in this conference [9,10], and astronomical data are now available which cover both lines. Figure 8 shows the spectra of several white dwarf stars compared with theoretical models that include unified theory calculations such as described here [13].

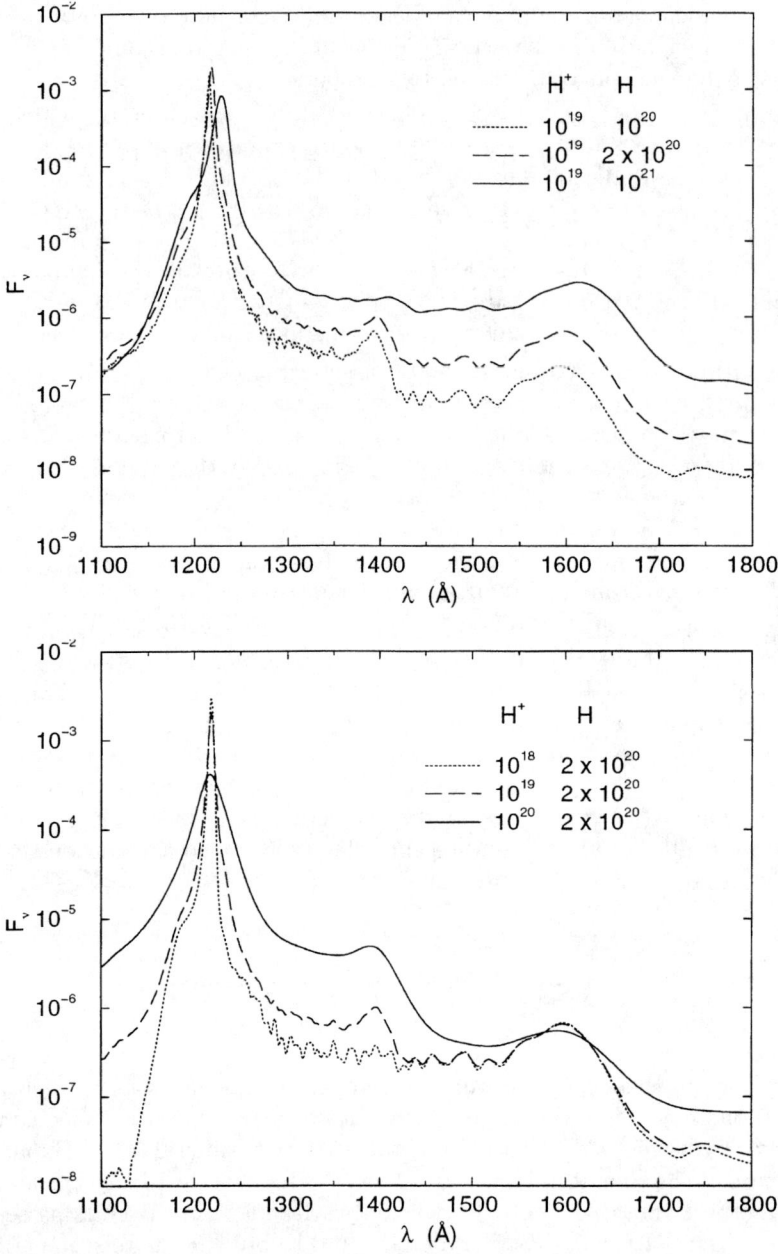

FIGURE 4. Theoretical models of Lyman α for different neutral atom and proton perturber densities at 10,000 K.

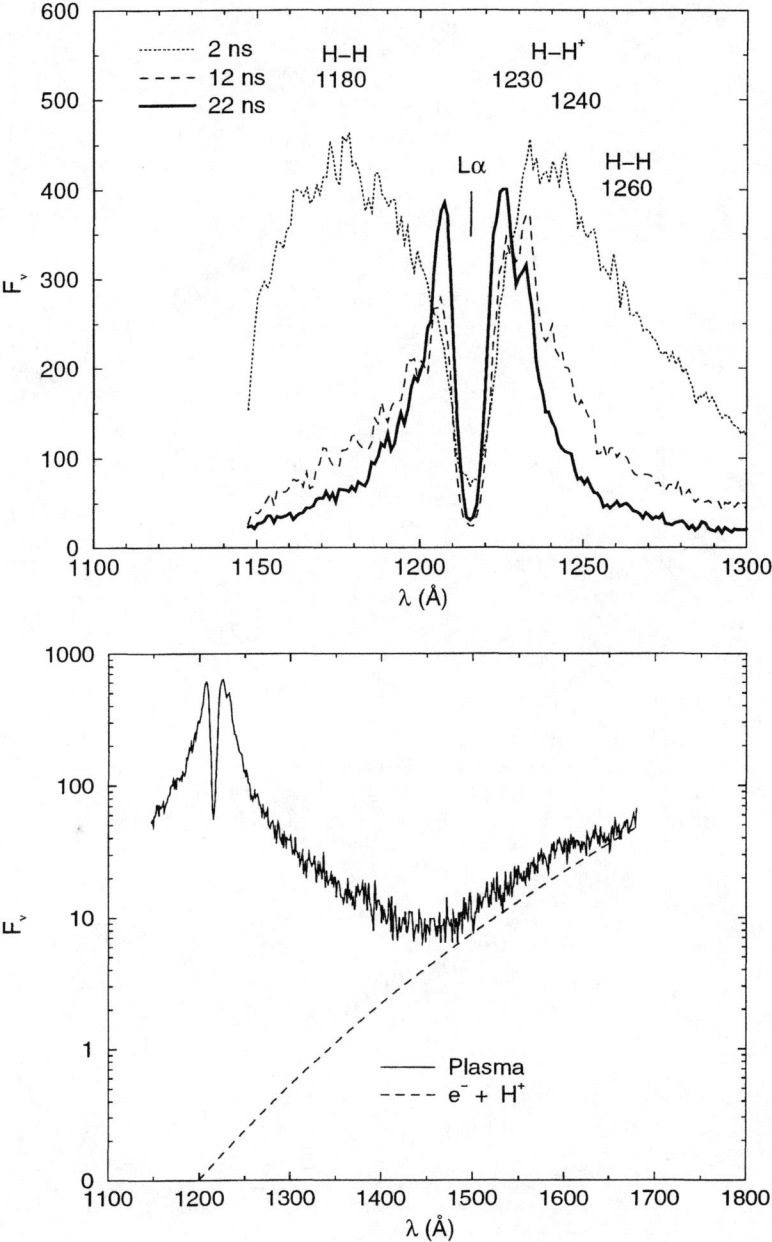

FIGURE 5. Lyman α in the laser-produced plasma. The line center (top) is self-reversed with satellites due to proton and neutral interactions in the near wing. At longer wavelengths the free-free and bound-free continuum due to electron-ion collisions boosts the observed line wing.

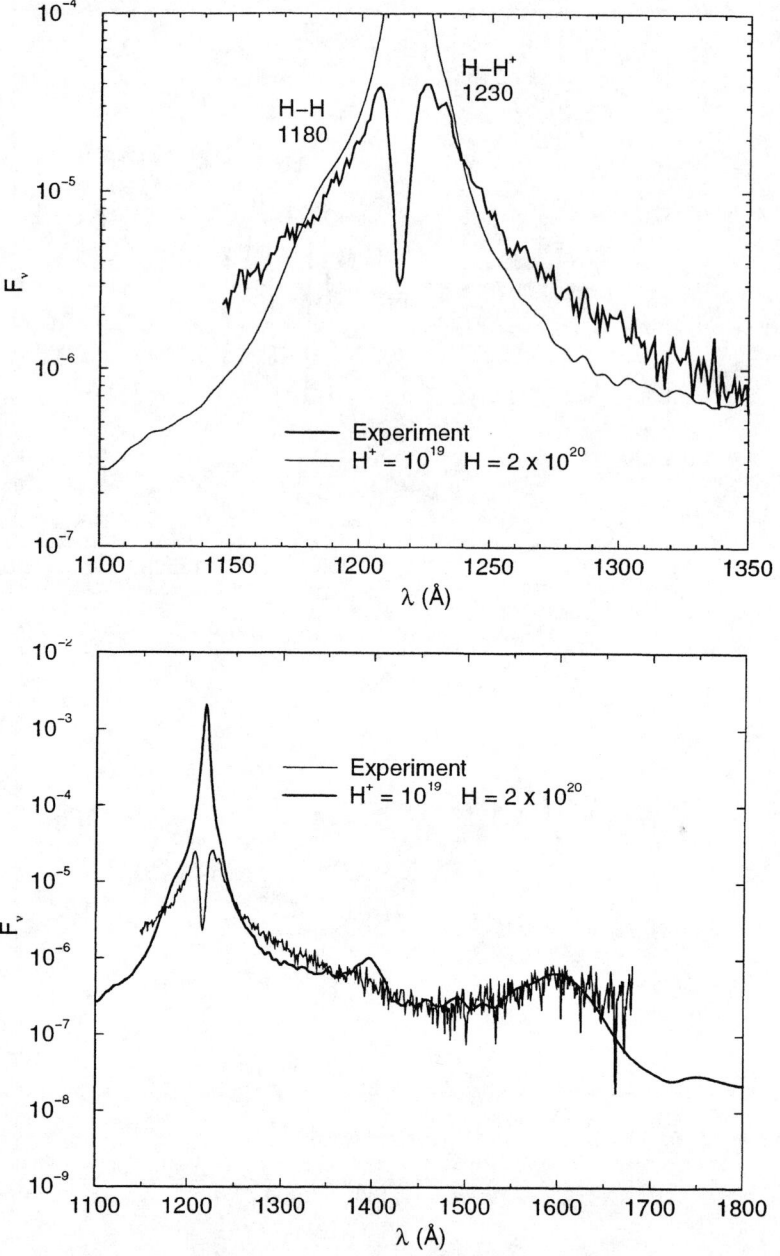

FIGURE 6. Comparison of near line wing after subtracting bremsstrahlung with theoretical profile for $N_{H^+} = 10^{19}$ ions/cm^3 $N_H = 2 \times 10^{20}$ atoms/cm^3.

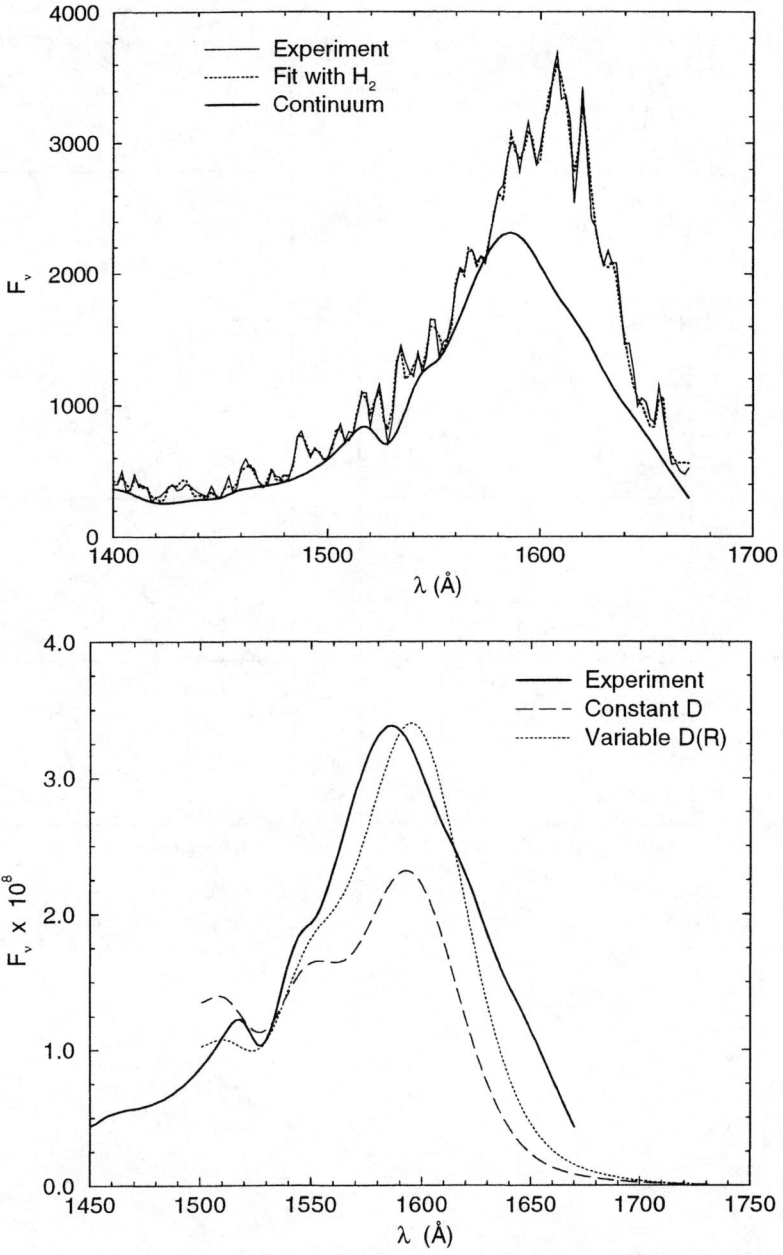

FIGURE 7. Far wing continuum with bound-bound molecular emission removed compared with theoretical models.

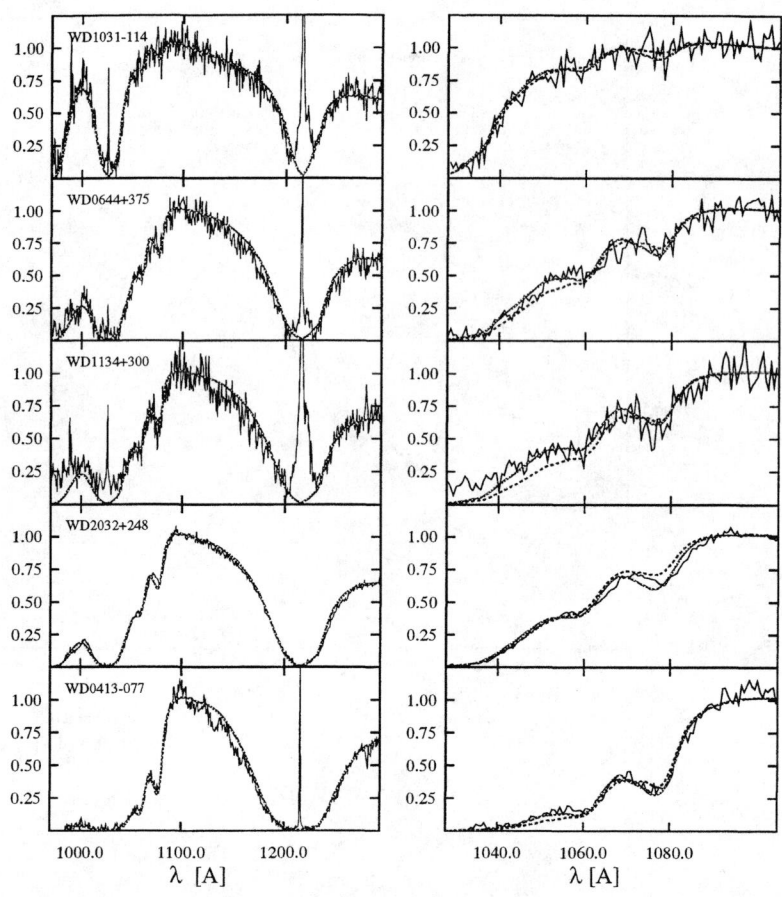

FIGURE 8. Comparison of observed white dwarf spectra with theoretical models [13].

V CONCLUSIONS

Satellites on Lyman α predicted 30 years ago [12,11] have now been observed in stars [2] and in the laboratory [6]. The theory has been developed to permit calculation of the combined profiles of Lyman α and β simultaneously perturbed by H and H$^+$, and to include the effect of the variation of the electric dipole transition moment during collision [4,8]. These advances open directions for new work. Observations of the Lyman β and Lyman γ regions in stellar spectra are expected from new generations of space-based observatories. High pressure gas targets are being used in the laboratory to explore many-body effects in the far wing of Lyman α, and pulsed gas jet targets have opened the windowless Lyman β and γ regions to experiments. As comprehensive potentials and transition dipole moments for the excited states of H$_2$ and H$_2^+$ become available, unified line shape theory can be used to calculate the profiles we need to interpret these new data.

ACKNOWLEDGMENTS

The work at the University of Louisville is supported by a grant from the U.S. Department of Energy, Division of Chemical Sciences, Office of Basic Energy Sciences, Office of Energy Research. We thank D. Koester for the data from ORFEUS observations in Fig. 8.

REFERENCES

1. J. F. Kielkopf and N. F. Allard, Ap. J. **450**, L75 (1995).
2. D. Koester and G. Vauclair, in *White Dwarfs*, edited by J. Isern, M. Hernanz, and E. Garcia-Berro (Kluwer, Dordrecht, 1997), p. 429.
3. N. F. Allard, J. F. Kielkopf, and N. Feautrier, Astr. Astrophys. **330**, 782 (1998).
4. N. F. Allard *et al.*, Astr. Astrophys. (1998), in press.
5. J. Kielkopf, Phys. Rev. E **52**, 2013 (1995).
6. J. F. Kielkopf, and N. F. Allard, preprint (1998).
7. J.F. Kielkopf, preprint (1998).
8. N. F. Allard, A. Royer, J. F. Kielkopf, and N. Feautrier, in preparation.
9. N. Allard, I. Drira, R. Faraggiana, M. Gerbaldi, J. Kielkopf, R. Kurucz, 14th International Conference on Spectral Line Shapes in State College, Pennsylvania, June 1998.
10. N. Allard, J. Kielkopf, I. Drira, 14th International Conference on Spectral Line Shapes in State College, Pennsylvania, June 1998.
11. J. C. Stewart, J. M. Peek, and J. Cooper, Ap. J. **179**, 983 (1973).
12. K. Sando, R. O. Doyle, and A. Dalgarno, Ap. J. **157**, L143 (1969).
13. D. Koester, U. Sperhake, N. F. Allard, D. S. Finley, and S. Jordan, Astr. Astrophys. (1998), in press.

Spectral Line Shapes for Atmospheric Work: Problems and Solutions

J. R. Drummond, A. D. May, R. Berman and S. Dolbeau

Department of Physics, University of Toronto
Toronto, Canada, M5S 1A7

ABSTRACT

We discuss the application of spectral line shapes to atmospheric physics. Two problems are identified, one of which requires detailed knowledge of spectral profiles over a limited region of the infrared. A simple master transport relaxation equation is proposed for the off-diagonal elements of the density matrix, the solution of which leads to a band profile. Exemplary solutions are then given, wherein a matrix technique is used to treat speed dependent broadening, shifting and line mixing. With reasonable computer power, detailed spectra may be constructed over regions containing tens if not hundreds of lines.

INTRODUCTION

The radiative properties of the Earth's atmosphere are significant in two respects. Firstly, the radiative budget and transfers within the atmosphere are a significant part of the larger earth system which determines the weather and climate of the planet. Secondly, the use of space-based instrumentation for monitoring the planet, both its atmosphere and surface, relies mainly on the detection and interpretation of radiation leaving the atmosphere.

These two fields, radiative transfer and remote sounding, have a great deal in common, since they both deal with radiative properties of the atmosphere. Furthermore, there is a strong link to gaseous spectroscopy which also investigates similar phenomena. However the situations which we meet in the laboratory are considerably different from those in the atmosphere, since in the former case most extrinsic quantities (e.g. concentration, pressure, temperature) are constant over the experimental volume, whereas in the latter case they are almost always spatially variable. It is acknowledged that by using a correct understanding of the physics of spectral formation, data may reliably be extrapolated from the laboratory case to the atmosphere. In practice, a full understanding of the physics is hard to obtain, particularly when the likely numbers of parameters are considered.

The volume of data required is likewise daunting. The active spectrum of the atmosphere covers the range from the far ultra-violet to the microwave region, encompasses dozens of gases and therefore millions of spectral lines. Producing a single model for the entire problem is likely beyond us at the moment, but work is progressing.

Radiative transfer and remote sounding also have differing requirements. The accurate computation of the radiative transfer requires the consideration of a large swath of the spectrum at a time, e.g. the infra-red spectrum for all absorbers between 300cm−1 and 2000cm−1, whereas the remote sounding problem usually concentrates on a few absorbers in a limited spectral region. In the former case inclusion of all significant gases, bands, etc becomes the primary goal, in the latter case the precision of the representation is often more significant.

As our understanding of spectra and their formation becomes more complete, these two areas must converge. The final area of separation will be the computational time required for a given problem since radiative transfer computations are often inserted in more complex models which tax the largest available computers at any given time.

A more significant issue for this paper is the increasing measurement capability in the atmosphere, given to us by the advance in the technological state of the instrumentation. This increased capability is fueled by a desire for more and more accurate measurements of the atmospheric state. This in turn calls into question our understanding of the spectroscopy of these systems. For example, current measurement requirements for atmospheric temperature are of the order of 1K which in turn (without getting into precise instrument scenarios) requires an understanding of the radiance at the level better than 1%. It is at this level of accuracy that many effects which have so far been neglected become significant.

Historically, atmospheric radiative transfer computations have progressed in step with the computational resources available. In the 1960s, it was certainly not possible to deal with hundreds of spectral lines in a routine manner and "band models" were in widespread use. The rise of computers permitted computations with groups of lines and the so—called "line—by—line" (LBL) codes began to be used for specialized problems. The line shape used for these codes was initially either the Lorentz function, ignoring translational effects, or the Doppler function, ignoring collisional effects. This was followed by the Voigt function which is the simplest model which combines the two. A great deal of work was done in the 1960s and 1970s on efficient computation of the Voigt function or, equivalently, the complex probability function. As computer power increased, LBL codes came into more widespread use until today it is feasible to run them on a desk—top machine as a matter of routine. In concert, the databases of spectral lines were enlarged and improved and today there are a number of very comprehensive databases in existence of which two, HITRAN[1] and GEISA[2], are perhaps the most widely known. These databases catalogue the positions, strengths, widths and temperature dependences of many of the gaseous components of the earth's atmosphere and some for the other planets.

There are still a number of problems which must be addressed. On the basic level the databases are still being improved as more precise measurements are made. However, although the most accurate parameters are known to 1%, many parameters, even some for very significant gases such as CO_2, are still poorly known.

A more severe problem is that a number of effects remain unaccounted for. Several of these are concerned with the line shape and some with other line parameters. Some of the line shape effects are: pressure shifts, line mixing, Dicke narrowing and speed dependence. All of the effects can produce changes of up to 2–3% of the absorption coefficients and these effects become significant as the accuracy and precision of space instrumentation becomes greater and the general requirements on modeling become more severe. As a specific example, ignoring the line mixing terms in the 4.3μm band of CO_2 can produce an error in the retrieved atmospheric temperature of 2–3K[3].

There is therefore a need for a comprehensive understanding of these phenomena. Although it is possible to add each of the additional effects piecemeal, the result rapidly becomes computationally cumbersome and the physics represented more opaque. However the historical path which has led to our current theories has led to a fractured approach in which a number of competing models, with apparently differing physical elements and processes, are used by different groups.

This paper presents a transport relaxation equation for the off–diagonal elements of the density matrix. In the case of electric dipole interaction a solution leads to an expression for the complex susceptibility, the imaginary part of which is directly related to the spectral profile. This unifies the treatment of spectral profiles as it is easy to see which specific set of approximations lead to specific models. However, what is more important, there is a well established matrix diagonalization and inversion technique which allows one to treat the general multi–line case, including speed dependent broadening and shifting and Dicke narrowing. This represents a significant step forward in our capabilities of modeling spectral profiles for atmospheric work.

A TRANSPORT RELAXATION EQUATION

In bath theory, the Liouville equation for the density matrix, ρ, for the entire system is contracted to one for the molecules of interest, plus some relaxation terms. The contracted form is often written symbolically as,

$$\partial \rho / \partial t = -(i/\hbar)[H,\rho] - \Gamma \rho \qquad (1)$$

where H consists of H_0, the Hamiltonian of an isolated atom or molecule, plus, H_1, the interaction with any applied (optical) field. In an elementary treatment, such as that used in laser theory, the matrix elements of Γ are constants and have the form that forces the populations (diagonal elements of the density matrix) towards their steady state values and the coherences (off–diagonal elements) towards zero. As an introduction, we make the same simplifying assumptions about the relaxation terms and treat the case of electric dipole interaction ($H_1 = -\mu \mathscr{E}$), with the field, $\mathscr{E}$, oscillating at a single frequency, ω. First we solve for ρ as a power series in the applied field. Hiding an $\hbar$ in $\mu \mathscr{E}$, it can be shown that the nth order jk component of ρ is given by,

$$[(\partial/\partial t) + i(\omega_j - \omega_k) + \gamma_{jk}](\rho^n)_{jk} = +i \mathscr{E} \sum_l [\mu_{jl}(\rho^{n-1})_{lk} - (\rho^{n-1})_{jl}\mu_{lk}] \qquad (2)$$

where ω_j is the energy of the jth level of the free molecule and γ_{jk} is the relaxation rate of the off–diagonal element, ρ_{jk}. For linear spectroscopy all we need is the first order solution. Thus only the zeroth order equilibrium populations (diagonal elements of ρ) will appear on the right hand side of eq.2. The zeroth order coherences (off–diagonal elements) are zero.

We now solve eq.2 for a two level molecule. We confine our treatment to steady state spectroscopy. Let the field be given by, $\mathscr{E} = E\exp[-i(\omega t - kz)] + $ cc. There are two levels, a the lower and b the upper with thermal equilibrium populations, $(\rho^0)_{aa} = n_a$ and $(\rho^0)_{bb} = n_b$, molecules per unit volume, respectively. Eq.2 for the component $(\rho^1)_{ba}$ then becomes,

$$[(\partial/\partial t)+i(\omega_b-\omega_a)+\gamma_{ba}](\rho^1)_{ba} = i\left\{E\exp[-i(\omega t-kz)] + cc\right\}[n_a-n_b]\mu_{ba} \qquad (3)$$

where, in the interest of simplicity, we have ignored the vector nature of the interaction. If we look for steady state solutions of the form,

$$(\rho^1)_{ba} = \left\{\rho^- E\exp[-i(\omega t-kz)] + \rho^+ E^*\exp[+i(\omega t-kz)]\right\} \qquad (4)$$

it is easy to show that only the ρ^- term may resonate and is given by,

$$\rho^- = [n_a - n_b]\mu_{ba}/[(\omega_{ba} - \omega) - i\gamma_{ba}] \qquad (5)$$

where ω and $\omega_{ba} = (\omega_b - \omega_a)$ are positive numbers. Thus, within the rotating wave approximation, i.e. neglecting the anti-resonant term, we have for $(\rho^1)_{ba}$,

$$(\rho^1)_{ba} = \left\{[n_a-n_b]\mu_{ba}/[(\omega_{ba}-\omega)-i\gamma_{ba}]\right\}E\exp[-i(\omega t-kz)] = \rho_{ba}E\exp[-i(\omega t-kz)] \qquad (6)$$

We have, at this point, solved our elementary relaxation equation. Now we make the connection to the spectral profile. First we calculate the macroscopic polarization (dipole moment per unit volume), P, using $P = Tr[\rho^1\mu]$. The dipole and density matrices are $\begin{bmatrix} 0 & \mu_{ab} \\ \mu_{ba} & 0 \end{bmatrix}$ and $\begin{bmatrix} 0 & \rho_{ab} \\ \rho_{ba} & 0 \end{bmatrix}$ so that P is just $\rho_{ab}\mu_{ba} + \rho_{ba}\mu_{ab}$. (For conciseness we have dropped the superscript, 1 on ρ.) In order to see the phase shift between the applied field and the polarization it is simpler to calculate the complex material polarization in the form $P = \rho_{ba}\mu_{ab}$ and since it is linear in the field, $\mathcal{E}$, to write it as $\epsilon_0 \chi \mathcal{E}$ where $\mathcal{E}$ and χ are both complex. Thus we find the complex susceptibility, χ, for our two level system, is given by,

$$\chi = \rho_{ba}\mu_{ab}/\hbar\epsilon_0 = A/\left\{(\omega_{ba} - \omega) - i\gamma_{ba}\right\}. \qquad (7)$$

The real part of χ is related to the index of refraction by $n^2 = 1 + \chi_r$ while the absorption coefficient, for $\chi_r \ll 1$, is proportional to χ_i. Thus our relaxation equation has led to a dispersion curve for the index of refraction and a Lorentzian profile, $A\gamma_{ba}/[(\omega_{ba} - \omega)^2 + (\gamma_{ba})^2]$, for the absorption. The half width at half maximum (HWHM) of the absorption profile, measured in radians per second is the relaxation rate γ_{ba}. If γ_{ba} is complex there will also be a frequency shift.

A GENERAL RELAXATION TRANSPORT EQUATION

It is clear from the preceding sections that a Lorentzian line shape follows directly from the form assumed for eq.3. Thus we need to consider the physics that is omitted from the simple relaxation equation used above. As was recognized by Rautian and Sobelman[4], one must think in terms of molecular distribution functions (of which ρ is an example) and some generalization of the Boltzmann equation. Thus we must glue together the off-diagonal elements of the density matrix, which deals with the internal states, and the Boltzmann distribution function, which describes the external or translational states. In other words, what we are seeking, is an equation for the off-diagonal element of a semi-classical density matrix as a function of **r**, **v** and, t, where **r** is a position vector and **v** the velocity. Tip[5] developed such an equation in the case of degenerate states, i.e. at the level of the Waldmann-Snider equation[6]. Since we will ignore effects dependent upon the polarization of the optical field, the complications associated with spatial degeneracy may be ignored. Technically, it can be said that we are seeking a transport relaxation equation at the Wang-Chang Uhlenbeck level[7]. This is proposed on physical arguments, much in the spirit of the original derivation of the Boltzmann equation.

Let ρ_{ba} be a general off-diagonal element with level b above level a. It is a function of position and velocity. Thus we must add the free streaming term, $-\mathbf{v}\cdot\nabla\rho_{ba}$, for the translational motion, to $-(i/\hbar)[(H_0 + H_1),\rho]$, the free streaming term for the internal motion. We saw already that a travelling wave field produced an off-diagonal element that varied spatially as exp(ikz). It is the gradient term acting on such a spatial term that produces the familiar kv_z term found in any elementary treatment of Doppler broadening or in a derivation of the Voigt profile.

In addition to the free streaming translational term, we must add the temporal evolution due to collisions with bath molecules. Two translational relaxation (transport) terms are to be added. The first is $-\nu\rho(\mathbf{r},\mathbf{v},t)_{ba}$ where ν is the kinetic collision frequency or the rate at which molecules leave a velocity class, $\rho(\mathbf{r},\mathbf{v},t)_{ba}$. The second is the rate of return to the velocity class from all the other velocity classes. We write this as $+\int A(\mathbf{v}\leftarrow\mathbf{v}')\rho(\mathbf{r},\mathbf{v}',t)_{ba}d\mathbf{v}'$. In equilibrium, the distribution function is constant. This does not imply that the velocity of a molecule is constant, only that the rate at which molecules leave a velocity class is dynamically balanced by a counter flow from other velocity classes. Neither of the forward/backward translational relaxation terms should come as a surprise to a reader familiar with the Boltzmann equation. Note, in the dynamical equation for ρ_{ba}, that the translational terms involve different velocity classes but only a single component of the optical (state) coherence, ρ_{ba}.

Now we consider the local decay terms. There is the usual term, $-\gamma_{ba}\rho(\mathbf{r},\mathbf{v},t)_{ba}$, describing the decay of the coherence. In addition there is the return contribution from other components of the optical coherence. Being discrete, this term is written as $+\Sigma W(ba\leftarrow dc)\rho(\mathbf{r},\mathbf{v},t)_{dc}$ where the sum over dc does not include ba. Here ba is to be read as a single index and is commonly (misleadingly) designated "line" in the literature. It is the return of coherence that is the source of line mixing. Note that the local decay terms involve only a single velocity but all of the optical coherences. For the moment we omit terms that involve both a change in the components of the optical coherence and a change in the velocity.

This completes the presentation of the model transport relaxation terms to be added to an equation for the off-diagonal elements of the density matrix. In the case of linear spectroscopy and dipole absorption, the final master equation for this distribution function is,

$$\left[(\partial/\partial t)+i(\omega_b-\omega_a) + \mathbf{v}\cdot\nabla\right]\rho_{ba} = -\gamma_{ba}\rho_{ba} + \sum_{dc} W(ba\leftarrow dc)\rho_{dc} - \nu\rho_{ba}$$
$$+ \int A(\mathbf{v}\leftarrow\mathbf{v}')\rho_{ba}d\mathbf{v}' + i\left\{E\exp[-i(\omega t-kz)] + cc\right\}[\rho_{aa}-\rho_{bb}]\mu_{ba} \qquad (8)$$

Here the equilibrium populations, ρ_{aa} and ρ_{bb} are to be written as $n_a f_0(v)$ and

$n_b f_0(v)$ where $f_0(v)$ is the Maxwellian distribution function, normalized to unity, and the n's are the number of molecules per unit volume.

In passing we note that most of the calculations of broadening and shifting of spectral lines deal with a calculation of the γ's or some Maxwellian average of γ, assuming that the real and imaginary parts give the width and shift of an isolated line. However, as we saw above such a relationship only follows from a simple relaxation equation, such as eq.3. Since the relaxation rates depend upon the relative velocity of the active molecule and the perturber they still depend upon the speed of the active molecule, even after averaging over the perturber motion[8]. If the speed dependence is significant for both the relaxation and the transport terms, then the relaxation of the optical coherence and the translational motion are coupled. If the motions are coupled, the usual simple relationship between the rates and widths or shifts is broken and the profiles, even for isolated lines, are no longer expected to be Lorentzian.

With the aid of the master equation (eq.8) we now address the problem of an isolated line. Here we may drop the line mixing terms involving the W's. Following the same path we used to solve eq.3, we find that the equation for ρ_{ba} (the quantity we actually need to compute the complex susceptibility) is given by

$$[\omega_{ba} - \omega + kv_z - i\gamma_{ba}]\rho_{ba} = i\nu\rho_{ba} - i\int A(\mathbf{v}\leftarrow\mathbf{v}')\rho_{ba}(\mathbf{v}')d\mathbf{v}' + n_a f_0(v)\mu_{ba} \quad (9)$$

where γ_{ba} and ν may both be functions of the speed of the active molecule and $\omega_{ba} = \omega_b - \omega_a$ is a positive number.

The Doppler profile, the speed dependent or independent Voigt, the speed dependent or independent Lorentz and the speed dependent or independent hard collision profile may all be recovered in a few lines from eq.9. For example, the simple Lorentz model we started with is obtained by making the mathematical, but not very physical, approximation of dropping all translational terms, i.e. those involving ν, $A(\mathbf{v}\leftarrow\mathbf{v}')$ and kv_z. If γ_{ba} is independent of v then the total susceptibility is found by multiplying by $(\mu_{ab}/\hbar\epsilon_0)d\mathbf{v}$, where $d\mathbf{v}$ is the volume element in velocity space, and integrating. (One observes contributions from all molecules, not just those in a single velocity class.) Below, when discussing a new treatment of Dicke narrowing, we will return to eq.9.

At this point we demonstrate how the well known expression for line mixing may be re-derived in a few lines from the master equation. For multi-line spectra or bands, the usual treatment of line mixing[9] suppresses all aspects of the translational motion including the speed dependence of the γ's and the W's. This allows us to integrate the master equation over the velocity distribution to arrive at an equation in which ρ_{ba} is now simply per unit volume, not per unit volume per unit velocity space. In the steady state, all parts of the optical coherence oscillate at the driving frequency, ω, not their natural frequencies, ω_{ba}, ω_{dc}, etc. Thus we can write all off-diagonal elements, ρ_{od}, as ρ_{od} times the complex field. As a result we find the equation,

$$(\omega_{ba} - \omega)\rho_{ba} = i\gamma_{ba}\rho_{ba} - \sum_{dc} iW(ba\leftarrow dc)\rho_{dc} + n_a\mu_{ba}. \quad (10)$$

There is a similar equation for each component or "line", ρ_{ba}. Thus we have a set of coupled linear equations, which in matrix notation may be written

$$[\omega_0 + i\mathbf{W} - \omega\mathbf{I}]\cdot\rho = \mathbf{N}\cdot\mu \quad (11)$$

where ω_0 is a diagonal matrix of free molecule transition frequencies, **W** is a relaxation matrix, **I** is the unit matrix, ρ is a column vector of the components of the optical coherence, **N** is a diagonal matrix of populations and μ is a column vector of transition dipoles. As defined, the off–diagonal elements of **W** are positive and just W(ba←dc). As pointed out earlier, they describe the back flow of the coherence ρ_{dc} to ρ_{ba}. The diagonal elements of **W** are negative and just the relaxation rates, γ_{ba}, for the transitions. This represent the outflow of coherence from ρ_{ba}. To find the total complex susceptibility for the band, eq.11 must be multiplied by $\mu_t/\hbar\epsilon_0$ where μ_t is the transpose of μ. The imaginary part of the result is identical to the well known expression for the absorption profile of a band experiencing line mixing[9]. It is usual to simplify the computation of the spectrum using well known matrix diagonalization and inversion techniques. We do not go into those details. Rather we now return to a single line case and consider the problem of Dicke narrowing[10]. We will give a new treatment of Dicke narrowing, establishing at the same time its equivalence to line mixing.

The spectral profile of an isolated line, at densities so low that all collisional relaxation may be neglected, has the well known Doppler profile. This arises because each velocity class maintains its integrity (no transitions to other velocity classes) and has its own Doppler shifted frequency. Of course, the Doppler shift arises from the free streaming kv_z term in eq.9. As the density increases, the translational relaxation rates, ν and A(**v**←**v**′), become important. Physically, at high densities, the molecules perform a random walk or diffusive motion. Thus one expects, perhaps naively, a line width (decay rate) which varies as k^2D, where D is the mass diffusion constant. Since D varies inversely with density, ρ, the contribution of the translational motion to the width decreases. As first noted by Dicke[8], if the broadening is sufficiently small an isolated line may actually narrow. This effect is well documented in the literature[11]. Two models of the translational motion are commonly used to describe the narrowing, all the way from the Doppler limit to the diffusion narrowed regime. These are, the soft collision (SC)[12] and the hard collision (HC)[13], model. We now turn to the question of how Dicke narrowing can be treated using the transport relaxation equation.

Eq.9 is the general equation for an isolated line. Since ρ_{ba} represents a distribution over velocity then $\rho_{ba}d\mathbf{v}$ is the amount of optical coherence per unit volume that lies between **v** and **v** + d**v**. If we multiply by d**v**, the three dimensional element of velocity space, eq.9 may then be written as

$$(\omega_{ba} - \omega + kv_z)\rho_{ba} = i\gamma_{ba}\rho_{ba} + i\nu\rho_{ba} - i\Sigma W(\mathbf{v}\leftarrow\mathbf{v}')\rho_{ba}(\mathbf{v}') + n_a f_0(\mathbf{v})\mu_{ba} \quad (12)$$

where now $\rho_{ba}(\mathbf{v})$ is to be interpreted as the number of molecules per unit volume with an optical coherence, ρ_{ba}, that lie in a velocity "cell" centered around **v**. More important is the fact that we have written the usual integral over **v**′ as a sum over velocity classes **v**′ and replaced A(**v**←**v**′)d**v** by W(**v**←**v**′), the rate at which the coherence is transferred from a "**v**′–cell" to a "**v**–cell". This rate is analogous to W(ba←dc), the rate of transfer between discrete components of the optical coherence in the case of line mixing. That the analogy is complete may be seen by comparing eq.12 (which is for a single speed class) to eq.10 (which is for a single component of the optical coherence). Note in particular that the discrete resonant frequencies ω_{ba} in eq.10 for line mixing are replaced in eq.12 either by the Doppler shifts kv_z or by the Doppler shifted resonant frequencies, $\omega_{ba} + kv_z$, where ω_{ba} is a constant for the isolated line. Thus we have established that line mixing and Dicke narrowing are not merely related but are in fact mathematically equivalent. Furthermore, they are almost physically identical since most of the quantities appearing in the equations have the same or a similar physical meaning.

Writing the equation for <u>one line</u> in terms of discrete velocity groups allows us to use the same diagonalization technique as that commonly used to solve the line mixing problem. Just as line mixing leads at high densities to a collapsed band, here we expect the exchange between the velocity groups to lead to a collapse of the Doppler profile, a profile which can always be considered as a band with a continuous distribution of "discrete" lines.

To illustrate this new treatment of Dicke narrowing we require an expression for the collision kernel, $A(\mathbf{v} \leftarrow \mathbf{v}')$. We chose the hard collision model since it is the only known case where an analytical expression for a spectral profile has been found when the relaxation rate of the optical coherence γ_{ba} is speed dependent[4]. The hard collision model takes a collision kernel that distributes all $\mathbf{v}'$ classes over the same Maxwellian in v. While not very physical the hard collision model does have the correct low and high density limit, satisfies detailed balance, is well behaved mathematically, and what is important for us, provides a benchmark, against which we can compare solutions obtained by velocity discretization and matrix inversion. We consider the case of pure Dicke narrowing (negligible broadening) from the Doppler to the Dicke limit.

Pure Dicke Narrowing (almost)

To reiterate, we wish to illustrate Dicke narrowing in the absence of collision broadening, from the low density Doppler profile to the Dicke narrowed profile at high densities. However, for the Doppler limit, it is not possible simply to set γ_{ba} and ν to zero in eq.9. In the absence of spontaneous radiation, setting the relaxation rate for the optical coherence to zero results in zero absorption. A correct treatment of Doppler broadening results from setting the kinetic collision frequency, ν, equal to zero, calculating the profile and taking the limit as γ_{ba} approaches zero [14]. For convenience, we shall take γ_{ba} as finite but small and speed independent. We could regard a small constant γ_{ba} as the natural width of the line.

For the hard collision model, ν is speed independent and the collision kernel is given by $A(\mathbf{v}\leftarrow\mathbf{v}') = \nu f_0(v)$ where $f_0(v)$ is the normalized Maxwellian $(1/\sqrt{\pi}v_0)^3 \exp-(v/v_0)^2$ and v_0 is the mean speed $(2kT/m)^{1/2}$. With $A(\mathbf{v}\leftarrow\mathbf{v}')$ given by the HC model and γ_{ba} a constant, eq.9 can be integrated over the x and y components of the velocity. This leads to an equation for a one dimensional distribution function, $\rho_{ba}(v_z) = \int \rho_{ba} dv_x dv_y$. The normalized Maxwellian $f_0(v)$, that appears in $A(\mathbf{v}\leftarrow\mathbf{v}')$ and the term involving n_a, also become one dimensional. When discretized, the elements of the relaxation matrix, off–diagonal in v_z, can be written as,

$$W_{ij} = W(v_{zi} \leftarrow v_{zj}) = (\nu K/\sqrt{\pi}v_0)\exp-(v_{zi}/v_0)^2 \qquad (13)$$

where K is a normalizing constant, determined by the size of the velocity cell. The subscripts i and j identify post and pre collisional values. As stated above, and shown by eq.13, the kernel for the hard collision model is independent of the velocity of the active molecule (**v**′) before a collision, i.e. independent of the subscript j. Thus the off–diagonal elements in the same row of the relaxation matrix are equal. The diagonal elements are written,

$$W_{ii} = -\gamma_{ba} - \nu[1 - K/\sqrt{\pi}v_0)\exp-(v_{zi}/v_0)^2] \qquad (14)$$

The final set of coupled linear equations is established by inserting eqs.13 and 14 into eq.11. This completes the main part of the transformation from a continuous distribution to a discrete distribution. The remainder of the problem may be solved by borrowing the standard matrix techniques from line mixing[14].

Fig.1 shows the computed line profile for γ_{ba} equal to 3 MHz and the kinetic collision frequency, ν, equal to 0, 15, 300 and 1500 MHz (30 GHz = 1 cm^{-1}). The mean speed v_0 and the Doppler width (kv_0 = 85 MHz) were chosen to represent a line in the fundamental band of CO at room temperature. We have deliberately chosen the spacing between the "v_z–cells" large so that discrete but fictitious "lines" appear when the kinetic collision frequency is low. A total of 25 speed groups were used, although Fig.1 is plotted over a frequency range that displays the presence of only seven of them. At ν equal to zero, the widths of the spectral components (HWHM) is just the constant relaxation rate, γ_{ba} = 3 MHz. We see, as the collision frequency is increased, that the fictitious lines in our band, broaden, overlap and collapse to a narrow single line. These details of Dicke narrowing, which are identical to those observed in line mixing would not have been apparent if we had binned the v_z speed classes on a scale fine compared to the collisional width γ_{ba}.

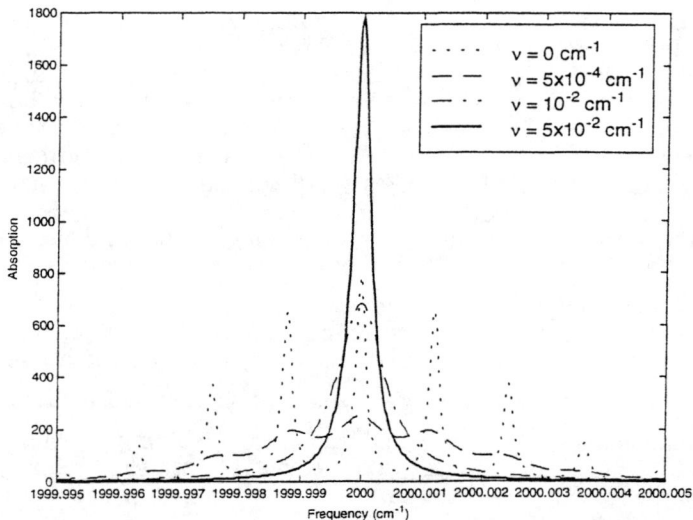

Figure 1 Speed independent profile with 25 z components of the velocity and a constant collisional width.

The curves for the HC model, calculated numerically, but using the standard analytic solutions[4], while not shown in the figure, pass nicely through the "bumps" associated with the coarse graining of the speed, in the matrix solution. What is surprising is how quickly the coarse grained matrix solutions approach the full solution as the density (collision frequency) is increased. This has a practical consequence. The larger the value of either ν or γ_{ba}, the coarser the speed graining can be. This reduces the size of the matrix that must be diagonalized, thus saving valuable computer time.

To further convince the reader of the viability of this new approach to the computation of spectral profiles of an isolated line, we show in Fig.2, a plot of the width (HWHM), as a function of the kinetic rate ν, for ν varying from 0.3 to 90 GHz. The HWHM were measured directly from the profiles. We see the $1/\nu$ behaviour which is characteristic of narrowing and the asymptotic approach to the "natural" width.

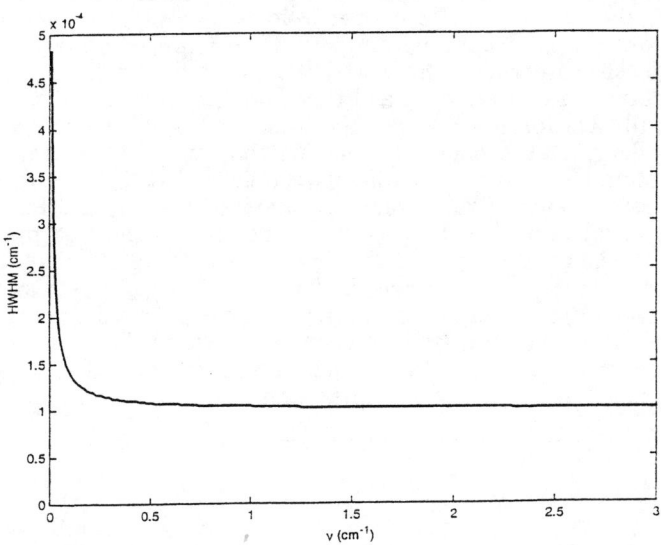

Figure 2 Half width at half maximum as a function of the kinetic collision frequency; other conditions as in Fig.1.

In the examples above, the relaxation rate of the optical coherence was kept constant and small. This was convenient for illustrating the well known properties of Dicke narrowing, but not realistic. Physically, both γ_{ba} and ν scale with density and maintain a fixed ratio, one with respect to the other. The same procedure for establishing the matrix equation as outlined above may be used to generate spectra, even when both γ_{ba} and ν are speed independent. We have computed spectral profiles for several ratios of $<\gamma_{ba}(v)>/\nu$ over a range of the kinetic collision frequency or what is equivalent a range of density. Here $<\gamma_{ba}(v)>$ is the thermal average of $\gamma_{ba}(v)$. For the speed dependence of the broadening and shifting we used the model given in[15] with q = 6, in their terminology. In all cases there was agreement between the matrix and standard profile, provided the binning of the v_z classes was fine enough. In the case of $<\gamma_{ba}(v)>$, small compared to ν a plot of the HWHM of the profiles as a function of kinetic collision frequency (density) showed the characteristic Dicke minimum[11].

So far we have shown that discretizing the velocity distribution allows us to recover results obtainable by more conventional techniques. However, the new approach can do much more. Given any model or set of numerical values that specify the speed dependence of the relaxation rates, γ_{ba}, ν and $W(\mathbf{v}\leftarrow\mathbf{v}')$ one can generate the spectrum for any isolated line. This is a major step forward. It transfers the problem from one of calculating a spectrum to one of determining input parameters, viz. the speed dependence of the relaxation rates. This problem may be solved, by carrying out microscopic calculations using reasonable interaction potentials.

It should be evident that the general approach may be extended to include line mixing. The master equation for the off–diagonal elements of the density matrix, eq.8 is already discretized in the components of the optical coherence. Discretizing the distribution ρ_{od} over velocity simply enlarges the size of the relaxation matrix. The input parameters, then required are the rates of transfer between the cells into which the distribution function has been partitioned. Each cell is characterized by say, v, v_z and the coherence component or "line".

As an illustration Fig.3 shows the computed profiles for a two line spectrum with speed dependent broadening and shifting and line mixing[16]. Because the master equation neglects relaxation between cells that differ both in velocity and "line", the components of the relaxation matrix, off–diagonal in coherence, were taken as diagonal in **v**. The hard collision model and the treatment of the speed dependence of the broadening and shifting was the same as above for a single line. The ratio $<\gamma_{ba}(v)>/\nu$ was taken as 2.5, while $<\gamma_{ba}(v)>$ was taken as 0.05 cm^{-1}/atm. These are values not uncharacteristic of infrared lines. In the top panel the two lines are Dicke narrowed. In the next panel, the lines have broadened and there is weak mixing. In the last panel, the spectrum has collapsed to a single speed dependent, Dicke narrowed line. This the first theoretical profile of this nature.

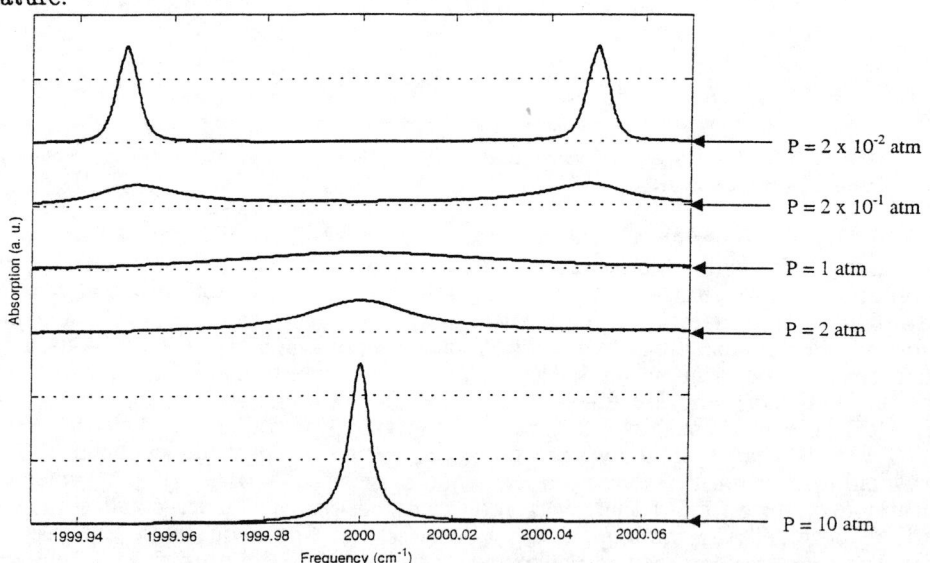

Figure 3 Line mixing with Dicke narrowing and speed dependent broadening and shifting. 5 cells in speed and 5 cells in z component of the velocity were used.

Finally, it should also be evident that the master equation and the technique of discretizing the distribution over velocity can be extended to include relaxation between cells, off–diagonal in **v** and "line". Thus not only would coupling between the internal and translation motion, which arises from their mutual speed dependence, be included, but so would the statistical correlation between the two motions. Again, microscopic calculations could be used to establish realistic models for the various rates. For atmospheric work, this must include the variation with temperature.

SUMMARY

We have introduced a relaxation transport equation for the off–diagonal elements of a semi–classical density matrix. Spectral profiles may be constructed from solutions linear in an applied optical field. This approach unifies the treatment of spectral line shapes. By discretizing the distribution function over the velocity the master equation is reduced to a set of coupled linear equations that may be solved using standard matrix techniques. For an isolated line the mathematical equivalence of Dicke narrowing to line mixing was demonstrated. We have shown how speed dependent effects may be included even in the case of line mixing. We have shown that the discretization over the velocity may be coarse and yet yield an accurate description of the profiles. Thus the numerical technique is not computer intensive and lends itself to the atmospheric problem of determining accurate and physically meaningful profiles for groups of lines. It appears that the remaining problem is to establish realistic models for the speed dependence of the various relaxation rates. This is a problem to be tackled using existing microscopic theories of the rates.

REFERENCES

1. L. S. Rothman et al, J. Quant. Spectrosc. Radiat. Transfer, **48**, 469 (1992).
2. N. Husson et al, Ann. Geophys. **4A**, 185 (1986).
3. L. Larabee et al, App. Opt. **27**, 872 (1988).
4. S. G. Rautian and I. I. Sobelman, Sov. Phys. Usp. **9**, 701 (1967).
5. A. Tip, Physica **52**, 492 (1971).
6. L. Waldmann Z. Naturforsch. **13a**, 600 (1958), R. F. Snider, J. Chem. Phys. 1051 (1960).
7. C. S. Wang–Chan et al, in Studies in statistical mechanics, North Holland Pub. (Amsterdam) vol II, 243 (1964)
8. G. Nienhuis, J. Quant. Spectrosc. Radiat. Transfer, **20**, 275 (1978).
9. M. Baranger, Phys. Rev. **111**, 481, 494, 855 (1958).
10. R. H. Dicke, Phys. Rev. **89**, 472 (1953).
11. J. W. Forsman et al, J. Chem. Phys. **97**, 5355 (1992).
12. L. Galatry, Phys. Rev. **122**, 1218 (1961).
13. M. Nelkin and A. Ghatak, Phys. Rev. A **135**, 4 (1964).
14. A paper giving a fuller discussion of the master equation and its use to recover standard results has been submitted for pulication along with a paper dealing specifically with Dicke narrowing.
15. J. Ward et al, J. Quant. Spectrosc. Radiat. Transfer 20 275 (1978.
16. S. Dolbeau, undergraduate report, Univ. of Toronto and Univ. de Rennes, "Cycle Optronique" (1998). A copy is available by writing to, ENSSAT, 6 rue de Kerampont, BP 447, 22305 Lannion, France.

Accurate Determination of Electron Densities from Hydrogen Line Widths in Active and Quiescent Prominences: the Mid-Infrared Advantage

Edward S. Chang and Donald W. Blair
Department of Physics & Astronomy, University of Massachusetts, Amherst, MA 01003

and

Drake Deming
Code 693, NASA/Goddard Space Flight Center, Greenbelt, MD 20771

Abstract. We have measured the widths of all hydrogen lines in the MIR (8 to 20 micron) spectra of a quiescent and two time-frames of an active prominence. Accounting for instrumental and Doppler broadening, we attribute the remaining width to the sum of Stark contributions, which are shown to be due to protons in the quasi-static and to electrons in the impact limits. We obtain electron densities, N_e=1.3(0.2), 5.2(0.8), and 2.2(0.4) in units of 10^{10} cm^{-3} for the above 3 prominences respectively. For the quiescent prominence, our value is consistent with the lower end of the accepted range (Bommier) but not with those at the higher end (Hirayama, Nikolsky).

1. What is a prominence?

In the Quiet Sun photosphere, infrared lines of atomic hydrogen are broadened by optical thickness effects as well as by the Stark Effect. Therefore, information on the electron density can only be extracted from detailed modelling with radiative transfer computations. On the other hand, prominences when viewed at the solar limb have a column density of only 10^{-5} g cm^{-2}, so infrared lines are usually optically thin (Gouttebroze et al. 1993) and their widths bear direct information on the electron density. Prominence temperatures are usually known to within a few hundred degrees, being about 10000K (Chang and Deming 1996), and therefore much cooler than the million degree corona surrounding them. However, the electron densities N_e are uncertain by a factor of five (Tandberg-Hanssen 1995). Appying Griem's theory of line broadening (Griem 1960; 1967), Chang and Deming (1998, referred as CD) show that accurate values of N_e can be conveniently derived from the observed line widths of highly excited hydrogen lines.

2. Solar Data

The solar prominence spectra were taken at Kitt Peak Observatory with Fourier transform spectroscopy (FTS) in the mid-infrared mode. Our data set consists of a quiescent prominence (Q) (Brault and Noyes 1983) and two active prominences (A1 and A2), which are two time-frames of the same prominence taken one and a half hours apart. In the spectal range of 524 to 1227 cm^{-1}, we have found a dozen hydrogen lines, with the principal quantum number ranging from 6 to 16. Details of the solar prominence spectra can be found in CD, including the study of three observed He I lines.

3. Line widths and electron densities

The hydrogen line shape fits a gaussian well, with the damping constant (1/e half width σ) varying from 25 to 171 mK (1000 mK=1 cm^{-1}, see Figures 1 and 2 in CD). We ascribe the width to Doppler, Stark and instrumental broadening in quadrature, since the hydrogen intrinsic line width due to atomic fine structure and radiative damping is negligible. Thus we have

$$\sigma^2 = \sigma_I^2 + \sigma_D^2 + \sigma_E^2. \tag{1}$$

The instrumental resolution of 0.0306 cm^{-1} implies an instrumental width (the e^{-1} half-width) σ_I=0.0214 cm^{-1}. So it is dominant in the narrowest observed widths, but becomes negligible for the broadest. Following previous works, we attribute the Doppler width to the turbulent velocity v_T and the temperature T,

$$\sigma_D^2 = \left(\frac{\nu}{c}\right)^2 \left(v_T^2 + \frac{2kT}{M}\right), \tag{2}$$

where k is Boltzmann's constant and M the atomic mass. In CD, values for σ/ν are found to be 36, 79 and 47 ppm (part per million) for Q, A1, and A2 respectively. The upper bound temperature T_{max} corresponds to the temperature when v_T vanishes.

Following the theory of Griem (1967), we have determined that Stark width of the hydrogen lines is the sum of two terms, one due to protons in the quasistatic approximation σ_p and the other to electrons in the impact limit σ_e. Explicit expressions for the hydrogen transition $\Delta n+n$ to n are derived in the appendix of CD and presented below.

$$\sigma_p = 4.32 \times 10^{-11} N_e^{2/3} \Delta n(2n + \Delta n). \tag{3}$$

and

$$\sigma_e = 6.14 \times 10^{-17} \frac{N_e}{T^{1/2}} \left[log\left(\frac{4.02 \times 10^6 T}{N_e^{1/2}(n+\Delta n)^2}\right) - 0.125\right][(n+\Delta n)^5 + n^5]. \tag{4}$$

We take the temperature T as consistent with but somewhat lower than T_{max}, i.e. 6000K, 30000K, and 10000K for Q, A1 and A2 respectively. Then we vary N_e and calculate the widths for a set of hydrogen lines and compare them with the observed widths in Table 1. The best set of calculated widths are also shown and they correspond to N_e = 1.3, 5.2 and 2.2 in units of 10^{10} cm^{-3}.

TABLE I

Observed and calculate hydrogen line widths. ν_c is the frequency calculated by Casini. σ is the 1/e-halfwidth.

Line n m	ν_c cm^{-1}	PromB σ mK	PromB σ_c mK	Prom1 σ mK	Prom1 σ_c mK	Prom2 σ mK	Prom2 σ_c mK
8 7	524.607	25(3)	28				
14 9	794.466	53(2)	57	115(7)	125	72(4)	77
11 8	807.287	40(1)	40	82(1)	82	49(1)	52
7 6	808.283	37(1)	35	71(1)	68	49(1)	44
15 9	866.589	74(7)	69	160(7)	153	87(8)	94
9 7	884.275	39(1)	38	73(1)	77	43(1)	49
16 9	925.617	66(10)	83	171(19)	187	127(16)	114
12 8	952.064	50(2)	48	116(4)	102	70(2)	63
10 7	1141.544	51(1)	47	103(2)	100	63(1)	62
14 8	1154.134	65(9)	66	141(9)	148	89(5)	90
15 8	1226.258			134(17)	175	101(11)	106

4. Conclusions

We have demonstrated the MIR advantage in measuring accurate electron densities from the widths of high excitation hydrogen lines in both quiescent and active prominences. Our method relies on calculations of Stark broadening by protons in the quasi-static and by electrons in the impulse approximations. We have corrected the widths in CD by adding the proton and the electron widths linearly, increasing by about 30% over our former values. Thus the electron densities are found to be $N_e=1.3(0.2)\times 10^{10}$ cm^{-3} for a quiescent prominence and $N_e=5.2(0.8)$ and $2.2(0.4)$ in units of 10^{10} cm^{-3} for an active prominence when the spectra are taken 1.5 hours apart. We have assumed temperatures of 6000K, 33000K and 10000K for the above prominences respectively, which are just below but consistent with the upper limits from Eq.(2). For the quiescent prominence, our value coincides with the Bommier et al. (1986) mean electron densities $N_e=1.3(0.2)\times 10^{10}$ cm^{-3} of 14 prominences observed at Pic-du-Midi, but not with the much higher values measured by Hirayama (Hirayama 1971; Hirayama 1978) and by Nikolsky (Nikolsky et al. 1971).

Acknowledgements

We thank Peter Foukal for sending us all the prominence spectra analyzed here, and for useful discussions. E.S.C. and D.W.B. are grateful to the Universities

Space Research Association for temporary appointments to the Goddard Visiting Scientist Program. This research was supported by NASA through NAS-5-32484 and RTOP 344-12-53-10.

References

Bommier, V., Leroy, J.-L., and Sahal-Brechot, S.: 1986, *Astron. Astrophys.* **156**, 90.
Brault, J. and Noyes, R.: 1983 *Astrop.J.* **260**, L61.
Chang, E. S., and Deming, D.: 1996, *Solar Physics* **165**, 257.
Chang, E. S., and Deming, D.: 1998, *Solar Physics* **179**, 89.
Griem, H. R.: 1960, *Astrophys. J.* **132**, 883.
Griem, H. R.: 1967, *Astrophys. J.* **148**, 547.
Gouttebroze, P., Heizel, P., and Vial, J. C.: 1993, *Astron. Astrophys. Suppl.* **99**, 513.
Hirayama, T.: 1971, *Solar Physics*, **17**, 50.
Hirayama, T.: 1978, in E. Jensen, P. Maltby, and Orrall (eds.) *Physics of Solar Prominences*, Inst. Theor. Astrophys., Oslo, p.4.
Nikolsky, G. M., Gulyaev, R. A. and Nikolskaya, K. I.: 1971, *Solar Physics* **21**, 332.
Tandberg-Hanssen, E.: 1995, *The Nature of Solar Prominences*, Kluwer Acad. Press.

Polarized Lines of Hydrogen in Magnetized Stellar Atmospheres

S. Brillant*, G. Mathys[†], C. Stehlé*

*DARC, Observatoire de Paris, 5 Place Jules Janssen F-92195 Meudon
[†]European Southern Observatory, Casilla 19, Santiago, Chile

Introduction.

Stellar magnetic fields are frequently diagnosed through the analysis of spectra of the I and V Stokes parameters for test elements. The following relation defined by Landstreet and al. [1] based on the formalism of Unno [2] allows one to determine the mean longitudinal magnetic field (i.e. the magnetic field along the line of sight averaged over the stellar disk),

$$V = -\bar{g}\,\Delta\lambda_B\,B\,\frac{dI}{d\lambda} \quad \text{with} \quad \Delta\lambda_B = \frac{e}{4\pi m_e c^2}\lambda_0^2. \tag{1}$$

The usual test elements are either heavy elements in the photographic method [3] or hydrogen (Balmer lines) for the photopolarimetric method [1]. These two methods give in average the same order of magnitude for the field strength, but the exact values that are obtained may differ from each other in an apparently non-systematic manner [3]. There are reasons to suspect that the discrepancy arises from an improper interpretation of the hydrogen line observations. In order to investigate this, we present below a new model for the formation of hydrogen lines.

Hydrogen Line Formation in Magnetized Atmospheres.

Let us consider a radiation of frequency ω, propagating through a plane parallel atmosphere along direction $\vec{k}$. Let $\hat{e}_1$ and $\hat{e}_2$ define arbitrary mutually orthogonal directions of polarization in the plane perpendicular to $\vec{k}$. The equation of transfer of polarized radiation is

$$\frac{dS}{dz} = \chi S - \epsilon, \tag{2}$$

where $S = (I, Q, U, V)^t$ is the Stokes vector, χ is the opacity matrix, and ϵ is the emissivity vector [4]. The latter two can be written as a function of the elementary absorption matrix b.

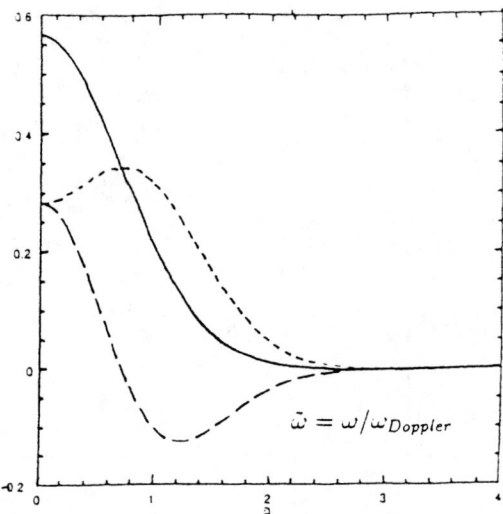

FIGURE 1. g_{Dopp} (solid line), g_0 (short dashes) and g_{cross} (long dashes)

We consider an absorbing hydrogen atom moving with the velocity $\vec{v}$ in the atmosphere. It is submitted to the actions of the magnetic field (parallel to the z axis) and of the Lorentz electric field ($\vec{v} \otimes \vec{B}$, parallel to the x axis). θ is the angle between $\vec{k}$ and $\vec{B}$, and ϕ the angle between $\hat{e}_1$ and the Lorentz field. Taking into account the symmetry of the system one proves that b can be written as [5]:

$$b(\omega, v_\perp, \theta, \phi) = b_0(\omega, v_\perp, \theta) + \cos(2\phi)\, b_{cross}(\omega, v_\perp, \theta)\,, \qquad (3)$$

where the b_{cross} matrix results from the departure from the cylindrical symmetry around $\vec{B}$ due to the Lorentz electric field.

To obtain the absorption matrix b, we need to average the terms b_0 and b_{cross} over $\vec{v}$, taking into account the Doppler shift. The average is computed over two components of the velocity ($v_\parallel$ parallel to z, $v_\perp$ parallel to y) and the angle ϕ.

Due to the dependence of b on $v_\perp$ and ϕ, a correlation appears between b and the Doppler effect. Thus one can not use the standard convolution by a Doppler profile. However in the case where

$$b(\omega, v_\perp, \theta, \phi) = b^{(0)}(\omega) + v_\perp^2\, b^{(2)}(\omega)\,, \qquad (4)$$

we proved that the velocity average reduces to:

$$b_{0(cross)} = g_{Dopp} * b_{0(cross)}(\omega) + g_{0(cross)} * b^{(2)}_{0(cross)}(\omega)\,, \qquad (5)$$

where g_{Dopp} is a standard Doppler profile and g_0 and g_{cross} are modified quasi-Doppler profiles (see Figure 1) [5].

The elements of the absorption matrix b are derived through calculation of the evolution operator $T(\Delta\omega)$ in the Unified Theory:

$$T(\Delta\omega) = -\frac{i}{\pi}[L_Z + L_m - \Delta\omega I) - i\gamma(\Delta\omega)]^{-1}\,, \qquad (6)$$

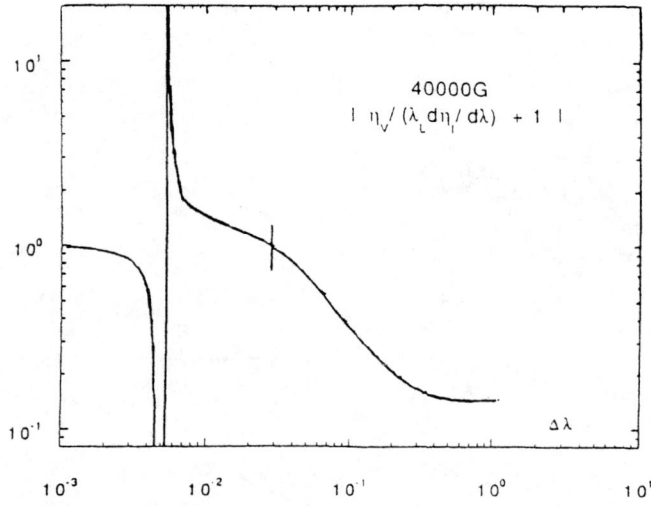

FIGURE 2. $| K(\Delta\lambda) + 1 |$ for an electronic density $N_e = 10^{14} \text{cm}^{-3}$, $T = 10^4 \text{K}$ and $B = 40 \text{kG}$.

where L_Z and L_m are the Liouville operators for the interaction with the magnetic field and with the Lorentz electric field respectively. Plasma microfields effects are included through the damping rate $\gamma(\omega)$, taking into account the ion dynamics [6].

We proved that the quadratic approximation (equation 4) is valid as long as $\gamma(\omega)$ is larger than the energy perturbation induced by the Lorentz field. This yields for Lyman α

$$\log B \leq .-3.8 + 0.4 \log N_e \quad (\text{G, cm}^{-3}). \tag{7}$$

This allows us to use (5) for the velocity average in the condition of Ap stars.

We also computed the function $K(\Delta\lambda) = (\lambda_L)^{-1} \times \eta_V d\lambda / d\eta_I$ for the line Lyα, in order to test the validity of relation (1)(Figure 2). η_I and η_V are elements of the b matrix [4].

Conclusions.

We showed that the relation (1) is qualitatively correct in the line wings of Lyman α, but $K(\Delta\lambda)$ differs from unity and varies with the density, the temperature and the magnetic field. We also presented a new formulation for the transfer coefficients and for averaging over the atom velocities including the Doppler effect.

REFERENCES

1. Landstreet, J.D., Borra, E.F., Angel, J.R.P., Illing, R.M.E., ApJ **201**, 624 (1975).
2. Unno, W., Pub. Astr. Jap. Vol 8, 108 (1956).
3. Mathys, G., A&AS. **89**, 121 (1991).
4. Mathys, G., J.Q.S.R.T. **44**, 143 (1990).
5. Brillant, S., Mathys, G., Stehlé, C., Submitted to A&A.
6. Stehlé, C., A&AS. **104**, 509 (1994).

The search for the bound-free emission from the LiH C $^1\Sigma^+$ state

H. Skenderović, T. Ban and G. Pichler
Institute of Physics, POB 304, HR-10000 Zagreb, Croatia

In spite of great interest in LiH molecule, because of its importance for astrophysics and other applications there is almost no knowledge about any other higher potential curves beside two lowest singlet states.[1] Recent ab initio theoretical calculations[2] revealed a very complex structure of LiH singlet potential curves, where ionic-covalent interaction causes many avoided crossings. Thus LiH C $^1\Sigma^+$ state appears as a double minimum state with its inner minimum lying entirely above the dissociation limit. The outer minimum is below the dissociation limit, but its equilibrium position is at internuclear distances where lower LiH A $^1\Sigma^+$ and X $^1\Sigma^+$ states are almost horizontal. Spectral transitions C→A and C→X in the region of the outer minimum of the C $^1\Sigma^+$ state are of bound-free type in the infrared and green part of the spectrum, respectively. Both spectra should be governed by the strong enhancement due to the extremum in the corresponding difference potential curves C-A and C-X.

In our experimental search for the bound-free emission from the LiH C $^1\Sigma^+$ state we employed a crossed heat-pipe oven with cylindrical electrodes inserted in two arms. The electrodes were isolated from the main oven body. The electric discharge in a Li+Li$_2$+H$_2$ mixture produced LiH molecules in large quantity so that the A $^1\Sigma^+$ → X $^1\Sigma^+$ could be easily observed. By increasing the partial pressure of hydrogen gas we succeeded to enhance LiH A $^1\Sigma^+$ → X $^1\Sigma^+$ emission and to entirely eliminate emission spectra of Li$_2$ dimers. Possible C-A and C-X features in the observed spectra will be discussed in connection with spectral simulations. Quite recently we learned about the successful optical-optical double resonance depletion fluorescence experiment in which LiH C $^1\Sigma^+$ was found.[3]

References:
1. W. C. Stwalley and W. T. Zemke *J. Phys. Chem. Ref. Data* **22** (1993) 87.
2. A. Boutalib and F. X. Gadea *J. Chem. Phys.* **97** (1992) 1144.
3. W. C. Lin, J. J. Chen and W. T. Luh, *J.Phys.Chem.* **101** (1997) 6709.

Transition Probabilities of Si II Lines from Laser Induced Breakdown Spectroscopy on solid target

P. MATHERON*, A. LESAGE** and J. RICHOU*

* Laboratoire d'Optoélectronique, Université de Toulon et du Var, BP 132, 83957 LA GARDE, France
** Observatoire de Paris-Meudon, DASGAL, URA 335, 92195 MEUDON Cedex, France

Silicon is among the most abundant element in stellar atmospheres. As underlined by Th. Lanz and M. C. Artru transition probabilities and Stark broadening parameters of Si II lines are still needed, especially for faint lines and for intercombination transitions for which calculations are not yet possible. This lack of data affect particularly the interpretation of the observations of chemically particular Ap silicon stars. The present experiment starts a series of measurements devised to provide such data.

The plasma is produced by a Nd:YAG laser operating at 1064 nm which generates pulses of 280 mJ maximum energy and 10 ns duration. Such method has already been used to study NI lines from this plasma source [1].

The laser beam is focused onto a pure silicon solid target, with a 7,56 cm focal length lens. The light flux is equal to 2.10^{11} W/cm2. The silicon sample is put inside a vacuum cell filled with argon gas. The silicon spectra are observed with a 2400 gr./mm grating Chromex spectrometer and recorded with an Optical Multichannel Analyzer. Time resolution of the emission spectra is achieved.

Relative transition probabilities measurements of some Si II multiplet lines are performed. We've studied here three Si II multiplets (1, 3 and 23) of visible range corresponding to $4p\ ^2P - 3p^2\ ^2D$, $4f\ ^2F - 3d\ ^2D$ and $4d''\ ^4F - 4p'\ ^4D$ transitions.

In this method, all transitions from the same level are considered [2]. Absolute density number, absolute line intensity and temperature are not needeed. In a situation where no autoionization and no de-excitation takes place, transition probability are determined with combining line branching ratio and upper state transition lifetime.

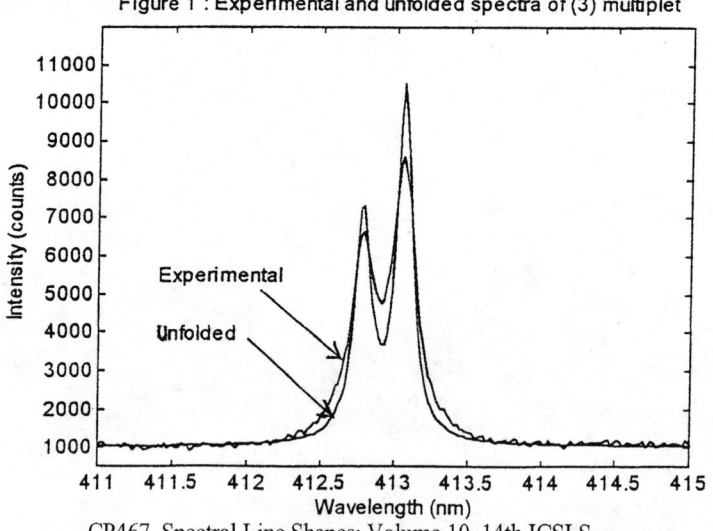

Figure 1 : Experimental and unfolded spectra of (3) multiplet

Table 1: Atomic transition probabilities of SII with experimental branching ratios.

Multiplet N°	Transition levels		Line (Å)	Branching ratios	Atomic transition probabilities (10^8 s^{-1})			
	Upper	Lower			This work	Ref. [4]	Ref. [6]	Ref. [7]
1	$4p\ ^2P_{3/2}$	$3p^2\ ^2D_{3/2}$	3853,7	0,054	0,059 ± 10 %	0,046 ± 16 %	0,054 ± 15 %	-
	$4p\ ^2P_{3/2}$	$3p^2\ ^2D_{5/2}$	3856,0	0,446	0,49 ± 10 %	0,42 ± 13 %	0,51 ± 15 %	-
	$4p\ ^2P_{1/2}$	$3p^2\ ^2D_{3/2}$	3862,6	0,318	0,35 ± 10 %	0,35 ± 14 %	0,33 ± 15 %	-
3	$4f\ ^2F_{5/2}$	$3d\ ^2D_{3/2}$	4128,1	0,424	1,45 ± 12 %	-		1,56
	$4f\ ^2F_{7/2}$	$3d\ ^2D_{5/2}$	4130,9	0,576	1,69 ± 12 %	-		1,67
23	$4d'\ ^4F_{3/2}$	$4p'\ ^4D_{1/2}$	5181,9	0,12	0,44 ± 25 %	-		0,97
	$4d'\ ^4F_{5/2}$	$4p'\ ^4D_{3/2}$	5185,25	0,19	0,68 ± 25 %	-		1,04
	$4d'\ ^4F_{7/2}$	$4p'\ ^4D_{5/2}$	5192,9	0,22	0,81 ± 25 %	-		1,19
	$4d'\ ^4F_{9/2}$	$4p'\ ^4D_{7/2}$	5202,4	0,4	1,42 ± 25 %	-		1,38
	$4d'\ ^4F_{5/2}$	$4p'\ ^4D_{5/2}$	5219,4	0,03	0,12 ± 25 %	-		0,33
	$4d'\ ^4F_{7/2}$	$4p'\ ^4D_{7/2}$	5240,3	0,047	0,09 ± 25 %	-		0,19

Line intensity analysis is carried out by unfolding the experimental profile from the apparatus function, which is determined with a thin mercury spectral line. The unfolded spectra, by resolving the overlapped lines, allows to improve continuum and line intensity estimation. The deconvolution result is shown in figure 1 for the multiplet (23).

Temperature is given by line intensities ratios from two adjacent ionization states elements [3]. The selected lines are choosen from Si II and Si III spectra for multiplet (1) and (23). For the multiplet (3), line intensities ratios from two same ionization states of Ar II are used. Calculated electronic temperatures are 2.10^4 K for multiplet (1) spectrum, $1,6.10^4$ K for multiplet (3) spectrum and $2,3.10^4$ K for multiplet (23) spectrum. Electron density of the plasma, derived from Stark broadening parameter of Si III and Ar II line, is $\sim 4.10^{16}$, $\sim 1,7.10^{17}$ and $\sim 3,3.10^{17}$ cm^{-3} respectively. Local Thermodynamic Equilibrium is then assumed by Mc Whirther criterion [4].

For the self-absorption correction of the studied lines, plasma homogeneity is assumed and the optical depth of each multiplet line is calculated.

Si II multiplets transition probabilities have been measured from available lifetime data for their upper states [5], [8]. Results are shown in table I and compared with previous measurements of F. Blanco [4], Schulz-Gulde [6] and with computation of M. C. Artru [7].

The uncertainties listed in table I have been estimated by taking into account the following contributions : a) systematic uncertainties of the line profile after deconvolution method,
b) uncertainty in the lifetime data,
c) possible deviation from linearity of the photoelectric detection system, estimated to be less than 1 % in our wavelength interval,
d) uncertainty of autoabsorption correction.
Integrated intensity computation error is also negligible.

As can be shown, present results agree with previous ones. That is a proof that laser produced plasmas can be used for such measurements.

References :

[1] P. Matheron et al., XXIII ICPIG Toulouse, **IV** 116 (1997)
[2] M. C. Huber and R. J. Sandeman, Physica Scripta, **22**, 373 (1980)
[3] H. R. Griem, Principles of Plasma Spectroscopy, University Press (1997)
[4] F. Blanco, Botho and J. Campos, Physica Scripta, **52**, 628 (1995)
[5] S. Bashkin and al., Physica Scripta, **21**, 820 (1980)
[6] Schulz-Gulde, J.Quant.Spectrosc.Radiat.Transf., **9**, 13 (1969).
[7] M. C. Artru et al., Astronomy & Astrophys. Suppl. Series, **44**, 171 (1981)
[8] H. G. Berry and al. Physica Scripta, **3**, 125 (1971)

Quasi-molecular satellites in the Lyman β profile: application to the white dwarf stars

N.F. Allard[1,2], I. Drira[1], J.F. Kielkopf[3]

[1] Observatoire de Paris-Meudon, Département Atomes et Molécules en Astrophysique, 92195 Meudon Principal Cedex, France
[2] CNRS Institut d' Astrophysique, 98 bis Boulevard Arago, 75014 Paris, France
[3] Department of Physics, University of Louisville, Louisville, KY 40292, USA

Satellite features in hydrogen lines are not limited to Lyman α, which is of course the only Lyman-series line that was accessible to the Internationall Ultraviolet Explorer (IUE) or is now accessible to the Hubble Space Telescope (HST). Observations made with the Hopkins Ultraviolet Telescope (HUT) of Lyman β in the spectrum of the DA white dwarf Wolf 1346, with T_{eff} close to 20000 K, have revealed a line shape very different from the expected simple Stark broadening, with a very steep red wing and satellites near 1078 and 1060 Å (Fig. 1). The theory of Allard and coworkers is easily extended to Lyman β (Koester et al., Astrophys. J. **463**, L93 (1996); Allard et al. Astron. Astrophys. **333**, 782 (1998)), and gives an excellent fit to the new observational data.

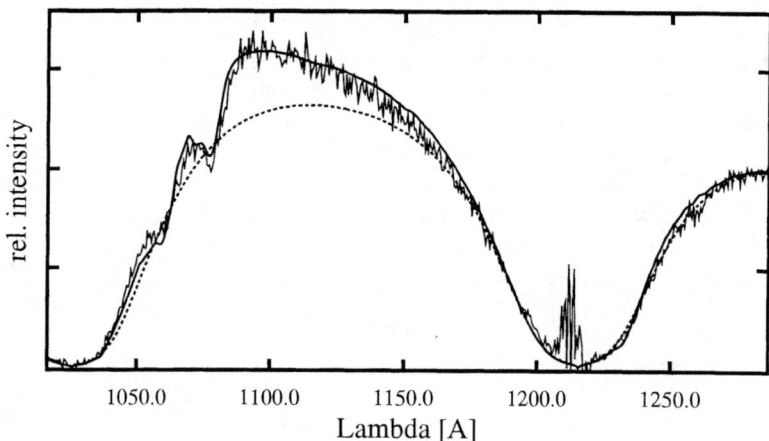

Fig. 1. HUT (Hopkins Ultraviolet Telescope) spectrum of the DA white dwarf Wolf 1346. The thick lines are synthetic spectra, computed with the new Lyman β profiles of Koester et al. 1996 (continuous), and with standard VCS Stark broadening (dotted).

Here we present new theoretical calculations of the total profile of the atomic hydrogen Lyman β spectral line. The variation of the radiative dipole moment during the collision is now included. This leads to a significant increase in the amplitude of the main spectral line satellites. As a consequence, when theoretical Lyman series profiles are to be used in the diagnostics of stellar parameters, the variation of the dipole moment must be included in order to obtain a reliable analysis.

CP467, Spectral Line Shapes: Volume 10, 14th ICSLS,
edited by Roger M. Herman
© 1999 The American Institute of Physics 1-56396-754-5/99/$15.00

Effect of the variation of electronic dipole moment on theoretical spectra: application to the λ Bootis stars

N.F. Allard[1,2], I. Drira[1], R. Faraggiana[3], M. Gerbaldi[4], J.F. Kielkopf[5], R. Kurucz[2,6]

[1] Observatoire de Paris-Meudon, Département Atomes et Molécules en Astrophysique, 92195 Meudon Principal Cedex, France
[2] CNRS Institut d' Astrophysique, 98 bis Boulevard Arago, 75014 Paris, France
[3] Dipartimento di Astronomia dell' Università degli Studi di Trieste, via Tiepolo 11, 34131
[4] Institut d' Astrophysique, Université de Paris Sud XI, 98 bis Boulevard Arago, 75014 Paris, France
[5] Department of Physics, University of Louisville, Louisville, KY 40292, USA
[6] Smithsonian Institution Astrophysical Observatory, Cambdidge, MA 02138, USA

Close collisions with neutral hydrogen and protons are responsible for the formation of the quasi-molecules H_2 and H_2^+ with the appearance of satellite features at 1600 Å and 1405 Å in the Lyman α wing. Previous theoretical calculations of the line profiles were based on the assumption that the radiative dipole moment is constant during the collision.

The theoretical results are improved by including the effect of the variation of the radiative dipole moment during the collision. The theoretical approach is described by Allard *et al.* in Astron. Astrophys. **200**, 58 (1994), and Astron. Astrophys. **333**, 782 (1998). Numerical calculations show that large changes in the intensity of the satellites may occur when the variation of the dipole moment is important in the region of internuclear distance where the satellite is formed. The new theoretical Lyman α line profiles have been included in a stellar atmosphere program for the computation of model stellar atmosphere spectra and synthetic spectra of λ Bootis stars. A comparison of the new calculations with observations made with the International Ultraviolet Explorer (IUE) as shown in Fig. 1 demonstrates that these last improvements are of fundamental importance for obtaining a better quantitative interpretation of the spectra and for determining stellar atmospheric parameters.

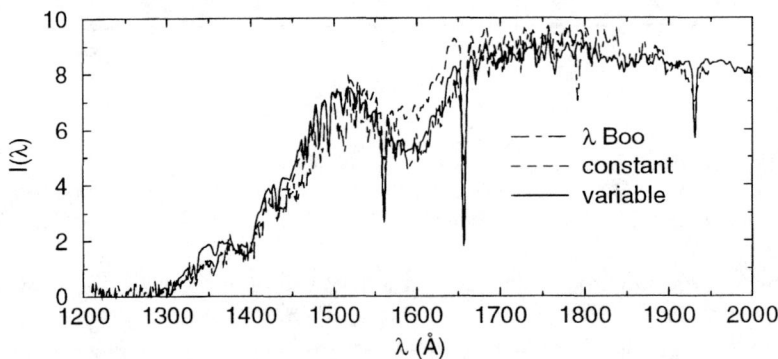

Fig. 1. *A comparison of synthetic spectra with and without the variation of dipole moment and the IUE spectrum of λ Bootis.*

INNOVATIVE TECHNIQUES
FOR LINESHAPE MEASUREMENT
AND ANALYSIS

Spectroscopy with Non-classical Light and Non-classical Atoms

E. S. Polzik, J. Hald, J. L. Sørensen

Institute of Physics and Astronomy, Aarhus University
Aarhus, 8000, Denmark

In the last few years a new field, atomic physics and spectroscopy with non-classical, squeezed light, has emerged. The frequency tunable sources of such light utilize tunable lasers and frequency doubling and/or parametric down conversion to create quantum correlations between frequency components of radiation [1-5]. The non-classical features of light result in manifestly non-classical features of spectral lines, namely in the rate of atomic processes, in the line shapes, and in the level of quantum noise in atomic spectroscopy. Moreover, the non-classical features of light can be transferred onto atoms leading to their manifestly non-classical behaviour. We will briefly review some theoretical proposals and experimental results in this new area.

We concentrate here on the non-classical light produced in the process of parametric downconversion, although other methods are also available [2,3]. Fig. 1 illustrates how phase and amplitude correlated fields are generated in the process of the parametric downconversion.

Parametric Generation of Squeezed Light

• Phase correlations for classical fields

$E_+ = \beta \, i \, E_p \, E_-^*$

$\varphi_p + \pi/2 = \varphi_- + \varphi_+$

$\omega_p = \omega_- + \omega_+$

• Phase correlations for quantum fields

spontaneous parametric emission

e.-m. vacuum

Squeezed vacuum

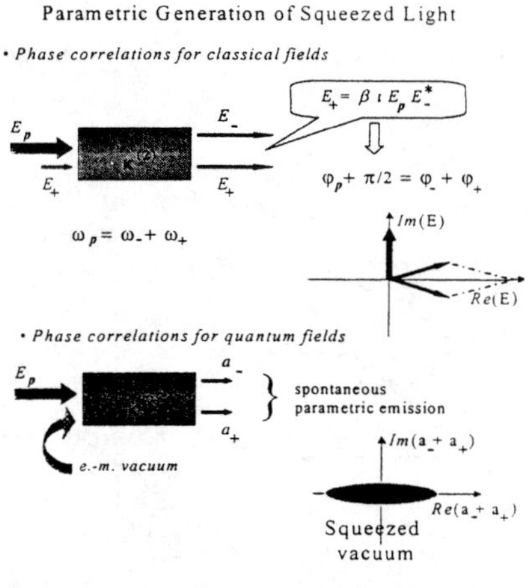

Figure 1. Classical and quantum correlations in the process of parametric downconversion

In the upper half of the figure classical signal and idler fields are shown to be phase correlated due to the momentum conservation for the photons involved in the process. In the lower part of the Figure a similar field distribution is shown for the amplification of the vacuum, i.e. for the process of the spontaneous parametric downconversion. Quantum theory of such a process has been developed in detail by various authors (see, for example, [6], and references therein). Parametrically amplified vacuum, i.e. squeezed vacuum, is shown as the ellipse in the phase diagram for the e.-m. field in Fig. 1. The principal difference between the squeezed fluctuations and classical phase sensitive fluctuations is in that squeezing corresponds to the fluctuations less than that for a coherent or vacuum state of the e.-m. field. The latter correspond to equal uncertainties in the two quadratures of the fields in the minimum uncertainty state [6].

Pairwise photon-photon correlations between the fields $a(\omega_+)$ and $a(\omega_-)$ created in the process of the parametic downconversion are described by correlation functions $\langle a(\omega_+), a(\omega_-)\rangle = M(\omega_+)\delta(2\omega - \omega_+ - \omega_-)$, $\langle a^+(\omega_+), a(\omega_-)\rangle = N(\omega_+)\delta(\omega_+ - \omega_-)$. For non-classical light $N^2 + N \geq |M|^2 \geq N^2$ [7]. In principle, any pair of modes $a(\omega_+), a(\omega_-)$ with the frequencies symmetrical around the half-pump frequency ω can be quantum correlated. In practice, the degree of correlations depends on the spectral profile of the parametric gain. To enhance the efficiency of generation of correlated photons the non-linear medium is often placed in an optical resonator. The resulting device is called an Optical Parametric Oscillator (OPO) or an Optical Parametric Amplifier (OPA). Due to the resonator mode structure significant quantum correlations occur between symmetric pairs of frequencies within the resonator mode centered around ω, and for frequencies which belong to the resonator modes symmetrical around ω. In the former case quantum correlations occur for the photons separated in frequency by several MHz up to a few GHz depending on the resonator configuration. These correlations allow for the suppression of the quantum noise of light in the radio frequency domain which is most relevant for sub-shot-noise atomic spectroscopy [1-4, 8]. In the latter case quantum correlations may occur between photons separated in

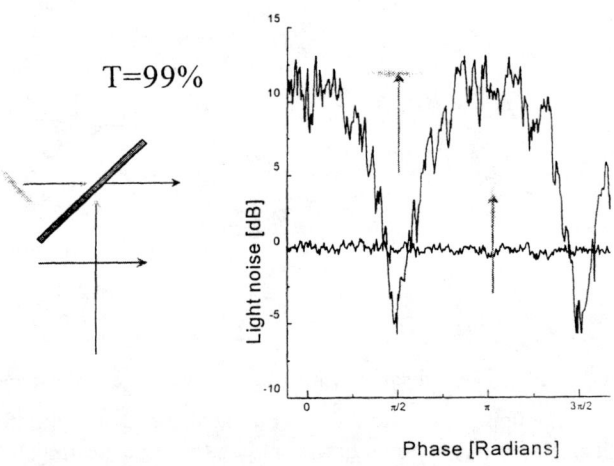

Figure 2. Mixing squeezed vacuum and a coherent beam on a beamsplitter (left). Quantum noise of light (right). Shot noise – the horizontal trace. Phase sensitive noise going below (squeezing by 5 db) and above the shot noise level is shown.

their wavelength by up to tens of nm. Such correlations allow to perform two- and four- photon spectroscopy with non-classical light [9,10].

Squeezed vacuum is characterized by the fluctuations of one of the quadrature phase components below the e.-m. vacuum level. When mixed with a coherent field on a beam splitter (Fig. 2) squeezing leads to reduced or enhanced amplitude noise of the light emerging from the beam splitter. A typical picture of the quantum noise of such light as a function of the phase between the squeezed vacuum and the coherent beam is shown in Fig. 2 (zero corresponds to the shot noise level in the figure) [4]. Apparently, when the squeezed vacuum is appropriately phased (with the ellipse orthogonal to the coherent component) the amplitude noise of the light is lower than the shot noise limit, which corresponds to the absence of the squeezed vacuum. This feature allows to reach the sub-shot-noise sensitivity in the atomic absorption spectroscopy [1].

By means of mixing of the squeezed vacuum with an orthogonally polarized coherent component, polarization squeezed probe can be generated [11]. Such a probe has been recently used to performed sub-shot-noise polarization spectroscopy of atoms [8]. In this case two quantum correlated modes $a(\omega_+), a(\omega_-)$ correspond to σ_+ and σ_- polarized components of the probe. The experimental set-up of the sub-shot-noise polarization interferometer is shown in Fig. 3. The polarization squeezed probe is prepared by mixing the coherent beam and the squeezed vacuum on a polarizing beamsplitter PBS1. The probe at 917 nm passes through the atomic sample whoes spin polarization has to be measured. The sample in our case is a collection of cold atoms trapped in a magneto-optical trap (MOT). The weak probe measures the spin polarization of the $6P_{3/2}$ state.

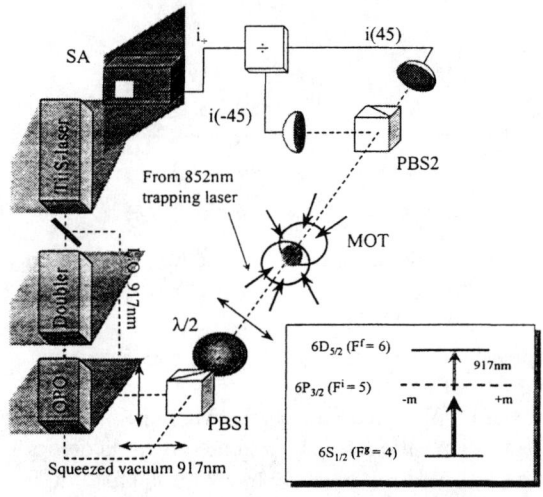

Figure 3. Setup for the polarization measurement of the quantum noise of the MOT

To avoid the influence of technical noise the spin polarization is modulated at 3 MHz by means of the modulation of the light exciting the $6P_{3/2}$ state.

The polarization rotation component at this frequency is then analyzed by means of the PBS2, balanced detectors and the spectrum analyzer SA. When squeezed light is used as a passive probe for absorption or polarization spectroscopy, the fundamental quantum noise which limits the sensitivity of spectral line measurements can be reduced beyond the shot noise level Φ_{shot}. The minimal noise is set by $\Phi_{squeezed} = \Phi_{shot}(1 + 2N - 2|M|)$. For the ideal squeezing $|M| = \sqrt{N^2 + N}$ and $\Phi_{squeezed} \to 0$ for a large number of photons, $N \to \infty$. The comparison of the shot-noise limited and sub-shot-noise polarimetry is shown in Fig. 4 reproduced from [8]. For both traces the probe laser is scanned across the atomic resonance and the Faradey rotation signal is measured. The upper trace corresponds to the probe in the coherent state (no squeezed vacuum applied. Applying the squeezed vacuum and thereby generating the polarization-squeezed probe we lower the noise floor by about 2 dB with the subsequent improvement in the signal to noise ratio (lower trace).

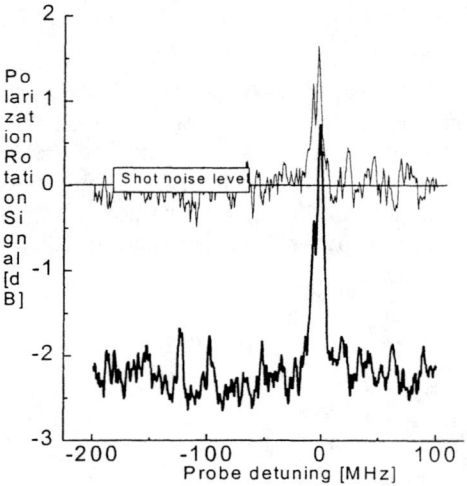

Figure 4. Shot noise limited polarization rotation signal from cold atoms (upper trace). Sub-shot noise polarization rotation signal obtained with the polarization squeezed probe (upper trace).

Besides precision measurements non-classical light can furthermore be used to alter fundamental properties of atomic dynamics, such as, e.g. the spontaneous decay. If an atom is illuminated with squeezed light from all directions matching its dipole pattern the spontaneous decay process is altered as shown theoretically in [7]. Instead of the familiar Lorentz-shaped fluorescence line with the natural linewidth γ, the spectral profile consists of two features, one of which has the the sub-natural width $\gamma_{squeezed} = \gamma(\frac{1}{2} + N - |M|)$. The experimental difficulties with the experimental observation of this alteration of the spontaneous emission process by squeezed light are mainly due to the necessity to provide squeezed vacuum from the full 4π solid angle and to localize an atom within a volume of a cubic wavelength. A route to partly overcome these difficulties has been investigated in [12] where an atom in a cavity illuminated with squeezed vacuum has been studied. Due to the preferential coupling of an atom to the cavity mode it is not necessary to provide squeezing of all the vacuum modes. Such a "one-dimentional" atom is not expected to have any sub-natural features in its spectrum. Nevertheless the effect of the squeezing of the cavity mode should be in narrowing of the spectral line of an atom-cavity system as compared to the "normal" vacuum input.

If the non-classical correlations are present between the photons corresponding to the frequencies of the two steps of a ladder two-photon transition the two-photon excitation rate becomes very different from what is expected for the excitation with regular laser light. In the two-photon excitation process the non-classical correlations of light lead to the experimentally observed [9] *linear* rather than quadratic dependence of the excitation rate on the intensity of light. This dependence stems from the pairwise character of the photon statistics described above.

The analysis of the fundamental noise limiting the accuracy of polarization spectroscopy has revealed that the two quantum noise sources contributing towards the limitations of the accuracy are the noise of the probe light and the noise of atomic medium itself [8,13]. To see that we write the differential photocurrent measured by the polarization interferometer as $i_- = 2\Theta i_{1,2} e^{-\alpha} + (i_1 - i_2)$ where $i_{1,2}, \alpha, \Theta$ are correspondingly the photocurrents of the detectors in the absence of the polarization rotation, the optical depth of the atomic medium and the angle of the polarization rotation caused by the atoms. In the absence of atoms and for balanced detectors the average values of the photocurrents are equal, however, the quantum noise of the difference $i_1 - i_2$ is non-zero and corresponds to the shot noise for the probe in the coherent state. We discussed above how this fundamental quantum noise of the probe light can be reduced beyond the vacuum state limit with the help of squeezed light. Let us now concentrate on the first term in the equation for i_-. Even in the absense of any collective spin polarization of the atomic ensemble, i.e. for the average angle $\Theta = 0$ quantum fluctuations of the collective atomic spin polarization will lead to the non-zero quantum noise of Θ. The angle of rotation contains two contributions: $2\Theta = (\phi^- - \phi^+) + (k^\Leftrightarrow - k^\Updownarrow)$. The first term corresponds to the circular birefringence (orientation) of the medium. It is observed mostly off resonance and is proportional to the projection of the collective atomic orientation vector on the axis of the probe propagation F_z. The second term is the linear dichroism proportional to the alignment of the ensemble. This term is maximal at resonance. Concentrating on the first term only for the sake of brievity we write the variance of the photocurrent fluctuations as $Var(i_-) \propto i_0^2 e^{-2\alpha} Var(\sum_{i=1}^{N} F_z^i)$+alignment term+shot noise. The sum is taken over all the atoms of the ensemble. Clearly, the statistical distribution of individual atomic spin orientations determines the size of this contribution to the fundamental noise of the atomic spin polarization measurement. The significance of the quantum noise of the collective atomic spin in achieving the ultimate sensitivity of the atomic polarization spectroscopy has been analyzed in [8, 13].

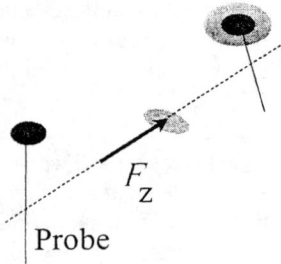

Figure 5. Transformation of the probe quantum noise due to the atomic spin noise. See comments in the text.

A pictorial view of the noise transformation described above is shown in Fig. 5. A probe with its quantum noise (e.g. shot noise) propagates through an atomic ensemble with the spin F_z. Due to the quantum noise of the spin δF the probe acquires an additional noise on top of its own shot noise. The ultimate precision of the probe

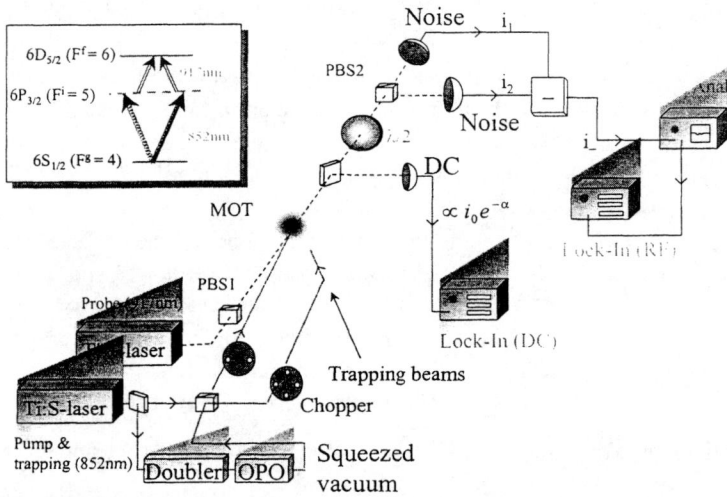

Figure 6. Experimental setup for generation of the spin squeezed state of cold atoms

polarization rotation measurement is determined by the total noise of the transmitted probe. The amount of the addditional noise acquired by the probe from the atomic medium depends on the quantum state of the collective atomic spin. For an ensemble of uncorrelated atoms the quantum spin noise cannot be lowered beyond a certain quantum limit analogous to the shot noise limit for uncorrelated photons of light. This quantum limit has been observed for an atomic spin polarization measurement in [8]. It is quite similar to the quantum projection noise described in [15].

A reduction of the atomic noise below the quantum limit is possible via introducing quantum correlations between individual atoms of the ensemble. One type of such an ensemble is a spin squeezed state (SSS). Various mechanisms has been proposed for generation of the SSS [14-17]. We are currently pursuing an experiment along the lines of the proposal [16]. According to the proposal SSS can be generated via mapping of the squeezed light on an ensemble of atoms in the process of the complete absorption. The experimental set-up is shown in Fig.6.

The squeezed pump resonant with the $6S_{1/2} \rightarrow 6P_{3/2}$ transition introduces quantum correlations between pairs of atoms in the magnetic substates of the $6P_{3/2}$ state with magnetic quantum numbers m_i and $m_{i\pm2}$. The resulting atomic state in $6P_{3/2}$ is characterized by the reduced or enhanced quantum noise of the collective atomic spin F_z depending on the phase of the polarization-squeezed pump. The spin state is analyzed by the polarization analysis of the probe at the transition $6P_{3/2} \rightarrow 6D_{5/2}$ as shown in the Fig. 6.

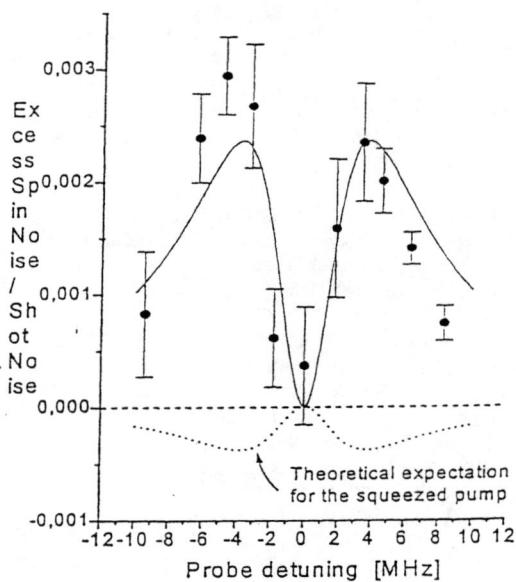

Figure 7. Anti-squeezing of the atomic spin noise. Spectral profile of the excess quantum spin noise of cold atoms produced by mapping of anti-squeezed light onto collective atomic spin. See comments in the text.

In a recent series of experiments we have observed the effect of mapping of quantum fluctuations of light on atoms in accordance with the prediction in [16]. In Fig.7 we present the observation of the anti-squeezed phase of the spin squeezed atomic ensemble. The polarization changes in the probe scanned across the resonance follow the square of the dispersion-like shape corresponding to the anti-squeezed noise of F_z. Further attempts to observe a much smaller effect of squeezing of the atomic spin are in progress. Squeezed phase of the spin will correspond to the data below the dashed line of the quantum limit, somewhere around the theoretical dotted line.

Besides the importance of SSS of atoms for overcoming quantum limits in frequency standards and atom interferometers, these states are also of significant interest to the newly emerging field of quantum information. For example, mapping of the non-classical light on atoms may pave the road towards storage of quantum information and quantum memory.

In summary, the area of atomic physics and spectroscopy with non-classical light has developed into an active experimental field. Experiments demonstrating various atomic measurements with sub-shot-noise sensitivity, as well as those demonstrating very unusual atomic dynamics brought about by the non-classical driving light have been carried out. Very recently attention has been attracted to the non-classical states of atomic systems where quantum correlations between separate atoms or ions occur. With an outburst of activity in the area of quantum information and quantum computing new and exciting developments involving quantum correlated photons and atoms can be foreseen.

REFERENCES

1. E. S. Polzik et al, *Phys. Rev. Lett.*, **68**, 3020 (1992); *Appl.Phys.***B 55**, 279 (1992).
2. S. Kasapi et al, *Opt. Lett.* **22**, 478 (1997).
3. F. Marin et al, *Optics Comm.*, **140**, 146 (1997).
4. Y. Li et al, *Phys. Rev. Lett.*, **78**, 3105 (1997).
5. J. L. Sørensen et al, *Quantum and Semiclass. Optics*, **9**, 239 (1997).
6. D. F. Walls, G. J. Milburn, *Quantum Optics*, Springer-Verlag, Berlin, 1994.
7. C. W. Gardiner, *Phys. Rev. Lett.*, **56**, 1917 (1986).
8. J. L. Sørensen et al, *Phys. Rev. Lett.*, **80**, 3487 (1998).
9. N. Ph. Georgiades et al, *Phys. Rev. Lett.*, **75**, 3426 (1995).
10. N. Ph. Georgiades et al, *Phys. Rev. A*, **55**, R1605 (1997).
11. P. Grangier et al, *Phys. Rev. Lett.*, **59**, 2153 (1987).
12. Q. Turchette et al, to appear in *Phys. Rev. A*.
13. J, L, Sørensen et al, *J. of Modern Opt.*, **44**, 1917 (1997).
14. M. Kitagawa et al, *Phys. Rev.* A**47**, 5138 (1993).
15. D. J. Wineland et al, *Phys. Rev.* **46**, R6797 (1992).
16. A. Kuzmich et al, *Phys. Rev. Lett.*, **79**, 4782 (1997).
17. A. Kuzmich et al, *Europhys. Lett.*, **42**, 481 (1998).

Single-Mode Cavity Ring-Down Spectroscopy for Line Shape Measurements

J. Patrick Looney, Roger D. van Zee & Joseph T. Hodges

Process Measurements Division
Chemical Science and Technology Laboratory
National Institute of Standards and Technology
Gaithersburg, MD 20899

Abstract. The precise measurement of absorption line shapes and line strengths is demonstrated using single-mode cavity ring-down spectroscopy. This technique utilizes a pulsed laser to excite a single transverse electromagnetic mode of a high-finesse, stable resonator. From the measurement of the decay time constant of this mode, the optical losses can be determined. As the cavity mode is tuned through an absorption resonance, the variation of the decay time constant yields a measure of the absorptive losses in the cavity. Spectra of the weak transitions of the molecular oxygen A-band are used to demonstrate the capability of this technique to measure absorption profiles and absolute line strengths. Repeated measurements of the line strength have a standard deviation <0.3 %.

INTRODUCTION

Cavity ring-down spectroscopy (CRDS) is a relatively new absorption spectroscopy that holds promise for sensitive and accurate absorption measurements with quantifiable uncertainties (1-3). Such measurements could enable quantitative, absolute partial pressure measurements for applications where traditional sensors are inadequate (4). This objective is one goal of our ongoing research. Any absorption-based pressure measurement, however, requires accurate absorption line shape and line strength measurements. Fulfilling this requirement is the subject of this paper.

The experimental approach to this problem is based on a variant of CRDS we refer to as single-mode cavity ring-down spectroscopy (5,6). Using this approach, absolute measurements of line strengths could eventually be made with an accuracy of <0.1 %, if all systematic effects in the measurement have been taken into account (6). In this paper, we will demonstrate the use of single-mode CRDS to measure absorption line shapes with a resolution of about 5 MHz. These measurement also demonstrate a standard deviation <0.3 % in the line strength determination.

SINGLE-MODE CAVITY RING-DOWN SPECTROSCOPY

The conventional implementation (1-3) of CRDS uses a pulsed laser to excite a linear cavity formed from two highly reflective, spherical mirrors (a Fabry-Pérot étalon or stable resonator). Usually the cavity design is not particularly special (*i.e.* not confocal or otherwise degenerate), and the cavity is not length-stabilized. Nevertheless, such a cavity can store the optical field for as long as 10^{-4} s, during which time the measured intensity decays, with a time constant called the ring-down time. This decay is measured at the exit of the cavity, and the ring-down time is extracted from the recorded signal. For a cavity filled with an absorbing gas under conditions when the Beer-Lambert law is valid, the ring-down time constant for a *single* cavity mode is given by the relationship (5,7),

$$\tau(\omega) = \frac{\ell}{c \cdot [(1-R) + \alpha(\omega)\ell]}, \qquad (1)$$

where ℓ is the cavity length, R is the intensity reflectivity of the mirror, $\alpha(\omega)$ is the frequency dependent absorptivity coefficient, and c is the speed of light. CRDS has the seductive feature that quantitative, high sensitivity absorption measurements are possible using pulsed laser sources. A difficulty with using the conventional approach to making high sensitivity measurements, however, arises from the modal structure of the cavity. The radiation field exiting the cavity is a linear superposition of the fields of the excited modes (7,8). The amplitude and relative phase of these excited cavity modes, in turn, depends on the overlap between the spatial profile and field spectrum of the incident laser field and the spatial profile and spectrum of the cavity modes (see Fig. 1a and 1c). If the excitation spectrum of the laser is broad relative to the longitudinal mode spacing of the cavity ($c/2\ell$), then periodic beating between the excited longitudinal and, perhaps, transverse modes will be evident as a modulation superimposed on the overall decay envelope of the ring-down signal (see Figs. 1c and 1d). The overall decay in this case will represent the superposition of many exponentially decaying signals and Eq. (1) may, at best, approximate the envelope of the observed decay.

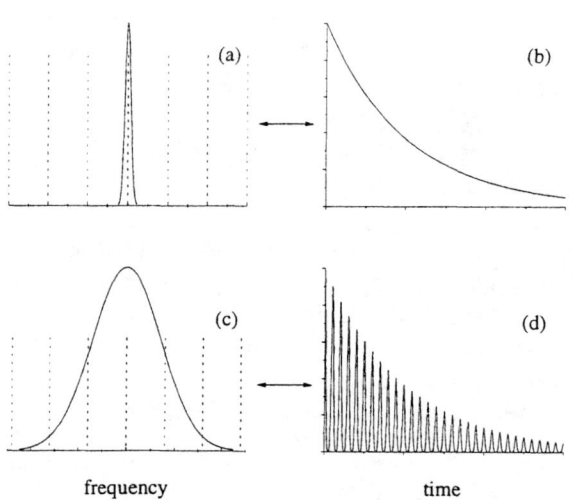

FIGURE 1. This figure illustrates excitation of a ring-down cavity by a transform-limited, Gaussian pulse in the frequency (left) and time (right) domains. On the left, the longitudinal mode structure (dashed lines) and excitation field spectrum (solid lines) are shown for a short (a) and a conventional, long (c) ring-down cavity. On the right, panels (b) and (d) depict the corresponding ring-down signals.

Fluctuations in the spectrum of the laser and variations in the cavity length further complicate the observed signals. For pulsed lasers, the central frequency and spectrum can vary considerably from pulse-to-pulse when compared to the longitudinal mode spacing. Furthermore, if the cavity is not length-stabilized, small length changes from vibrations or thermal fluctuations change the frequency of the longitudinal modes. These two variations conspire to change which cavity modes are excited from shot-to-shot. This effect, averaged over time, masks the cavity mode structure and limits the ultimate spectral resolution to that of the time-averaged power spectrum of the excitation laser. For absorption profiles that are comparable to or narrower than the laser's spectrum, this leads to a loss of spectral resolution and non-exponential ring-down signals (9,10). These factors place practical limits on the precision in the determination of the ring-down time constant, distort absorption profiles and reduce the sensitivity far from the shot-noise limit (9).

Excitation of a single transverse electromagnetic mode (TEM_{qmn}) (11) of a ring-down cavity eliminates these limitations (see Figs. 1a and 1b) and constitutes the special case where Eq. (1) is strictly valid. By constructing length-stable cavities with a mode spacing that is large compared to the excitation field spectrum, single cavity modes can be excited, and the ring-down signal is a single exponential decay. The precision with which the time constant can be extracted from a single exponential is greatly increased when compared to a multi-exponential signal. If a pulsed laser with good spatial and frequency stability is used, the same mode of the cavity can be excited repeatedly. Through synchronized scanning of the excitation laser and the ring-down cavity mode frequency position by varying the cavity length, absorption spectra can be measured. The ultimate spectral resolution of this approach would then be the cavity mode line width, typically 10 - 100 kHz. In practice, the ultimate resolution is limited to the accuracy with which one can synchronously scan the excitation laser and ring-down cavity.

For an excitation of a single mode of a cavity by a laser with a transform-limited, Gaussian spectrum, the power of the radiation exiting the cavity is (5,6,8)

$$\dot{\mathcal{E}}(t;\omega) = \mathcal{E}_p \left[\left(\sqrt{\frac{\pi}{2}} \frac{\Gamma_{qmn}}{\sigma_\omega} e^{-(\omega_{qmn} - \omega_c)^2/2\sigma_\omega^2} \right) \left(\frac{1-R}{1 - Re^{-\alpha(\omega)\ell}} \right)^2 \right] f_{mn} \times \exp\left(-\frac{t}{\tau(\omega_{qmn})} \right) \Big/ \tau(\omega_{qmn}) \qquad (2)$$

In this expression, $\mathcal{E}_p$ is the pulse energy incident on the cavity. The second term in this expression represents the ratio of the areas of the excited (qmn) cavity mode, a Lorentzian of width Γ_{qmn} [$\Gamma_{qmn} = \{2\tau(\omega_{qmn})\}^{-1}$] centered at ω_{qmn}, to the area of the excitation pulse centered at ω_c for an incident field with an $e^{-\frac{1}{2}}$ power spectrum width of σ_ω. The spatial coupling factor, f_{mn}, is the fraction of the incident energy that couples into the (mn) transverse mode series. Together, the product of these terms represent the fraction of the incident energy that is transmitted through the cavity. The last term represents the evolution of the light exiting the cavity with time, a single exponential decay. The time constant of the decay is given in Eq. 1. This decay constant depends on the optical losses within the cavity and varies across an absorption profile. The measurement of the decay time as a function of cavity detuning therefore provides a straight forward measurement of the absorption spectrum.

By synchronized scanning of the excitation laser and cavity ($\omega_{qmn} = \omega_c$), the spectral distribution of cavity losses can be measured. From Eq.(1), we arrive at the following simple expression for the $\alpha(\omega)$,

$$\alpha(\omega) = \frac{\tau(\omega_{qmn})_{empty} - \tau(\omega_{qmn})}{c\,\tau(\omega_{qmn})_{empty}\,\tau(\omega_{qmn})} \quad . \quad (3)$$

where $\tau(\omega_{qmn})_{empty}$ is the empty or off-resonant cavity decay constant. From Eq.(3), one can see that the absorptivity measurement does not depend upon knowing the cavity length. In principle, an absorptivity measurement simply involves measuring two time constants, $\tau(\omega_{qmn})_{empty}$ and $\tau(\omega_{qmn})$, when cavity mode (qmn) is centered at the peak of the absorption profile. If the functional form of the profile is unknown, then the absorptivity as a function of frequency can be measured and the line strength can be extracted by integrating the absorptivity over the entire profile.

EXPERIMENT

As discussed above, a tunable, length-stable optical resonator with a proper mode spacing and a frequency-stable pulsed laser are required for single-mode CRDS measurements. The cavity was constructed within a sample cell made of invar. The cavity was fashioned from two spherical mirrors with radius of curvature of 20 cm, separated by ~10.5 cm. This cavity length and mirror curvature combine to give a longitudinal mode spacing of 1500 MHz and a transverse mode spacing of 500 MHz. One of the mirrors was fixed to the sample cell housing, and the other mirror was affixed to a tubular piezo-electric element in the sample cell. The piezo-electric provided continuous tuning of the cavity length and thus the mode frequency. The optical cavity was enclosed in a sample cell that had fused silica windows which were sealed using o-rings. A platinum resistance thermometer was used to measure the temperature of the sample cell with an inaccuracy of <10 mK. The entire sample cell was contained in a temperature-regulated enclosure. The temperature of the sample cell was found to be constant to ±10 mK during the course of a day and ±2 mK over an hour. It is estimated that the length of the cavity did not change by more than 0.25 nm due to thermal effects during a one hour period, corresponding to frequency shifts of the cavity modes of less than 1 MHz. All samples were natural isotopic oxygen samples. The sample cell pressures were measured with capacitance diaphragm gauges, calibrated against NIST primary pressure standards. The absolute pressure measurement inaccuracy is estimated to be ~0.5%.

An injection seeded optical parametric oscillator (OPO) was used to generate single mode tunable radiation (6). The spectral profile of the OPO is approximately Gaussian with a full width half maximum frequency bandwidth of ~115 MHz in a ~4.5 ns pulse. The OPO seed laser was locked to length-stabilized Fabry-Pérot étalon for frequency stability. It was estimated that the central frequency jitter of the pulsed laser was ~3 MHz. The output of the OPO was spatially filtered and mode matched into the ring-down cavity. The mode excitation spectrum was verified by visual inspection using a CCD camera. With careful alignment, typically about three quarters of the total radiation that coupled into the ring-down cavity was in the lowest order transverse mode. The equivalent power levels at the beginning of a ring-down signal were 20-50 µW, corresponding to $10^8 - 10^9$ photons in a single ring-down trace.

Synchronized scanning of the cavity and the OPO was achieved by first stepping the pulsed laser by frequency shifting the seed laser with an acousto-optic modulator, then bringing the ring-down cavity into resonance by maximizing the ring-down signal at the beginning of the decay. This method of scanning leads to an imprecision in tuning the cavity resonance of ~5 MHz, which is the principle limitation in our line shape measurements. Future efforts will be devoted to developing a higher resolution tuning process.

The transmitted radiation was focused onto a Si:PIN photodiode, and the photocurrent was amplified with a transresistance gain of 2.2×10^3 Ω. The signals were processed by a 12-bit vertical resolution digitizer sampling at 10^8 Hz. Digitized ring-down data were processed shot-by-shot and in real time. The DC-offset was determined from the baseline preceding the decay curve and subtracted from the decay signal. The natural logarithm of the signal was taken, and a weighted linear regression was used to extract the ring-down time (5,6). The weighting factors were taken to be the reciprocal of the sum of the variance in the baseline and the shot-noise variance. Two factors contributed to the baseline variance: the finite digitizer resolution and noise in the photodiode/amplifier and digitizer (5,6). This last source was the dominant component of the variance, and hence these measurements must be considered technical-noise limited.

RESULTS

Time-Decay Measurements

Figure 2 shows a ring-down for an empty cavity. This data is equivalent to when the laser and cavity were detuned far from any optical absorption at low pressures. The measured ring-down time was 13.18 µs ± 0.004 µs, corresponding to single pass losses of 26.33×10^{-6} and a cavity finesse of 1.2×10^5. The uncertainty in a time constant extracted from a single ring-down trace was typically 0.03 % – 0.06 % for this detection/digitizer system (6). Ensemble measurements of ring-down times yielded normally distributed data with a standard deviation equivalent to the uncertainty for a single-shot time constant determination. This level of imprecision is approximately a decade above the shot-noise level.

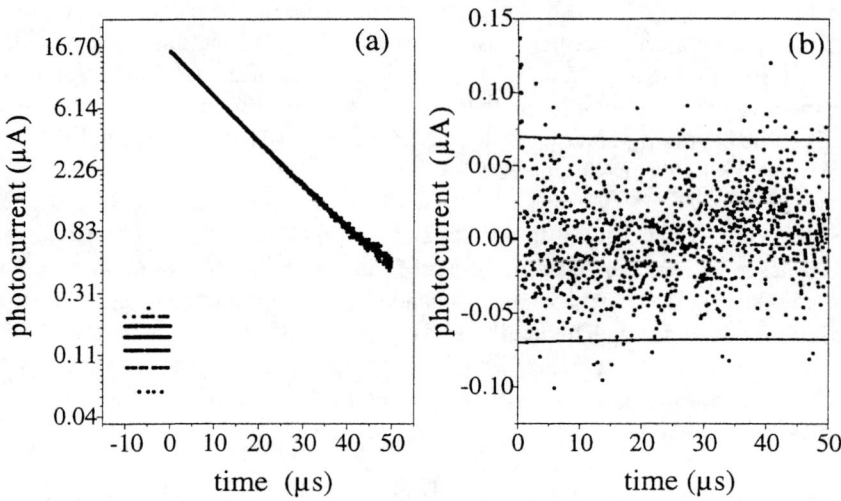

Figure 2. (a) A ring-down signal for a single excited cavity mode. (b) The residuals to a fit of an exponential decay to the measured curve. The solid line in (b) shows the 2-σ uncertainty associated with the measured technical noise.

Spectral Line Shape Measurements

To demonstrate the feasibility of using single-mode CRDS for detailed line shape measurements, absorption profiles for the $b\,^1\Sigma_g^+(v=0) \leftarrow X\,^3\Sigma_g^-(v=0)$ transition (the A-band) of molecular oxygen were recorded. These transitions were chosen for two reasons: (I) the transitions are very weak and therefore measurements made at a modest pressure serve as a good test of the sensitivity of measurement capability, and (ii) this spectrum has been extensively studied, and good line shape and line strength information are available for comparison (12).

Specific examples relevant to line shape measurements are given below. First, to test our ability to record repeatable scans, multiple spectra across a one rovibronic transition are shown. Then, pressure broadened line profiles of this transition and precise measurement of its line strength are demonstrated. Finally, the ability to discern small differences between line shape profiles is demonstrated.

Doppler Profiles

Figure 3 shows three repeat measurements or scans of the $^PQ(9)$ line recorded at an oxygen pressure of 199.98 Pa. The average absorptivity is plotted as a function of detuning. At each detuning for each scan ten ring-down traces were recorded and fit, and the average of the measured time decays was used to calculate the absorptivity using Eq. 3. At this pressure, the transition is expected to very nearly Doppler broadened (12). The line is a best-fit Gaussian profile to the entire data set. Below, the residuals to the Gaussian fits are shown. The residuals are largest in the region where the line shape is steepest and are a result of our laser/cavity tuning procedure, that is, the finite spectral resolution of our measurements. In the far wings of the transitions the residuals have a standard deviation of $\sim 3\times 10^{-9}$ cm^{-1}. In addition, there is a small variation in the empty cavity transmission, the empty cavity mirror transmission changes from about 35.8×10^{-6} to 35.6×10^{-6} over the 4 GHz, red-to-blue interval of this scan. This variation corresponds to an absorptivity of $\sim 1\times 10^{-8}$ cm^{-1}. The line profile data have been fit in two different ways. Fitting individual Doppler profiles (Gaussian) to the data results in an average fitted line width (half-width at half maximum–HWHM) of 428.9 MHz ± 1.6 MHz, about 1.9 MHz or 0.5% larger than the calculated Doppler width for this transition of 427.02 MHz.

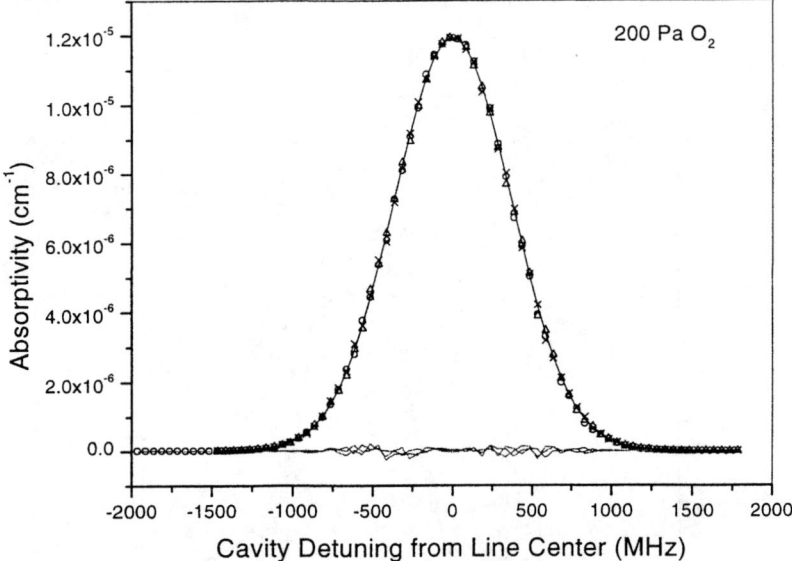

Figure 3. The absorption profile for the $^PQ(9)$ transition of $^{16}O_2$ recorded at a total pressure of ~200 Pa. Three separate scans are shown (solid symbols), along with a Gaussian fit to the data and the residuals of the fit to the data (obs. - calc.).

At this pressure, the estimated self-broadening contribution should be ~2.9 MHz, and a small amount of motional narrowing is also expected, as well. In fact, by fitting a Galatry or soft-collision profile to these data, fixing the broadening and narrowing parameters (calculated using the reported coefficients of Ritter and Wilkerson (12) and measured sample pressures), the best-fit Doppler width is indeed 427 MHz (HWHM).

Using the measured temperature of the cell, fitting the profiles with the line width fixed to the theoretical Doppler width yields line strength 7.387×10^{-24} cm^{-1} (molecule cm^{-2})$^{-1}$ with a standard deviation of 0.003×10^{-24} cm^{-1} (molecule cm^{-2})$^{-1}$. This value is about 1.2 % smaller than the value previously reported by Ritter and Wilkerson (12). It is important to recognize that the standard deviation quoted here is the statistical or random uncertainty with which the line strength is determined; the absolute accuracy requires an assessment of all potential systematic errors. This notwithstanding, the ability to make sensitive line shape and line strength measurements demonstrate the fundamental precision of single-mode cavity ring-down spectroscopy.

Pressure Broadening Measurements

Another test of the measurement approach is demonstrated in Fig. 4 where nitrogen foreign-gas broadening of the oxygen $^PQ(9)$ line was measured for a fixed oxygen pressure

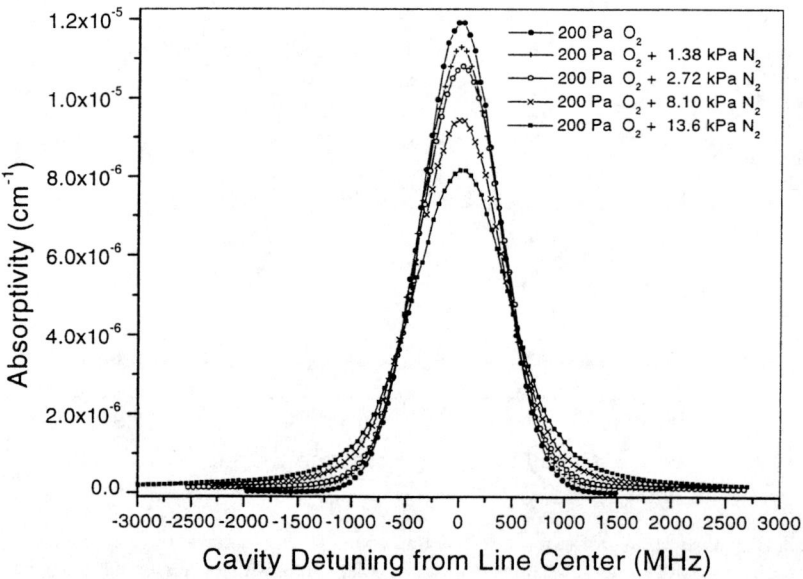

Figure 4. The effect of added molecular nitrogen on the absorption profile of the $^PQ(9)$ transition of $^{16}O_2$.

of ~200 Pa. The average absorptivity at each laser detuning is plotted as a function of the ring-down cavity detuning. As before, at each detuning for each scan ten ring-down traces where recorded and fit, and then the average of the measured time decays was used to calculate the absorptivity. The relative frequency axis for each trace was not adjusted and, within the estimated resolution of our experiment, we do not observe a pressure shift for this transition. The nitrogen pressure shift therefore would be <3.8 MHz Pa^{-1}. The HWHM was extracted for each of the traces in Fig. 4, and the results plotted versus nitrogen pressure in Fig. 5. A linear regression to the data is also displayed. The best-fit to the data has a slope of 13.7 kHz Pa^{-1}. The best-fit intercept has a value of 427 MHz, in excellent agreement of the calculated Doppler-limited value of 427.02 MHz.

The line strength can be determined from this data by fitting a Galatry or soft collision profile to the data using the line shape parameters of Ritter and Wilkerson (12). Mole fraction-weighted values of the line-broadening and collisional narrowing parameters where used in these fits. From the fitting of all of these scans, we find that the line strength determination has a standard deviation of 0.28 %. Again, assessment of the overall accuracy will involve the assessment of all potential systematic errors, which was not undertaken in this analysis.

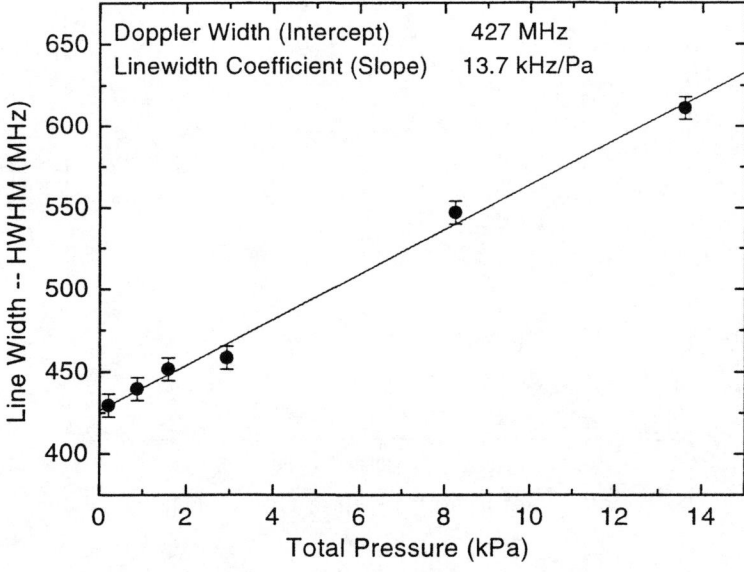

Figure 5. The measured line width (HWHM) for the $^PQ(9)$ absorption profile as a function of total pressure.

Tests of Line Shape Models

As a final demonstration of this measurement capability, we show in Fig. 6 the $^PP(9)$ line shape measured at 10 kPa total oxygen pressure. At these pressures, the line shape is expected to deviate slightly from a Doppler-broadened profile (12). These data were fit to two line shape models. First, they were fit using a Galatry line shape and the broadening and narrowing coefficients reported by Ritter and Wilkerson (13). The off-resonance ring-down cavity loss and the transition line strength were varied. For each datum, the weighting coefficient was taken to be the reciprocal of the sum of the measured variance in cavity loss and the variance associated with the uncertainty in the frequency axis mention earlier. Except in the far wings and near the line center, the maximum contribution to the uncertainty in each datum is associated with the imprecision in the frequency axis. This fit gives a line strength of $8.56 \cdot 10^{-24}$ cm^{-1} (molecule cm^{-2})$^{-1}$, about 0.9 % larger than that reported by Ritter and Wilkerson (12). The fit to the line profile is quite good and yields a reduced-$\chi^2 = 0.81$. Figure 6 also shows a best-fit Doppler profile. The line strength extracted from this fit is ~0.5 % larger than the Ritter and Wilkerson value; however, the quality of this fit is inferior (reduced-$\chi^2 = 31$). The discrepancies

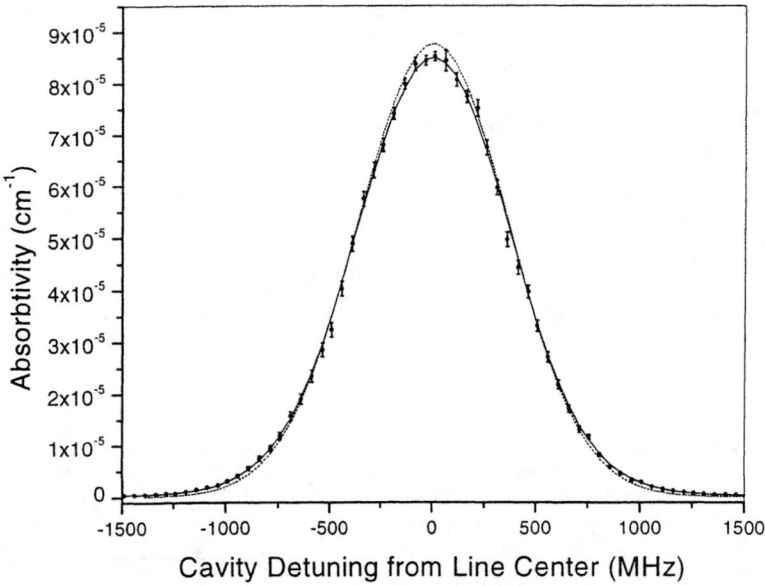

Figure 6. The absorption profile of the $^PP(9)$ transition of $^{16}O_2$. The 2-σ uncertainty in the measured absorptivity are also shown. The solid line is a Galatry fit to the data, the dashed line is a Doppler profile.

between the measured profile and best-fit Doppler profile are most notable near the line center. In the far wings, the discrepancies are also notable. Here, the best-fit Doppler profile gave empty-cavity losses of ~34 $\times 10^{-6}$ per pass, ~10 % larger than the measured value of $31\pm.1\times10^{-6}$ per pass. The ability to distinguish between line shapes demonstrates that the inherent frequency selectivity of optically short cavities can be used to make sensitive and accurate measurements.

CONCLUSIONS

The motivation behind single-mode cavity ring-down spectroscopy was discussed and spectral profiles were measured to demonstrate this technique. This variant has all the inherent advantages of CRDS; to wit, absorption measurements involve simple voltage versus time measurements, the time constant determination is laser intensity independent, the time decay is not impacted by absorption outside of the cavity and noise sources, and potential sources of error can be easily traced. Additionally, the signal are single exponential, which makes it easy to extract quantitative information from the signal and improves the sensitivity of the measurement.

It is clear from these considerations and the quality of the data that single-mode cavity ring-down spectroscopy can be a powerful tool for measurement of weak absorption features and detailed tests of line shape theories. One of the long-standing difficulties in traditional absorption spectroscopies has been the accurate measurement of the "baseline absorption". In the approach discussed here, the baseline can be determined very accurately, better than 0.05% in most cases. This precision should facilitate the measurement of weak continuum absorption or accurate measurements in the far wings of spectral transitions where subtle effects, such as finite duration of collisions, dominate the absorption. These attributes make single-mode CRDS an ideal approach to quantitative absorption and line shape measurements.

REFERENCES

1. O'Keefe, A.; Deacon, D. A. G., *Rev. Sci. Instrum.* **59** 2544-2551 (1988).
2. Romanini, D.; Lehmann K. K., *J. Chem. Phys.* **99**, 6287- 6301 (1993).
3. Zalicki, P. and Zare, R.N., *J. Chem. Phys.* **102**, 2708-2717 (1995).
4. van Zee, R.D., Hodges, J.T., and Looney, J.P., Proc. SPIE **3535** (1999) (in press).
5. Looney, J.P., Hodges,J.T., and van Zee, R.D., "Quantitative absorption measurements using cavity-ringdown spectroscopy with pulsed lasers", in *Cavity Ringdown-Spectroscopy: A New Technique for Trace Absorption Meaurements,* K. A. Busch and M. A. Busch eds., American Chemical Society, 1998.
6. van Zee, R.D., Hodges, J.T., and Looney, J.P., Applied Optics (submitted).
7. Lehmann, K.K. and Romanini, D., *J. Chem. Phys.* **105**, 10263-10277 (1996).
8. Hodges, J.T., Looney, J.P., and van Zee, R.D., *J. Chem. Phys.* **105**, 10278-10288 (1996).
9. Hodges, J. T. Looney, J.P., and van Zee, R.D., *Appl. Opt.* **35**, 4112-4116 (1996).
10. Jongma, R.T. *et al.*, *Rev. Sci. Instrum.* **66** 2821-2828 (1995).
11. Boyd, G.D. and Gordon, J.P., *Bell Syst. Tech. J.* **40** 489-508 (1961).
12. Ritter, K.J. and Wilkerson, T.D., *J. Mol. Spectrosc.* **121**, 1-19 (1987).

Quantum Oscillations: Collisional Broadening of Rydberg States by Alkali Perturbers

Megan E. Henry and Roger M. Herman

Department of Physics, The Pennsylvania State University, University Park, PA 16802

Abstract

This paper qualitatively and quantitatively develops a theory which allows the calculation of the widths and shifts of the two photon Doppler free transitions from the ground states of alkali atoms to Rydberg n^2S and n^2D states. The widths and shifts result from collisions between Rydberg atoms and ground state alkali atoms, either of the same of other species. Impact theory calculations of the broadening and shift calculations applied to $K^{**}(nS) - K$, $K^{**}(nS) - Rb$, $Rb^{**}(nS) - Rb$, $Rb^{**}(nS) - K$ show substantial agreement with experiment.

I. INTRODUCTION

An outstanding problem of fundamental importance in the theory of spectral lines is that of systematically predicting pressure induced widths and shifts for transitions from the ground state to various Rydberg states of alkalis though an impact theory of lineshapes. To date, the best model [1-2] utilizes the ^3P free electron-neutral alkali scattering resonance as if it were a stable state whose energy dynamically crosses that of the Rydberg level in question, leading to coherence destruction. Notwithstanding its success in predicting widths, there seems to be no systematic theory emerging from this model as is illustrated in part by the fact that there is no obvious way to extend the theory to include the line shifts. Furthermore, Borodin and Kazansky treat the non-resonant part of the scattering between the alkali perturber and the Rydberg atom with the free electron theory which limits the validity of the calculation to large n.

In this paper, we shall qualitatively and quantitatively develop a theory first suggested by Herman [3] which predicts the pressure induced broadening for transitions from the ground state to various Rydberg states of alkalis through an impact theory of lineshapes. In describing the interaction between a neutral ground state perturber and the Rydberg electron, we derive a time and kinetic energy dependent Fermi type pseudohamiltonian generalized to include free electron scattering phase shifts for all quasifree electron partial waves relative to the perturbing atom. Utilizing H_{eff} we shall then calculate the widths and shifts for alkali broadening through standard impact theoretical method.

I. QUALITATIVE CONSIDERATIONS

Within the context of the present impact theory the oscillations in the width and shift of Rydberg absorption lines due to collisions with ground state alkali atoms can be understood to result in large part from the ^3P resonance between the Rydberg electron and the alkali perturber. The free electron scattering resonance occurs at an electron energy $E_r \approx 0.02 eV$ for both K and Rb [4]. The Rydberg electron has kinetic energy E_r at a radial distance

$$r_{res} = \left(\frac{1}{2n^{*2}} + E_r \right)^{-1} \tag{1}$$

relative to the ionic core, where $n^* = n - \mu$, μ being the quantum defect. As we shall show in the next section, because the free electron scattering resonance is a p-resonance it is the *gradient* of the Rydberg electronic wavefunction which is associated with the interactions involving this resonance. A maximum in the optical interruption process should therefore occur when the radial derivative is large at r_{res}. For $n \lesssim 15$, this radius lies beyond the outermost extremum in the electronic wavefunction and the resonance therefore causes a monotonic increase in the optical interruption cross section with n. For $n \gtrsim 15$, the derivative of the Rydberg wavefunction at r_{res} passes through maxima and minima as n increases, which leads to the oscillatory behavior of the widths and shifts. For $n > 30$, as r_{res} is further removed from the classical radius of the atom, the distances between maxima in the radial derivative decrease and the oscillations, though still present, become much less pronounced.

This understanding of the relationship between the energy of the ^3P resonance and the collisional broadening suggests that it might be possible to predict with good accuracy the maxima and minima in the width as a function of the principle quantum number n. These results for the self broadening of the $K\ 4^2S - n^2S$ two photon lines are shown in Fig. 1 in which the square of the n^2S wave function radial derivative at r_{res} multiplied by n^{*6}, in arbitrary units,

$$\gamma \sim n^{*6} \left[\frac{dR_{n0}(r)}{dr}\right]^2_{r_{res}} \qquad (2)$$

is compared with the experimental widths. The factor n^{*6} arises from the physical cross section proportionality to n^{*4} multiplied by the collision duration proportionality to n^{*2}. The correlation between them is indeed quite good for the resonance energy $E_r = 19 meV$ [4]. These arguments can be repeated for other alkali combinations in self and foreign atom broadening. While the actual calculation depends upon far more detailed considerations, the idea outlined above provides an intuitive picture of the phenomenon often referred to as the quantum oscillation in lineshape parameters.

III. THEORY

The fundamental problem to be addressed is the description of quasi-free electron scattering in a bound system. Specifically, we shall derive a general pseudohamiltonian H_{eff} which represents the effect of the interaction between a free electron and a neutral perturber, and then use this pseudohamiltonian to describe the interactions between the weakly bound electron and the perturber. Since the speed of ground state neutral atoms at thermal energies is small compared with the average speed of the Rydberg electron in the states under consideration, the Rydberg electron interacts with the perturber as if it were static. This is equivalent to the Born-Oppenheimer separation of slow and fast motions. Once H_{eff} is established for the fast motions involved in the electron perturber interactions, holding the perturber fixed, it can then be used in the description of the effects of slower motions, including that of the perturber. We then carry out this latter step by utilizing the position and kinetic energy dependent H_{eff} into the Anderson-Tsao-Curnutte impact theory of line broadening [5-6] to calculate the elastic and inelastic contributions to the width and shift of the Rydberg spectral lines. In all relevant experiments, two photon laser spectroscopy is used to eliminate Doppler broadening. Therefore, the measured widths and shifts are half those of the complete transition. In our use of the parameters γ and Δ, we shall mean twice the measured value, corresponding to the width and shift of the full transition [1].

A. Derivation of H_{eff}

The pseudohamiltonian describing Rydberg electron-perturber interactions will be written in terms of theoretically calculated free electron-alkali atom elastic scattering phase shifts. These phase shifts describe not the detailed information about what goes on during the encounter but rather its equivalent effect upon the wavefunctions at large distances (compared with the perturber size). Electron-alkali atom elastic phase shifts have been calculated by Sinfailam and Nesbet using variational methods [7] and by Fabrikant, for energies less than $0.1eV$, using the effective range theory [4].

Let us begin with Schrodinger's equation for a free electron in the presence of a potential $V(\vec{r})$,

$$\left[\nabla^2 + k^2 - 2V(\vec{r})\right]\psi(\vec{r}) = 0. \tag{3}$$

This potential represents the actual potential of the ground state perturbing atom and will eventually be replaced with the *effect* of the potential, H_{eff}. Solving for $\psi(\vec{r})$ in (3) for large r, we get the integral equation [8]

$$\psi(\vec{r}) \underset{r\to\infty}{\sim} \exp\left(i\vec{k}_0 \cdot \vec{r}\right) - \frac{\exp(ikr)}{2\pi r} \int d^3r' \exp\left(-i\vec{k}_s \cdot \vec{r}'\right) V(r')\psi(\vec{r}') \tag{4}$$

---

[1] There has been some confusion regarding this factor of two in past literature. Specifically, the experimental results recorded in [10-11] are approximately a factor of two greater than those recorded in [12]. We have verified through a direct correspondense with the authors of [10-11] that in these articles, the full transition was recorded without any note to that effect. Consequently, experimentalists [12] and theorists [20-23] who referred to this article assumed that the measured lines of the half transition were recorded and doubled the results incorrectly. The corrected results of [10-11] now agree in magnitude with the results in [12].

where $\vec{k}_0$ and $\vec{k}_s$ are, respectively, the incident and scattered wavevectors. Since we want to find an expression for the potential which uses the perturbing atom's free electron scattering phase shifts, we now expand the total wavefunction in terms of its partial waves

$$\psi(\vec{r}) = \sum_{l=0}^{\infty} i^l \sqrt{4\pi(2l+1)} \exp(i\delta_l) \frac{u_l(kr)}{kr} Y_l^0(\theta) \qquad (5)$$

where $u_l(kr)$ satisfies the radial equation,

$$\frac{d^2 u_l}{dr^2} + \left[k^2 - 2V(\vec{r}) - \frac{l(l+1)}{r^2} \right] u_l(r) = 0. \qquad (6)$$

We have chosen for convenience the incident wave vector to be directed along the $z - axis$, though eventually we shall generalize the results to handle one or more incident waves traveling in arbitrary directions. Here $u_l(r)$ is normalized, and satisfies a condition of finiteness at $r = 0$. For large r it behaves as

$$u_l(kr) \rightarrow \sin\left(kr - \frac{l\pi}{2} + \delta_l\right). \qquad (7)$$

In order to relate this expression to the atom's potential $V(r)$, we approximate $\psi(\vec{r})$ on the right side of (4). For weak potential scattering at large energies, the scattering phase shifts go to zero and the usual Born approximation can be made for the scattered wavefunction $\psi(\vec{r})$. At low energies, according to Levenson's theorem [9], $\delta_l \rightarrow \nu_l \pi$ where ν_l is equal to the number of bound states of appropriate symmetry within the scattering center. Since the scattering amplitude is small in that case, the scattered wave can be approximated with a free wavefunction as in the Born approximation [8]. However, in order that the radial wavefunction be consistent with its limiting behavior as $r \rightarrow \infty$ and $\delta_l \rightarrow \nu_l \pi$, u_l must behave as

$$u_l(kr) \stackrel{\sim}{\rightarrow} (-1)^{\nu_l} \sin\left(kr - \frac{l\pi}{2}\right) \qquad (8)$$

that is, a factor of $(-1)^{\nu_l}$ must multiply the radial wavefunction so that

$$u_l(kr) \approx (-1)^{\nu_l} kr j_l(kr) \qquad (9)$$

whenever $V(\vec{r}) \to 0$ or $k \to 0$. On the other hand, above a scattering resonance δ_l is more nearly given by $\delta_l \approx n_l \pi$, where $n_l = \nu_l + 1$, the scattering amplitude is small and therefore the wavefunction is largely unperturbed except for an additional 180° phaseshift. Again, using the asymptotic expression to ensure the correct phase of the wavefunction,

$$u_l(kr) \to (-1)^{n_l} \sin\left(kr - \frac{l\pi}{2}\right), \tag{10}$$

the radial wavefunction for small scattering amplitude can be approximated more generally as

$$u_l(kr) \approx (-1)^{n_l} kr j_l(kr), \tag{11}$$

which goes to zero at the origin, is consistent with the asymptotic behavior of the exact radial wavefunction for small δ_l, and remains properly normalized for all values of the phaseshift. Equation (11) therefore serves as the Born approximation to u_l in all cases, where now n_l is the nearest integer multiple of π to δ_l and the complete wavefunction can be approximated as

$$\psi(\vec{r}) \approx \sum_{l=0}^{\infty} (-1)^{n_l} i^l \sqrt{4\pi(2l+1)} \exp(i\delta_l) j_l(kr) Y_l^0(\theta)\,^2. \tag{12}$$

Within the Born approximation (4) becomes, after rearranging,

$$\sum_l \sqrt{4\pi(2l+1)} \frac{\sin\delta_l}{k} (-1)^{n_l+1} Y_l^0(\theta) = \frac{1}{2\pi} \int d^3r' \exp\left(-\vec{k}_s \cdot \vec{r}\right) V(r') \exp(ikz), \tag{13}$$

which is consistent with results found in previous literature [8].

The form of the potential $V(\vec{r})$ representing the interaction between the Rydberg electron and the foreign atom must be sufficiently localized that the electronic wave function acts as a plane wave with a single momentum within the range of $V(\vec{r})$. Therefore the pseudohamiltonian H_{eff} is most conveniently written as a delta function potential. Furthermore, we wish to expand the potential in a way such that each term acts only upon the partial electronic wave (relative to the perturber location) having angular quantum numbers l and m. Therefore, following Fermi, the $l = 0$ term, H_{00} will be a delta function multiplied by a constant so that only the $l = 0, m = 0$ parts

---

[2] We retain the factor of $\exp(i\delta_l)$ here because it is to be cancelled out exactly when used in (4).

of the incident and scattered waves will be seen in the pseudohamiltonian representation of (13); by way of extension of this idea, we now let the $l = 1$ terms be delta functions times operators which pick out the appropriate first derivatives of the wavefunctions. In this way, H_{1m} will isolate the scattering effects of the Y_1^m–dependent contributions to the wavefunctions in the vicinity of the perturbing atom. In general, let us then choose H_{lm} such that it will pick out only the Y_l^m terms. A Hamiltonian which satisfies these criteria is

$$H_{eff}\left(\vec{r}, \vec{R}\right) = \sum_{l=0}^{\infty} \sum_{m=-l}^{+l} H_{lm}\left(\vec{r}, \vec{R}\right)$$

$$= \sum_{l=0}^{\infty} \sum_{m=-l}^{+l} C_l(k) \vec{T}_{lm}^* \delta^3\left(\vec{r} - \vec{R}\right) \vec{T}_{lm} \qquad (14)$$

where the components of each spherical tensor component of rank l, $\vec{T}_{lm}$, satisfies the defining equation

$$T_{lm} \exp\left(i\vec{k} \cdot \vec{r}\right) = Y_l^m(\theta_k, \phi_k) \exp\left(i\vec{k} \cdot \vec{r}\right). \qquad (15)$$

They are shown in Table I for the lowest l values ($l = 0, 1, 2$). While the T_{l0} terms have essentially been derived by Omont for small $|\sin \delta_l|$ [10], the T_{lm} operators defined in (15) allow $H_{eff}\left(\vec{r}, \vec{R}\right)$ to be applied in completely general situations, in which the electronic wavefunctions could have arbitrarily complicated forms in the vicinity of $\vec{r} \cong \vec{R}$.

Inserting $H_{eff}(\vec{r})$ into (13), we find the following relation between the free electron scattering phase shifts δ_l and the coefficients of the collisional Hamiltonian

$$C_l(k) = (-1)^{n_l+1} \frac{8\pi^2}{k} \sin \delta_l(k) \qquad (16)$$

where, of course

$$k = k(r) = \sqrt{-\frac{1}{n^{*2}} + \frac{2}{r}}. \qquad (17)$$

The final expression for the pseudohamiltonian of the Rydberg electron interacting with an intruding ground state atom, written in terms of its phase shifts, is therefore

$$H_{eff}\left(\vec{r}, \vec{R}\right) = \frac{8\pi^2}{k} \sum_{l=0}^{\infty} \sum_{m=-l}^{+l} (-1)^{n_l+1} \sin \delta_l \vec{T}_{lm}^* \delta^3\left(\vec{r} - \vec{R}\right) \vec{T}_{lm}. \qquad (18)$$

It should be noted that this Hamiltonian is valid only within matrix formulations of quantum mechanics, inasmuch as the $\vec{T}_{lm}^*$ operator must operate to the left in forming matrix elements of H_{eff}.

We now show that the pseudohamiltonian (18) is compatible with Fabrikant's *ab initio* calculation of the phase shifts. Fabrikant [4] used the effective range theory to extrapolate low energy scattering phase shifts, $E < 0.1 eV$, from scattering phase shifts at higher energies for the scattering of electrons by the heavy alkali atoms K, Rb and Cs. Matching the external and internal wavefunctions of the scattered electron, he found the free electron scattering phase shifts in terms of the polarizability of the alkali atom α, and two coefficients β and η which he determined from scattering data for each partial wave. He showed the ^3P phase shifts for the alkali atoms to be given as a function[3] of electron energy E in the form

$$\tan \delta_1 (E) = \frac{\frac{-5}{3\pi}\sqrt{\frac{E\alpha}{13.6}} + \frac{\pi \alpha E}{204} + \beta + \eta E}{1 + \frac{204}{\pi \alpha}\left(\frac{\beta}{E} + \eta\right)}. \qquad (19)$$

If the width of the resonance is small, $(\Gamma \ll E_r)$, the resonance occurs at $\delta \approx \frac{\pi}{2}$ which, according to (19), corresponds to an energy $E_r \approx \frac{-\beta}{(\pi\alpha/204 + \eta)}$. However, when the width of the resonance is significant, the background scattering acts to lower the effective resonance due to the dependence of ^3P scattering amplitudes on k^{-3}. Since the resonance between electrons and ground state alkali atoms is broad, $\Gamma_r \approx E_r$ for potassium and rubidium, Fabrikant determined the resonance energy by maximizing the derivative of the phase shift with respect to energy, $\left(\frac{d\delta_1(E)}{dE}\right)_{E_r} = \max$, where the effect of the resonance on the scattering phase shifts is the greatest. A maximum in H_{eff} should occur at the resonance energy. Therefore, to verify the phase shift and energy dependence of the pseudohamiltonian, we compare the value of E_r as found by maximizing the change in the phase shift with that found by maximizing $H_{eff} \propto \frac{\sin \delta_1(E)}{k^3}$ for various parameters β and η as shown in table 2. Good agreement is found between the E_r calculated by the two methods, and as compared with the energy at which $\delta = \frac{\pi}{2}$.

---

[3] In the following, δ_1 indicates the *triplet* P phase shift. Also note that β and η depend on the angular quantum number l and the spin state of the electronic partial wave in scattering from the perturbing spin 1/2 alkali atom; the polarizability α is a characteristic of the perturbing atom by itself.

B. Application of Impact theory to the present problem

We now use the Hamiltonian derived in the previous section in standard Anderson-Tsao-Curnutte impact theory of line broadening [5-6] to calculate the widths and shifts of the collisionally broadened transition spectra. However, we need first to modify the Hamiltonian for the purpose of describing inelastic scattering. Specifically, free electron scattering phase shifts corresponding to this type of inelastic transition are not defined, and the wavevector k in the above expression for H_{eff} (18) is defined only for elastic collisions. Inelastic transitions contribute significantly in the present problem because of the strong collision interaction. The energy difference between Rydberg states which contribute measurably is much smaller than the estimated width of the alkali scattering resonances ($\Gamma_r = 0.016 eV$ for potassium and $\Gamma_r = 0.025 eV$ for rubidium [4]), and are small compared with the resonance (kinetic) energies. We therefore approximate H_{eff} by making the replacement in H_{eff},

$$\frac{\sin \delta_l (k)}{2k} \rightarrow \frac{\sin \delta_l \left(\tilde{k}\right)}{2\tilde{k}} \qquad (20)$$

where $\tilde{k}$ is the geometric mean

$$\tilde{k} = \sqrt{k^2 + k'^2} \qquad (21)$$

and k' is the propagation constant associated with the Rydberg electron at $\vec{R}$ for the inelastically coupled state $\phi_{n'l'm'}$, representing an averaging of the effect of both states. This replacement should introduce negligible error [5].

There are several remarkable aspects of the impact theory as applied to these transitions. The optical interruption cross sections are quite probably the largest of all atomic transitions known to date, characteristically of order $10^5 A^2$. Since these cross sections are significantly larger than interaction cross sections of the atoms in their ground states, in effect only the upper state is perturbed. This situation is possibly unique in all of neutral species spectral line broadening studies to date, and leads to great computational simplifications, in that the dynamics only of the upper state are important. Moreover, for such large impact parameters, classical straight line constant velocity trajectories can be assumed exact.

Since the quantum number of the Rydberg state n is large enough that on average the perturber does not interact with the Rydberg electron and the Rydberg ionic core simultaneously, we ignore the interaction between the Rydberg electron and the dipole moment of the perturber due to its polarization by the Rydberg core Coulomb field. The total collision Hamiltonian is then comprised of two terms,

$$V = H_{pol} + H_{eff} \qquad (22)$$

where

$$H_{pol} = -\frac{\alpha}{2R^4} \qquad (23)$$

is the long range polarization interaction between the Rydberg core and the perturber and H_{eff} represents the complete collisional interaction, including short and long range effects, between the Rydberg electron and the ground state neutral atom as described above. We calculate the orbital of the Rydberg electron using the quantum defect method [11]. We calculate the adiabatic broadening to all orders and explicitly calculate the inelastic contributions to the broadening to second order terms, utilizing the Murphy-Boggs procedure for approximating the cumulative effects of all higher order terms [12].

IV. CONCLUSIONS

The results of the calculations applied to $K^{**}(nS) - K$, $K^{**}(nS) - Rb$, $Rb^{**}(nS) - Rb$, $Rb^{**}(nS) - K$ are shown in Figs. 2-5. Our theory is generally able to predict the overall magnitudes of the widths and shifts, as well as the existence and, in most cases, the period of oscillations as a function of Rydberg quantum number, n, and to some extent the depths of oscillation. In many cases the oscillations are out of phase, and magnitudes differ by factors ranging up to two.

It is difficult to estimate the source of the errors arising from our calculations. Possibly the most significant arises from the fact that we can only estimate the form of the Hamiltonian for inelastic terms; there is no real theory for handling this. We have neglected any contributions coming from the perturber crossing the classically forbidden regions. While the scattering effects presumably can be understood via the analytic continuation of the elastic scattering amplitudes, these effects should be small and will not alter significantly the calculations of the quantum oscillations which one would hope to improve.

There could be errors of an unanticipated nature arising through the Murphy-Boggs procedure for handling those cases in which the second order term of the collisional efficiency, $S_2(b)$, approaches or exceeds unity and possibly higher order terms become significant before the second order terms $S_2(b)$ become very large (which might happen, say, if H_{eff} causes the energy difference to become substantially altered. Furthermore, the effect of polarization of the perturbing atom by the Rydberg atom core may affect the electron scattering phase shifts, and there may exist possible improvements in H_{eff} near the scattering resonances. All of the above lie beyond the present calculation but should eventually be considered.

Lastly, there may be considerable error in the free electron scattering phase shifts used to calculate the broadening of Rydberg lines. In particular, the low energy phase shifts ($E \lesssim 0.1 eV$) cannot presently be verified by beam experiments due to the difficulty in measuring the scattering cross sections at such low energies. In addition, Fabrikant's calculation of scattering parameters through extrapolation from scattering data at higher energies predicts a ^3P bound state for cesium which experimentally does not exist. Until free electron scattering experiments can be performed at lower energies, the best way to verify the low energy free electron phase shifts may well ultimately be through an impact theory of lineshapes which uses free electron scattering phase shifts to predict the width and shifts of Rydberg transitions, whose upper state electrons have kinetic energies in the desired region. The present theory, if sufficiently improved, might eventually be used to verify and improve one's knowledge of the free electron scattering phase shifts.

TABLE I. The $l = 0$, $l = 1$ and $l = 2$ terms in T_{lm}.

l, m	T_{lm}
$0, 0$	$T_{00} = \frac{1}{\sqrt{4\pi}}$
$1, 0$	$T_{10} = \frac{1}{ik}\sqrt{\frac{3}{4\pi}} \nabla_z$
$1, \pm 1$	$T_{11} = -T_{1,-1}^* = -\frac{1}{ik}\sqrt{\frac{3}{8\pi}} (\nabla_x + i\nabla_y)$
$2, 0$	$T_{20} = \frac{-1}{k^2}\sqrt{\frac{5}{4\pi}} \left(\frac{3}{2}\nabla_z^2 - \frac{1}{2}\nabla_r^2\right)$
$2, \pm 1$	$T_{21} = -T_{2,-1}^* = \frac{1}{k^2}\sqrt{\frac{15}{8\pi}} \nabla_z (\nabla_x + i\nabla_y)$
$2, \pm 2$	$T_{22} = -T_{2,-2}^* = \frac{-1}{4k^2}\sqrt{\frac{15}{2\pi}} (\nabla_x + i\nabla_y)^2$

TABLE II. A comparison between the energy when $\delta_1 = \frac{\pi}{2}$ and the energy resonance found by maximizing $\frac{d\delta}{dE}$ and $\frac{\sin \delta_1}{k^3}$. All energies are in eV.

Atom	β	$\eta\,(eV^{-1})$	$E\left(\delta=\frac{\pi}{2}\right)$	$E_r\,(\max \frac{d\delta}{dE})$	$E_r\,(\max \frac{\sin \delta_1}{k^3})$
K	-0.1065	0.1877	0.0219	0.0193	0.0185
Rb	-0.1507	0.1562	0.0289	0.0234	0.0226
	-0.01507	0.1562	0.00289	0.00283	0.00282
	-0.0351	-0.0365	0.00700	0.0066	0.00655
Cs	-0.0655	0.3574	0.0100	0.0092	0.0090
	-0.0923	0.4021	0.0140	0.0125	0.0122
	-0.00593	0.2584	0.000920	0.00091	0.00091

REFERENCES

1. V. M. Borodin and A. K. Kazansky, J. Phys. B **25**, 971 (1992).

2. V. M. Borodin and A. K. Kazansky, Opt. Spectrosc. (USSR) **83**, 664 (1997).

3. T. F. Gallagher, *Rydberg Atoms* (Cambridge University Press, Cambridge, Massachusetts 1994).

4. I. I. Fabrikant, J. Phys. B **19**, 1527 (1986).

5. P. W. Anderson, Phys. Rev. **76**, 647 (1949).

6. C. J. Tsao and B. Curnutte, J. Quant. Spectrosc. Radiat. Transfer **2**, 41 (1961).

7. A. L. Sinfailam and R. K. Nesbet, Phys. Rev. A **7**, 1987 (1973).

8. E. Merzbacher, *Quantum Mechanics* (John Wiley & Sons, New York 1961).

9. B. H. Bransden, *Atomic Collision Theory* (Benjamin/Cummings, Reading, Massachusetts, 1983).

10. A. Omont, J. Phys. (Paris) **38**, 1343 (1977).

11. C. Fabre and S. Haroche, in *Rydberg States of Atoms and Molecules*, edited by R. F. Stebbing and F. B. Dunning (Cambridge University Press, Cambridge, Massachusetts 1983).

12. J. S. Murphy and J. E. Boggs, J. Chem. Phys. **47**, 691 (1967).

13. D. C. Thompson, E. Kammermayer, B. P. Stoicheff, and E. Weinberger, Phys. Rev. A **36**, 2134 (1987).

14. H. Heinke, J. Lawrenz, K. Niemax, and K. H. Weber, Z. Phys. A **312**, 329 (1983).

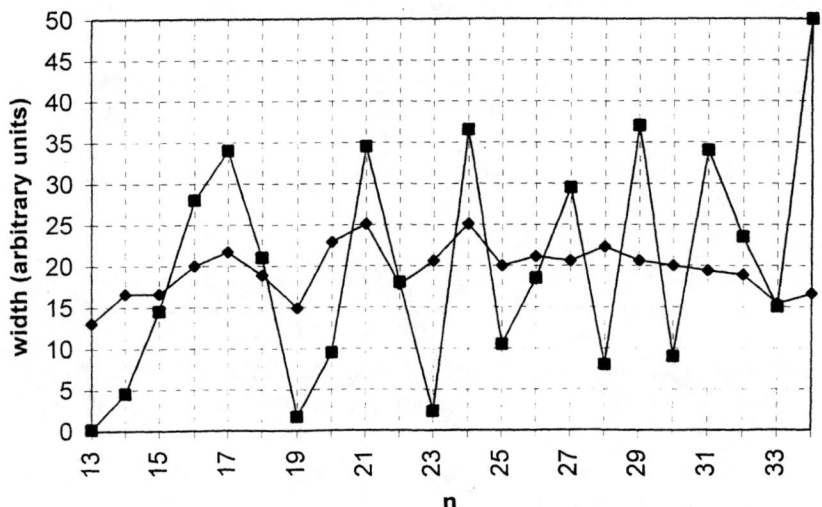

Fig. 1. Estimated widths (arbitrary units), solid squares, based on (2) and observed widths [13] versus n for $K^{**}(ns) - K$.

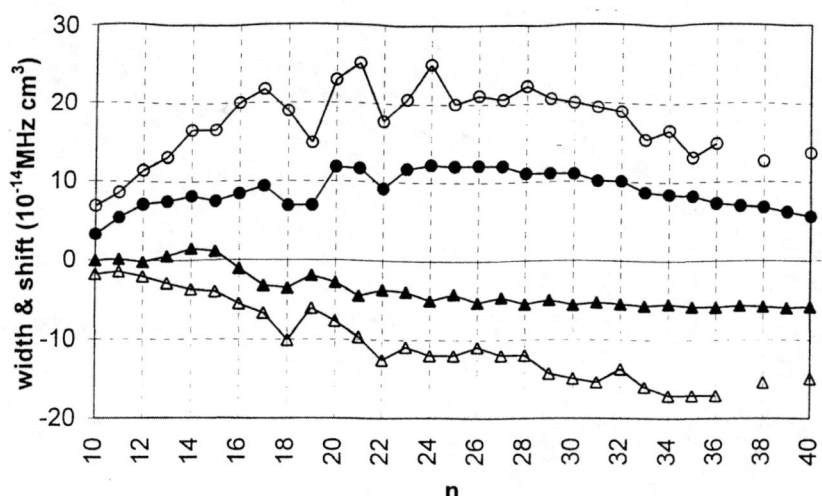

Fig. 2. Theoretical (closed circles) and observed (open circles) widths and theoretical (closed triangles) and observed (open triangles) shifts for alkali transitions to n^2S states versus n for $K^{**}(ns) - K$. Experimental results are taken from [13].

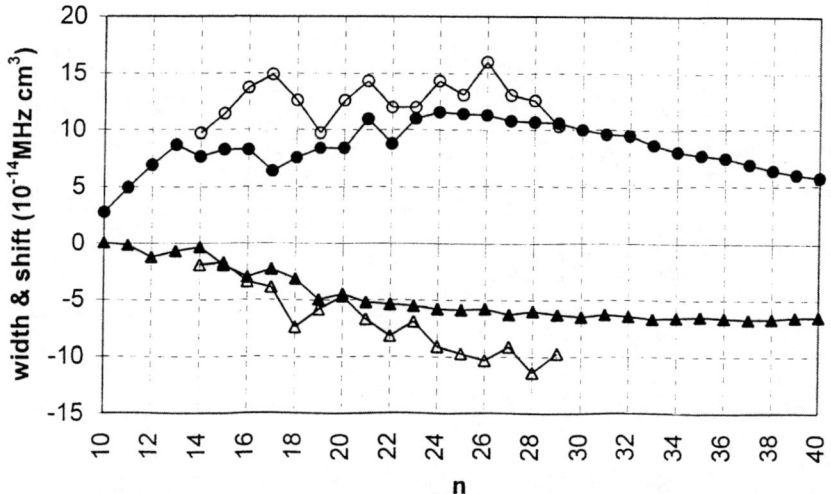

Fig. 3. Same as Fig. 2 except for $K^{**}(ns) - Rb$.

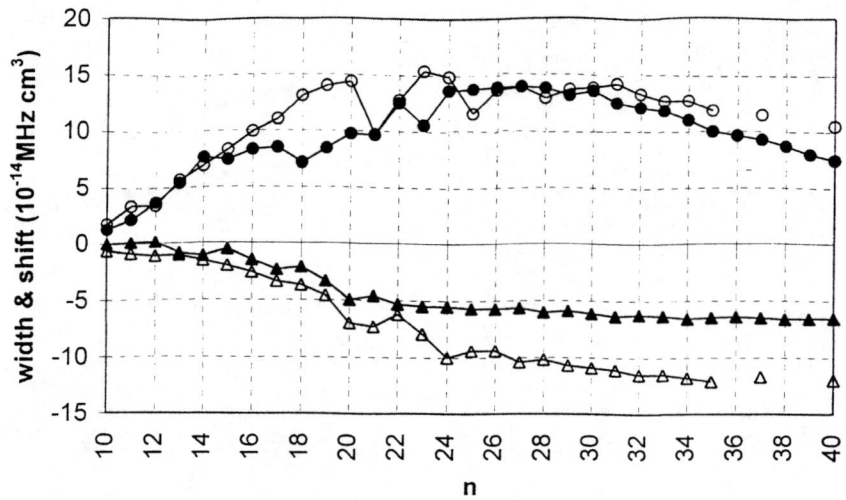

Fig. 4. Same as Fig. 2 except for $Rb^{**}(ns) - Rb$.

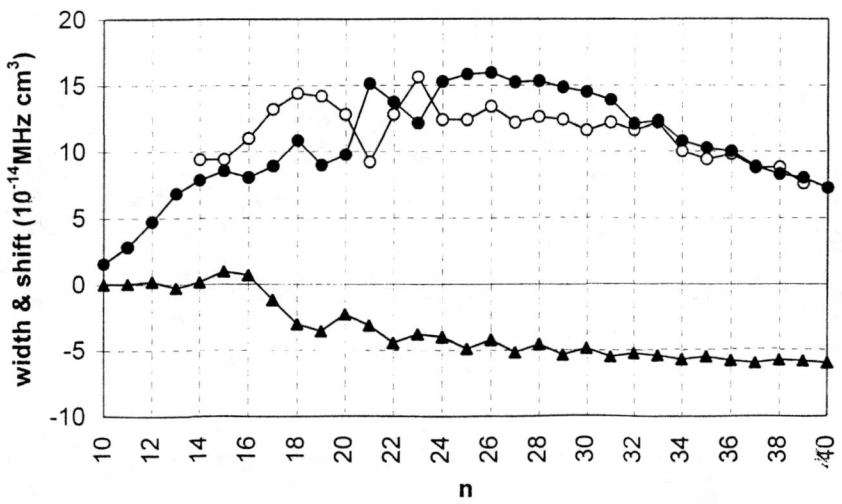

Fig. 5. Same as Fig. 2 except for $Rb^{**}(ns) - K$. Experimental results are taken from [14].

Spectral Boltzmann Distribution: an "Infrared Catastrophe"

Andrey N. Starostin[*], Alexei G. Leonov[†], Dmitriy I. Chekhov[†], Andrey Yu. Sechin[*], Yuriy K. Zemtsov[*]

[*]*Troitsk Institute for Innovation & Fusion Research 142092, Troitsk, Moscow Region, Russia*
[†]*Moscow Institute of Physics & Technology 141700, Dolgoprudnyi, Moscow Region, Russia*

Abstract. The review of recent achievements in the problem of resonance radiation transfer in dense highly absorptive media is presented. It is shown, in particular, that a spectral distribution of a resonance level population is proportional to $\exp(-\hbar\omega/T)$, where $\hbar$ is Planck constant, ω – current frequency, T - temperature. The availability of this factor can result in occurrence of the second maximum on a far-red wing of a resonant line. According to the theoretical prediction in a luminescence spectrum for sodium resonance line a maximum in the region of 2-3 micrometers must be observed. Intensity of a radiation in this maximum by some orders of magnitude exceeds the value, obtained from a standard theory of radiation transfer. This phenomenon can be conditionally named as "infrared catastrophe". The experimental measurement of the spectra and radiation intensity in IR and visible regions are in the agreement with the advanced theory.

1. INTRODUCTION

In recent years increasingly more attention is paid to a problem of the luminescence spectra formation and the resonance radiation transfer processes in dense plasmas and gaseous media [1-7]. Substantial progress in this field demonstrates the indispensability to essentially review certain conventional assumptions in the standard theory of resonance radiation transfer [8,9] developed for the case of low gaseous pressure and, in particular, the necessity to abandon a number of adopted approximations. First of all, it refers to the known Biberman's applicability criterion of the standard theory $\varepsilon'-1\ll 1$ (ε' is the real part of dielectric permeability of medium: $\varepsilon=\varepsilon'+i\varepsilon''$), and to the approximation of a narrow line $\Delta\omega=\omega-\omega_o \ll\omega_o$, where ω_o is the frequency of resonance transition, ω is the frequency of emitted photon. Taking these circumstances into account, the generalized theory of resonance radiation transfer, free from the above limitations, has been first developed in [10,11]. In particular, it was shown in [11,12] that in the interior of the dense highly absorptive medium the intensity of equilibrium radiation J_ω within the line profile may essentially exceeds its Plank value. Besides, it has been found in [10] that under certain conditions the resonance line contour may have the second wide maximum displaced from ω_o to the low frequency region, in which case the radiation intensity on this "red" wing of the line may exceed by many orders of magnitude the intensity of near-resonance part of the line. This phenomenon was referred to in [13] as an infrared "catastrophe".

The effects considered may manifests themselves to a large extent in the observation of thermal radiation emitted by a dense nonuniformly heated medium, which permits an experimental verification of the theory outlined in [10,11]. The present work is concerned with the development of theory and the experimental observation of the effects predicted in [10,11] on the example of thermal radiation of the dense sodium vapors steadily heated up to temperatures 600-1200 K, comparing the experimental data with the results of numerical modeling. In the experiments a structure of sodium resonance doublet 3P - 3S has been studied. It should be specially noted that a pure thermal luminescence of the vapors has been registered in the absence of their excitation by electric field or external source of radiation.

2. THEORETICAL CONSIDERATION

A problem of resonance radiation transfer is solved using the equation for the Fourier components of correlation functions of electromagnetic field, which can be formulated in terms of kinetic Green functions [11]. It permits to introduce generalized spectral "intensity" of radiation $J(\omega,\mathbf{k})$ in which the frequency ω and the wave vector $\mathbf{k}$ are independent variables. Thus, the spectral intensity of radiation J_ω, with which one usually deals in measurements and in the standard theory of radiation, is connected with $J(\omega,\mathbf{k})$ by the relation [11]:

$$J_\omega(\Omega) = \frac{2c^2}{\omega} \int_0^\infty kJ(\omega,\mathbf{k}) \frac{k^2 dk}{(2\pi)^5}, \qquad (1)$$

where Ω is the unit vector in the direction $\mathbf{k}$. Under conditions of the thermodynamic equilibrium in infinite medium the function $J(\omega,\mathbf{k})$ can be found from fluctuation-dissipation theorem for fluctuations of spectral density of transversal electromagnetic field strength:

$$J(\omega,\mathbf{k}) = \frac{8\pi\hbar\omega^4 n(\omega)\varepsilon''}{\left|\omega^2\varepsilon - c^2 k^2\right|^2}, \qquad (2)$$

where n(ω) are the equilibrium photon occupation numbers given by the Planck formula:

$$n(\omega) = (\exp(\hbar\omega/T) - 1)^{-1}. \qquad (3)$$

If we substitute (2) in (1) we obtain (v_T is the thermal velocity of atom):

$$J_\omega \approx \frac{\hbar\omega^3 n(\omega)}{4\pi^3 c^2} \frac{\varepsilon'}{2}\left\{1 + \frac{2}{\pi}\arctg\frac{\varepsilon'}{\varepsilon''} + \frac{4\varepsilon''}{\varepsilon'}\ln\left[\frac{\Gamma}{\omega v_T/c}\right]\right\} \equiv J_\omega^{Pl}\psi(\omega)\varepsilon', \qquad (4)$$

where J_ω^{Pl} is the Planck intensity of black body radiation in vacuum. Since at large medium density N the width Γ is determined by the resonance collisions according to Vlasov-Fursov mechanism [14] ($\Gamma \propto N$), then due to logarithmic term in (4), growing with increasing density, the equilibrium intensity in absorbing medium can exceed the intensity of radiation in transparent medium determined by the Klausius formula $J_\omega^{Pl}\varepsilon'$ almost by the order of magnitude. It should be noted that the numerical calculations of J_ω for sodium vapors with account of the dependence ε on ω and **k** show satisfactory accuracy of the approximation (4) [11]. The calculation of intensity of such radiation should be also executed using the function $J(\omega,k,\Omega,\mathbf{r})$ introduced above. In steady state case the generalized spectral "intensity" of radiation $J(\omega,k,\Omega,\mathbf{r})$ should satisfy simultaneously two equations. One of them has a form of kinetic equation:

$$(\Omega,\Delta)J = -k_\omega J + \tilde{\varepsilon}(\omega,k,\Omega,\mathbf{r}), \qquad (5)$$

where k_ω is the absorption coefficient and $\tilde{\varepsilon}$ is the generalized spectral intensity of volume spontaneous emission [11]:

$$k_\omega = \frac{\omega^2 \varepsilon''(\omega,k)}{c^2 k} =$$

$$= \frac{4}{3}(g_2/g_1)\frac{(\pi d\omega)^2}{\hbar c^2 k}a(\omega,k)\{Ñ_1 - (g_1/g_2)Ñ_2 \exp[-\hbar(\omega-\omega_o)/T]\}; \qquad (6)$$

$$\tilde{\varepsilon} = \frac{4}{3}\frac{d^2\omega^3\hbar\omega}{\hbar c^2 k}\frac{\omega^2\varepsilon''}{|\omega^2\varepsilon - c^2 k^2|^2}(2\pi)^3 a(\omega,k)Ñ_2 \exp[-\hbar(\omega-\omega_o)/T]. \qquad (7)$$

In (6) and (7) $a(\omega,k)$ is the generalized contour of line, $g_{1,2}$ are the statistical weights of the ground and excited states, $N_{1,2}$ are their populations (see below) which in equilibrium obey the Boltzmann relation and in the absence of equilibrium can be found from the equations of kinetics given in [10]. Integrating (7) over k and neglecting spatial dispersion (thereby $a(\omega,k) \approx a(\omega)$) one can obtain an obvious expression for the spectral intensity of spontaneous radiation in highly absorptive medium:

$$\varepsilon_\omega = (1/4\pi)\hbar\omega A_o \left(\frac{\omega}{\omega_o}\right)^3 \text{Re}(\sqrt{\varepsilon})a(\omega)\tilde{N}_2 \exp[-\hbar(\omega-\omega_o)/T]. \tag{8}$$

It follows from the equation (5) that in the general case for the spectral intensity J_ω determined by (1) does not exist a closed equation and, besides the equation with partial derivatives (5), the function $J(\omega,k,\Omega,\mathbf{r})$ should also satisfy a nonuniform wave equation which in the stationary case has a form:

$$\left[-\frac{c^2}{2}\Delta + 2(c^2k^2 - \omega^2\varepsilon')\right]J =$$
$$= \frac{8\pi\hbar\omega^4\varepsilon''}{|\omega^2\varepsilon - c^2k^2|^2} \frac{2(c^2k^2 - \omega^2\varepsilon')}{\{(N_1/N_2)(g_2/g_1)\exp[-\hbar(\omega-\omega_o)/T]-1\}}. \tag{9}$$

In the uniform case the same result (2) follows from (9) as from kinetic equation (5). The boundary conditions for intensity on the limiting surface can be obtained considering a problem of radiation emission by heated medium into vacuum. In particular, for intensity of radiation $J_\omega^{ex}(\Omega)$ escaping the heated medium which is characterized on the boundary by the temperature T_o, the reflection coefficient $R = R_o$, the dielectric permeability $\varepsilon = \varepsilon_o$, and the function $\psi = \psi_o$, one obtains:

$$J_\omega^{ex}(\Omega) = \frac{J_\omega^{(-)}(\Omega)|_{z=0}}{\varepsilon_o(\omega)\psi_o(\omega)}[1 - R_o(\omega,\Omega)], \tag{10}$$

where $J_\omega^{(-)}$ is the intensity of radiation from medium onto a boundary.

Occurrence of the exponential factor $\exp[-\hbar(\omega-\omega_o)/T]$ in (6)-(8) (the necessity of which was suggested formerly in [15,16]) is connected with generalization of the standard theory of radiation transfer [8,9] for the case of wide lines typical for dense media and is rigorously substantiated in [10] in the framework of L.V.Keldysh's theory using the kinetic Green functions [17-20]. Let us consider, for instance, the case of equilibrium of two-level atoms. The equilibrium condition implies an equality of the radiation decay rate for spectral density of exited atoms (described by Green function $G_2^{+-}(E,\mathbf{p})$, so that $N(\omega,\mathbf{p}) = -(iG^{+-}/2\pi)$, [18]) and the rate of photoabsorption from the ground state (cf. [17,18]):

$$G_2^{+-}\Sigma^{+-} = G_2^{-+}\Sigma^{-+} \tag{11}$$

Here the mass operators Σ^{+-} and Σ^{-+} correspond to the rates of decay (including spontaneous and stimulated radiation) and excitation under the effect of radiative processes. From equilibrium condition (11) one obtains:

$$\tilde{N}_2(\mathbf{p})a_2(\varepsilon)\int d\omega_k d\Omega \omega_k^3 (1+n_k)[1-\tilde{N}_1(\mathbf{p}-\hbar\mathbf{k})] \times$$
$$\times a_1\{\varepsilon - \omega_k + \omega_o + [E(\mathbf{p})-E(\mathbf{p}-\hbar\mathbf{k})]/\hbar\} =$$
$$= [1-\tilde{N}_2(\mathbf{p})]a_2(\varepsilon)\int d\omega_k d\Omega \omega_k^3 n_k \tilde{N}_1(\mathbf{p}-\hbar\mathbf{k}) \times$$
$$\times a_1\{\varepsilon - \omega_k + \omega_o + [E(\mathbf{p})-E(\mathbf{p}-\hbar\mathbf{k})]/\hbar\}, \tag{12}$$

where $\omega_o = \omega_2 - \omega_1$. $a_i(\varepsilon)$ is the spectral density (contour) of the state i:

$$a(\varepsilon) = \frac{\gamma_i/2\pi}{\varepsilon^2 + (\gamma_i/2)^2}, \tag{13}$$

γ_i is the width of the state. In (12) and (13) we denoted:

$$\varepsilon = \omega - \omega_i - E(\mathbf{p})/\hbar + \mu. \tag{14}$$

Here ω_i is the frequency (energy) of the state i, $E(\mathbf{p}) = p^2/2m$ is the kinetic energy of atom as a whole, μ is the chemical potential.

The condition (12) becomes identity if $\tilde{N}_i(\mathbf{p})$ has a form (cf. [10]):

$$\tilde{N}_i(\mathbf{p}) = \frac{1}{\exp\{[\hbar\varepsilon + \hbar\omega_i + E(\mathbf{p}) - \mu]/T\}+1} \equiv \frac{1}{\exp(\hbar\omega/T)+1} \tag{15}$$

If we introduce auxiliary populations $\tilde{N}_i$ determined as:

$$\tilde{N}_i = \exp(-\omega_i/T)[\exp(\mu/T)/\lambda_T^3] \tag{16}$$

(where the statistical weights g_i are put, for simplicity, equal to unity; λ_T is the thermal de Broglie wavelength of the atom) and neglect the degeneration, then the rate of radiative decay of atom in a state 2 with momentum $\mathbf{p}$ and detuning ε will be proportional to $R_\downarrow$:

$$R_\downarrow \cong \tilde{N}_2 \lambda_T^3 \exp[-E(p)/T - \varepsilon/T](1+n_k)a_2(\varepsilon). \tag{17}$$

Here $n_k = [\exp(\hbar\omega_k/T)-1]^{-1}$, and the quantities ω_k and ε - detuning from a position of level 2 are connected by relation:

$$\varepsilon = \omega_k - \omega_o - [E(\mathbf{p}) - E(\mathbf{p}-\hbar\mathbf{k})]/\hbar. \tag{18}$$

Similarly, a rate of photoexcitation of atom from the state i = 1, with momentum $\mathbf{p}$ - $\hbar\mathbf{k}$ in the same notations has a form:

$$R_\uparrow \cong \tilde{N}_1 \lambda_T^3 \exp[-E(\mathbf{p}-\hbar\mathbf{k})/T] n_k a_2(\varepsilon). \tag{19}$$

Setting (17) equal to (19) one obtains (taking into account (16), (18)):

$$\frac{1+n_k}{n_k} = \exp(\omega_k/T) \equiv \exp\{[\hbar\varepsilon + \hbar\omega_o + E(\mathbf{p}) - E(\mathbf{p}-\hbar\mathbf{k})]/T\}. \tag{20}$$

The relation (18) for the frequency of photon ω_k means that for a broadened atomic state 2 the photon emitted on a wing of line ($\varepsilon \neq 0$) has an energy which is determined not only by the frequency of transition and the recoil energy of atom $E(\mathbf{p}) - E(\mathbf{p}-\hbar\mathbf{k})$ but also by the value of detuning ε.

Integrating the spontaneous decay rate over momenta of emitting atoms (and omitting a term with n_k in (17)) one can obtain:

$$\langle R_\downarrow \rangle_{spont} = \tilde{N}_2 \exp[-\hbar(\omega_k - \omega_o)/T]\varphi(\omega_k - \omega_o) \tag{21}$$

where $\varphi(\omega_k - \omega_o) \equiv \varphi(\Delta)$ is the Voigt profile of the emission line. One must note that in deducing the above relations the particular mechanism of homogeneous broadening is not important, so that γ_2 may be even determined by pure radiative mechanism of broadening. Qualitatively the occurrence of the exponential factor may be explained in the following way. In the approximation of narrow line $\Delta\omega \ll \omega_0$ and in a rarefied medium the spectral intensity of volume radiation for a resonance medium is described by expression [8] (cf. (10)):

$$\varepsilon_\omega = (1/4\pi)\hbar\omega_o A N_2(\omega), \tag{22}$$

where $N_2(\omega)$ is the spectral density of resonance level population, i.e., the density of atoms capable to emit a photon with frequency ω (in the approximation of complete redistribution of absorbed and emitted photons over frequencies, it is assumed that $N_2(\omega) = N_2 a(\omega)$, where N_2 is the total population of the resonance level). However, assuming for equilibrium $N_2 = (g_2/g_1) N_1 \exp(-\omega_0/T)$ [8], one can not obtain from (22) a correct expression for the equilibrium Planck intensity J_ω^{Pl}, since the photon occupation numbers $n(\omega)$ would contain a resonance frequency ω_0, rather than a current frequency ω. Moreover, as can be easily seen, the integral of ε_ω over all frequencies, determining a total intensity of line radiation, in this approximation will diverge. To obtain a correct formula for J_ω^{Pl}, and ensure a convergence of corresponding integral, one must modify the expression for $N_2(\omega)$, introducing in it the correction factor:

$$\widetilde{N}_2(\omega) = \widetilde{N}_2 a(\omega) \exp(-\hbar(\omega - \omega_0)/T). \qquad (23)$$

In equilibrium one has the Boltzmann spectral distribution:

$$\widetilde{N}_2(\omega) = \widetilde{N}_1 a(\omega) \exp(-\hbar\omega/T) \qquad (24)$$

(in the absence of equilibrium the effective populations $\widetilde{N}_i(\omega)$ are found from equations of kinetics given in [10]).

The introduction of factor $\exp[-\hbar(\omega-\omega_o)/T]$ leaves the core of the line $\Delta\omega \sim \Gamma$ practically unaffected (though, for sufficiently wide line even more enhances an asymmetry of absorption and emission lines in a dense medium), whereas at $T \ll \omega_o$ it may lead to the second maximum of the line contour on its far "red" wing in the region of several T. Being so, the most part of the energy emitted in line may just pertain to its non-resonance red wing, rather than to the central near-resonance part, which was called conventionally in [13] an infrared "catastrophe".

3. NUMERICAL MODELING PROCESSES OF RESONANCE RADIATION TRANSFER IN DENSE SODIUM VAPORS

A numerical solution of equation (5) with boundary conditions (10) was employed to calculate the intensity J_ω of the resonance line radiation both inside and outside the nonuniformly heated slab of sodium vapors with length L in the presence of buffer gas - argon at pressure $P \sim 1$ atm. In calculations it was assumed that in near-resonance region the broadening is determined by impact mechanism with a dominant role of resonance collisions in the region of high temperatures. The results of calculations are presented in Figs.1a,b. In Fig.1a the frequency dependence is shown for the radiation intensity in the center of a slab with thickness L = 15 cm for the temperature in center T = 1000 K. The temperature profile through the slab corresponds to the conditions of experiment (see

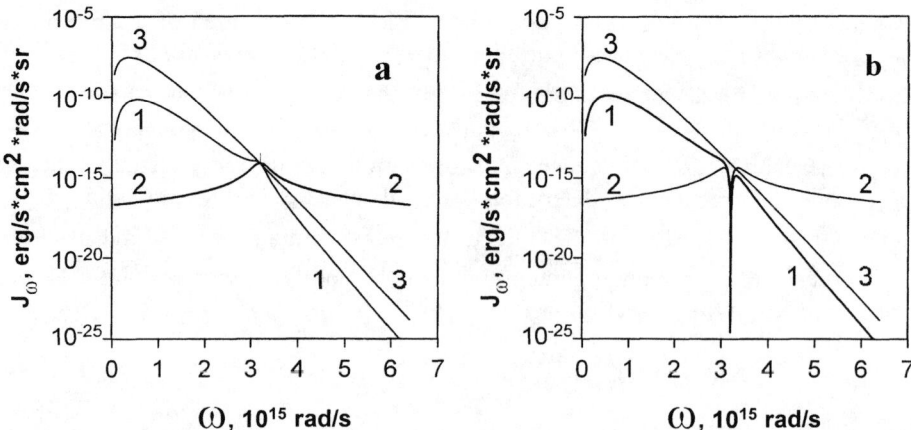

FIGURE 1. Frequency dependence of spectral intensity of radiation in the center of sodium vapor slab (a) and outgoing intensity of radiation (b). T = 1000 K, L = 15 cm. 1 - calculations by present theory, 2 - calculations by standard theory of radiation [8], 3 - Planck intensity.

below), the density of Na atoms was assumed to be corresponding to saturation vapor pressure at a given temperature [28]. The curve 1 corresponds to the theory presented above, the curve 2 - to the standard theory of spectral line radiation [8], the curve 3 - to the Planck intensity. It is easily seen from the graph that on the curve 1 a narrow peak reveals distinctly itself, exceeding the Planck intensity, that owes its origin to strong dispersion and absorption in sodium vapors, which was discussed in detail above. In the IR-region of spectrum (at $\omega = \omega_{max} \sim 5 \cdot 10^{14}$ s^{-1}) the calculations demonstrate the appearance of wide maximum with radiation intensity exceeding by several orders of magnitude the value obtained for the same frequency according to the standard theory. On the "blue" wing the situation reverses and, which is most important, the integral of intensity over frequency calculated by the standard theory diverges at high frequencies. From the presented data it follows that the integral intensity of resonance line spectrum is predominantly determined by a "red" maximum, whereas the contribution of the central part of line is relatively small. It should be noted, that according to calculations, the optical depth of slab τ in the "red" maximum is comparatively small (under given conditions $\tau \sim 0.1$), so that a soft photon can escape a limited medium.

In Fig.1b the spectral dependencies are shown for the intensity of thermal radiation outgoing from the medium under the same conditions. From the curves presented it is seen that the same features manifest themselves in the escaping radiation far from resonance. However, at small detunings a peak at the resonance frequency converts into a dip due to self-absorption of line; nevertheless, as the calculation shows, in this case the intensity at the center of line still exceeds the intensity calculated by standard theory by several orders of magnitude.

4. RADIATION OF PLASMA. ESTIMATES FOR INTENSITIES OF RESONANCE LINE AND CONTINUUM DUE TO PHOTORECOMBINATION AND BREMSSTRAHLUNG

In previous section a spectral structure of pure thermal radiation of resonance medium has been considered on the example of sodium vapors in the absence of ionization. In plasma the intensity of radiation in a far line wing, high as it may be, will compete with intensity of radiation determined by other mechanisms, in particular, Bremsstrahlung and photorecombination. Let us consider, as an example, the radiation of line L_α in plasma containing hydrogen atoms or hydrogenlike ions, comparing the volume emissivity ε_ω on the line wing with the intensity of continuum. For multiply charged ions the Bremsstrahlung and the recombination radiation were taken into account (in the Kramers' approximation [29]), the calculations were carried out for plasma consisting of "bare" nuclei, hydrogenlike and heliumlike ions (ionic composition was determined by Saha formula). For hydrogen the Bremsstrahlung of electrons on atoms was also taken into account. The line contour for large detuning from resonance was taken as the sum of Stark and dispersion wings. The results of calculations for atom of hydrogen and hydrogenlike ions are presented in Figs.2a,b.

Fig.2a shows the emissivity of hydrogen atom in an equilibrium isothermal plasma due to various mechanisms of radiation at the electron density $N_e = 6.4 \cdot 10^{13}$ cm^{-3} and the temperature $T = 0.55$ eV. The parameters mentioned correspond

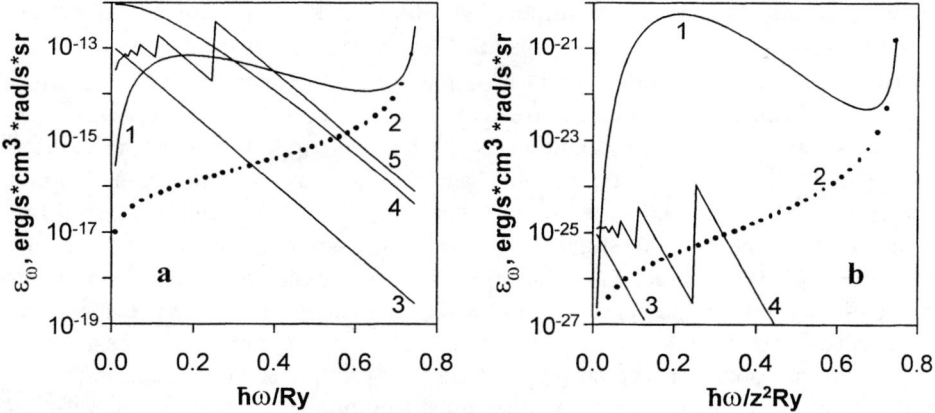

FIGURE 2. Spectral emissivity of hydrogenic plasma.
(a) $z = 1$, $N_e = 6.4 \cdot 10^{13}$ cm^{-3}, $T = 0.55$ eV. 1,2 - radiation of line L_α (1 - calculation by present theory, 2 - by standard theory). 3 - Bremsstrahlung on atoms, 4 - Bremsstrahlung on ions, 5 - photorecombination radiation.
(b) $z = 20$, $N_e = 10^8$ cm^{-3}, $T = 150$ eV. 1,2 - radiation of line L_α, 3 - Bremsstrahlung on ions, 4 - photorecombination radiation. 1 - calculation by present theory, 2 - by standard theory.

to conditions in the photosphere of the solar disk, where the plasma consists mainly of hydrogen (Harward-Smithson standard model of solar atmosphere - HSRA [16]). On the graph the curve ε_ω corresponding to the radiation in line does not exceed the continuum in its "red" maximum. Nevertheless, at the energies $\hbar\omega > \hbar\omega_1 = 0.412$ Ry ($\lambda_1 = 221.3$ nm) the intensity of radiation in the wing of line L_α may appreciably exceed the intensity of continuum.

Fig.2b represents the results of calculations for multicharged hydrogenlike ion for plasma parameters ($N_e = 10^8$ cm^{-3}, and T = 150 eV) typical for the solar corona where the element calcium (Z = 20) is abundant. Under such conditions the intensity of line wing is much higher than that of continuum and the intensity calculated by the standard theory.

5. RESULTS OF MEASUREMENTS AND COMPARISON OF EXPERIMENTAL DATA WITH RESULTS OF CALCULATIONS

The spectral measurements of pure thermal luminescence, fulfilled for verification of the above theory, were carried out with the aid of setup consisting of a cylindrical "heat-pipe" cell filled with sodium vapor and a complex of high-sensitivity recording system.

The most essential feature displayed both in the theory and in the experiment is a pronounced asymmetry of thermal radiation spectrum of resonance medium, which can not be obtained in the standard transfer theory. Far from resonance the effects of radiation surplus over the Planck level become unimportant and the optical depth of vapors drops much below unity. However, at $\Delta\omega \gg \Gamma$ the exponential factor $\exp[-\hbar(\omega-\omega_o)/T]$ begins to play an essential role for the intensity of luminescence in IR-region (see section 2). The results of measuring the absolute intensity of thermal luminescence of the dense sodium vapors in the infrared (band 2-3 μm) and in the visible (band 0.5-0.6 μm) spectral regions are presented in Fig.3. Notice, that a luminescence in such a far IR-wing of spectral line has not been hitherto investigated (as a rule, measurements were limited to detunings ~ 1000 cm^{-1}, see, e.g. [31]); much less accessible were the measurements of pure thermal luminescence). From the data presented in Fig.3 it is evident that the intensity of IR luminescence is by several orders of magnitude greater than that of near-resonance spectral region, with well-pronounced maximum in the IR-region, which confirms the theoretical considerations. Fig.3 presents also the results of the integral intensity calculations by the theory outlined above for the spectral regions corresponding to experiment. As follows from the graphs, the theoretical curve for the region 0.5-0.6 μm is in good agreement with experiment. As for the IR-region, the experimental and computational data are in agreement only within an order of magnitude. Such discrepancy may be partially explained

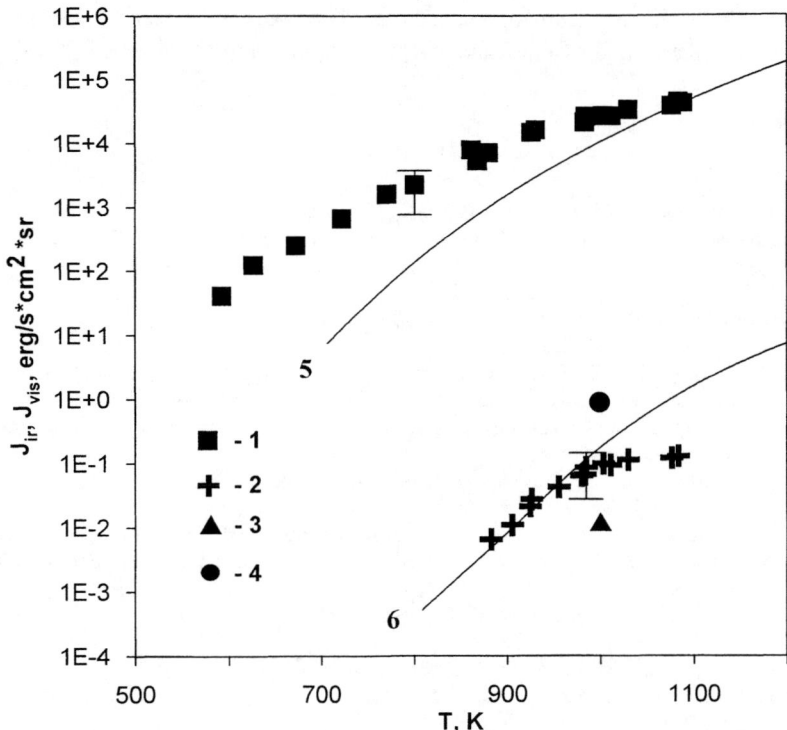

FIGURE 3. Dependence of integral intensity of luminescence for sodium vapors upon temperature at the center of heat tube for IR (1,3,5) and visible (2,4,6) regions of spectrum. 1,2 - experiment, 3,4 - standard theory, 5,6 - theory [10,11]. L = 30 cm.

by limitations of theory approximately describing the formation of statistical wing at such large detuning.

The integral intensities obtained by the standard theory differ drastically from observed ones, being several times higher in the visible region and almost by six orders of magnitude lower in the IR-region of spectrum. Thus, the experimental data are in fair agreement with the theory developed in [10,11], assuming the Boltzmann distribution for spectral population of the resonance level proportional to $\exp[-\hbar(\omega-\omega_0)/T]$, which provides an evidence of the infrared "catastrophe" in the luminescence of resonance medium.

ACKNOWLEDGEMENTS

The authors express their gratitude to A.M.Dykhne, P.D.Gasparian, Yu.K.Kochubey, and A.A.Panteleev for interest to work and stimulating discussions, J.Cooper, A.Gallagher, H.-J.Kunze, R.More, A.Osterheld, A.Phelps,

A.Sureau and A.Szöke for fruitful discussions. The work was carried out with support of the Russian Fund of Fundamental Investigations (grant # 97-02-17796).

REFERENCES

1. Stamm R., Talin B., Pollock E. and Iglesias C., *Phys. Rev.* A **34**, 4144 (1986).
2. Calisti A., Khelfaoui F., Stamm R. and Talin B., *Spectral Line Shapes*, v.6, ed. Frommhold L. and Kato J.W., New York: AIP, 1990, p.3.
3. Khelfaoui F., Calisti A., Stamm R. and Talin, ibid., p.102.
4. Calisti A., Stamm R. and Talin, *Europhys. Lett.* **4**, 1003 (1987).
5. Anufrienko A.V., Godunov A.L., Demura A.V. et al., *Zh. Eksp. Teor. Fiz,* **98**, 1304 (1990).
6. Anufrienko A.V., Bulyshev A.E., Godunov A.L. et al., *Zh. Eksp. Teor. Fiz* **103**, 417 (1993).
7. Bulyshev A.E., Demura A.V., Lisitsa V.S. et al., *Zh. Eksp. Teor. Fiz.* **108**, 212 (1995).
8. Biberman L.M., Vorobjov V.S., Yakubov I.T., *Kinetics of Nonequilibrium Low-Temperature Plasmas*, New York: Plenum Press, 1987.
9. Apresian L.A., Kravtsov Yu.A., *Teorija perenosa izluchenia*, Moscow: Nauka, 1983.
10. Starostin A.N., Zemtsov Yu.K., *Zh. Eksp. Teor. Fiz.* **103**, 345 (1993).
11. Sechin A.Yu., Starostin A.N., Zemtsov Yu.K., *Zh. Eksp. Teor. Fiz.* **110**, 1654 (1996).
12. Sechin A.Yu., Starostin A.N., Zemtsov Yu.K., et al., *Pis'ma v Zh. Eksp. Teor. Fiz.,* **65**, 20 (1997).
13. Sechin A.Yu., Starostin A.N., Zemtsov Yu.K., et al., *Pis'ma v Zh. Eksp. Teor. Fiz.* **65**, 807 (1997).
14. Sobel'man I.I., *Vvedenie v teoriju atomnykh spektrov*, Moscow: Fizmatgiz, (1963).
15. York G. and Gallagher A., *Power Gas Laser in Alkali-Dimers A-X Band Radiation,* Boulder, Colorado: JILA Rpt.114, Univ. of Colorado, 1974.
16. Mihalas D., *Stellar atmospheres*, ed. W. H. Freeman & Co., S. Francisco, 1978.
17. Keldysh L.V., *Zh. Eksp. Teor. Fiz.* **47**, 1515 (1964).
18. Lifshitz E.M., Pitaevskij L.V., *Fizicheskaja kinetika*, Moscow: Nauka, 1979.
19. Veklenko V.A., Tkachuk G.B., *Optika i spectroskopija* **38**, 1132 (1975).
20. Makhrov V.A., Sechin A.Yu., Starostin A.N., *Zh. Eksp. Teor. Fiz.* **97**, 1114 (1990).
21. Galitskij V.M., Yakimetz V.V., *Zh. Eksp. Teor. Fiz.* **51**, 957 (1966).
22. Kasianov V.A., Starostin A.N., *Zh. Eksp. Teor. Fiz.,* **48**, 295 (1965).
23. Borisov J., Frommhold L., in *Phenomena Induced by Intermolecular Interaction*, ed. Birnbaum G., New York: Plenum, 1985, p.67.
24. Gallagher A., in *Eximer Lasers*, ed. Rhodes Ch.K., Berlin-Heidelberg-New York: Springer-Verlag, 1979.
25. Carrington C.G., Stacey D.N., and Cooper J., *J. Phys. B: Atom. Mol. Phys.* **6**, 417 (1973).
26. Vainshtein L.A., Sobel'man I.I., Yukov E.A., *Vozbuzhdenie atomov i ushyrenie spektral'nykh linij,* Moscow: Nauka, (1979).
27. Chatham R.H., Gallagher A., and Levis E.L., *J. Phys. B: Atom. Mol. Phys.* **13**, 47 (1980).
28. *Fizicheskie velichiny*, Reference book, Moscow: Energoatomizdat, 1986.
29. Zel'dovich Ya.B., Raizer Yu.P., *Fizika udarnykh voln i vysokotemperaturnykh gidrodinamicheskikh yavlenij*, Moscow: Nauka, 1966.
30. Chekhov D.I., Ph.D. Thesis, Moscow: MIPT, 1994.
31. Jongerius M.J., in *Spectral Line Shapes*, ed. B.Wende, Berlin, 1980, p.963.
32. Ratzig A.A., Smirnov B.M., *Parametry atomov i atomnykh ionov*, Moscow: Energoatomizdat, 1986.

Theory of optical shielding in cold atom beams

Vladimir A. Yurovsky and Abraham Ben-Reuven

School of Chemistry, Tel Aviv University, 69978 Tel Aviv, Israel

Abstract. Experiments were conducted recently (by Weiner and co-workers) on the optical shielding (suppression) of atomic collisions in cold-atom beams, and its variation with the angle between the polarization direction of the shielding light and the direction of approach of the beam. This case is shown here to be a typical example of an optical collision in which quantum interference may persevere between two incident collision partial waves leading to the same output state. This effect depends on the relative collisional phaseshift of the two interfering channels, as well as on the angle of approach, and will vanish when averaging over the latter (as with collisions in the bulk).

The extent of variation of this interference effect with the relative phaseshift is quite broad, and may lead, under favorable conditions, to almost complete shielding at a finite value of the shielding-laser power. The latter observation leaves open the possibility of exerting coherent control over the interference effect in order to optimize the shielding.

The need to shield cold atoms (in the mK regime and below) from loss-inducing collisions has drawn recently some attention (see [1,2,4] and the references therein). The laser radiation, tuned to the blue of the atomic resonance transition, was supposed to couple the ground electronic state of the collision pair to an excited (repulsive) state. According to the simple Landau-Zener theory of curve crossing, the probability of the atoms approaching each other (to within range of loss-inducing processes) was supposed to decrease exponentially as a function of the laser power. Experiments show that, instead of decreasing exponentially, the penetration probability rather tends to level off, producing incomplete shielding at the higher power levels.

A simple model, explaining the main reason for incomplete shielding in cold traps of metastable Xe atoms, has been suggested recently by the authors [1]. This model of multiple curve crossing gives a three-dimensional description of the shielding process. The atoms, on approaching the shielding Condon point at the ground electronic state in a given partial wave, may end up penetrating the inner zone again in the ground electronic state, but in a higher partial wave, along the diverted path of the kind $g, J \to e, (J+1) \to g, (J+2)$, while g and e are, respectively, the ground and excited electronic states, and J is the angular momentum (including the relative motion of the atoms). Unlike the direct path going straight on through all crossing points, the diverted path is not suppressed even at high laser powers. The treatment of this model was based on a theory of multiple curve crossing processes [3], drawn from the similarity to a well-known exactly-soluble problem of linear potentials.

The work of Weiner and co-workers on cold atom beams [4] has drawn the attention to the question of optical shielding in beams vs. shielding in traps, with their isotropy of atomic motion. The experiments show that the shielding, using linearly polarized light, produces significant dependence on the polarization direction (relative to the direction of approach of the beam).

The present report treats the main results of the analysis of these experiments based on the approach [1] (for complete presentation see [2]). The dependence of the penetration probability on the angle ϑ between the electromagnetic field polarization and the direction of the electron-beam is given by

$$P(\vartheta) = \frac{2}{(J_{\max}+1)(J_{\max}+2)} \sum_{J=0}^{J_{\max}}{}' \sum_{M=-J}^{J} \left[\tilde{T}_{JM}^d \left(P_J^{(M)}(\cos\vartheta) \right)^2 \right.$$
$$\left. + \tilde{T}_{JM}^u \left(P_{J-2}^{(M)}(\cos\vartheta) \right)^2 + 2\sqrt{\tilde{T}_{JM}^d \tilde{T}_{JM}^u} P_J^{(M)}(\cos\vartheta) P_{J-2}^{(M)}(\cos\vartheta) \cos\chi_J^{(M)} \right] \quad (1)$$

where

$$\tilde{T}_{JM}^d = (2J+1)\frac{(J-|M|)!}{(J+|M|)!} \exp\left(-2\pi\lambda_{RJ+1}^{(M)} - 2\pi\lambda_{QJ}^{(M)} - 2\pi\lambda_{PJ-1}^{(M)}\right) \quad (2)$$

$$\tilde{T}_{JM}^u = (2J-3)\frac{(J-2-|M|)!}{(J-2+|M|)!}\left[1-\exp\left(-2\pi\lambda_{RJ-1}^{(M)}\right)\right]\left[1-\exp\left(-2\pi\lambda_{PJ-1}^{(M)}\right)\right] \quad (3)$$

Here $P_J^{(M)}$ are the Legendre functions, the Landau-Zener exponents $\lambda_{QJ}^{(M)}$, $\lambda_{PJ}^{(M)}$ and $\lambda_{RJ}^{(M)}$ are given in [1], and $J_{\max}$ is the maximal (even) angular momentum for which the collision energy surpasses the centrifugal barrier.

The three terms within the square brackets in Eq. (1) correspond, respectively, to the penetration along a direct path, a diverted path, and the interference of such paths, respectively. The interference terms vanish on averaging over an isotropic distribution of the angle of approach (ϑ), and are therefore absent from the penetration probability in the case of atomic traps, as considered in [1]. They nevertheless must be taken into account in analyzing atomic-beam experiments [4].

The amplitudes representing the direct and diverted paths may have different phases. The phaseshift difference $\chi_J^{(M)}$ between these amplitudes can be expressed in the semiclassical theory as

$$\chi_J^{(M)} = \left[\chi\left(\lambda_{RJ-1}^{(M)}\right) - \chi\left(\lambda_{PJ-1}^{(M)}\right)\right] + \left[\phi_{\text{WKB}}(J-2) - \phi_{\text{WKB}}(J)\right] \quad (4)$$

It includes contributions at the crossing points [5]

$$\chi(\lambda) = \arg \Gamma(i\lambda) - \lambda \ln \lambda + \lambda + \pi/4 \tag{5}$$

and the contributions of the WKB phaseshifts $\phi_{\text{WKB}}(J)$ produced by the adiabatic collision potentials. These potentials are obtained by diagonalizing the potential matrix, including the diabatic potentials, the centrifugal potentials, and the radiative couplings.

At low shielding-laser intensities we have $\tilde{T}^d_{JM} \gg \tilde{T}^u_{JM}$ and the interference term is negligible in comparison to the first term in Eq. (1) corresponding to the direct penetration path. The interference vanishes in the limit of high intensity, as well; i.e., when $\tilde{T}^d_{JM} \ll \tilde{T}^u_{JM}$. In this limit,

$$P(\vartheta) \approx P_\infty = 1 - 2\frac{2J_{\max} + 1}{(J_{\max} + 1)(J_{\max} + 2)} \tag{6}$$

does not depend on ϑ. The angle-averaged penetration probability, considered in [1], has the same "hangup" value P_∞. For the conditions prevailing in the Na beam experiment, [4] $J_{\max} = 4$ and $P_\infty = 0.4$. For the Xe bulk experiment, considered in [1], $J_{\max} = 2$ and $P_\infty = 1/6$.

The interference terms substantially depends on the phaseshifts $\chi_J^{(M)}$. Unfortunately, their evaluation, in the adiabatic case, should involve an infinite number of interacting channels. Because of this, and in order to shows the full range of variation, the penetration probabilities presented below include not only results of the phaseshifts of Eq. (4), but also those of the phaseshifts $\pi/2$ and π.

Figure 1 presents the penetration probability as a function of the shielding-laser intensity, for several choices of $\cos\chi_J^{(M)}$, calculated by using parameters appropriate to the Na beam experiment [4]. The results are compared with the experimental data of [4]. This figure shows a strong influence of the interference on the penetration probability. The interference correction has the same sign as $\cos\chi_J^{(M)}$ if the atomic beam and the electric field are parallel and the opposite sign if they are perpendicular.

A better agreement with the experimental data may be obtained by choosing M-dependent phaseshifts, such as $\cos\chi_J^{(0)} = 0$ and $\cos\chi_J^{(2)} = -1$. Phaseshifts for other M values have no effect, since $P_J^{(M)}(0) = P_J^{(M)}(1) = 0$ for odd M.

The present model predicts non-monotonic dependence of the penetration probability on the shielding-laser intensity, discernible on letting the laser intensity go just beyond the maximum value used in the experiments [4]. The extent of this non-monotonic variation, showing an optimization dip at a finite value of the laser intensity, is less pronounced in the bulk probabilities, as calculated for Xe in [1].

Thus, we have demonstrated here how channel interference can affect the optical shielding in the case of beams interacting in the presence of polarized radiation. In particular, we have pointed out the non-monotonic dependence of the shielding

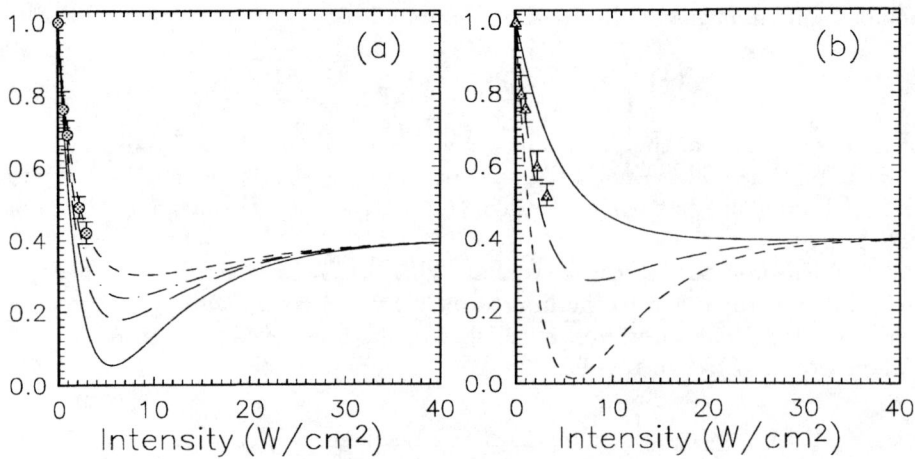

FIGURE 1. Penetration probability for Na as a function of the shielding-laser intensity at perpendicular (a) and parallel (b) polarization angles. Lines denote calculations using the semiclassical phase shifts (solid line), $\cos\chi_J^{(M)} = 0$ (long-dashed line) or -1 (dashed line) for all J and M and $\cos\chi_J^{(0)} = 0$, $\cos\chi_J^{(2)} = -1$ (dash-dotted line) for all J. (The dash-dotted and long-dashed lines overlap at the parallel polarization.) The experimental data of Weiner and co-workers are shown for comparison, using the same laser detuning (250 MHz) and collision energy ($60mK$) for all plots.

efficiency on the laser intensity, and especially the way it depends on the polarization angle and on the phaseshift difference between the two interfering channels. It would be therefore desirable to use techniques of coherent control over this phaseshift, exploiting the dip in the intensity dependence in order to optimize the shielding at a reasonably low value of the laser intensity.

Further experimental investigations, reaching slightly higher laser intensities than those used previously, will help clarify and establish the points considered here.

The authors are most grateful to John Weiner for prepublication results, and to him as well as to Paul Julienne and Moshe Shapiro for helpful discussions.

REFERENCES

1. Yurovsky, V. A., Ben-Reuven, A., *Phys. Rev.* A **55**, 3772 (1997).
2. Yurovsky, V. A., Ben-Reuven, A., *J. Phys. Chem.* A (1998) (submitted).
3. Yurovsky, V. A., Ben-Reuven, A., *J. Phys.* B **31**, 1 (1998).
4. Tsao, C.-C., Wang, Y., Napolitano, R., Weiner, J., *European Physical Journal* D (1998) (in press).
5. Child, M. S., *Semiclassical Mechanics with Molecular Applications*, Oxford: Clarendon Press, 1991.

Coherent transients as an effective technique to distinguish different collisional relaxational channels.

N.N. Rubtsova, L.S. Vasilenko and E.B. Khvorostov

*Institute of Semiconductor Physics, Siberian Branch RAS,
acad. Lavrentyev ave., 13, Novosibirsk, 630090, Russia*

Abstract.
Precise knowledge of collisional relaxation channels is in close relation with the research of both specific features of interaction potential and details of lineshape. When using nonlinear laser technique of spectral domain, it's not so easy to distinguish between different channels of relaxation because of their integral contribution to the lineshape of rarefied gas. This report intends to demonstrate ability of coherent transient phenomena to overcome this difficulty.

Comparative study of several photon echo modifications, performed in molecular gas SF_6 has permitted to choose definite type of coherent transient phenomena, which is the most sensitive to peculiar channel of collisional relaxation. The data on relaxation rates of rotationally inelastic, elastic small angle scattering and depolarizing collisions in SF_6 and its mixtures with buffer gases are examined.

I INTRODUCTION

The shape of isolated spectral line ot neutral gas is controlled by translational motion (nonhomogeneous Doppler broadening or Dicke narrowing), particles interactions with external electromagnetic fields (spectral line shift or splitting due to dynamic Stark effect and radiation broadening). Nonlinear Doppler-free laser spectroscopy methods are able to reduce the influence of translational motion while experiments in weak electromagnetic fields are able to decrease power broadening till natural broadening. So the Doppler-free spectroscopy allows to concentrate efforts on homogeneous line broadening investigation.

This work is performed by photon echo (PE) and stimulated photon echo (SPE) techniques. Coherent nature of these phenomena gives some advantages in comparison with saturated spectroscopy technique [1]. These are significant signal-to-noise increase and the absence of power broadening distortion. The goal of this work is to analyze the abilities of coherent spectroscopy methods to determine different channels of collisional relaxation. Elastic small angle scattering, depolarizing collisions and inelastic scattering (which is rotational relaxation in our case) are mainly considered.

II PHOTON ECHO TECHNIQUES AS AN EFFECTIVE METHOD TO INVESTIGATE DIFFERENT KINDS OF COLLISIONS

Simplified picture of the photon echo formation in the gas of two-level particles allows to reveal qualitatively the role of inelastic and elastic small angle scattering collisions. Really, the action of first exciting resonant pulse creates the macroscopic polarizations in sub-ensemble of particles within longitudinal velocities range controlled by the pulse width. Just after finish of this pulse this sub-ensemble emits spontaneously. Due to Doppler dephasing the macroscopic polarization vanishes while microscopic polarization of individual particles exists for a long time. Arrived after time delay T_{12} the second exciting pulse is able to reverse the sign of accumulated Doppler phase in the PE case or to convert the polarization (non-diagonal components of coherence matrix $\rho_{ij}, i \neq j$) into resonant levels populations (diagonal components of coherence matrix ρ_{ii}) in the SPE case. In the last case the application of additional third exciting pulse with time delay T_{23} converts these nonequilibrium level populations back into polarization with the same Doppler phases reversal. PE and SPE signals arise correspondingly at time moments $t_{PE}=2T_{12}$ and $t_{SPE}=2T_{12}+T_{23}$.

Therefore the PE signal is sensitive to the decay of nondiagonal components ρ_{ij}, which may be provided by inelastic and elastic collisions both. However, the elastic small angle scattering may be neglected in PE experiments till time delays $T_{12} \leq T_c$, where $T_c \approx \pi/(k\delta v_z)$, k is the wave vector and δv_z is average change of longitudinal projection of relative velocity per one collision. Hence the initial part of PE decay curve corresponds to inelastic relaxation with rate $\gamma^{(0)}$ while the part with $T_{12} > T_c$ characterizes the combined action of elastic (velocity changing) and inelastic relaxation channels with rate $\Gamma_{tot} = \gamma^{(0)} + \Gamma_{vcc}$.

Under condition of small and constant T_{12} the stimulated photon echo kinetic curve reflects only dynamics of populations decay. However the levels degeneracy leads not only to specific polarization properties of SPE but gives unique possibility to study collisional decay of polarization moments induced on magnetic sublevels by polarized resonant pulses. Null polarization moment corresponds to the total level population and decays with rate $\gamma^{(0)}$, first polarization moment or orientation of the level corresponds to macroscopic magnetic moment induced in the gas sample and decays with rate $\gamma^{(1)}$, second polarization moment or alignment of level corresponds to macroscopic electric qudrupole moment and decays with rate $\gamma^{(2)}$ [2]. Study of relaxation of orientation and alignment gives data about asymmetry of interaction potential.

The above mentioned features of coherent transient phenomena allows to understand the contribution of each relaxation channel.

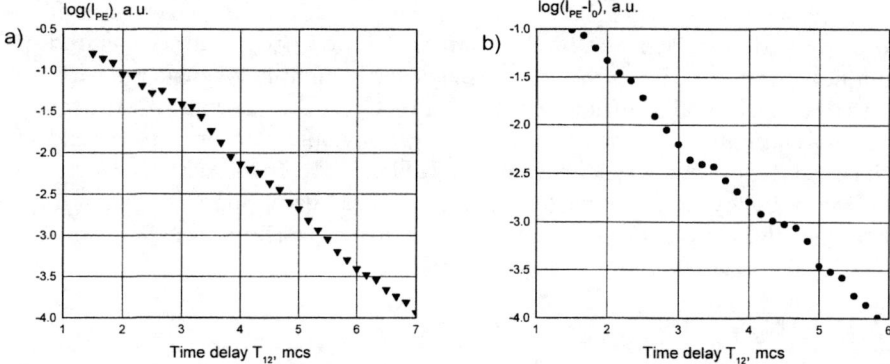

FIGURE 1. Deacrease of the photon echo intensity due to collisions with different partners: a) SF_6–SF_6 collisions in pure SF_6 at pressure 2.5 mTorr, b) SF_6–He collisions in mixture of 2.5 mTorr SF_6 with 3.5 mTorr He. I_{PE} – photon echo intensity. I_0 – photon echo intensity at the same SF_6 pressure without buffer gas.

III EXPERIMENTAL RESULTS

PE was formed in SF_6 gas and its mixtures with buffers He and Xe for time delays T_{12}=1.5–10 mcs under excitation by electro-optically shuttered pulses of CW CO_2-laser radiation with its frequency locked to the center of P(33) A_2^1 line of ν_3 vibrational band of SF_6. Nonlinear character of Fig. 1a shows typical [3] nonexponential PE decay in pure SF_6. Fig. 1b show influence of SF_6–He collisions only on PE decay in the mixture. Curve at Fig. 1b may be treated as linear. Omitted curve for SF_6–Xe collisions demonstrates intermediate behaviour.

Let us estimate the scattering angle as diffractional one [4] via relation $\theta = \lambda_D/\rho_W$, where de Broglie wavelength $\lambda_D = \hbar/(\mu v_{rel})$, μ is the reduced mass of collisional partners, v_{rel} is average relative velocity. Weisskopf radius ρ_W may be determined from experimental data on Γ_{tot}. These values of θ may be applied to T_c estimate by accounting for $\delta v_z \approx \theta v_{rel}$. (See Table I.)

TABLE I.

Colliding partners	$\gamma^{(0)}$, 10^6 s^{-1}Torr^{-1}	Γ_{tot}, 10^6 s^{-1}Torr^{-1}	λ_D, Å	θ, rad	δv_z, cm/s	T_c, 10^{-6} s
SF_6–SF_6	(36 ± 4.5)	(49.8 ± 6.3)	0.03	$3.6 \cdot 10^{-3}$	145	3.4
SF_6–Xe	not measured	(27.6 ± 4.7)	0.029	$4.8 \cdot 10^{-3}$	199	2.5
SF_6–He	not measured	(44.2 ± 6.7)	0.13	$3.5 \cdot 10^{-2}$	4995	0.1

Comparison of T_c, estimated for different colliding partners, may explain characteristic features of curves at Fig. 1a–1b.

Stimulated photon echo was generated at the same transition under action of three pulses with specially chosen linear polarizations [5]. Decay rates of orientation and alignment are shown in the Table II. Since in SF_6 the decay rates for both orientation and alignment are higher than population decay rate (measured by PE technique), the $\gamma^{(1)}$ and $\gamma^{(2)}$ values serve as over estimates for inelastic rates $\gamma^{(0)}$ in mixtures with Xe and He (not measured directly). This permits to estimate rate of velocity changing collisions $\Gamma_{vcc} = \Gamma_{tot} - \gamma^{(0)}$ by substitution $\gamma^{(1)}$ instead of $\gamma^{(0)}$ for these mixtures. As it's clear from the Table II, the cross sections σ_{vcc} of elastic scattering of heavy active particle SF_6 is larger for heavy buffer Xe, than for small mass buffer He. The same statement is true for inelastic processes cross sections σ_{rot}.

TABLE II.

Colliding partners	$\gamma^{(1)}$, $s^{-1}Torr^{-1}$	$\gamma^{(2)}$, $s^{-1}Torr^{-1}$	Γ_{vcc}, $s^{-1}Torr^{-1}$	σ_{vcc}, $Å^2$	σ_{rot}, $Å^2$
SF_6–SF_6	$(32 \pm 3) \cdot 10^6$	$(38 \pm 3) \cdot 10^6$	$13.8 \cdot 10^6$	160	390
SF_6–Xe	$(16 \pm 2) \cdot 10^6$	$(16 \pm 2) \cdot 10^6$	$\approx 11.6 \cdot 10^6$	$120 \leq \sigma \leq 280$	≤ 160
SF_6–He	$(36 \pm 4) \cdot 10^6$	$(47 \pm 4) \cdot 10^6$	$\approx 8.2 \cdot 10^6$	$20 \leq \sigma \leq 105$	≤ 85

IV CONCLUSION

The investigation of photon echo decay kinetics in a wide range of time delays T_{12} demonstrates ability to distinguish elastic and inelastic scattering channels, especially for heavy buffers. Heavy buffers proved more effective for both elastic and inelastic scattering of SF_6. The application of stimulated photon echo polarization technique gives additional data on collisional decay of population, orientation and alignment. These data permit to estimate inelastic channel contribution, even when it was not measured directly.

V ACKNOWLEDGEMENTS

This research was supported by grants of Russian Foundation for Basic Research (No 97-02-18496 and 98-02-16390) and by State program "Laser Physics".

REFERENCES

1. L.S. Vasilenko, N.N. Rubtsova, Laser Phys., 1997, v. 7(3), p. 1021–1031.
2. I.V.Yevseyev, V.M.Yermachenko, V.A.Reshetov, ZhETF, 1980, v. 78(6), p. 2213–2221.
3. L.S. Vasilenko, N.N. Rubtsova, Optika i Spektroskopiya, 1985, v. 58, p. 697–699.
4. V.P. Kochanov, S.G. Rautian, A.M. Shalagin, ZhETF, 1977, v. 72(4), p. 1358–1374.
5. L.S. Vasilenko, N.N. Rubtsova, E.B. Khvorostov, ZhETF, v. 113(3), p. 827–836.

Ionic spectra under strong laser field in plasma

D.A. Shapiro, M.G. Stepanov

*Institute of Automation and Electrometry
Siberian Branch, Russian Academy of Sciences
1 Acad.Koptjug Ave, Novosibirsk, 630090, Russia*

Abstract. An expression is found for the profile of Bennett hole with different level lifetimes under strong field and weak collisions. It is shown that the square of the total width being equal to the sum of the squares of the diffusion and field widths. Probe-field spectrum of the three-level system is obtained in the presence of the strong field at the adjacent transition. The Autler – Townes doublet components are shown to broaden and repel each other under the collisions. The main reason of both the effects occurs to be the diffusion of phase due to quadratic frequency shift.

The weak (or "soft") collision model [1,2] supposes a small velocity change in each collision act. It describes heavy molecules moving in a gas of light buffer particles or Coulomb ion-ion scattering in plasma [3]. In nonlinear spectroscopy weak collisions transforms the density matrix equations from algebraic to differential set. The weak saturation allows one to calculate the ionic density matrix as a series of the perturbation theory in field intensity. Meanwhile, several experiments were carried out last 15 years in plasma spectroscopy by powerful continuous ion and dye lasers [4–7] beyond the perturbation theory.

The present paper is devoted to strong field effects in probe-field spectra of two- and three-level systems. The weak collisions result in the diffusion wandering of

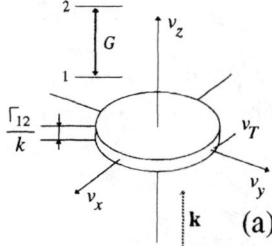 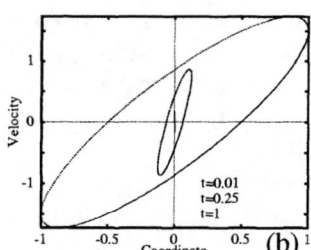

FIGURE 1. (a) Domain of interaction in velocity space. (b) The diffusion of ion cloud in the phase space (arb. units).

atom in the velocity space with diffusion coefficient D and then the corresponding broadening. Let us consider the interaction between a two-level system and the strong monochromatic wave. The domain, Fig. 1 (a), has the shape of a disk of the radius about thermal velocity v_T and the height about the Bennett hole width $\delta v_0 = \Gamma_{12}/k$, where Γ_{12} is the homogeneous linewidth, $\mathbf{k}$ is the wavevector. The significant diffusion broadening occurs when velocity changing $\delta v \sim (Dt)^{1/2}$ exceeds δv_0.

Effect of diffusion dephasing, Fig. 1 (b), also results in the line broadening but have a different nature. Atoms start at the origin of coordinates. The diffusion over velocity blurs the point to a segment δv extended along the velocity axis. The level lines of equal distribution density acquire and keep the elliptic shape. Atoms gather speed, and start drifting in the positive/negative direction in the upper/lower half-plane, then the ellipse is rotated clockwise. The scatter of coordinate $\delta x \sim D^{1/2} t^{3/2}$ grows up with time faster than of the velocity. When δx gets the order of wavelength, atoms occurs out of phase with radiation and the line broadens. The characteristic dephasing width is $\delta \Omega \sim (Dk^2)^{1/3}$. In contrast to the diffusion width $\delta \Omega \sim k\delta v \sim (Dk^2/\Gamma_j)^{1/2}$ it is independent of relaxation constants.

Consider the resonance interaction of the intense traveling electromagnetic wave with two excited atomic states $|1\rangle$, $|2\rangle$ of energies $E_1 < E_2$. Suppose that one level L is long-lived and the diffusion width of the Bennett hole in the short-lived level S being small: $v_T (\nu/2\Gamma_S)^{1/2} \ll \Gamma_{12}/k$, $\Gamma_S = \max_{j=1,2} \Gamma_j \gg \Gamma_L = \min_{j=1,2} \Gamma_j$. The population of long-lived level satisfies the equation

$$\left(1 + \frac{A^2}{W^2 + x^2}\right) y = \frac{d^2}{dx^2} y + \frac{A^2}{W^2 + x^2}, \tag{1}$$

where parameter $A = w_F/w_D$ is the ratio of the field width w_F and the diffusion width w_D in the long-lived level

$$w_F = \sqrt{2\Gamma_{12}|G|^2/k^2 \left(\Gamma_1^{-1} + \Gamma_2^{-1} - A_{21}/\Gamma_1\Gamma_2\right)}, \quad w_D = v_T\sqrt{\nu/2\Gamma_L}.$$

Here $G = \mathbf{E}\mathbf{d}_{21}/2\hbar$ is the interaction between wave $\mathbf{E}$ and dipole moment $\mathbf{d}_{21}$, A_{21} is the Einstein coefficient of transition 1-2, $W^2 = (\Gamma_{12}^2 + 2\Gamma_{12}|G|^2/\Gamma_S)/k^2 w_D^2$, $x = (v_z - \Omega/k)/w_D$ is the normalized projection v_z of the velocity $\mathbf{v}$ onto wavevector $\mathbf{k}$, Ω is the detuning, $y = (\rho_L - \rho_L^{(0)})/(\rho_S^{(0)} - \rho_L^{(0)})$, $\rho_j^{(0)}$ is the unperturbed population of level $j = L, S$. At small Γ_{12}

$$y = A^2 \sqrt{i|x|} \left(c \cdot K_\alpha(|x|) - S_{-3/2,\alpha}(i|x|)\right), \tag{2}$$

$$c = \frac{e^{-i\alpha\pi}}{2^{3/2}\pi} \Gamma\left(-\tfrac{1}{4} - \tfrac{\alpha}{2}\right) \Gamma\left(-\tfrac{1}{4} + \tfrac{\alpha}{2}\right) \cos\left(\tfrac{\pi}{2}\left(\alpha + \tfrac{3}{2}\right)\right),$$

where $K_\nu(z)$, $S_{\mu,\nu}(z)$ are McDonald and Lommel's functions, $\alpha = (A^2 + 1/4)^{1/2}$. Solution (2) involves exponential function $e^{-|x|}$ and Lorentzian $(1 + x^2/A^2)^{-1}$ as

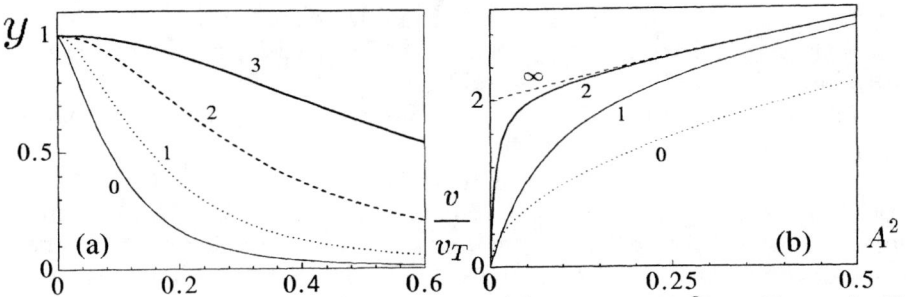

FIGURE 2. (a) The Bennett hole profile $y(x)$. $\Gamma_L = 10^{-3}kv_T$, $\Gamma_S = 4 \cdot 10^{-2}kv_T$, $\Gamma_{12} = (\Gamma_L + \Gamma_S)/2$, $\nu = 10^{-5}kv_T$, $\Omega = 0$. Curve $n = 0, 1, 2, 3$ corresponds to $|G| = 10^{n/3-2}kv_T$. (b) The absorbed power as a function of intensity A^2. Curve number n corresponds to the width $W = 10^{-n}$ (curve ∞ is plotted for $W = 0$). Curve 0 describes the collisionless case.

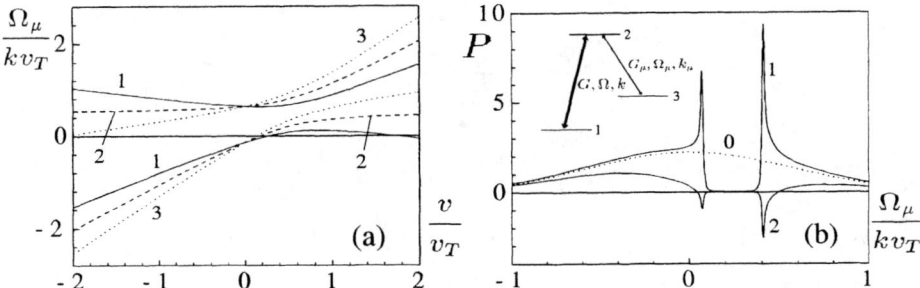

FIGURE 3. (a) Hyperbole 3 for $k_\mu/k = 0.75$ (solid line); 1 (dashed line); 1.25 (dotted line). $\Omega = 0.5kv_T$, $|G| = 0.3kv_T$, $\gamma_\pm = 0$. (b) Probe field spectrum $P_\mu(\Omega_\mu)$ at transition 1-3, $\Gamma_j, \Gamma_{ij} = 10^{-3}kv_T$, $\nu = 5 \cdot 10^{-4}kv_T$, $\Omega = 0.3kv_T$, $|G| = 0.2kv_T$. Curve 0: $G = 0$; 1: $N_1 = N_2$; 2: $N_1 = 3N_3$. *Insert*: Three-level system interacting with strong field G and probe field G_μ.

limiting cases $A \ll 1$ or $A \gg 1$, respectively. Fig. 2 (a) illustrates a transformation from exponential function to Lorentzian (from curve 0 to curve 3). Half a width at half a maximum (HWHM) $v_{1/2}$ of the hole has a good approximation $v_{1/2} \simeq (w_F^2 + w_D^2/2)^{1/2}$. The HWHM was calculated earlier [5] within the variational approximation. The profile of probe function $y(x)$ considered as Lorentzian, but yielded the correct HWHM. Examples of saturation curves, the dependence of absorbed power on the intensity of falling wave, are shown in Fig. 2 (b). The smooth transition occurs from a steeply slopping curve, corresponding to the homogeneous saturation, to the less slopping asymptote describing the inhomogeneous saturation. The limit of homogeneous saturation agrees with measurements [7].

Let us consider the probe-field spectrum modified by the strong field at the adjacent transition, Fig. 3 (insert). Choosing the copropagating probe wave we obtain in collisionless case the Autler – Townes doublet

$$\Omega_\mu^{(1,2)}(v_z) = k_\mu v_z + \frac{\Omega - kv_z + i\gamma_+}{2} \pm \sqrt{\frac{(\Omega - kv_z + i\gamma_-)^2}{4} + |G|^2}, \quad (3)$$

where $\gamma_\pm = \Gamma_{13} \pm \Gamma_{23}$, Ω_μ, k_μ are the detuning and wavenumber of the probe field, Γ_{13}, Γ_{23} are the nondiagonal relaxation constants. The positions of a component is defined by the real part of Ω_μ, while the width equals to its imaginary part.

The number of atoms occurring in resonance δn_r is proportional to $\delta n_r \propto \gamma/|d\Omega_\mu^{(1,2)}/dv_z|_{v_z=v_r}$, where $\Omega_\mu = \text{Re}\Omega_\mu^{(1,2)}(v_r)$, γ is the width of resonance. At $k_\mu > k$ the dependence $\Omega_\mu^{(j)}(v_z)$ is a monotonic function, whereas at $k_\mu < k$ it has points of extremum, Fig. 3 (a). In the vicinity of a stationary point the narrow peak appears in the probe-field spectrum, because at this detuning the probe field interacts with the highest number of atoms $\propto \gamma^{1/2}$. If the Rabi splitting is strong $|G| \gg \nu^{1/3}(kv_T)^{2/3}$, two resonances in the probe-field spectrum are independent, Fig. 3 (b), and

$$P_\mu(\Omega_\mu) = 2\hbar\omega_\mu|G_\mu|^2 \sum_{j=1}^{2}(-1)^j \text{Re}\int dv \frac{(\Omega_\mu^{(j)} - \Omega_{\mu B}^{(1)})(\rho_2 - \rho_3) + G^*\rho_{21}}{\Omega_\mu^{(1)} - \Omega_\mu^{(2)}} \cdot \int_0^\infty \frac{dt}{\sqrt{\text{ch}\,\tau}} \exp\left(-i(\Omega_\mu - \Omega_\mu^{(j)})t + h_j\left(\frac{\Gamma_{D1}^{(j)}}{\Gamma_{D2}}\right)^3(\tau - \text{th}\,\tau)\right), \quad (4)$$

$$\Gamma_{D1}^{(1,2)}(v) = \left(\frac{\nu}{2}\left(v_T\frac{d\Omega_\mu^{(1,2)}}{dv}(v)\right)^2\right)^{1/3}, \quad \Gamma_{D2}(v) = v_T\left(\frac{2\nu k_\mu|k - k_\mu|}{\Omega_\mu^{(1)}(v) - \Omega_\mu^{(2)}(v)}\right)^{1/2};$$

$$h_{1,2} = \exp\left(\mp i\,\text{sgn}(k - k_\mu)\frac{\pi}{4}\right), \quad \tau = h_{1,2}\Gamma_{D2}t, \quad \Omega_{\mu B}^{(1)} = k_\mu v_z + (\Omega - kv) + i\Gamma_{13}.$$

In experiment [4] the profile of nonlinear resonance was observed in three-level scheme with $\lambda = 488$ and 514 nm at adjacent transitions of ArII. There was no strong Coulomb broadening of the resonance, that could be interpreted as a consequence of the quadratic shift.

This work supported by RFBR (96-02-19052, 00069, 96-15-96642,), R&D Programs "Optics. Laser physics" (1.53), "Fundamental Spectroscopy" (08.02.32), Soros graduate student's program (M S, a98-674), and by INTAS (M S, 96-0457) within the research program of the ICFPM.

REFERENCES

1. L. Galatry, Phys. Rev. **122**, 1218 (1961).
2. S. Rautian, Zh. Eksp. Teor. Fiz. **51**, 1176 (1966).
3. G. Smirnov, D. Shapiro, Zh. Eksp. Teor. Fiz. **76**, 2084 (1979).
4. V. Lebedeva et al., Zh. Prikl. Spektr. **41**, 385 (1984).
5. S. Babin, V. Donin, D. Shapiro, Zh. Eksp. Teor. Fiz. **91**, 1270 (1986).
6. A. Feitish, D. Schnier, T. Müller, B. Wellegehausen, IEEE JQE **24**, 507 (1988).
7. A. Apolonsky et al., Phys. Rev. A **55**, 661 (1997).
8. O. Bykova, V. Lebedeva, N. Bykova, A. Petukhov, Opt. Spektr. **53**, 171 (1982).

Towards a 'rule of thumb' for the wings of forbidden transitions

A.Devdariani

Department of Optics and Spectroscopy St.Petersburg University
Ulianovskaia 1, St. Petersburg 1998904, Russia

Abstract. Our calculations of dipole moments and potential energy curves lead to the conclusion that collision induced quasimolecular radiation can be understood in the two-level approach. Results leads to a simple rule for the formation of a satellite near the position of the forbidden atomic transition.

It is widely accepted that there is no useful 'rule of thumb' to predict *a priory* the shapes of spectral lines produced by atomic collisions in the vicinity of atomic transitions which are strictly forbidden in an isolated atom, see, e.g., (1). In any case, the results of earlier calculations (2-4) have supported this conclusion. In a rigorous approach one should first solve a quantum chemistry part of the problem, that is to calculate dipole transition moments taking into consideration the configuration interaction. The profile shapes are then obtained by solving the coupled equations which take into account non-adiabatic transitions. There is however a simpler approach.

Consider two close atomic states such that optical transitions from one of them to the ground state are forbidden due to symmetry. For example, this situation can be recognized in the (nsnp, $^3P_2 \rightarrow ns^2$, $^1S_0$), (n_0p^5ns, $^3P_2 \rightarrow n_0p^6$, $^1S_0$) transitions or in the parity forbidden transitions from low excited states of light atoms. The interaction with a buffer gas atom produces quasimolecular states from the non-radiating atomic state and reduces symmetry so that some adiabatic quasimolecular states can in principle radiate. The closeness of the initial excited atomic states allows to apply a simple few - or even a two-state model when calculating the dipole moment and the profile shape. The advantage of this approach is the possibility to include the interaction between atoms at intermediate interatomic distances. In this region the atomic wave function basis is switched over to the molecular one due to the influence of the dominant exchange interaction, and induced dipole moments attain their maximum values. The disadvantage is that the approach neglects the influence of other states not included in the model. However, the multi-configuration perturbation theory cannot be applied in the switching region.

As a test case, the calculations of potential energy curves and dipole moments for the Ca(4s4p,4s3d), Mg(3s3p)-He quasimolecules have been performed. The results demonstrate a reasonable agreement between the multi-configuration and the two-level approaches. Thus, the two-level results could be used for profile shape calculations based on quasistatic or path trajectories approaches. However, both of these approaches are in general inaccurate because they do not take into account dynamical nature of the atomic interaction during collisions or, in other words, nonadiabatic transitions. The resulting spectrum is not simply the sum of optical transitions from two non-interacting quasimolecular states treated separately, rather optical and nonadiabatic transitions must be treated simultaneously.

A useful tool to solve the dynamic problem is the well-known Demkov model. It has been extensively used in atomic collision physics to determine the probabilities of non-adiabatic transitions between two states which correspond to parallel diabatic potential energy curves and an exponential interaction. To calculate the profile we require the time-dependent solutions of the model for the whole collision rather than only its large-time asymptotic values normally used. We then obtain the profiles double averaged analytically over the impact parameters and the Maxwellian distribution (6).

Our conclusions can be summarized as follows:

i. the resulting profile depends on the Massey adiabatic parameter, i.e.,

$$\xi^* = \frac{\pi \Delta \varepsilon}{2\alpha \sqrt{\frac{2kT}{\mu}}}$$

and also on the state which is populated initially;

ii. the interaction leads to the formation of a satellite shifted to the blue side from the position of the atomic forbidden transition if the energy of the allowed atomic transition is less than the energy of the forbidden one;

iii. the Lorentzian distribution at the position of the allowed transition and the satellite give rise to red and blue wings respectively; the intensities of these wings decrease as $1/\omega$ where ω is a detuning from the frequency of the allowed or the forbidden transitions.

Fig. 1 qualitatively summarizes the case $\xi^* > 1$.

The calculations are performed (5) for the Ca(4s4p,$^3P_1$,$^3P_2$ $\rightarrow$ 4s^2, $^1S_0$)-He optical transitions which corresponds to $\xi^*=0.79$, the Ca(4s4p,$^1P_1$, 4s3d^{1}D$\rightarrow$4s^2,$^1S_0$)-He transitions with $\xi^*=20$, and the Mg(3s3p,$^3P_2$, $^3P_1\rightarrow$3s^2,$^1S_0$) transitions with $\xi^*=0.22$. They confirm a simple thumb rule: *the interaction between two nearest quasimolecular states leads to the formation of a satellite. The distance between its maximum and the position of the nearest allowed transition is larger than that between the allowed transition and the forbidden one.*

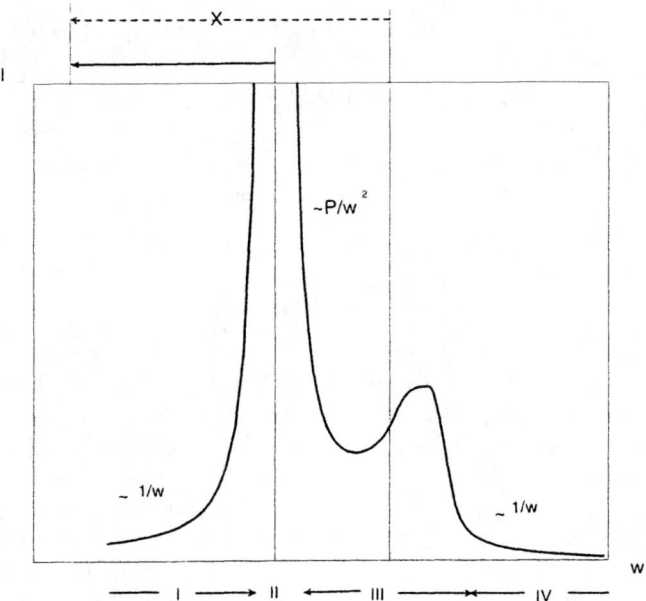

FIGURE 1. The qualitative sketch of the intensity of radiation as function of frequency when the non-radiating state is initially populated. The frequency of the allowed transition is less than the frequency of the forbidden one. Regions: I, IV are the wings, II is the Lorentzian distribution near the position of the allowed transition, III is the satellite, and P is for the probability of the non-adiabatic transition between two states.

ACKNOWLEDGMENTS

It is my pleasure to thank Prof. Y.Sato for numerous stimulating discussions, communications and warm hospitality during my stay at Tohoku University. I gratefully acknowledge the assistance of my colleagues: PhD student E.Bichutskaia, Drs.Yu.Sebyakin, and A.Zagrebin. Support by the Russian Found for Basic Research under the grant 96-03-33679a is gratefully acknowledged.

REFRENCES

1. Kleiber, P.D., Sando, K.M., *Phys.Rev.A* **35**, 3715-3718 (1987).
2. Julienne, P.S., Krauss M., Stevens W., *Chem.Phys.Lett.* **38**,374-381 (1976)
3. Pascale,J., *J.Chem.Phys.* **67**, 204-209 (1977).
4. Gallagher A., Holstein T., *Phys.Rev.A* **16**, 2413-2429 (1977)
5. Bichutskaia,E.N., Devdariani, A.Z., Zagrebin,A.L.,and Sebyakin, Yu.N., *Opt.Spectrosc.*, submitted for publication
6. Bichutskaia, E.N., Devdariani, A.Z., and Sebyakin, Yu.N., *Opt.spectrosc.*, **85**, n. 2 (1998)

Asymmetric Rautian-Sobelman profile

Roman Ciuryło and Józef Szudy

Institute of Physics, Nicholas Copernicus University,
87-100 Toruń, ul. Grudziądzka 5, Poland

I. INTRODUCTION

In their pionieering work [1] on the influence of velocity- and phase-changing collisions on spectral lines Rautian and Sobelman have derived formulae for line profiles using classical kinetic equation method. They have modeled the effect of velocity-changing collisions in two limits denoted, respectively, as soft and hard collisions. On the other hand, in their analysis the effect of phase-changing collision was taken into account assuming the impact approximation, which predicts a symetric (Lorentzian) distribution. The first-order correction term to the Lorentzian profile resulting from the finite duration of a collision has a dispersion form and thus gives rise to an asymmetry in the near-wing region [2,3]. The dispersion asymmetry can also be caused by other effects such as the mixing of overlapping lines[4]. In the present work we have generalized the Rautian-Sobelman treatment on cases when the dispersion asymmetry and the correlation between the Doppler and pressure broadening [5,6] occur. The intensity distribution in a line perturbed due to simultaneous action of velocity-changing and phase-changing collisions can be writen as real part $I(\omega) = \text{Re } \mathcal{I}(\omega)$ of the complex line-shape function

$$\mathcal{I}(\omega) = \frac{1}{\pi} \int_0^{+\infty} dt \int d^3\vec{v}_E \int d^3\vec{r}\, f(\vec{r}, \vec{v}_E, t) \exp\left[i(\omega - \omega_0)t - i\vec{k}\vec{r} - Ng(t)\right], \quad (1)$$

where N is the density number of perturbers, $f(\vec{r}, \vec{v}_E, t)$ is the distribution function giving the relative number of emitters at time t with a velocity between $\vec{v}_E$ and $\vec{v}_E + d\vec{v}_E$ which have moved from their initial position $\vec{r} = 0$ at time $t = 0$ to position between $\vec{r}$ and $\vec{r} + d\vec{r}$. Here $\vec{k}$ ($|\vec{k}| = \omega_0/c$) is the wave vector of the emitted radiation, ω_0 is the unperturbed frequency of a line. The function $g(t)$ which, in general, has complicated form describes the effect of phase-changing collisions and plays a crucial role in the line shape analysis.

II. NEAR WINGS

In the near-wing region, which corresponds to the limit $t \to \infty$ the function $g(t)$ can be approximated using the following equation:

$$Ng(t) = \xi(v_E) + i\beta(v_E) + [\Gamma(v_E) + i\Delta(v_E)]t, \quad (2)$$

where $\Gamma(v_E)$ and $\Delta(v_E)$ are the emitter-speed-dependent Lorentzian width and shift of the line, respectively. The speed-dependent quantities $\xi(v_E)$ and $\beta(v_E)$ first of all depend on the collision duration time. Including the term $i\beta(v_E)$ we obtain the dispersion-shaped correction to the Lorentzian distribution which is responsible for the occurence of a so-called "collision-time" asymmetry in the near-wing regime.

III. SOFT COLLISIONS

In the soft collision model the emitter motion is described in terms of the theory of Brownian motion. The line shape can then be derived from Eq. (1) using the distribution function $f(\vec{r}, \vec{v}, t) \equiv W_{\nu_S}(\vec{r}, t; \vec{v}) f_{m_E}(\vec{v}_E)$, where $f_{m_E}(\vec{v}_E)$ is the Maxwellian distribution of emiter velocities, and

$$W_{\nu_S}(\vec{r}, t; \vec{v}) = \left(\frac{A}{\pi}\right)^{3/2} \exp\left\{-A\left[\vec{r} - \vec{v}(1 - e^{-\nu_S t})/\nu_S\right]^2\right\}, \qquad (3)$$

is the distribution function which describes the probability to find after time t the emitter with the initial velocity $\vec{v_E}$ at a possitiom $\vec{r}$ in the form given by Chndrasekhar [7]. Here $A = [m_E \nu_S^2/(2k_B T)]/[2\nu_S t - 3 + 4e^{-\nu_S t} - e^{-2\nu_S t}]$ and ν_S is the coefficient of the dynamical friction undergone by the moving emitters or, more precisely, the frequency of "soft" velocity-changing collisions. There are two possibilities:

(i) We assume that ξ, β, Γ and Δ are speed-independent. The line profile $I_{AG}(\omega)$ obtained from Eq. (1) in such a case will be reffered to as asymmetric Galatry profile (AGP). This profile becomes identical to the ordinary (i.e. symmetric) Galatry profile [8] denoted by $I_G(\omega)$ in the limits $\xi = 0$ and $\beta = 0$.

(ii) We consider the collision parameters to be dependent on the emitter speed $\xi(v_E)$, $\beta(v_E)$, $\Gamma(v_E)$ and $\Delta(v_E)$. The profile $I_{SDAG}(\omega)$ which results in this case from Eqs. (1) – (2) will be referred to as the speed-dependent asymmetric Galatry profile (SDAGP). For $\xi(v_E) = 0$, $\beta(v_E) = 0$ this profile becomes identical to the profile derived in our previous paper [9], where it was called the *speed-dependent Galatry profile* (SDGP). It should be noted that in a very recent paper by Duggan *et al.* [10] the profile derived in [9] is referred to as *correlated SDGP* (c-SDGP) to distinguish it from another version of the SDGP proposed by Duggan *et al.* [11].

IV. HARD COLLISIONS

In the hard collision model one assumes that the emiter velocity after every collision is independent of the velocity prior to collisions. The line shape in such a case was first derived by Nelkin and Ghatak (NGP) [12], who analyzed the profiles of Doppler-broadened Mössbauer lines. Using speed independent parameters: ξ, β, Γ and Δ one can obtain the profile $I_{ANG}(\omega)$ which will be refered to as asymetric Nelkin-Ghatak profile [13]. On the other hand for speed-dependent parameters: $\xi(\vec{v}_E)$, $\beta(\vec{v}_E)$, $\Gamma(\vec{v}_E)$ and $\Delta(\vec{v}_E)$ one arrives at the speed-dependent asymmetric Nelkin-Ghatak profile (SDANGP), which reduces to the speed-dependent Nelkin-Ghatak profile (SDNGP) if the dispersion asymmetry of the line is neglected ($\xi(\vec{v}_E) = 0$, $\beta(\vec{v}_E) = 0$). The SDNGP has recently been used to analyse line shapes of C_2H_2 pertubed by Xe [14]. Robert *et al.* [15] were the first who used the hard collision model in their analysis of the influence of simultaneous effect of the Dicke narrowing and the Doppler-collision correlation on the shape of the H_2 lines perturbed by Ar.

V. SOFT AND HARD COLLISIONS

In real atomic or molecular systems the assumptions of the soft- and hard-collision models are not fulfilled completely. Following Rautian and Sobelman [1] we can assume that some of the collisions can be treated as soft-collisions and their frequency will be denoted as ν_S. The remaining part is treated as hard-collisions and their efective frequency will be denote as ν_H. The effective frequency $\nu = \nu_S + \nu_H$ of all velocity-changing collisions is connected with the diffusion coefficient D for the emitter with mass m_E in the gas of temperature T by the relation $\nu = k_B T/(m_E D)$. In this case the kinetic equation for the function $f(\vec{r}, \vec{v}, t)$ should be taken in the form proposed by Rautian and Sobelman [1]:

$$\frac{\partial f(\vec{r}, \vec{v}, t)}{\partial t} = -\vec{v}\vec{\nabla}_r f(\vec{r}, \vec{v}, t) + \hat{D}f(\vec{r}, \vec{v}, t) - \nu_H f(\vec{r}, \vec{v}, t) + f_{m_E}(\vec{v}) \int d^3\vec{v}' \, \nu_H f(\vec{r}, \vec{v}', t), \quad (4)$$

where $\hat{D}f(\vec{r}, \vec{v}, t) = \nu_S \vec{\nabla}_v [\vec{v} f(\vec{r}, \vec{v}, t)] + (v_{m_E}^2 \nu_S/2)\Delta_v f(\vec{r}, \vec{v}, t)$. To obtain the profile for the soft- and hard-collisions in speed-independent case, we multiply both sides of Eq. (4) by $\exp[i(\omega - \omega_0 - \Delta + i\Gamma)t - i\vec{k}\vec{r} - \xi - i\beta]/\pi$ and integrate by parts over t and $\vec{r}$. After some manipulations one obtains the asymmetric Rautian-Sobelman profile (ARSP) $I_{ARS}(\omega)$ which is identical with Rautian-Sobelman profile (RSP) given in paper [1] if dispersion asymmetry is neglected. The complex function of the asymmetric Rautian-Sobelman profile can be written in the following form

$$\mathcal{I}_{ARS}(\omega) = \frac{\mathcal{I}_{AG}(\omega)}{1 - \pi\nu_H \mathcal{I}_G(\omega)}, \quad (5)$$

where

$$\mathcal{I}_{AG}(\omega) = \frac{1}{\pi} \int_0^{+\infty} dt \int d^3\vec{v}_E \, f_{m_E}(\vec{v}_E) \int d^3\vec{r} \, W_{\nu_S}(\vec{r}, t; \vec{v}_E) \\ \exp\left\{i(\omega - \omega_0)t - i\vec{k}\vec{r} - (\xi + i\beta) - (\nu_H + \Gamma + i\Delta)t\right\} \quad (6)$$

and $\mathcal{I}_G(\omega) = \mathcal{I}_{AG}(\omega)$ for $\xi + i\beta = 0$. We have undertaken an attempt to derive a speed-dependent ARSP (SDARSP) but we failed. Nevertheless, in analogy to Eq. (5) we suggest to write the SDARSP formula in the following form [16]

$$\mathcal{I}_{SDARS}(\omega) = \frac{\mathcal{I}_{SDAG}(\omega)}{1 - \pi\nu_H \mathcal{I}_{SDG}(\omega)} \qquad (7)$$

Here $\mathcal{I}_{SDAG}(\omega)$ is given by Eq. (6) where $\xi = \xi(v_E)$, $\beta = \beta(v_E)$, $\Gamma = \Gamma(v_E)$, and $\Delta = \Delta(v_E)$. It must be emphasized that Eq. (7) is just a hypothetical formula and as such it needs a theoretical justification.

All four "elementary" profiles descibing the Doppler and pressure effec on spectral lines that means: Voigt profile (VP), Galatry profile (GP), Nelkin-Ghatak profile (NGP) and Rautian-Sobelman profile (RSP) as well as asymmetric and speed-dependent profiles can be optained from the speed-dependent asymmetric Rautian-Sobelman profile (SDARSP).

Recent studies on the of Ar-broadened lines of CO [10] and HF [16,17] clearly indicate that the inclusion of the correlation between the velocity-changing and dephasing collisions in some case seems to be necessary.

REFERENCES

1. S. G. Rautian and I. I. Sobelman, Usp. Fiz. Nauk **90**, 209 (1966) [Sov. Phys. Usp. **9**, 701 (1967)].
2. N. Allard and J. Kielkopf, Rev. Mod. Phys. **54**, 1103 (1982).
3. J. Szudy and W. E. Baylis, Phys. Rep. **266**, 127 (1996).
4. M. Baranger, Phys. Rev. **111**, 494 (1958).
5. P. R. Berman, J. Quant. Spectrosc. Radiat. Transf. **12**, 1331 (1972).
6. J. Ward, J. Cooper, and E. W. Smith, J. Quant. Spectrosc. Radiat. Transf. **14**, 555 (1974).
7. S. Chandrasekhar, Rev. Mod. Phys. **15**, 1 (1943).
8. L. Galatry, Phys. Rev. **122**, 1218 (1961).
9. R. Ciuryło and J. Szudy, J. Quant. Spectrosc. Radiat. Transf. **57**, 411 (1997).
10. P. Duggan, P. M. Sinclair, R. Berman, A. D. May, and J. R. Drummond, J. Mol. Spectrosc. **186**, 90 (1997).
11. P. Duggan, P. M. Sinclair, R. Berman, A. D. May, and J. R. Drummond, Phys. Rev. A **51**, 218 (1995).
12. M. Nelkin and A. Ghatak, Phys. Rev. **135**, A4 (1964).
13. R. Berman, P. Duggan, P. M. Sinclair, A. D. May, and J. R. Drummond, J. Mol. Spectrosc. **182**, 350 (1997).
14. B. Lance, G. Blanquet, J. Walrand, and J. P. Bouanich, J. Mol. Spectrosc. **185**, 262 (1997).
15. D. Robert, J. M. Thuet, J. Bonamy, and S. Temkin, Phys. Rev. A **47**, R771 (1993).
16. R. Ciuryło, Phys. Rev. A (to be published).
17. A. S. Pine, J. Chem. Phys. **101**, 3444 (1994).
18. A. S. Pine, (submitted to J. Quant. Spectrosc. Radiat. Transf.).

Moyal Quantum Dynamics: Atomic Scattering and Line Shapes

B. R. McQuarrie, T. A. Osborn, M. F. Kondrat'eva, and
G. C. Tabisz

Physics and Astronomy Department, University of Manitoba
Winnipeg, Manitoba, Canada R3T 2N2

Abstract. For the first time, a computational version of Moyal quantum mechanics is used to treat a three dimensional system having physical relevant parameters, namely atom–atom scattering. The application of the technique to spectral line calculation is outlined.

INTRODUCTION

Calculations of collision–broadened line shapes frequently treat the average over the collision through a classical trajectory approximation. Often, though, a quantum mechanical approach is necessary. Semi–classical approximations can be useful in that they play the dual role of providing computational tools when full quantum calculations are too difficult or unnecessary and also are a source of insight and intuition even when numerical solutions are available.

In 1949, Moyal [1] published a complete description of quantum mechanics set in classical phase space. Within this formalism there is a natural semi–classical expansion which is based on classical dynamics. Recently, Osborn and Molzahn [2] reformulated Moyal quantum mechanics (MQM) through a graph cluster expansion. This approach provides representations of the semi–classical expansion that are much more computationally accessible than the original.

The Moyal formalism is based on the Wigner–Weyl transformation σ that maps a QM operator $\hat{A}$ on Hilbert space to a function or symbol A_w on classical phase space and vice versa [3,4]

$$A_w(z) = (\sigma \hat{A})(q,p) = \int dx \, e^{-ip \cdot x/\hbar} \langle q + \frac{x}{2} | \hat{A} | q - \frac{x}{2} \rangle, \tag{1}$$

where $z = (q,p)$ are conjugate phase space variables and the angular brackets denote a Dirac matrix element of $\hat{A}$ in coordinate space. An operator $\hat{A}$ with a symbol A_w, which has a regular asymptotic expansion about $\hbar = 0$, is called semi-classically admissible. Commonly, Hamiltonian symbols $H(z)$ are $\hbar$ independent and observables A_w are semi-classically admissible.

Denote by $\hat{A}(t)$ the Heisenberg picture time evolution of $\hat{A}$ generated by Hamiltonian $\hat{H}$. The corresponding evolution in the Moyal formalism is Γ_t, which maps the A_w into the symbol $\sigma\hat{A}(t)$. Expectation values are realized as integrals over z. For an initial state ψ with a density matrix $\hat{\rho} = |\psi\rangle\langle\psi|$ one needs the Wigner function $w_\psi(z) \equiv h^{-3}\sigma\hat{\rho}(z)$. Exact evolution and the semi-classical expansion are, respectively,

$$\langle \hat{A}(t)\rangle_\psi = \mathrm{Tr}\hat{A}(t)\hat{\rho} = \int dz\, w_\psi(z)\Gamma_t A_\mathrm{w}(z)\,, \tag{2}$$

$$= \sum_{n=0}^{\infty} \frac{\hbar^n}{n!} \int dz\, w_\psi(z)(\gamma_t^{(n)} A_\mathrm{w})(z) = \sum_{n=0}^{\infty} \frac{\hbar^n}{n!} \langle \gamma_t^{(n)} A_\mathrm{w}\rangle_\psi\,. \tag{3}$$

Like Γ_t, the semi-classical flow operators $\gamma_t^{(n)}$ map symbols to symbols. The first of these, $\gamma_t^{(0)}$, is constructed solely in terms of classical motion. Let $g(t|z)$ be the solution of Hamilton's equations $\dot{g}(t|z) = J\nabla H(g(t|z))$ (here J denotes the standard symplectic matrix and $g(0|z) = z$) then $\gamma_t^{(0)} A_\mathrm{w}(z) = A_\mathrm{w}(g(t|z))$. For a Hamiltonian $H(z)$ which is $\hbar$ independent, all $n = $ odd operators $\gamma_t^{(n)}$ vanish and only even powers of $\hbar$ appear in expansion (3).

The higher order correction $\gamma_t^{(2)}$ depends on two basic ingredients—the notion of a quantum trajectory and the stability of the classical flow. The quantum trajectory $Z_\nu(t,\hbar;z)$ is defined by the action of Γ_t on the linear coordinate symbol $\pi_\nu(z) = z_\nu$

$$Z_\nu(t,\hbar;z) \equiv \Gamma_t \pi_\nu(z) = g_\nu(t|z) + \frac{\hbar^2}{2} z_\nu^{(2)}(t|z) + O(\hbar^4)\,. \tag{4}$$

As one sees from (4) the classical trajectory $g(t|z)$ is the first order approximation to the quantum trajectory. Classical flow stability is defined by the Jacobi field $\nabla g(t|z)$, which measures the shift of the trajectory at t under small changes of the initial data z. In terms of these quantities, the graph analysis of [2] obtains the $O(\hbar^2)$ operator as

$$\gamma_t^{(2)} A_\mathrm{w}(z) = z_\alpha^{(2)}(t|z)(A_\mathrm{w})_{;\alpha}(g(t|z)) - \frac{1}{8}w_{\alpha\beta}(t|z)(A_\mathrm{w})_{;\alpha\beta}(g(t|z))$$
$$+ \frac{1}{12}w_{\alpha\beta\gamma}(t|z)(A_\mathrm{w})_{;\alpha\beta\gamma}(g(t|z))\,, \tag{5}$$

where the coefficient functions are

$$w_{\alpha\beta}(t|z) = -\mathrm{tr}\nabla\nabla g_\alpha(t|z) J \nabla\nabla g_\beta(t|z) J\,, \tag{6}$$
$$w_{\alpha\beta\gamma}(t|z) = -J\nabla g_\alpha(t|z) \cdot \nabla\nabla g_\beta(t|z) J \nabla g_\gamma(t|z)\,. \tag{7}$$

SCATTERING OF A WAVE PACKET

As a practical application, we compute Heisenberg evolution under the influence of a spherical Lennard–Jones potential $v(r)$ in three dimensions. In this case the Hamiltonian symbol is $H(z) = p^2/2m + v(r)$. The initial state ψ is a Gaussian. The method consists first of all of solving the differential equations for classical motion. The motion is confined to a plane which takes all orientations with respect to space–fixed axes. It can be shown that the quantum corrections $z^{(2)}(t|z)$ also remain in the scattering plane [5,6]. By "squeezing" the Gaussian Wigner function, $w_\psi(z)$, in the momentum direction it is possible to reduce the six dimensional phase space integrals in (3) to integrals over d^3q. This is achieved by an asymptotic expansion about the $w_\psi(z)$ mean value of the momentum.

Figure 1 shows the classical and quantum trajectories for various impact parameters. The quantum trajectories do not differ appreciably from the classical except where the potential is significantly different from zero. The trajectories near rainbow scattering ($b = 3.62$ Å) produce the largest quantum corrections.

FIGURE 1. (a) Classical and (b) quantum trajectories for He at v = 450 m/s. The circle $r = 1.8$ Å indicates the inner core region of the potential.

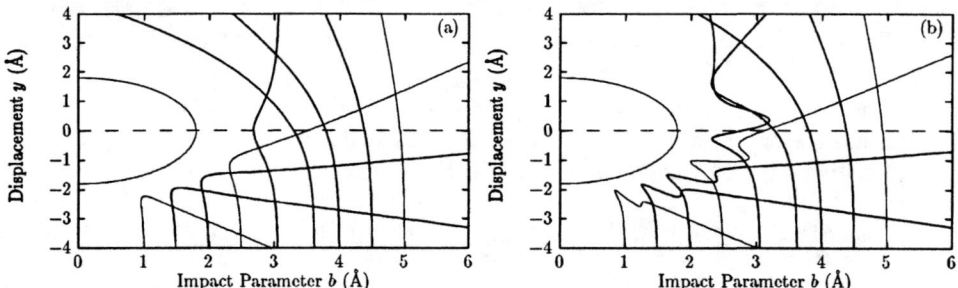

In general we find that the $\langle \gamma_t^{(2)} A_w \rangle_\psi$ correction to $\langle \gamma_t^{(0)} A_w \rangle_\psi$ is small for a wide range of initial Gaussian wave packets and observables $\hat{A}$. The exception to this rule occurs for initial states containing trajectories that are near unstable fixed points of the classical system—i.e. near metastable states. The $\gamma_t^{(2)}$ corrections are largest for helium, while those for argon and neon are smaller but comparable to each other.

LINE SHAPE FORMALISM

To apply the Moyal semi–classical expansion to line shape problems, our approach is to treat the actual molecule and perturber as a composite system characterized by two independent degrees of freedom: I) internal molecular coordinates $\hat{Z} = (\hat{Q}, \hat{P})$ and II) external coordinates $\hat{z} = (\hat{q}, \hat{p})$ suitable for describing the relative motion of the molecule and perturber. Note $[\hat{z}, \hat{Z}] = 0$.

The Hamiltonian $\hat{H}_1 = H_1(\hat{Z})$ and its associated eigenvalue problem $\hat{H}_1|\Phi_n\rangle = E_n|\Phi_n\rangle$ determine the energy spectrum and wave functions of the molecule. Relative motion and the scattering process is generated by $\hat{H}_2 = H_2(\hat{z})$. Normally $\hat{H}_2$ is the sum of a kinetic energy term plus a Lennard–Jones potential. The full system dynamics is given by

$$\hat{H} = \hat{H}_1 + \hat{H}_2 + \hat{H}_{12}. \tag{8}$$

Here $\hat{H}_{12} = H_{12}(\hat{z}, \hat{Z})$ is the anisotropic part of the intermolecular potential. It mixes both degrees of freedom.

It is to be expected that dynamics generated by $\hat{H}$ will be semi–classical in external variables $\hat{z}$, but strictly quantum in the internal coordinates $\hat{Z}$. We incorporate these features into the formalism by defining a (new) mixed Weyl symbol associated with an operator $A(\hat{z}, \hat{Z})$,

$$A_{w1}(z; \hat{Z}) \equiv \int dx \, e^{-ip \cdot x/\hbar} \langle q + \frac{x}{2}|A(\hat{z}, \hat{Z})|q - \frac{x}{2}\rangle. \tag{9}$$

The value of this partial Wigner transform remains an operator in $\hat{Z}$ Hilbert space and continuously depends on $z = (q, p)$.

The dipole auto-correlation function for the description of collision broadening with dipole operator $\hat{M}^r$ and initial density matrix $\hat{\rho} = |\Phi_1\rangle\langle\Phi_1| \otimes e^{-\beta \hat{H}_2}$, is

$$C(t) = \text{Tr}\left(\hat{M}^r \exp\left(\frac{i}{\hbar}\hat{H}t\right)\hat{M}^r \exp\left(-\frac{i}{\hbar}\hat{H}t\right)\hat{\rho}\right). \tag{10}$$

In terms of the mixed Weyl symbol this correlation function becomes

$$C(t) = \sum_n \int dz \, \langle\Phi_n|(\hat{X}^\dagger(t)\hat{M}^r\hat{X}(t))_{w1}(z;\hat{Z})|\Phi_1\rangle\langle\Phi_1|(\hat{\rho}\hat{M}^r)_{w1}(z;\hat{Z})|\Phi_n\rangle. \tag{11}$$

Here $\hat{X}(t) = e^{-\frac{i}{\hbar}\hat{H}t}e^{\frac{i}{\hbar}\hat{H}_2 t}$ is the H_2 interaction picture evolution. We evaluate (11) by obtaining an equation of motion for $\langle\Phi_n|(\hat{X}^\dagger(t)\hat{M}^r\hat{X}(t))_{w1}(z;\hat{Z})|\Phi_k\rangle$. The lowest order $O(\hbar^0)$ approximation of this system recovers the classical path approximation and employs the trajectory $g(H_2; t|z)$. In addition, the mixed Weyl symbol formalism systematically defines higher order corrections, the first of which is $O(\hbar^1)$.

Numerical calculations for the HD-He infrared spectrum are underway.

REFERENCES

1. Moyal, J. E., *Proc. Cambridge Philos. Soc.* **45**, 99 (1949).
2. Osborn, T. A., and Molzahn, F. H., *Ann. Phys.* **241**, 79 (1995).
3. Wigner, E. P., *Phys. Rev.* **40**, 749 (1932).
4. Weyl, H., *Zeitschr. f. Phys.* **46**, 1 (1927).
5. McQuarrie, B. R., Ph.D. thesis, University of Manitoba (1997).
6. McQuarrie, B. R., Osborn, T. A., and Tabisz, G. C., submitted for publication.

The "on the energy shell" simplification of the Impulse approximation in Rydberg atom-neutral collisions

D. HOANG-BINH and H. VAN REGEMORTER
LAM, OBSERVATOIRE DE PARIS, F-92195 MEUDON CEDEX, FRANCE

Abstract. Three types of calculations are compared according to the approximations used for the (e+B) scattering amplitude $f_e(q \to q')$: (i) Exact impulse approximation (without any further approximation); (ii) Impulse approximation + "on the energy shell" approximation; (iii) Impulse approximation + "on the energy shell" approximation + "scattering length" approximation. Preliminary results have been obtained for Na(n,l)+Ar collisions, suggesting that the "on the energy shell" approximation breaks down for large momentum transfers Q.

Within the impulse approximation (IA), the scattering amplitude for inelastic collisions between a Rydberg atom $A(\alpha)$ and a neutral perturber B is given by

$$f(\alpha k \to \alpha' k') = \mu \iiint G_\alpha(q) f_e(q \to q') G_{\alpha'}^*(q') dq, \qquad (1)$$

where α denotes the atomic Rydberg state, k is the momentum of (A,B) relative motion, μ is the reduced mass, $G_\alpha(q)$ is the atomic wave function in momentum space, and $f_e(q - q')$ is the free Rydberg electron-perturber scattering amplitude. For simplicity, we assume below that B is a rare gas staying in its ground state. Atomic units are used.

Conservation of momentum implies

$$Q = k - k' = q' - q, \text{ and } |q - Q| \leq q' \leq q + Q. \qquad (2)$$

Therefore, a knowledge of the amplitude $f_e(q \to q' = q + Q)$ is required both "on the energy shell" ($q'^2 = q^2$) and "out of the energy shell" ($q'^2 \neq q^2$), $q^2/2$ being the kinetic energy of the free electron e.

In all applications of the IA, the additional "on the energy shell approximation" (ESA) has been used, according to which the exact $f_e(q \rightarrow q')$ in (1) is replaced by $f_e(q, \cos\theta_{qq'})$ for $q'=q$, the latter being given by known results of elastic electron scattering theory. When the e+B interaction can be represented by a central field $U(x)$, the exact amplitude is given by

$$f_e(q \rightarrow q') = (-1/4\pi) \iiint e^{-i q' \cdot x} U(x) \Psi_q^+(x) dx, \qquad (3)$$

where $\Psi_q^+(x)$ is the outgoing wave function for the initial channel. From (3), qualitative arguments show that the ESA should break down for large momentum transfers Q, when $Q x_0 \geq 1$, x_0 being of the order of

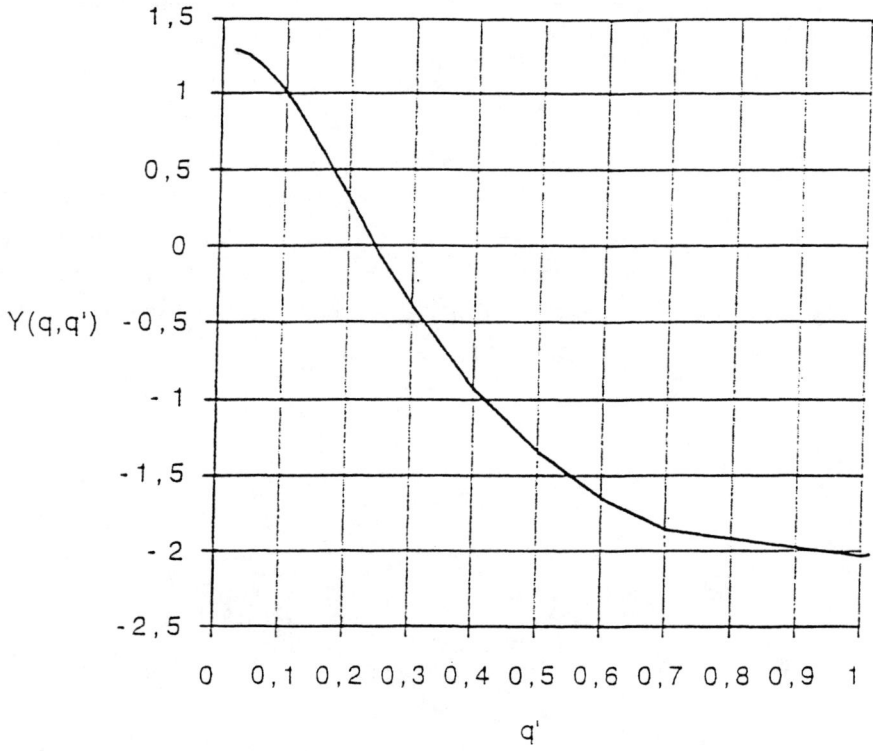

Figure 1. Ratio $Y(q,q') = f_e^S(q,q')/f_e^S(q)$, for q= 0.10.

the range of U(x). Therefore, we have undertaken detailed calculations of excitation cross-sections $\sigma(\alpha \rightarrow \alpha')$, using the exact IA amplitude (1), in order to check the accuracy of the ESA.

Three types of calculations are compared according to the approximations used for the (e+B) scattering amplitude $f_e(q \rightarrow q')$. For simplicity, we consider s-waves only.

1/ **Exact IA** (without any further approximation). From (3),
$$f_e(q \rightarrow q') = f_e^S(q,q'). \tag{4}$$

2/ **IA+ESA**. q=q' in (3), in the interval of interest (2),
$$f_e(q \rightarrow q') = f_e^S(q). \tag{5}$$

3/ **IA+ESA+SLA**, corresponding to the usual "scattering length approximation" (SLA), in which
$$f_e(q \rightarrow q') = f_e^S(q\text{->}0) = L. \tag{6}$$

Preliminary results have been obtained for Na(n,l)+Ar collisions, for 10d-10f transitions. Using e+Ar potential U(x) given by Holtsmark, figure 1 displays the ratio Y(q,q') of the two amplitudes (4) and (5), for $q = <q> = 1/n = 0.10$, the momentum expectation value. It can be seen that, owing to relation (2), the ESA breaks down for large momentum transfers Q. Detailed calculations of the inelastic amplitude (1) and the corresponding cross-section will be published elsewhere.

Quantum Interference and Thermodynamic Equilibrium Between the Gas of Three-Level Atoms and the Photon Gas

V. I. Savchenko+, N. J. Fisch+, A. A. Panteleev*, A. N. Starostin*

+ *Princeton University, Plasma Physics Laboratory, N. J. 08543,*
* *TRINITI and Moscow Institute of Physics and Technology, Moscow Region, Russia*

Abstract.
We consider the influence of quantum interference on the state of thermodynamic equilibrium between atoms and the photon gas. We show, that this interference leads to nonzero non-diagonal elements of the atomic density matrix. These elements in turn modify the emissivity of the medium, which acquires a zero point at a certain frequency and has a red wing with higher intensity.

We consider quantum interference and its influence on thermodynamic equilibrium of atomic and photon gases. In three-level atoms different decay channels may interfere with each other [1], leading to non-diagonal elements in the relaxation operator. Here we will be concerned with the case of thermodynamic equilibirum, where the photon distribution is determined self-consistently with the electron distribution in atoms.

We use the diagram approach developed by Keldysh and Korenman. In this approach quantum interference can be taken into account by the non-diagonal mass operators $\Sigma_{32}^{\alpha\alpha'}$, $\Sigma_{23}^{\alpha\alpha'}$, which are shown in the Fig. 1.

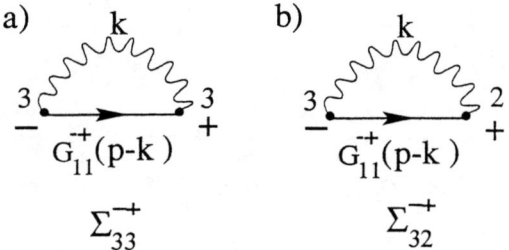

FIG. 1. Mass operators of a three-level atom

Other mass operators are represented by similar diagrams. Wavy lines on these diagrams correspond to the photon Green functions. These mass operators determine the collision integrals in the right-hand side of equations for the atomic Green functions:

$$\left[i\left(\frac{\partial}{\partial \tau} + \frac{\mathbf{p}}{m}\frac{\partial}{\partial \mathbf{R}}\right) - \omega_u + \omega_{u'}\right] G_{uu'}^{-+}(\mathbf{R}, \tau; \mathbf{p}, \omega) = \Sigma_{uu_1}^{-\alpha} G_{u_1 u}^{\alpha+} + G_{uu_1}^{-\alpha} \Sigma_{u_1 u'}^{\alpha+} \quad (1)$$

$$\left[\frac{i}{2}\frac{\partial}{\partial \tau} + \omega - \omega_u + \frac{1}{8m}\frac{\partial^2}{\partial \mathbf{R}^2} + \frac{i\mathbf{p}}{2m}\cdot\frac{\partial}{\partial \mathbf{R}} - \epsilon_{\mathbf{p}}\right] G_{uu'}^{R} = \delta_{uu'} + \Sigma_{uu_1}^{R} G_{u_1 u'}^{R} \quad (2)$$

By solving Eqs. (1), (2) with mass operators from the Fig. 1 we get:

$$G^{-+}_{uu'}(\omega) = -\frac{1}{e^{\omega/T}+1} 2\pi i\, a_{uu'}(\omega), \qquad a_{uu'}(\omega) = \frac{\sqrt{A_u A_{u'}}\,\Omega_u \Omega_{u'}}{(\Omega_2 \Omega_3)^2 + \frac{1}{4}(\Omega_2 A_3 + \Omega_3 A_2)^2} \qquad (3)$$

We see, from Eq. (3), that quantum interference leads to nonzero non-diagonal elements of G^{-+} and, therefore, atomic density matrix. This has to be compared with the usual case of thermodynamic equilibrium, when these elements are zero. Diagonal elements are the same Boltzman exponents in both cases. These elements determine the emissivity of the medium as is clear from the following equation:

$$\varepsilon(\omega) = \frac{4}{3}\frac{\omega^4}{4\pi c^3}\,(1 - N(0))\,N(\omega) \sum_{uu'} d_{u1} d_{u'1} a_{uu'}(\omega), \qquad (4)$$

where $N(\omega) = 1/(1 + e^{\omega/T})$ is the Fermi distribution function. We plot $\varepsilon(\omega)$ in the Fig. 2, 3, where thick and thin lines correspond to spectra with and without interference.

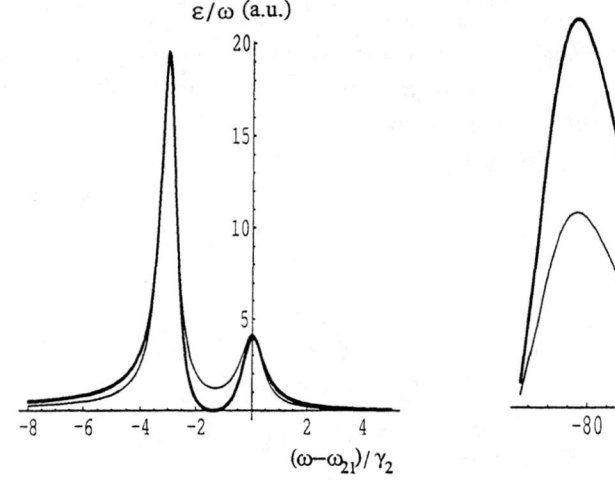
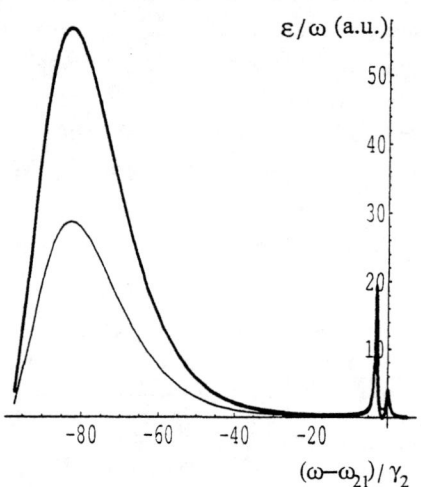

Fig.2. Zero point in the emission spectrum. Atomic parameters are: $\omega_{32}/\gamma_2 = -3$, $\omega_{21}/\gamma_2 = 100$, $T/\gamma_2 = 5$, $d_{31}/d_{21} = \sqrt{2}$, $\omega_{31}/\omega_{21} = 0.7$.

Fig. 3. Enhancement of the red wing due to quantum interference. All conventions are the same as in Fig. 2.

---

[1] V. I. Savchenko, A. A. Panteleev, and A. N. Starostin, in *Laser Interactions and Related Plasma Phenomena* (G. H. Miley and E. M. Campbell, eds.) AIP Conference Proceedings 406, AIP, Woodbury, NY (1997) 371-378.

Relative intensity of Brillouin lines resulting from quasi-transverse hypersonic acoustic waves

T. Błachowicz

Institute of Physics
Silesian Technical University
Krzywoustego 2, 44-100 Gliwice, POLAND

Observation of the acoustic wave in the hypersonic range consists in measuring changes of photon energy nonelastically scattered in annihilation and creation processes by phonons lying at the beginning of the first Brillouin zone. The intensities of these optical signals are very low with respect to the intensity of light incident on the investigated sample, so a single photon counting registration method must be applied.

The signal coming to the counter possessed a stochastic nature due to the thermal noise of the photomultiplier and the random nature of the low-intensity-level scattered light. For these reasons a numerical filtration must be performed, assuming that the noise is conformable to the Poisson distribution. So a weighted average of some neighboring values of counts, creating an "average window" was used. A separate problem is the choice of an appropriate width of the "filtration window". If the width is too large, important properties of the spectrum may be lost (Fig. 1). As we can see from Fig. 1, the best choice of a width of the "filtration window" is the "c" case. A filtration width equal to 60 loses physical information completely.

One of the physical quantities suitable to determine the intensity of light scattered on the acoustic wave is a scattering coefficient which dependence on the differential scattering cross-section can be written in the following form

$$R = \frac{1}{V} \cdot \frac{d\sigma}{d\Omega}, \qquad (1)$$

where V is the volume of a medium in which light scattering occurs. Calculations based on the classical theory of elasticity and classical electrodynamics leads to the following formula (the LiTaO$_3$ piezoelectric crystal belonged to the trigonal system was taken as an example)

$$R_{1/2} = \frac{\pi^2 k T n_o^8}{2\lambda^4 X_{1/2}} \left[p_{41}\gamma_1 - \left(p_{66} + \frac{r_{22}e_{16}}{\varepsilon_{11}}\right)\gamma_2 - \left(p_{14} + \frac{r_{22}e_{15}}{\varepsilon_{11}}\right)\gamma_3 \right]^2, \qquad (2)$$

where k is the Boltzmann constant, T is the temperature of the crystal, λ is the light wavelength, n_0 is the refractive index of the medium, p_{ik}, r_{ik}, and e_{ik} are elements of the elasto-optic Pockels tensor, the electro-optic tensor, and the piezoelectric tensor, respectively. $X_{1/2}$ are the eigen-values and γ_i is the eigen-vector element (describes states of polarization of the acoustic wave) of a so-called "characteristic matrix" defined as $Q_{ik} = c_{ijkl}^{ef} \chi_j \chi_l$, where the c_{ijkl}^{ef} are elastic constants of the medium and where the unit vector $\vec{\chi}$ informs us about direction of propagation of the acoustic wave. The eigen-values $X_{1/2}$, divided by a density of the medium, inform us about velocities of the two quasi-transverse waves. Tab. 1 provides information about frequencies, scattering coefficients R_1 and R_2, for the first and second quasi-transverse waves, and their

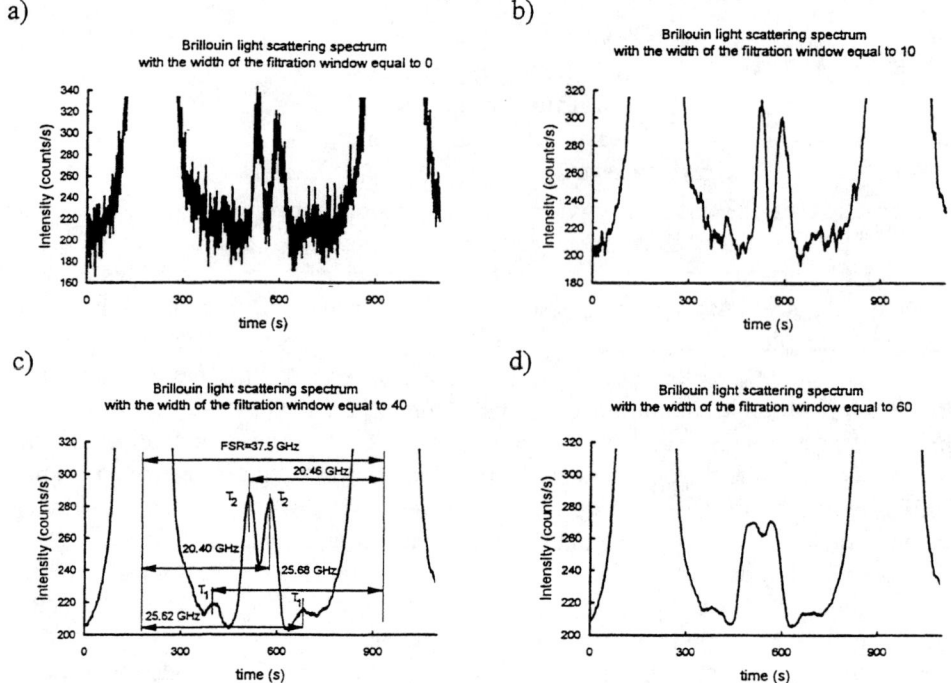

FIG. 1. Examples of the Brillouin spectra from the LiTaO$_3$ crystal after filtration by different widths of the filtration window. Each picture was created from 1100 data points. Descriptions: T$_1$, T$_2$ - quasi-transverse waves, FSR - full spectral range of the Fabry-Perot interferometer[1].

quotient R_1/R_2. The quotient is equivalent in order to make measurements of a relative intensity of the two Brillouin signals. The table provides the experimental results[2], as well as make a comparison with calculated values.

Tab. 1. Comparison of theory with experimental values of basic acoustic wave propagation parameters. Descriptions: v - velocity, f - frequency, T$_1$, T$_2$ - quasi-transverse waves.

	v [T$_1$] (m/s)	v [T$_2$] (m/s)	f [T$_1$] (GHz)	f [T$_2$] (GHz)	R$_1$ (m^{-1})	R$_2$ (m^{-1})	R$_1$/R$_2$
theory	4172.36	3338.10	25.35	20.28	$1.72\cdot10^{-6}$	$1.26\cdot10^{-5}$	0.14
experiment	4213.9±25.6	3352.3±25.2	25.60±0.08	20.36±0.03			0.76±0.01

It is easy to notice that there exists discordance between classical theory predictions and light intensity measurements. It means that quantum level phenomenon (photon-phonon interaction) require more advanced quantum calculations.

References
1. T. Błachowicz, R. Bukowski, Z. Kleszczewski, Fabry-Perot interferometer in Brillouin scattering experiments, Rev. Sci. Instrum. 67, 4057, 1996.
2. T. Błachowicz, Ph. D. Thesis, Silesian Technical University in Gliwice, Poland 1997.

COLD ATOM INTERACTIONS

Optical Spectra of Atoms in Liquid Helium and Cold Helium Gas

Michio Takami

The Institute of Physical and Chemical Research, Wako, Saitama 351-0198, Japan.

Abstract. Lineshapes associated with the absorption and emission spectra of neutral atoms in liquid helium and cold helium gas are discussed. A neutral atom in liquid helium forms a microcavity due to a strong short-range repulsive force between the helium and impurity atoms, producing very broad and largely blueshifted absorption bands and relatively narrow emission lines with much smaller shifts. A broad and weak emission band observed by exciting the D2 line of Ag atoms in liquid helium has confirmed the hypothesis that quenching of D2 emission lines for heavy group 1a and 1b atoms are due to the formation of exciplex. In cold helium gas, sudden decrease of the emission lifetimes below 5.2 K by the excitation of the D2 line of Ag atoms is attributed to the condensation of He atoms on the metastable $^2D_{5/2}$ state. Very narrow absorption lines of Eu innershell electron transitions have revealed that superfluid helium behaves as a static and highly homogeneous low temperature matrix.

INTRODUCTION

Liquid helium has unique properties as a solvent. Its boiling point, 4.21 K, is the lowest among all liquids except liquid ^{3}He. The liquid becomes superfluid below 2.18 K by Bose condensation, and shows no apparent viscosity and almost infinite thermal conductivity. However, the most important property of liquid helium that makes this liquid unique arises from the smallest polarizability of He atom among all elements. This leads to two characteristics of liquid helium. First, the liquid hardly dissolves materials spontaneously because the solvent He atoms do not stabilize solute atoms by wetting. Second, if a neutral atom is implanted into the liquid by some means, the atom forms a microcavity by the short range repulsive force between the helium and implanted atoms [1]. The coupling of trapped neutral atoms with the bubbles (or soft cages) produces 1 ~ 2 nm broadening and shifts for the emission lines, while the absorption lines show large blueshifts and broadening [2–4]. This unusual behavior of the absorption lines originates from the Franck-Condon principle that, at the moment of optical excitation, the excited atom still exists in the bubble equilibrated with the ground state atom, causing very strong interaction with the liquid. Such shift and broadening has been extensively studied for Ba, Rb, and Cs in a pressure range of up to 30 atm [5–7], and reproduced reasonably well by a bubble model with Ba-He and Cs-He pair potentials [6].

As more experimental data became available, many new phenomena came up which cannot be understood within the framework of the bubble model. We will present three such examples from our experimental results; the formation of impurity atom-helium exciplex in liquid helium and cold helium gas, exciplex of light alkaline earth atoms, and innershell electron transitions of Eu atoms in liquid helium. In all these studies, analyses of the observed lineshapes are crucial for correct understanding of the properties and dynamics of impurity atoms in liquid helium.

EXPERIMENTAL METHOD

All experiments were carried out in a double bath Pyrex or metal Dewar [8,9]. Superfluid helium (He II) was produced simply by reducing the He vapor pressure above the liquid. The lowest temperature attainable was 1.6 K in our setup. A piece of metal sample immersed in He II is irradiated with a focused pulsed YAG laser beam (2nd harmonics, 10-30 mJ/pulse, 10 Hz). Plasma of sample material produced by the laser ablation is rapidly cooled by the surrounding liquid, whereby producing a large number of small sample particles (clusters). Neutral atoms and molecules are produced by dissociating these particles with a second YAG laser pulse (10-30 mJ, 2-4th harmonics, 10 Hz) [10]. Pulsed dye lasers are used to measure absorption and emission spectra of atoms and molecules in liquid helium. Optical emission from the produced atoms and molecules is detected with a photomultiplier through a 25 cm grating monochromator. The detected signals were averaged by a Boxcar integrator and stored in a computer.

FORMATION OF EXCIPLEX IN THE LIQUID

From the early stage of the work, light alkali atoms (Li, Na, and K) were known to show no laser induced fluorescence (LIF) from the lowest p state. Even for heavy alkali atoms like Rb and Cs, LIF in the D2 line is absent or very weak despite the fact that the D2 line is clearly observed in the absorption spectra [7]. To explain these unusual observations, formation of exciplexes with the surrounding He atoms was proposed [11]. According to this model, a light alkali atom in the p state forms an exciplex with several helium atoms bonded to the ion core of the alkali atoms in the nodal plane by the induction effect. This produces largely redshifted and broadened bound-free emission band, or even quenches the emission by a non-radiative relaxation to the ground state. For heavy alkali atoms, on the other hand, the interaction should be considered for each fine structure component because the fine structure splitting is larger than the atom-He interaction energy. The $P_{1/2}$ state has a spherical electron density distribution and shows only the repulsive interaction with the helium atoms. However, the electron density distribution for the $P_{3/2}$ state has a doughnut-like structure, and can accommodate two helium atoms along the symmetry axis. This model predicts rapid relaxation of the atomic $P_{3/2}$ state to the exciplex and a bound-free emission band from the exciplex with a large redshift from the atomic D2 line.

Recently, we accidentally found a broad and weak emission band on the red side of the D1 line when the D2 line of Ag atom was excited in liquid helium [12]. In order to find the origin of this broad emission band, absorption spectra were measured by monitoring the atomic D1 line and the broad emission band. The absorption spectrum taken by monitoring the D1 line shows two absorption bands corresponding to the D1 and D2 lines, with blue shifts and asymmetric band shapes as expected for valence electron transitions in liquid helium. On the other hand, the absorption spectrum taken by monitoring the broad emission shows only the D2 absorption line. Since the exciplex is formed only when the D2 line is excited, we concluded that this weak band is the emission from the AgHe$_2$ exciplex. An *ab initio* calculation of the He-Ag-He potential reproduced the observed exciplex emission band satisfactorily. Recently, a similar exciplex emission band has been observed also for Rb [13].

FORMATION OF EXCIPLEX IN LOW TEMPERATURE HE GAS

The above measurement suggests the presence of AgHe exciplex as well. In liquid helium, however, density of He is too high to produce this unsaturated exciplex with enough concentration to be detected. In order to detect this unsaturated exciplex, we studied the emission spectra of Ag atoms with a similar experimental arrangement in cold helium gas at around 1 atm He pressure simply by lifting up the metal sample above the liquid. Surprisingly, the ablation/dissociation method worked perfect even in cold helium gas. We measured absorption/emission spectra in the temperature range of 7.5 to 25 K by changing the height of observation [14]. The observed emission spectra show two atomic lines ($5p\ ^2P_{1/2}$ and $4d^9 5s^2\ ^2D_{5/2}$) and three broad emission bands from AgHe$_2$ $5p\ ^2\Pi_{3/2}$, AgHe $5p\ ^2\Pi_{3/2}$ and $^2\Pi_{1/2}$ states (the final assignment has been made recently).

The observed spectrum shows an interesting temperature dependence. At 25 K, the emission from AgHe $5p\ ^2\Pi_{1/2}$ is predominant. As the temperature decreases to 17.5 K, this emission becomes weaker while the emission from AgHe $5p\ ^2\Pi_{3/2}$ and AgHe$_2$ $5p\ ^2\Pi_{3/2}$ become stronger. At 7.5 K, the emission from AgHe$_2$ $5p$ $^2\Pi_{3/2}$ becomes dominant as observed in liquid helium. However, most remarkable temperature dependence was observed in emission lifetimes. At higher temperature, all emission lines and bands have very long and nearly equal lifetimes (~ 350 ns). As the temperature decreases, the lifetimes of all lines and bands suddenly decrease to 10-20 ns at around 5.2 K, the critical temperature of helium gas. This unusual feature suggests the existence of a reservoir state. In fact, the $^2D_{5/2}$ state, which exists between $^2\Pi_{3/2}$ and $^2\Pi_{1/2}$ and its emission to the ground state is doubly forbidden by the selection rules of parity and angular momentum, shows very weak collision-induced emission line. An Ag-He *ab initio* potential [15] indicates that, as the Ag-He distance decreases, the $^2\Pi_{3/2}$ state is lowered while the $^2D_{5/2}$ state is pushed up, crossing at around $r = 3.5$ Å. Thus part of the atoms excited to $^2P_{3/2}$ is transferred to this reservoir state, and shows long emission lifetimes by collisional energy transfer to nearby emitting states.

A clue to understand the abrupt decrease of emission lifetimes around 5.2 K is in the Ag-He potential for the $^2D_{5/2}$ state, which has a shallow potential minimum with only one bound state (about 6 cm^{-1} bonding energy). Since this potential is almost isotropic (there are three different states, Σ, Π, and Δ but the differences among the three potential depths are within 30 %.), roughly 20 He atoms can be accommodated. As the temperature becomes lower, the number of bonded He atoms increases. Each time one He atom is bonded, the $^2D_{5/2}$ state is pushed down while the $^2P_{1/2}$ state below $^2D_{5/2}$ is pushed up. When about 16 He atoms are bonded, these states cross each other, draining the atoms in the $^2D_{5/2}$ reservoir state into the emitting $^2P_{1/2}$ state. This model explains most of the observed features semiquantitatively, but the profile of absorption spectrum is not well understood yet.

FORMATION OF EXCIPLEX FOR LIGHT ALKALINE EARTH ATOMS

Alkaline earth atoms are one of the best studied atomic species in liquid helium. However, Mg and Be were not well studied until recently because their absorption lines are in the UV region where it is not easy to access by a conventional laser. During our systematic study of alkaline earth atoms in liquid helium, we found Mg and Be show spectra, especially absorption spectra, quite different from those of Ca, Sr, and Ba. For Ca, Sr, and Ba, emission lines are narrow and close to their free atomic lines (with a few nm shift). On the contrary, Mg and Be show very broad and largely redshifted emission bands, suggesting that the dynamics of excited states are different from those of heavy alkaline earth atoms [16]. In general, one ns electron still left in the lowest $^1P_1$ state of an alkaline earth atom prevents the surrounding He atoms to bond with the central ion core. For light alkaline earth atoms, however, the $2s$ and $3s$ electron orbital sizes are small enough to allow for the He atoms to bond to the ion core. A model calculation shows that the $^1P_1$ state of a Mg atom bonded with 6 He atoms in its nodal plane reproduces the observed emission band reasonably well, confirming the formation of exciplex as well. Unfortunately the simulated band profile is strongly dependent on the pair potential used. In fact, the experimental profile exists between the two profiles calculated by our own potential and by Czuchaj *et al.* [17], showing that the available Mg-He pair potentials are not accurate enough to reproduce the experimental data satisfactorily.

INNERSHELL TRANSITIONS OF EU ATOMS

So far almost all studied atomic spectra were those of valence electron transitions. Since large shift and broadening of atomic transitions in liquid helium originate from the change of electron density distribution by optical excitation, innershell electron transition is expected to show much narrower lines. Recently such studies were reported for Tm [18] and Eu [19]. Stimulated by the work on Eu atoms in liquid helium droplet which showed quite anomalous spectral features, we studied Eu atom in bulk liquid helium. This atom has the [Xe] $4f^7$ ($^8S_{7/2}$) $6s^2$ $^8S_{7/2}$ ground state, and its electronic transitions are divided into two groups, the $6s7p \leftarrow 6s^2$ valence electron transitions and the $5d \leftarrow 4f$ innershell electron transitions. Since the former shows absorption spectra typical for a bubble atom, Eu atom is definitely trapped in a bubble in liquid helium. The innershell electron transition showed entirely different spectral profiles as expected. We studied absorption spectra to the $4f^65d6s^2$ $^8F_{5/2}$, $^8F_{7/2}$, and $^8G_{9/2}$ states. All lines have narrow zero phonon lines (ZPL, ~ 0.5 cm^{-1} HWHM) with phonon bands extending over ~ 20 cm^{-1} on the higher energy side of ZPL. The phonon band is well reproduced by a simple model assuming the presence of thermal phonons. One unexpected observation is that all ZPL's show redshifts from their free atomic lines. This is reasonable because, since the outer $6s^2$ electrons hardly change their density distribution by the excitation of one of the innershell electrons, the only change in the Eu-He interaction by the excitation is the increase of polarizability for excited Eu atoms by the increased state density. Thus the change of interaction by exciting a Eu atom is the increase of dispersion energy which slightly lowers the energy of the excited atom, being consistent with the observation.

During this analysis, an interesting observation came up. As stated above, we measured absorption lines to three different angular momentum states of the same electronic configuration. All three ZPL show almost equal linewidths, ~ 0.5 cm^{-1} (HWHM) while their shifts from free atomic lines range form -15 to -43 cm^{-1}. This is incompatible with the collisional line broadening theories (e.g. Anderson's theory [20]) which provide strong correlation between shifts and widths. This observation reminds us of extremely narrow optical absorption lines observed for an organic molecule embedded in low temperature organic matrix [21]. Although a solid matrix shows site effect, spectrum of a single molecule in the matrix shows very narrow linewidth (≤ 1 MHz) because the matrix is rigid and thermally inert. From this point of view, it is not surprising that Eu atoms in superfluid liquid helium show narrow absorption lines because, at 1.6 K, all He atoms are more or less delocalized and the liquid behaves as a highly homogeneous and static matrix. Then the observed width should reflect some relaxation process. One possible process is the relaxation of the transient state produced at the moment of excitation. Due to the Franck-Condon principle, the liquid in the vicinity of excited Eu atom is in a density-deficient state at the moment of excitation due to the increased polarizability of an excited Eu atom. The time period necessary to restore equilibrium is the time for rearrangement of the liquid, or the time for phonons to propagate over the density deficient region. If we assume 240 m/s for the velocity of phonon and estimate the size of the defect to produce the observed width, the defect should have about 2 nm radius, being consistent with the size of expected defect [22].

Recently Hartmann *et al.* reported the observation of roton/maxon peaks over a phonon band associated with an electronic transition of a glyoxial molecule embedded in a He cluster [23]. In the present measurement, however, the roton/maxon peaks are absent in spite of the fact that the laser excitation provides enough energy to produce a roton which needs finite energy and momentum to create (the observed phonon band extends well beyond the threshold energy for creating a roton). This will be explained by the fact that, in the optical excitation of a Eu atom, there is no mechanism to transfer a minimum linear momentum to the liquid for creation of a roton because, general motion of He atoms after the optical excitation is towards the central Eu atom and a photon momentum is well below the threshold momentum. It is interesting to note that a thermal neutron scattering spectra, carried out for measuring dispersion curve of superfluid helium and for confirming the presence of roton, has similar line widths [24]. Furthermore, temperature dependence of the width for these two entirely different experiments shows a similar trend, suggesting that the two linewidths have the same origin. Although many observations are still not well understood yet, liquid helium will provide unique environment for physics, chemistry, and engineering.

ACKNOWKLEDGEMENTS

I would like to thank Drs. J.H.M. Beijersbergen, Qin Hui, J. Persson, M.P. Coquard, Z.J. Jakubek, Y. Kasai, M. Yonekura, and M. Nakamura who contributed to this work. This work was supported by a Grant for International Joint Research Projects for the New Energy and Industrial Technology Development Organization (NEDO), Japan, and Special Grant for Promotion of Research from The Institute of Physical and Chemiacl Research (RIKEN).

REFERENCES

1. Hickman, A.P. and Lane, N.F., *Phys. Rev. Lett.* **26**, 1216 (1971).
2. Bauer, H., Beau, M., Friedel, B., Marchand, C., Miltner, K., and Reyher, H.J., *Phys. Lett.* A **146**, 134 (1990).
3. Takahashi, Y., Sano, K., Kinoshita, T., and Yabuzaki, T., *Phys. Rev. Lett.* **71**, 1035 (1993).
4. Beijersbergen, J.H.M., Hui, Q., and Takami, M., *Phys. Lett.* A, **181**, 393 (1993).
5. Kanorsky, S.I., Arndt, M., Dziewior, R., Weis, A., and Hänsch, T.W.Z., *Phys. Rev.* B, **49**, 3645 (1994).
6. Kanorsky, S.I., Weis, A., Arndt, M., Dziewior, R. and Hänsch, T.W.Z., *Z. Phys.* B, **98**, 371 (1995).

7. Kinoshita, T., Fukuda, K., Takahashi, Y., and Yabuzaki, T., *Phys. Rev.* A, **52**, 2707 (1995).
8. Hui, Q., Persson, J.L., Beijersbergen, J.H.M., and Takami, M., *Z. Phys.* B, **98**, 353 (1995).
9. Takami, M., *Comments At. Mol. Phys.*, **32**, 219 (1996).
10. Fujisaki, A, Sano, K., Kinoshita, T., Takahashi, Y. and Yabuzaki, T., *Phys. Rev. Lett.* **71**, 1039 (1993).
11. Dupont-Roc, J., *Z. Phys.* B, **98**, 383 (1995).
12. Persson, J.L., Hui, Q., Jakubek, Z.J., Nakamura, M., and Takami, M., *Phys. Rev. Lett.* **76**, 1501 (1996).
13. Kinoshita, T., Fukuda, K., Matsuura, T., and Yabuzaki, T., *Phys. Rev.* A, **53**, 4054 (1996).
14. Jakubek, Z.J., Hui, Q., and Takami, M., *Phys. Rev. Lett.* **79**, 629 (1997).
15. Jakubek, Z.J. and Takami, M., *Chem. Phys. Lett.* **265**, 653 (1997).
16. Hui, Q. and Takami, M., in preparation.
17. Czuchaj, E., Stoll, H., and Preuss, H., *J. Phys.* B **20**, 1487 (1987).
18. Ishikawa, K., Hatakeyama, A., Goshono-o, K., Wada, S., Takahashi, Y., and Yabuzaki, T., *Phys. Rev.* B **56**, 780 (1997).
19. Bartlet, A., Close, J.D., Federmann, F., Hoffmann, K., Quaas, N., and Toennies, J.P., *Z. Phys.* D, **39**, 1 (1997).
20. Anderson, P.W., *Phys. Rev.* **76**, 647 (1949).
21. See, for example, Moerner, W.E., *Science* **265**, 46 (1994).
22. Hui, Q. and Takami, M., in prparation.
23. Hartmann, M., Mielke, F., Toennies, J.P., and Vilesov, A.F. *Phys. Rev. Lett.* **76**, 4560 (1996).
24. Stirling, W.G. and Glyde, H.R., *Phys. Rev.* B **41**, 4224 (1990).

Far-Wing Excitation Studies on the Quasimolecular Transitions in the Hg–Rare-Gas, Simple-Molecule Half-Collisions

Kenji Ohmori

Research Institute for Scientific Measurements, Tohoku University
Katahira 2-1-1, Sendai 980-8577, Japan

Abstract. We have developed a bran-new half-collision approach to the transitory quasimolecular states in the cold and thermal atomic and molecular collisions and applied it to the Hg – rare-gas, simple-molecule system. This approach consists of (a) the far-wing excitation technique to trigger a thermal half-collision by the laser excitation of a collisional quasimolecule; and (b) the vdW technique to trigger a cold half-collision by the laser excitation of a van der Waals complex in a supersonic-jet expansion; the nascent products of those half-collisions are detected in a quantum-state-resolved fashion by another laser pulse. Measurements have been made in both the frequency and time domains. In the frequency-domain measurements, a pair of nanosecond pulses are used to measure the excitation spectra of the quasimolecule. Inspection of these spectral line shapes have realized for the first time the complete separation of the rotational- and translational-energy dependence of a quantum-state-resolved scattering probability with an extensive tuning range from an ultracold (50 mK) to a higher-thermal (1000 K) regions. Interaction potentials for the relevant colliding pairs are also extracted from those line shapes for a wide range of internuclear distances. In the time-domain measurements, a pair of femtosecond laser-pulses are used to clock the half-collision in real time, giving the important information that could never be extracted in the frequency domain. Selected results are presented to demonstrate that these half-collision techniques are combined to work as a useful tool to extract the dynamical features of the transit regions in atomic and molecular collisions.

INTRODUCTION

Photoexcitation of a quasimolecule triggers the second-half collision on the excited potential-surface. It is certainly nothing less than jumping into a "transit region" in an atomic or molecular collision. Here I use the term "transit region" as all intermediate configurations between reagents and products, including what we call a "transition state" at the saddle point of the potential-energy surface. We have developed a brand-new half-collision approach to that transit region aiming at detailed understanding of dynamical aspects in chemical reactions or energy transfers and of interaction potentials for the relevant colliding pairs. Our half-collision approach is subdivided into two categories ; one is a far wing approach, and the other is a van der Waals (vdW) approach. As is now well understood, a far-wing spectrum of a pressure-broadened atomic line is ascribed to an optical transition of an ultrashort-lived quasimolecule formed during a thermal binary-collision [1] and has been used by several groups to extract the potential curve and the transit-region dynamics [2]. We have developed our own far-wing approach [3-8]. It is based on a laser pump/probe and/or a double-beam absorption/dispersion methods applied to the far wings of a pressure-broadened atomic line. A half collision is thus triggered by excitation of thermal quasimolecules in this approach. The vdW approach, on the other hand, is based on laser excitation of a vdW complex generated in a supersonic jet expansion, and a half collision is triggered by excitation of rotationally and vibrationally cold quasimolecules. These two approaches are combined to be quite useful in understanding a nonadiabatic coupling induced by nuclear rotation and vibration and in determining an interaction potential for a wide range of nuclear configuration.

We have successfully applied our half-collision approach to (1) the chemical reactions of $Hg(6s6p^3P_1)$ with H_2, HD and D_2 [3]; (2) the fine-structure transitions of the excited Hg atom within $6s6p^3P$ manifold in collisions with N_2, CO and rare gases (RG's) [4]; and (3) the interaction potentials for the excited Hg, Ba, and Ca atoms with various RG's and diatomic molecules [5-7]. Here I present a selected number of topics that demonstrate our "combined half-collision" approach works nicely as a promising tool for a transit-region spectroscopy both in the time and frequency domains.

EXPERIMENTS

Two different half-collision approaches have been combined to realize an access to the transit regions of Hg – RG, simple-molecule collisions. They are the far-wing approach and the vdW approach. Details of the experimental setups and procedures have been given or will be given elsewhere [3-8], and only brief descriptions are given in the following subsections.

Far-wing approach

The Hg vapor was generated in a heat-pipe oven, and the perturber gas X was admitted into the oven to give a total pressure of 5 Torr (X = N_2, CO, and RG). The second or third harmonic of a pulsed Nd:YAG laser was used to pump two dye lasers simultaneously. One of their outputs was frequency doubled to generate the pump pulse around 253.7 nm by which the Hg–X quasimolecule was excited and a thermal half-collision was triggered. The other dye-laser output was used for the laser-induced-fluorescence (LIF) detection of the nascent products formed from the excited quasimolecules. These nanosecond pump and probe pulses were introduced collinearly into the oven with the probe pulse being delayed by 3 ns from the pump pulse. The effects of secondary collisions were negligible. The fluorescence signal was sampled from a viewing port of the oven and was detected by a photomultiplier attached to a 20 cm monochromator.

vdW approach

The Hg–X vdW complexes were produced by expanding a high-pressure Hg/Ar mixture (for Hg–Ar complex) or Hg/X/He mixture (for other complexes) into a vacuum chamber through a pulsed valve of 0.3 mm in diameter. Both the frequency- and time-domain measurements were made on the half collisions triggered by photoexcitation of those vibrationally- and rotationally-cold quasimolecules.

In the frequency-domain measurements, the laser systems were the same as those used in the far-wing experiment except that the two dye lasers were pumped independently by different YAG lasers. The pump and probe pulses were introduced into a vacuum chamber to intersect the jet perpendicularly with the probe pulse being delayed typically by 30 ns from the pump pulse. A rotationally-cold half collision was thus triggered by excitation of the Hg–X vdW complex with a pump pulse. The probe-pulse induced fluorescence from the complex or a nascent product was detected through suitable interference filters with a photomultiplier directly attached to a vacuum chamber.

In the time-domain measurements, the experimental system was the same as those used in the frequency-domain measurements except that the nanosecond laser system was replaced by an ultrashort-pulse one and a monochromator was used instead of the interference filters to discriminate true signals from background. The 80 fs output pulse of a mode-locked Ti:Sapphire laser pumped by Ar^+ laser was stretched and seeded into a regenerative amplifier pumped by the 2nd harmonic of a pulsed Nd:YAG laser, and its output was further amplified by another Nd:YAG laser and then recompressed to give the final output around 760 nm in 200 fs. It passed through a home-built wavelength-conversion system to yield ultrashort pump and probe pulses in UV region. They were introduced collinearly into the vacuum chamber to intersect the jet perpendicularly. Scanning the pump-probe delay time in a femtosecond time-scale, the half collision was clocked in real-time. A mercury-vapor cell is inserted in the pump and probe beam paths to remove a wavelength portion at 253.7 nm. Without this vapor cell, the true signal was smeared out by the large background attributed to Hg $^1S_0 \leftarrow\, ^3P_1$ atomic fluorescence.

SELECTED RESULTS AND DISCUSSION

Hg($^3P_2$) – RG, N$_2$, CO systems; usefulness of the combined approach

Figure 1 shows the quasimolecular states arising from the asymptotes Hg($6s6p^3P$)+X($^1S_0$ or $^1\Sigma$); they are subdivided into three groups corresponding to the spin-orbit components for $6s6p^3P$ configurations in the Hg atom. Although we have studied the half collisions associated with all of these components, here I refer to our most recent series of works on the Hg(6^3P_2) – RG, N$_2$, and CO systems.

At the very beginning, we have to find out a way to reach the quasimolecular states b^32, c^31, and d^30^- correlated to their dipole-forbidden asymptotes. None of them is optically accessible from the ground X^10^+ state at least in its asymptotic limit.

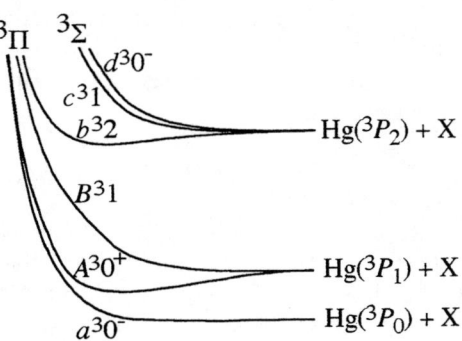

FIG. 1. Schematic representation of the Hg($6s6p\ ^3P_j$)–X($^1S_0$ or $^1\Sigma$) quasimolecular states.

The c state can be represented as a linear superposition of atomic wave function corresponding to $\Omega = 1$ and originating from the manifold of the states for 6^3P_j+X and 6^1P_1 +X systems. In the limit of large interatomic distances R, the c state is the pure $6^3P_{2,\Omega=1}$ atomic state. But with decrease in R, the wave function of the c state contains more and more admixture of two other atomic states, $6^3P_{1,\Omega=1}$ and $6^1P_{1,\Omega=1}$, representing the separated-atom limits of B^31 and C^11 molecular states. This mixing introduces the pure LS-coupling $^1P_1$-character into the adiabatic c-state wave-function and induces the $c-X$ dipole transition at a close approach of Hg and X, which we call a collision-induced-dipole (CID) transition. It is thus possible to make a direct approach to the c state by photoexcitation of quasimolecules in the X state. We use this CID transition as a tool to extract the interaction potential curve of the c state and half-collision dynamics triggered on this potential.

interaction potential

We start with the far-wing approach. The Hg–X quasimolecule (X = Ar, Kr, and Xe) is excited by a pump-laser pulse in the CID band in the far wing of the $^1S_0 - ^3P_2$ forbidden atomic transition under the thermal gas-cell condition, and a thermal half-collision is triggered on the c state. The half collision proceeds adiabatically on the c-state potential and yields the separated Hg atom in the $^3P_2$ state. This

Hg($^3P_2$) is probed by another laser-pulse with 7^3S_1 being the fluorescent state. By scanning the pump-laser wavelength, we measure the excitation spectrum.

Figure 2 presents the excitation spectra of the Hg – Ar, Kr, and Xe thermal quasimolecules. Each spectrum shows its own characteristic line-shape reflecting the potential curves of the c and X states and R dependence of the $c - X$ dipole moment. We have carried out semiempirical simulations of the spectra to obtain these information. Details of the simulating procedure have been given elsewhere [5], and only brief descriptions are given below. The adiabatic wave-function for the c state is expanded in terms of Hund's case (c)-type bases diagonalizing the spin-orbit Hamiltonian

$$|(^3P_2)1\rangle^a = c_1(R)|(^1P_1)1\rangle^{(c)} + c_2(R)|(^3P_2)1\rangle^{(c)} + c_3(R)|(^3P_1)1\rangle^{(c)}, \quad (1)$$

where they correlate to the separated atom limits of C^11, c^31, and B^31 molecular states, respectively. Of these three diabatic bases, the $|(^1P_1)1\rangle^{(c)}$ and $|(^3P_1)1\rangle^{(c)}$ contain the pure LS-coupling $^1P_1$-character and are optically connected with the X state, and their coefficients $c_1(R)$ and $c_3(R)$, functions of R, give the R dependence of the dipole moment $d(R)$. The potential-energy matrix for those (c)-type bases is diagonalized at each R by using empirical potentials for the A^30^+ and B^31 states, $V_A(R)$ and $V_B(R)$ [7], to give the adiabatic potential for the c state, $V_c(R)$, and the coefficients $c_1(R)$ and $c_3(R)$. $d(R)$ is then calculated from these $c_1(R)$ and $c_3(R)$.

These semiempirical $V_c(R)$ and $d(R)$ and the empirical $V_X(R)$ for the X state [9] are incorporated into the quasistatic theory (QST) to calculate $\gamma(R)$, which is the reduced absorption coefficient of the quasimolecule and should be proportional to the excitation spectrum,

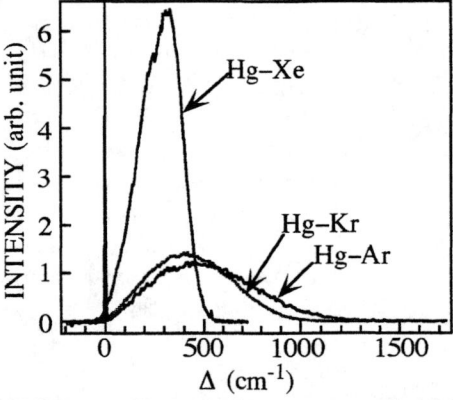 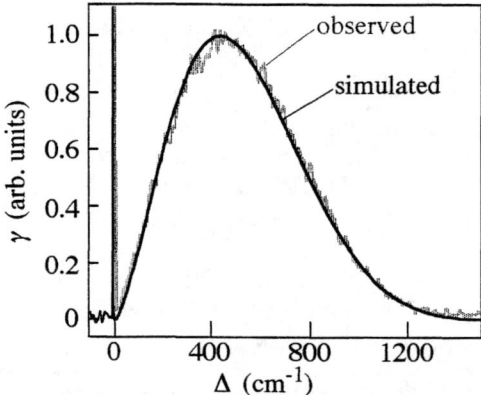

FIG. 2. $c-X$ excitation spectra of the Hg–RG thermal quasimolecules for the formation of Hg($^3P_2$). Δ denotes the detuning of the excitation wave-number from the Hg $^1S_0$–$^3P_2$ atomic transition.

FIG. 3. Comparison of the observed and simulated $c-X$ bands of the Hg–Ar thermal quasimolecule. Δ denotes the detuning of the excitation wave-number from the Hg $^1S_0$–$^3P_2$ atomic transition.

$$\gamma(\Delta) = \frac{(32\pi)^4 \nu}{3c} \cdot \frac{R_c^2 d^2(R_c)}{\Delta V'_{cX}(R_c)} \exp\left(-\frac{V_X(R_c)}{k_B T}\right). \tag{2}$$

R_c is the Hg–RG distance that satisfies the classical Franck-Condon relation

$$\Delta = V_c(R_c) - V_X(R_c) - \nu_0, \tag{3}$$

where ν_0 is the wave number for the $^1S_0 - ^3P_2$ atomic transition. And $\Delta V'_{cX}(R_c)$ is the first derivative of the potential difference between the c and X states. $V_A(R)$ and $V_B(R)$ are adjusted so that a better agreement is achieved between the calculated and measured spectra. Here I present the result for the Hg–Ar system as an example. An excellent agreement has been achieved between calculated and measured spectra, as shown in Fig. 3. The $V_c(R)$ finally obtained is fitted to the modified form of Morse-type function

$$V_c(R) = a[e^{-\beta_2(R-R_e)} - \frac{\beta_2}{\beta_1} e^{-\beta_1(R-R_e)}] \cdot f(R) \tag{4}$$

with

$$f(R) = 1 - b_1 \exp[-(\frac{R-b_2}{b_3})^2]. \tag{5}$$

The potential parameters thus obtained are $R_e = 4.70\text{Å}, D_e = 61.1\text{cm}^{-1}, \beta_1 = 1.03\text{Å}^{-1}, \beta_2/\beta_1 = 2.47, b_1 = 0.40, b_2 = 1.93\text{Å}$, and $b_3 = 0.699\text{Å}$. But, unfortunately, the present far-wing approach doesn't yield the precise information for the attractive branch of the potential curve. This attractive branch, however, can be determined precisely by the other half-collision approach, that is the vdW approach.

The Hg–Ar vdW complex is produced at the vibrational ground-level of the X state in a supersonic jet expansion. The complex is excited by a pump pulse to the bound levels of the c state. We tried to observe the fluorescence from these dark bound-levels down to the X state; it was impossible, however, since its fluorescence life-time could be longer than 1 ms. This is why the bound region of the c state has never been observed previously for any RG as a counterpart. Then we introduce another laser pulse, a probe pulse, to bring the complex from those dark bound-levels to another bright state, which is the E state correlated to $Hg(7^3S_1) + X$. A probe pulse is tuned so that the free–bound $E - c$ transition of the complex has a sufficient Franck-Condon (FC) overlap for all the bound levels of the c state. The complex thus excited by the probe pulse dissociates adiabatically on the E-state potential to produce the separated Hg atom in the 7^3S_1 state, which is now readily emissive to any of the lower 6^3P_2, 6^3P_1, and 6^3P_0 states, and this fluorescence is monitored.

Figure 4 presents the bound–bound $c-X$ excitation spectrum measured with the pump wavelength scanned and the probe wavelength fixed at 544.2 nm. As is seen in the figure, the typical anharmonic progression has been observed. And Figs. 5 (a)–(c) show the free–bound $E - c(v')$ excitation spectra measured with the probe wavelength scanned and the pump wavelength tuned to the lowest three members of the anharmonic progression seen in Fig. 4. Each of these three spectra shows its

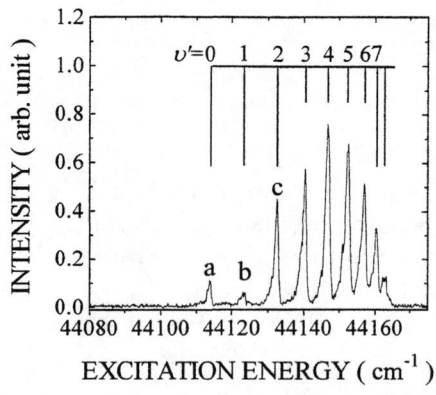

FIG. 4. $c(v')-X(v''=0)$ excitation spectrum of the Hg–Ar vdW complex.

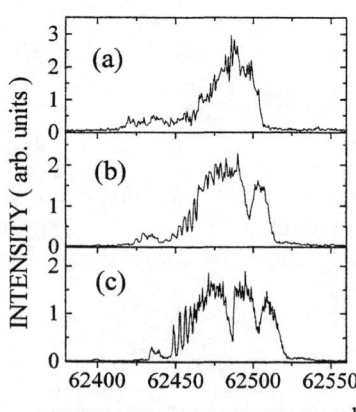

FIG. 5. Free–bound $E-c(v')$ excitation spectra of Hg–Ar measured with the pump laser tuned to the bands (a) a, (b) b, and (c) c in Fig. 4 plotted against the sum of the pump and probe wave-numbers.

own characteristic nodal structure which can be elucidated by the FC projection of the vibrational wave-function of the c state onto the hump of the E-state potential $V_E(R)$ around 5 Å. It is clearly seen that Figs. 5 (a), (b), and (c) are for $v'=0$, 1, and 2, respectively. The lowest member seen in Fig. 4 is thus assigned as the band origin.

The Birge-Sponer (BS) plot for the observed progression is well fitted by a straight line [6], suggesting that $V_c(R)$ is well approximated by the Morse function. The Morse potential-parameters are thus obtained for $V_c(R)$ by a linear fitting of the BS plot as $\omega_e = 11$ cm^{-1}, $\omega_e x_e = 0.54$ cm^{-1}, and $D_e = 56$ cm^{-1}, where D_e was calculated from the relation $D_e = \omega_e^2/(4\omega_e x_e)$. D_0 is then calculated to be 51cm^{-1} from the relation $D_0 = D_e - \omega_e/2 + \omega_e x_e/4$.

The bound region obtained by the vdW approach is joined to the repulsive branch obtained by the far-wing approach to give the whole feature of $V_c(R)$ in Fig. 6.

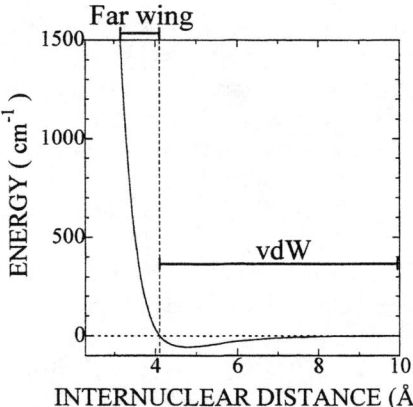

FIG. 6. Potential energy curve of the c state for Hg-Ar.

decay dynamics

Here I discuss the dynamics of the Hg–X half-collision triggered on the c-state potential. The possible pathways are adiabatic $Hg(^3P_2)$ formation and nonadiabatic $Hg(^3P_1)$ and $Hg(^3P_0)$ formation. The latter two processes occur respectively via the $c^31 \to B^31$ and $c^31 \to a^30^-$ nonadiabatic transitions of the quasimolecule.

We start with the far-wing excitation technique again. The Hg–X quasimolecule (X = Ar, Kr, Xe, N_2, and CO) is excited to the c state in the CID band under the thermal gas-cell condition, and the nascent $Hg(^3P_2)$, $Hg(^3P_1)$, and $Hg(^3P_0)$ atoms are probed by another laser pulse with 7^3S_1 being the fluorescent state. By scanning a pump-laser wavelength, we measure the $c - X$ excitation spectra of the thermal quasimolecule for the adiabatic and nonadiabatic scatterings.

Examples are plotted against Δ in Figs. 7 (a)–(c) for Ar, N_2, CO as the counterparts. For RG's, $Hg(^3P_2)$ is the only product. This is produced via an adiabatic scattering, and no trace of nonadiabatic transition is observed in this case. But for molecules, a considerable number of $Hg(^3P_1)$ and $Hg(^3P_0)$ are observed as the products of the $\tilde{c} \to \tilde{B}$ and $\tilde{c} \to \tilde{a}$ nonadiabatic transitions. It is clear that the molecular counterpart does something crucial on this nonadiabatic process. Figure 8 presents the $\tilde{c} \to \tilde{B}$ and $\tilde{c} \to \tilde{a}$ transition probabilities for a single approach of Hg and N_2 and Hg and CO, which are the nascent absolute ratios of the amounts of $Hg(^3P_1)$ and $Hg(^3P_0)$ to the sum of the amounts of $Hg(^3P_2)$, $Hg(^3P_1)$ and $Hg(^3P_0)$. The transitions are seen to be quite efficient; they occur every four or five approaches

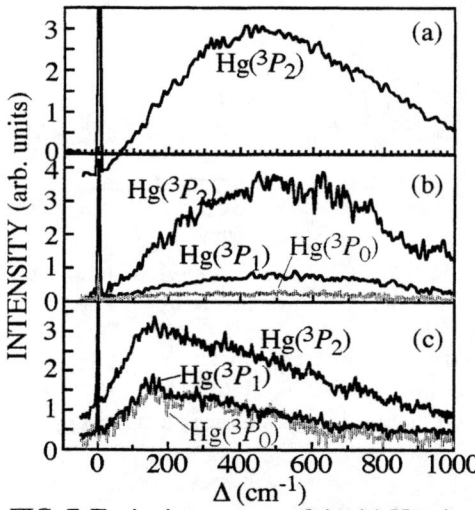
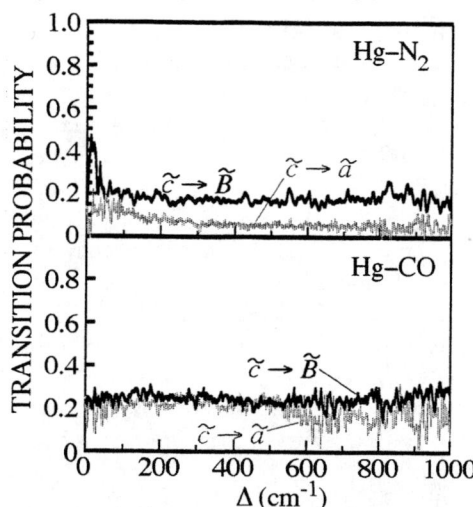

FIG. 7. Excitation spectra of the (a) Hg–Ar, (b) Hg–N_2, and (c) Hg–CO thermal quasimolecules for the formation of $Hg(6^3P_2)$, $Hg(6^3P_1)$, and $Hg(6^3P_0)$. Δ denotes the detuning of the excitation wave-number from the Hg $^1S_0$–$^3P_2$ atomic transition.

FIG. 8. $\tilde{c} \to \tilde{B}$ and $\tilde{c} \to \tilde{a}$ nonadiabatic transition probabilities in a single approach of Hg and N_2 and Hg and CO under the thermal conditions (see text). Δ denotes the detuning of the excitation wave-number from the Hg $^1S_0$–$^3P_2$ atomic transition.

of the colliding pair. These are much more efficient than the $^3P_1 \rightarrow {}^3P_0$ processes [4]. Moreover, the probabilities are almost irrespective of Δ, indicating that they are not influenced by the collision velocity. At this point, it might be reasonable to have an idea that the rotation of the molecular counterpart is important, as we have seen previously for the $^3P_1 \rightarrow {}^3P_0$ transition [4].

In order to check the validity of this idea, we have measured the corresponding probabilities at a much lower rotational temperature ($\sim$5K) of N_2 using the vdW approach. The Hg–N_2 vdW complex is generated in the vibrational ground-level of the X state in a supersonic beam and excited by a pump pulse to the free state above the dissociation limit on the c state potential to trigger a rotationally-cold half-collision. In this technique, collision energy is controllable with the excitation-laser wavelength between an ultracold ($\sim$50 mK) and a higher thermal ($\sim$1000 K) regions with a resolution of $\sim$0.05 cm^{-1}. The adiabatic and nonadiabatic atomic products are probed by another laser-pulse with 7^3S_1 being the fluorescent state. By scanning the pump-laser wavelength, the excitation spectra are measured. The corresponding probabilities are shown in Fig. 9. As is seen here, the result is completely against our expectation: no significant difference is observed at this extremely lower rotational temperature of N_2.

Now we understand that the rotation of N_2 is of no importance. The fact is that the vibration plays a crucial role in the $c \rightarrow B$ and $c \rightarrow a$ nonadiabatic transitions. But why do these $E - V$ transfers occur at those quite high probabilities? For Hg($^3P_1$) formation, there is a super resonance within 2 cm^{-1} between the asymptotes Hg($^3P_2$) + N_2 ($v = 0$) and Hg($^3P_1$) + N_2 ($v = 2$). This resonance could be the source of a high probability for $c \rightarrow B$ nonadiabatic transition. But there is no such a resonance for $^3P_0$ formation with any number of quanta in the vibration of N_2 ; while the probability is still high enough for this process. Theoretical efforts are needed in future.

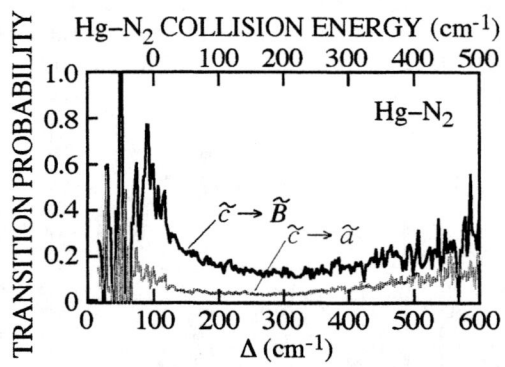

FIG. 9. $\tilde{c} \rightarrow \tilde{B}$ and $\tilde{c} \rightarrow \tilde{a}$ nonadiabatic transition probabilities in a single approach of Hg and N_2 under the rotationally cold conditions (see text). Δ denotes the detuning of the excitation wave-number from the Hg $^1S_0$–$^3P_2$ atomic transition.

Hg($^3P_1$)–N_2, CO systems; from the frequency domain to the time domain

As a final subject, I briefly refer to a bran-new half-collision approach we started recently. It is the time-domain half-collision approach based on femtosecond ultrafast spectroscopy, which I believe is complementary to the frequency-domain ones mentioned so far.

Here I present the results for the Hg(6^3P_1)–N_2, CO systems. The main pathway is the fine-structure transition to produce Hg(6^3P_0). This process proceeds via nonadiabatic quasimolecular transitions from the $\tilde{A}$ and $\tilde{B}$ states to the $\tilde{a}$ state.

First I present the results of the conventional frequency-domain measurement, aiming to demonstrate that how it works and how it fails to work for the present systems. The Hg–N_2 or CO vdW complex is generated in a supersonic beam and is excited by a nanosecond laser pulse to the bound levels of the $\tilde{A}$ or $\tilde{B}$ state. The complex then decays through a nonadiabatic transition to the $\tilde{a}$ state to produce Hg($^3P_0$) or fluoresces down to the $\tilde{X}$ state. The nascent Hg($^3P_0$) is probed by another nanosecond laser pulse with a pump-probe delay in a nanosecond timescale with 7^3S_1 being the fluorescent state. Scanning the pump-laser wavelength, we measure the bound-bound $\tilde{A}$–$\tilde{X}$ and $\tilde{B}$–$\tilde{X}$ excitation spectra of the complex for the fluorescence decay and $\tilde{A} \to \tilde{a}$ and $\tilde{B} \to \tilde{a}$ nonadiabatic transitions. Figure 10 shows the results for N_2 as a counterpart. The excitation switches from bound-bound to free-bound around 253 nm in going towards the shorter wavelength side, where Hg($^3P_1$) is predominantly produced via an adiabatic dissociation of the complex, and fluorescence is emitted by that isolated Hg($^3P_1$) atom. We are now concerned in the bound-bound region in the longer-wavelength side of 253 nm. It is seen, in this region, that both fluorescence decay and Hg($^3P_0$) formation are observed.

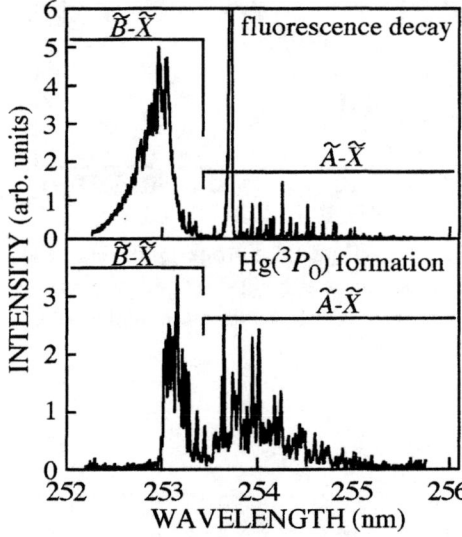
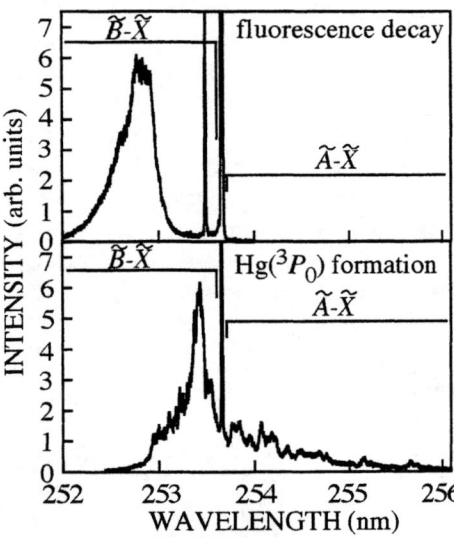

FIG. 10. $\tilde{A}$–$\tilde{X}$ and $\tilde{B}$–$\tilde{X}$ excitation spectra of the Hg–N_2 van der Waals complex for fluorescence decay and Hg($^3P_0$) formation.

FIG. 11. $\tilde{A}$–$\tilde{X}$ and $\tilde{B}$–$\tilde{X}$ excitation spectra of the Hg–CO van der Waals complex for fluorescence decay and Hg($^3P_0$) formation.

The branching ratio between these two gives the lifetime of the excited state to be 10-50 nanosecond in this case [10]. Or you could use a higher resolution laser to resolve the rotational structure of these spectra and measure the linewidth to know the lifetime. The frequency-domain approach works well for this system as a tool to extract the half-collision dynamics. However it does not work at all sometimes. Figure 11 shows the results for CO as a counterpart. It is clearly seen here that no fluorescence is observed in the bound-bound region and the spectrum is quite diffuse. There is no chance to extract dynamical information in the frequency domain.

Hence we move from the frequency domain to the time domain for this system. The Hg–CO complex is excited to the coherent superposition of the bound levels on the A- or B-state potential by a femtosecond laser-pulse to generate a wave packet moving around on the excited-state surface. Every time it reaches a particular point of nonadiabatic transition, a fraction hops the a-state surface, traveling away to the asymptote $Hg(^3P_0)$ + CO. The nascent $Hg(^3P_0)$ is probed by another femtosecond laser-pulse with 7^3D_1 being the fluorescent state. By scanning the delay in a femtosecond time-scale, the nonadiabatic decay is clocked in real-time. Figure 12 presents the real-time clocking of the $B \rightarrow a$ nonadiabatic transition obtained with the pump pulse tuned into the $B-X$ band. This bran-new measurement has shown for the first time that the bound Hg–CO complex in the B state decays in 19 ps through the $B \rightarrow a$ nonadiabatic transition. We are now investigating an early stage before $\Delta t = 10$ ps with a higher time-resolution to chase the wave-packet motion on the excited potential-surface by detuning the probe wavelength from Hg $6^3P_0 - 7^3D_1$ resonance-line.

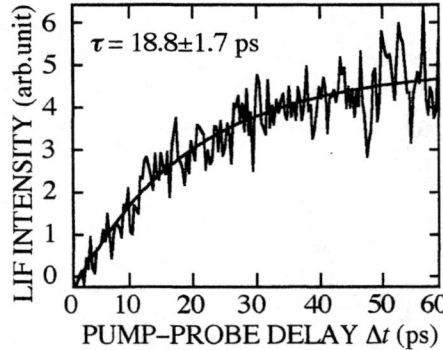

FIG. 12. Real-time clocking of $Hg(^3P_0)$ production in the Hg–CO half collisions triggered on the quasimolecular $\tilde{B}$-state potential surface.

ACKNOWLEDGMENT

The author thanks the members of the laboratory for UV spectroscopy at RISM (Tohoku University) : Mr. K. Amano, Mr. H. Chiba, Mr. M. Shoji, Dr. M. Okunishi, Dr. K. Ueda, and Professor Y. Sato. He also thanks Mr. T. Kurosawa (AIT), Mr. T. Takahashi (Tokyo Electron co.,ltd.), Professor Evgueni Nikitin (Technion - Israel Institute of Technology) and Professor Alex Devdariani (St. Petersburg University) and gratefully acknowledges the financial support from Ogasawara Foundation for Science and Technology.

REFERENCES

1. see, for example, (a) R.E.M. Hedges, D.L. Drummond, and A. Gallagher, Phys. Rev. A **6**, 1519 (1972) ; (b) N. Allard and J. Kielkopf, Rev. Mod. Phys. **54**, 1103 (1982).
2. see, for example, (a) P. Arrowsmith, S.H.P. Bly, P.E. Charters, and J.C. Polanyi, J. Chem. Phys. **79**, 283 (1983) ; (b) T.C. Maguire, P.R. Brooks, and R.F. Curl, Jr., Phys. Rev. Lett. **50**, 1918 (1983) ; (c) T.H. Wong and P.D. Kleiber, J. Chem. Phys. **102**, 6476 (1995); (d) J. Grosser, O. Hoffmann, C. Rakete, and F. Rebentrost, in *Photonic, Electronic, andAtomicCollisions*, XX ICPEAC, Vienna, Austria, 1997, edited by F. Aumayr and H.P. Winter (World Scientific, 1997), invited paper, p. 571.
3. (a) K. Ohmori, T. Takahashi, H. Chiba, K. Saito, T. Nakamura, M. Okunishi, K. Ueda, and Y. Sato, J. Chem. Phys. **105**, 7464 (1996) ; (b) K. Ohmori, T. Takahashi, H. Chiba, K. Saito, T. Nakamura, M. Okunishi, K. Ueda, and Y. Sato, *ibid.* **105**, 7474 (1996).
4. (a) K. Ohmori, T. Kurosawa, H. Chiba, M. Okunishi, K. Ueda, Y. Sato, and E.E. Nikitin, J. Chem. Phys. **102**, 7341 (1995) ; (b) K. Ohmori, T. Kurosawa, H. Chiba, M. Okunishi, and Y. Sato, *ibid.* **100**, 5381 (1994).
5. T. Kurosawa, K. Ohmori, H. Chiba, M. Okunishi, K. Ueda, Y. Sato, E.E. Nikitin, and A.Z. Devdariani, J. Chem. Phys. **108**, 8101 (1998).
6. K. Amano, K. Ohmori, T. Kurosawa, H. Chiba, M. Okunishi, K. Ueda, Y. Sato, E.E. Nikitin, and A.Z. Devdariani, J. Chem. Phys. **108**, 8110 (1998).
7. (a) K. Ohmori, K. Amano, M. Okunishi, H. Chiba, K. Ueda, and Y. Sato, Chem. Lett., 765 (1996) ; (b) Y. Sato, T. Nakamura, M. Okunishi, K. Ohmori, H. Chiba, and K. Ueda, Phys. Rev. A **53**, 867 (1996) ; (c) T. Maeyama, H. Ito, H. Chiba, K. Ohmori, K. Ueda, and Y. Sato, J. Chem. Phys. **97**, 9492 (1992).
8. K. Ohmori and Y. Sato, in *Photonic, Electronic, andAtomicCollisions*, XX ICPEAC, Vienna, Austria, 1997, edited by F. Aumayr and HP. Winter (World Scientific, 1997), invited paper, p. 593.
9. K. Fuke, T. Saito, and K. Kaya, J. Chem. Phys. **81**, 2591 (1984).
10. K. Yamanouchi, S. Isogai, and S. Tsuchiya, M. -C. Duval, C. Jouvet, O. B. d'Azy, and B. Soep, J. Chem. Phys. **89**, 2975 (1988).

Probing a Bose-Einstein Condensate by Near-Resonant Light Scattering

C. A. Sackett, J. M. Gerton, M. Welling, and R. G. Hulet[1]

Physics Department, MS 61
Rice University
Houston, TX 77251

Abstract. Bose-Einstein condensation (BEC) of atoms with attractive interactions is profoundly different from BEC of atoms with repulsive interactions. We describe experiments with Bose condensates of ^{7}Li atoms, which are weakly attracting at ultralow temperature. We measure the distribution of condensate occupation numbers occurring in the gas using a near-resonant optical probe to image the atom cloud. This data shows that the number is limited and demonstrates the dynamics of condensate growth and collapse due to the attractive interactions.

INTRODUCTION

The recent attainment of Bose-Einstein condensation (BEC) of dilute atomic gases [1–3] has enabled new investigations of weakly interacting many-body systems. ^{7}Li is unique among these gases in that the interactions are effectively attractive, which profoundly effects the nature of BEC. In fact, it was long believed that attractive interactions precluded the attainment of BEC in the gas phase [4,5]. It is now known that BEC *can* exist in a confined gas, provided the condensate number remains small [6]. These condensates are predicted to exhibit fascinating dynamical behavior, including soliton formation [7] and macroscopic quantum tunneling [8–11]. This paper reviews our work on BEC of ^{7}Li, including the measurement of limited condensate number, and the dynamics of condensate growth and collapse.

INTERACTIONS IN DILUTE GASES

One of the primary interests in dilute Bose-Einstein condensates is that the interactions are weak, facilitating comparison between theory and experiment. When

---

[1] This work is supported by the National Science Foundation, the Office of Naval Research, NASA, and the Welch Foundation.

TABLE 1. Singlet and triplet scattering lengths in units of a_o, for isotopically pure and mixed gases of lithium isotopes [15].

	^{6}Li	^{7}Li	^{6}Li/^{7}Li
a_T	-2160 ± 250	-27.6 ± 0.5	40.9 ± 0.2
a_S	45.5 ± 2.5	33 ± 2	-20 ± 10

the de Broglie wavelength Λ is much longer than the characteristic two-body interaction length, the effect of the interaction can be represented by a single parameter, the s-wave scattering length a [12]. The magnitude of a indicates the strength of the interaction, while the sign determines whether the interactions are effectively attractive ($a < 0$) or repulsive ($a > 0$). In the experiments, the density n is small enough that $n|a|^3 \ll 1$, so only binary interactions need be considered.

Photoassociative Spectroscopy

Although the interaction potentials for the alkali atoms lithium through francium are all qualitatively the same, in that they all have a repulsive inner-wall, a minimum that supports vibrational bound states, and a long-range van der Waals tail, their respective scattering lengths differ enormously in magnitude and in sign. This variation arises because of differences in the proximity of the least-bound vibrational state to the dissociation limit. As with the familiar attractive square-well potential, a barely bound or barely unbound state leads to collisional resonances that produce very large magnitude scattering lengths. Therefore, small changes in the interaction potential may result in a large change in the magnitude, or even change the sign of a. In the past few years, photoassociative spectroscopy of ultracold atoms has proven to be the most precise method for determining scattering lengths [13]. In one-photon photoassociation, a laser beam is passed through a gas of ultracold atoms confined in a trap. As the laser frequency is tuned to a free-bound resonance, diatomic molecules are formed resulting in a detectable decrease in the number of trapped atoms. The intensity of the trap-loss signal is sensitive to the ground-state wavefunction, providing useful information for determining the ground-state interaction potential. The value of the scattering length is found by numerically solving the Schrödinger equation using this potential. This method has been used to find the scattering lengths for Li, Na, K, and Rb [13].

A more precise method for finding scattering lengths is to probe the ground state molecular levels directly. We have used two-photon photoassociation to directly measure this binding energy for both stable isotopes of lithium, the bosonic isotope ^{7}Li [14] and the fermionic isotope ^{6}Li [15]. This technique has resulted in the most precisely known atomic potentials. Table 1 gives the triplet and singlet scattering lengths for both isotopes individually, as well as for mixed isotope interactions. Two-photon spectroscopy of the ground-state has also been used recently to find the scattering lengths of rubidium [16].

Mean-Field Theory

The effects of interactions on the condensate have been studied using mean-field theory and neglecting inelastic collisions [17]. In this approximation, the interaction part of the Hamiltonian is replaced by its mean value, resulting in an interaction energy of $U = 4\pi\hbar^2 an/m$, where n is the density and m is the atomic mass [12]. For a gas at zero temperature, the net result of the interactions and the confining potential can be found by solving the non-linear Schrödinger equation for the wave function of the condensate, $\psi(r)$ [18]:

$$\left(-\frac{\hbar^2}{2m}\nabla^2 + V(r) + U(r) - \mu\right)\psi = 0. \tag{1}$$

Here μ is the chemical potential, and $V(r)$ is the confining potential provided by the trap. In a spherically symmetric harmonic trap with oscillation frequency ω, $V(r) = m\omega^2 r^2/2$. The interaction energy $U(r)$ is determined by taking $n(r) = |\psi(r)|^2$.

Implications of $a < 0$

Limited Condensate Number

For a dilute gas with $a > 0$, corresponding to repulsive interactions, it was shown long ago that the condensate will be stable. The situation is drastically different for $a < 0$, since U decreases with increasing n so an untrapped (homogeneous) gas is mechanically unstable to collapse. It was therefore believed that BEC was not possible in the gas phase. In a system with finite volume, however, the zero-point kinetic energy of the atoms provides a stabilizing influence. A numerical solution to Eq. (1) is found to exist only when N_0 is smaller than a limiting value N_m [19]. Physically, this limit can be understood as requiring that the interaction energy U be small compared to the trap level spacing $\hbar\omega$, so that the interactions act as a small perturbation to the ideal-gas solution. This condition implies that N_m is of the order $l_0/|a|$, where $l_0 = (\hbar/m\omega)^{1/2}$ is the length scale of the single-particle trap ground state [20]. For condensate occupation numbers below N_m, ψ is determined using Eq. (1). It is found that for $N_0 \ll N_m$, ψ is closely approximated by the single-particle ground state, and as N_0 increases, the interaction energy causes the spatial extent of ψ to decrease.

A variational method has been used to study the decay of condensates with attractive interactions [21,9,10], which we discuss here following the development of Stoof [10]. The ground-state solution to Eq. (1), ψ_0, satisfies an extremal condition

$$\langle\psi_0|H|\psi_0\rangle \leq \langle\psi|H|\psi\rangle \tag{2}$$

for any other function ψ. The energy operator H is given by

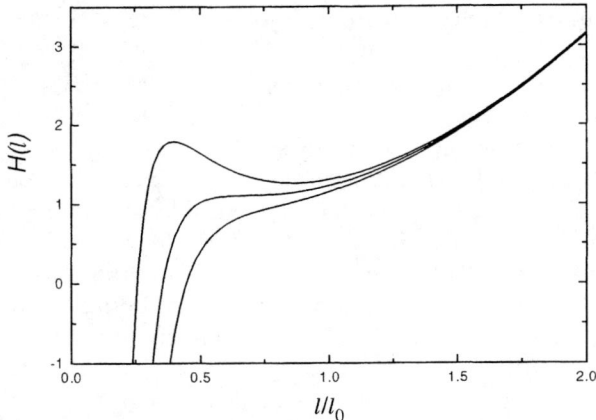

FIGURE 1. The condensate energy H, plotted in units of $N_0 \hbar^2/ml_0^2$. The upper curve corresponds to $N_0 = 0.48\, l_0/|a|$, the middle curve to $N_0 = 0.68\, l_0/|a|$, and the lower curve to $N_0 = 0.87\, l_0/|a|$. It is evident that a local minimum in H exists near $l = l_0$ if N_0 is sufficiently low, indicating that a metastable condensate can exist.

$$H = -\frac{\hbar^2}{2m}\nabla^2 + V(r) + \frac{U(r)}{2}, \qquad (3)$$

where the factor of $1/2$ in the interaction term arises from the dependence of U on ψ. Because the solution to Eq. (1) for the ideal gas is a Gaussian function, it is reasonable to minimize $\langle H \rangle$ using a Gaussian trial wavefunction with a variable Gaussian size l. Evaluating $\langle H \rangle \equiv H(l)$ yields the function plotted for three values of N_0 in Fig. 1. It is observed that for sufficiently small N_0, a local minimum exists near $l = l_0$, indicating that a metastable condensate is possible. For larger N_0, however, the minimum vanishes, and the system will be unstable. The condition for stability is $N_0 \leq 0.68\, l_0/|a|$, which is in reasonable agreement with the exact value obtained by numerical integration of Eq. (1), $N_m = 0.58\, l_0/|a|$ [19].

Condensate Collapse

Although a condensate can exist in a trapped gas, it is predicted to be metastable and to decay by quantum or thermal fluctuations [8–11]. The condensate has only one unstable collective mode, which in the case of an isotropic trap corresponds to the breathing mode [7,21]. The condensate therefore collapses as a whole, either by thermal excitation over, or by macroscopic quantum mechanical tunneling through the energy barrier in configuration space, shown in Fig. 1. The rates of decay for both quantum tunneling and thermal excitation can be calculated within the formalism of the variational calculation [10].

Condensate Growth and Collapse During Evaporation

Experimentally, the condensate is formed by evaporatively cooling the gas. As the gas is cooled below the critical temperature for BEC, N_0 grows until N_m is reached. The condensate then collapses spontaneously if $N_0 \geq N_m$, or the collapse can be initiated by thermal fluctuations or quantum tunneling for $N_0 \simeq N_m$ [8–11,22]. During the collapse, the condensate shrinks on the time scale of the trap oscillation period. As the density rises, the rates for inelastic collisions such as dipolar decay and three-body molecular recombination increase. These processes release sufficient energy to immediately eject the colliding atoms from the trap, thus reducing N_0. The ejected atoms are very unlikely to further interact with the gas before leaving the trap, since the density of noncondensed atoms is low. As the collapse proceeds, the collision rate grows quickly enough that the density remains small compared to a^{-3} and the condensate remains a dilute gas [22,23]. However, the theories are not yet conclusive as to what fraction of the condensate atoms participates in the collapse, and of those participating, what fraction is eventually ejected.

Both the collapse and the initial cooling process displace the gas from thermal equilibrium. As long as N_0 is smaller than its equilibrium value, as determined by the total number and average energy of the trapped atoms, the condensate will continue to fill until another collapse occurs. This results in a cycle of condensate growth and collapse, which repeats until the gas comes to equilibrium with some $N_0 < N_m$. We have modeled the kinetics of the equilibration process by numerical solution of the quantum Boltzmann equation, as described in Ref. [22]. Fig. 2 shows a typical trajectory of N_0 in time, for our experimental conditions. In this calculation we make the *ad hoc* assumption that N_0 is reduced to zero when a collapse occurs, because an accurate theoretical model of the collapse is not yet available.

EXPERIMENT

Magnetic Trap

The apparatus used to produce BEC of ^{7}Li is described most completely in Ref. [20]. Laser cooling using the Zeeman technique is used to slow an atomic beam of lithium atoms, which are then directly loaded into a magnetic trap. There is no magneto-optical trap used in the experiment. The magnetic trap is unique in that it is made from permanent magnets [24]. By exploiting the enormous field gradients produced by rare-earth magnets, the resulting trap potential was made nearly spherically-symmetric with a large harmonic oscillation frequency of ~150 Hz. N_m is limited by the tightest trap direction [20], so the condensate density is maximized for a spherically symmetric potential. In addition, by actively stabilizing the temperature of the magnets the fields are made highly stable, allowing for

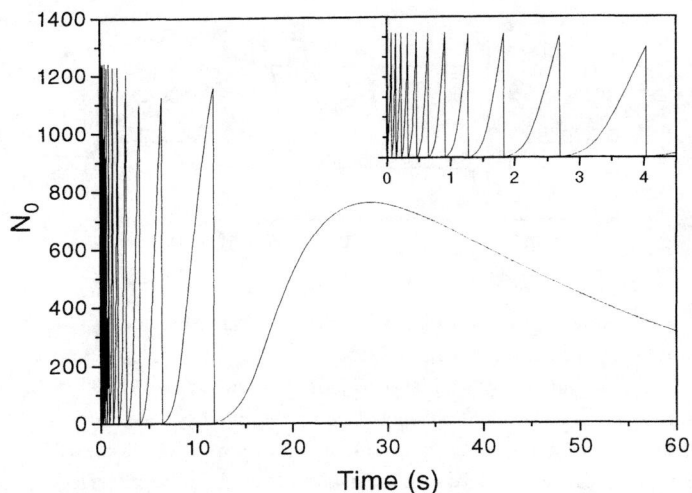

FIGURE 2. Numerical solution of the quantum Boltzmann equation, showing evolution of condensate occupation number. A trapped, degenerate ^{7}Li gas is cooled at $t=0$ to a temperature of about 100 nK and a total number of 4×10^4 atoms. The gas then freely evolves in time. The inset shows an expanded view of the early time behavior on the same vertical scale.

relatively repeatable and stable experimental conditions. The bias field at the center of the trap is 1004 G.

Evaporative Cooling

After about 1 s of loading, $\sim 2 \times 10^8$ atoms in the doubly spin-polarized $F = 2, m_F = 2$ state are accumulated. These atoms are then laser cooled to near the Doppler cooling limit of 200 μK. At this number and temperature, the phase space density, $n\Lambda^3$, is still more than 10^5 times too low for BEC. The atoms are cooled further by forced evaporative cooling [25]. The hottest atoms are driven to an untrapped ground state by a microwave field tuned just above the $(F = 2, m_F = 2) \leftrightarrow (F = 1, m_F = 1)$ Zeeman transition frequency of approximately 3450 MHz. As the atoms cool, the microwave frequency is reduced. The optimal frequency vs. time trajectory that maximizes the phase-space density of the trapped atoms is calculated ahead of time [26], and depends on the elastic collision rate and the trap loss rate. The elastic collision rate $n\sigma v$ is roughly 1 s^{-1}, with cross-section $\sigma = 8\pi a^2 \approx 5 \times 10^{-13}$ cm^2. The collision rate is approximately constant during evaporative cooling. We have recently measured the loss rate due to collisions with hot background gas atoms to be $< 10^{-4}$ s^{-1}, and the inelastic dipolar-relaxation collision rate constant to be 1.05×10^{-14} cm^3 s^{-1} [27]. From the low background collision loss rate, we estimate the background gas pressure in the apparatus to be $< 10^{-12}$ torr. Quantum degeneracy is typically reached after 200 seconds, with

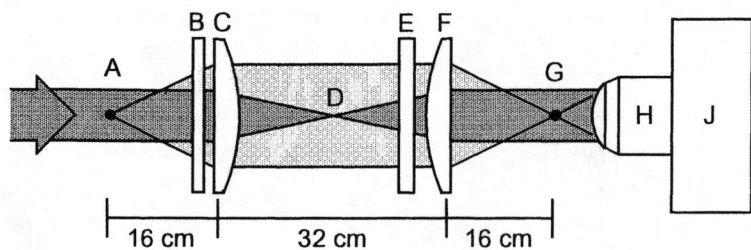

FIGURE 3. A schematic of the imaging system used for *in situ* phase-contrast polarization imaging. A linearly polarized laser beam is directed through the cloud of trapped atoms located at A. The probe beam and scattered light field pass out of a vacuum viewport B, and are relayed to the primary image plane G by an identical pair of 3-cm-diameter, 16-cm-focal-length doublet lenses C and F. The light is then re-imaged and magnified onto a camera J by a microscope objective H. The measured magnification is 14, and the camera pixels are 19 μm square. A linear polarizer E is used to cause the scattered light and probe fields to interfere, producing an image sensitive to the refractive index of the cloud.

$N \approx 10^6$ atoms at $T \approx 700$ nK. Lower temperatures are reached by extending the cooling time or by the application of a short, deep cooling pulse.

Phase-Contrast Imaging

After evaporative cooling, the spatial distribution of the atoms is imaged *in situ* using an optical probe. Since the single-particle harmonic oscillator ground state of our trap has a Gaussian density distribution with a $1/e$-radius of only 3 μm, a high-resolution imaging system is required. Because the optical density of the atoms is sufficiently high to cause image distortions when probed by near-resonant absorption [28], we instead use a phase-contrast technique with a relatively large detuning from resonance $\Delta = \pm 250$ MHz. Our implementation of phase-contrast imaging, shown schematically in Fig. 3, is both simple and powerful. It exploits the fact that atoms in a magnetic field are birefringent, so that the light scattered by the atoms is polarized differently from the incident probe light. A linear polarizer decomposes the scattered and probe light onto a common axis, which causes them to interfere. Since the phase of the scattered light is equal to $\alpha/4\Delta$, where α is the on-resonance optical density, the spatial image recorded on the CCD camera is a representation of the integrated atomic column density. Phase-contrast polarization imaging is described more fully in Ref. [20].

Fig. 4 shows two images obtained using phase-contrast polarization imaging. For these images, the trap symmetry is exploited by averaging the data around the cylindrical trap axis to improve the signal to noise.

FIGURE 4. Phase-contrast images averaged around the cylindrical axis of the trap. For both cases, $N \approx 23{,}000$ atoms and $T \approx 190$ nK. For the image on the right, $N_0 \approx 1050$, while for the image on the left $N_0 \approx 65$. These images demonstrates our sensitivity to a small number of condensate atoms on a background of a large number of non-condensed atoms.

Data Analysis

Image profiles are obtained from the averaged data. These profiles are fit with a model energy distribution to determine N, T, and N_0. If the gas is in thermal equilibrium, then any two of N, T, or N_0 completely determine the density of the gas through the Bose-Einstein distribution function. However, if the gas is undergoing the growth/collapse cycles shown in Fig. 2, it certainly is not in thermal equilibrium and a more complicated function is required. Using the quantum Boltzmann equation model, we find that atoms in low-lying levels quickly equilibrate among themselves and the condensate, and that high-energy atoms are well thermalized among each other. Therefore, a three parameter function, including two chemical potentials corresponding to the two parts of the distribution, and a temperature given by the high-energy tail of the distribution, is sufficient to describe the expected non-equilibrium distributions and to determine N_0 [29]. The fits yield an average reduced χ^2 of very nearly 1, indicating that the model is consistent with the data within the noise level. The procedure was tested by applying it to simulated data generated by the quantum Boltzmann model, and also by comparing the analysis of experimental images of thermalized clouds using both equilibrium and nonequilibrium models. From these tests, the systematic error introduced by the nonequilibrium model is estimated to be not more than ±50 atoms. The most significant uncertainty in N_0 is the systematic uncertainty introduced by imaging limitations. While the imaging system is nearly diffraction limited, the resolution is not negligible compared to the size of the condensate, and imaging effects must be included in the fit [28]. Imaging resolution is accounted for by measuring the point transfer function of the lens system and convolving this function with the images. Uncertainties in the resolution lead to a systematic uncertainty in N_0 of ±20% [29].

EXPERIMENTAL RESULTS

In this section, we give our experimental results on the observation of limited condensate number [6], and on the collapse of the condensate [29].

Limited Condensate Number

We have measured N_0 for several thousand different degenerate distributions with T ranging between 80 and 400 nK, and for N between 2,000 and 250,000 atoms. In all cases, N_0 is found to be relatively small. The maximum N_0 observed is between 900 and 1400 atoms, depending on the assumed imaging resolution. This measurement is in very good agreement with the mean-field prediction of 1250 atoms.

In the analysis we have assumed that the gas is ideal, but interactions are expected to alter the size and shape of the density distribution. Mean-field theory predicts that interactions will reduce the $1/e$-radius of the condensate from 3 μm for low occupation number to $\sim$2 μm as the maximum N_0 is approached [30,7,10,?,?]. If the smaller condensate radius is used in the fit, the maximum N_0 decreases by $\sim$100 atoms.

Condensate Collapse

To explore the predicted collapse of the condensate, evaporative cooling is continued well into the degenerate regime, to $N \sim 4 \times 10^5$ atoms at a microwave frequency 100 kHz above the trap bottom. The frequency is then rapidly reduced to $\sim$10 kHz and raised again, leaving approximately 4×10^4 atoms. The frequency is swept quickly compared to the collision rate of $\sim$3 Hz, so that this "microwave razor" simply eliminates all atoms above a cutoff energy. It thereby creates a definite energy distribution at a specified time whose relaxation to equilibrium can be followed. Fig. 2 shows the expected trajectory of N_0 in time, for our experimental conditions. For this calculation, we have assumed that N_0 is reduced to zero following a collapse [22].

Although phase-contrast imaging can in principle be nearly nonperturbative, it is not possible to reduce incoherent scattering to a negligible level and simultaneously obtain low enough shot noise to measure N_0 accurately. Each atom therefore scatters several photons during a probe pulse, heating the gas and precluding the possibility of directly observing the evolution of N_0 in time as in Fig. 2. This limitation cannot be overcome by repeating the experiment and varying the delay time τ between the microwave razor and the probe, because the evolution of N_0 is

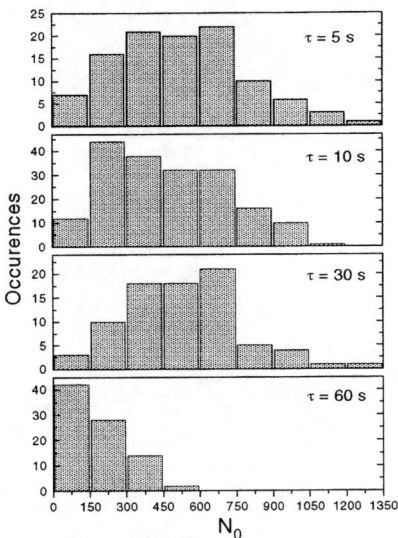

FIGURE 5. Frequency of occurrence of condensate occupation number. For each measurement, a nonequilibrium degenerate gas was produced, allowed to evolve freely for time τ, and then probed. The spread in N_0 values arises as the collapse/fill cycle is sampled at random points.

made unrepeatable by random thermal and quantum fluctuations in the condensate growth and collapse processes, as well as experimental fluctuations in the initial conditions. Because of this, however, the values of N_0 occurring at a particular τ are expected to vary as different points in the collapse/fill cycle are sampled. We have observed such variations by measuring N_0 for many similarly prepared samples at several values of τ. Their measured distribution are shown as histograms in Fig. 5. For small τ, N_0 ranges from near zero to about 1200 atoms, as expected if the condensate is alternately filling to near the theoretical maximum and subsequently collapsing. At longer time delays, the histograms change shape, narrowing somewhat at $\tau = 30$ s, and having only small N_0 values at $\tau = 60$ s. The variations in N_0 are uncorrelated with changes in N, T, probe parameters, imaging model parameters, and goodness of fit. To our knowledge, no other explanation for variations of this magnitude has been proposed, so we consider the observation of these variations to strongly support the collapse/fill model. In particular, the observed spread in observed N_0 values (250 atoms) is not simply noise because the statistical uncertainty in N_0 is only 65 atoms, as verified by obtaining a histogram at very long delay for which the measured N_0 values are consistent with zero.

The histogram data can be compared with the predictions of the quantum Boltzmann model. In the model trajectory shown in Fig. 2, three time domains can be discerned with which the data can be correlated. For $\tau \leq 20$ s, the condensate collapses frequently as the gas is equilibrating. Model histograms for delays of 5 and 10 seconds are similar to each other, and agree qualitatively with the experimen-

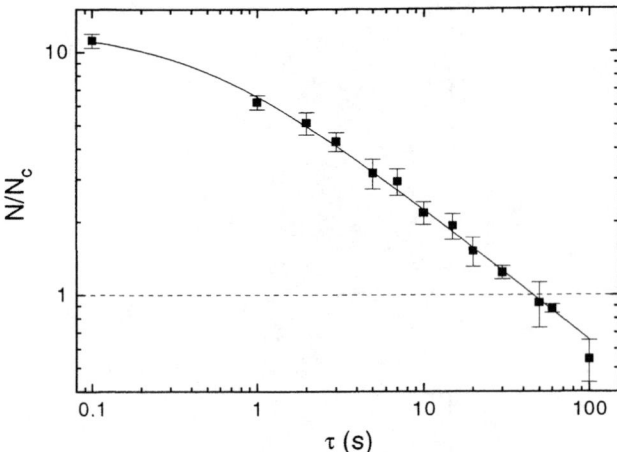

FIGURE 6. Relaxation of degenerate gas to equilibrium. Data were taken as in Fig. 5, but for each image the total number of atoms N and the temperature T were used to determine N/N_c, where $N_c = 1.2(kT/\hbar\omega)^3$. Points represent averages of several measurements, and error bars are standard deviations. The dashed line at $N/N_c = 1$ approximately denotes the point at which equilibrium is reached. The solid curve is a fit to the empirical form $A(1 + \kappa\tau)^\gamma$, yielding $A = 12 \pm 1$, $\kappa = 2.1 \pm .7$ s^{-1}, and $\gamma = -.55 \pm .03$.

tally observed distributions in being broadly spread between 0 and N_m. Around $\tau = 20$-40 s, equilibrium is reached and N_0 is stabilized for several seconds at a maximum value. As is observed in the data, N_0 declines at later times as atoms are lost through inelastic collisions.

The detailed shape of the measured histograms give us insight into the actual collapse process. The fact that small N_0 is not likely probably indicates that the collapse is not complete, but rather to a value near 100 atoms. Kagan *et al.* have observed the condensate to collapse to a nonzero value in numerical solutions of the NLSE [23]. However, while those authors found that close to 50% of the condensate was lost during a collapse, our data suggest that considerably smaller remainders are more likely, since a large fraction of our observations show $N_0 < 600$ atoms. We also observe the frequency of occurrence to drop steadily as N_0 increases. This may be the result of the decrease in the condensate energy with increasing N_0 due to the attractive interactions. This would likely lead to an acceleration of the rate of filling of the condensate, and therefore a reduction in the probability of observing condensates with large N_0.

The condensate growth and collapse cycle is driven by an excess of noncondensed atoms compared to a thermal distribution. This excess can be examined directly. From N and T, the critical number for the BEC transition, N_c, is calculated and the ratio N/N_c plotted as a function of delay time in Fig. 6. The ratio decays according to a power law, which signifies that a nonlinear process governs equilibration. This nonlinearity is reasonable since the rate of decay of the excess atoms should depend both on the excess number and on the collision rate, which in turn depends on N and T. Since $N_0 \ll N$, equilibrium is reached when $N/N_c \approx 1$, which occurs at $\tau \approx 40$ seconds. This time is consistent with the delay required to accurately fit the image data with an equilibrium model, and with the results of the quantum Boltzmann model. Comparison of Figs. 5 and 6 shows that the equilibration time is also consistent with the changing shape of the measured histograms. This further strengthens the conclusion that the variations in N_0 are related to the growth and collapse of the condensate during the equilibration process, since the distribution of N_0 values changes when the population imbalance driving condensate growth is eliminated.

CONCLUSIONS

These observations provide quantitative support for the applicability of mean-field theory to attractive gases. The measurements described here are the first indicator of the complex dynamics accompanying BEC in a gas with attractive interactions. We believe that they support the collective collapse/fill model as a useful framework for considering such systems. It is clear, however, that additional theoretical work is necessary to accurately describe the collapse in detail. Experimentally, we are pursuing more direct methods of observing the growth and collapse of the condensate By such means, we hope to further our understanding of this novel and interesting state of matter.

REFERENCES

1. M. H. Anderson, J. R. Ensher, M. R. Matthews, C. E. Wieman, and E. A. Cornell, Science **269**, 198 (1995).
2. C. C. Bradley, C. A. Sackett, J. J. Tollett, and R. G. Hulet, Phys. Rev. Lett. **75**, 1687 (1995).
3. K. B. Davis, M.-O. Mewes, M. R. Andrews, N. J. van Druten, D. S. Durfee, D. M. Kurn, and W. Ketterle, Phys. Rev. Lett. **75**, 3969 (1995).
4. N. Bogolubov, J. of Phys. **XI**, 23 (1947).
5. H. T. C. Stoof, Phys. Rev. A **49**, 3824 (1994).
6. C. C. Bradley, C. A. Sackett, and R. G. Hulet, Phys. Rev. Lett. **78**, 985 (1997).
7. R. J. Dodd, M. Edwards, C. J. Williams, C. W. Clark, M. J. Holland, P. A. Ruprecht, and K. Burnett, Phys. Rev. A **54**, 661 (1996).
8. Y. Kagan, G. V. Shlyapnikov, and J. T. M. Walraven, Phys. Rev. Lett. **76**, 2670 (1996).
9. E. Shuryak, Phys. Rev. A **54**, 3151 (1996).
10. H. T. C. Stoof, J. Stat. Phys. **87**, 1353 (1997).
11. M. Ueda and A. J. Leggett, Phys. Rev. Lett. **80**, 1576 (1998).
12. K. Huang, *Statistical Mechanics*, 2 ed. (John Wiley & Sons, New York, 1987).

13. J. Weiner, V. S. Bagnato, S. Zilio, and P. S. Julienne, review to be published.
14. E. R. I. Abraham, W. I. McAlexander, C. A. Sackett, and R. G. Hulet, Phys. Rev. Lett. **74**, 1315 (1995).
15. E. R. I. Abraham, W. I. McAlexander, J. M. Gerton, R. G. Hulet, R. Côté, and A. Dalgarno, Phys. Rev. A **55**, R3299 (1997).
16. C. C. Tsai, R. S. Freeland, J. M. Vogels, H. M. J. M. Boesten, B. J. Verhaar, and D. J. Heinzen, Phys. Rev. Lett **79**, 1245 (1997).
17. F. Dalfovo, S. Giorgini, L. P. Petaevskii, and S. Stringari, review to be published.
18. E. M. Lifshitz and L. P. Pitaevskii, *Statistical Physics, Part 2* (Butterworth-Heineman, Oxford, 1980).
19. P. A. Ruprecht, M. J. Holland, K. Burnett, and M. Edwards, Phys. Rev. A **51**, 4704 (1995).
20. C. A. Sackett, C. C. Bradley, M. Welling, and R. G. Hulet, Appl. Phys. B **65**, 433 (1997).
21. K. Singh and D. Rokhsar, Phys. Rev. Lett. **77**, 1667 (1996).
22. C. A. Sackett, H. T. C. Stoof, and R. G. Hulet, Phys. Rev. Lett. **80**, 2031 (1998).
23. Y. Kagan, A. Muryshev, and G. Shlyapnikov, Phys. Rev. Lett **81**, 933 (1998).
24. J. J. Tollett, C. C. Bradley, C. A. Sackett, and R. G. Hulet, Phys. Rev. A **51**, R22 (1995).
25. W. Ketterle and N. J. van Druten, in *Advances in Atomic, Molecular, and Optical Physics* (Academic Press, San Diego, 1996), No. 37, p. 181.
26. C. A. Sackett, C. C. Bradley, and R. G. Hulet, Phys. Rev. A **55**, 3797 (1997).
27. J. M. Gerton, C. A. Sackett, B. J. Frew, and R. G. Hulet, to be published in Phys. Rev. A (1998).
28. C. C. Bradley, C. A. Sackett, and R. G. Hulet, Phys. Rev. A **55**, 3951 (1997).
29. C. A. Sackett, J. M. Gerton, M. Welling, and R. G. Hulet, to be published.
30. F. Dalfovo and S. Stringari, Phys. Rev. A **53**, 2477 (1996).

Determination of Long-Range Interactions from Photoassociative Spectroscopy of Ultracold Atoms

William C. Stwalley

*Department of Physics
University of Connecticut
Storrs, CT 06269-3046 U.S.A.*

Abstract. The photoassociative spectroscopy of ultracold atoms provides understanding of long-range interatomic interactions at an unprecedented level of detail. We illustrate this technique for ^{39}K photoassociation at the 4s + 4p, 4d, 6s, 5d, 7s and 6d asymptotes. Such information should allow complete calculation of atomic line broadening between appropriate asymptotes.

INTRODUCTION

In the region below ionization, single photon two atom photophysical/photochemical processes can be broken down as summarized in Table I into bound-bound, bound-free, free-bound and free-free processes (1). The focus here will be on free → bound absorption (photoassociation) of ultracold (T ≤ 1mK) atoms, a new and exciting technique which is providing unprecedented understanding of photoprocesses and long range interactions. We will use as our primary illustrative example the potassium atom and molecule since K is perhaps the alkali atom with the best understood photoassociation spectra (2), its hyperfine splittings are relatively small, and much work on Li, Na and Rb has previously been reviewed (3,4).

In Figure I, we see a selection of the potential curves and corresponding asymptotes of the ^{39}K$_2$ molecule. The ordinary molecular absorption spectrum of K$_2$ (A$^1\Sigma_u^+ \leftarrow$ X$^1\Sigma_g^+$ and B$^1\Pi_u \leftarrow$ X$^1\Sigma_g^+$) has been known since 1874 (5). Laser spectroscopy of the A$^1\Sigma_u^+$ and B$^1\Pi_u$ states correlating to the 4s + 4p asymptotes is quite extensive (e.g. 6), and the spectroscopy of higher gerade states through the intermediate A$^1\Sigma_u^+$, B$^1\Pi_u$ and A$^1\Sigma_u^+ \sim$ b$^3\Pi_u$ mixed levels is also quite extensive (e.g. 7). All-optical triple resonance (AOTR) has been used to study the A$^1\Sigma_u^+ \sim$ b$^3\Pi_u$ mixed levels as well as pure b$^3\Pi_u$ levels inaccessible from the ground state (e.g. 8), and also to study state-selected photodissociation of vibrationally excited ground state molecules (e.g. 9). In addition, high quality electronic structure calculations are available for most of these states.

The connection of these short range molecular potentials with long range potentials and various atomic asymptotes is more problematic, with a few exceptions (e.g. in K$_2$, the 1$^1\Pi_g$ state (10, 11) now known out to 40Å). Fortunately, a new and complementary alternative for probing the long range potentials (typically ≥ 15Å) is now available, namely photoassociation of ultracold atoms. Figure I shows the examples of short range single photon excitation of the A$^1\Sigma_u^+$ state (laser L$_1$) and short range OODR excitation (lasers L$_1$ and L$_2$) of the 5$^1\Pi_g$ state of K$_2$ correlating to the 4s + 4d asymptote via the intermediate A$^1\Sigma_u^+$ state. Fig. I also shows the examples of long range A$^1\Sigma_u^+ \leftarrow$ X$^1\Sigma_g^+$ photoassociation (laser L_1') of two colliding ground state atoms and of long range OODR excitation (lasers L_1' and L_2') of the $5\,^1\Pi_g$ state near the 4s + 4d asymptote (in fact, the long range states are better described in Hund's case c notation; thus the photoassociation is labeled $O_u^+ \leftarrow O_g^+$ and the OODR excitation is $1_g \leftarrow O_u^+$). It should be clear that the long range ultracold atom photoassociation is in many ways the perfect complement to ordinary short range spectroscopy.

TABLE I. Single Photon Photoprocesses Involving Two Atoms (M) and No Ionization

Absorption

bound → bound	$M_2(v'', J'') + h\nu \rightarrow M_2^*(v', J')$
bound → free (photodissociation)	$M_2(v'', J'') + h\nu \rightarrow M^* + M$
free → bound (photoassociation)	$M + M + h\nu \rightarrow M_2^*(v', J')$
free → free	$M + M + h\nu \rightarrow M^* + M$

Emission (Spontaneous [or Stimulated])

bound → bound	$M_2^*(v', J')\,[+h\nu] \rightarrow M_2(v'', J'') + h\nu\,[+h\nu]$
bound → free	$M_2^*(v', J')\,[+h\nu] \rightarrow M + M + h\nu\,[+h\nu]$
free → bound	$M^* + M\,[+h\nu] \rightarrow M_2(v'', J'') + h\nu\,[+h\nu]$
free → free	$M^* + M\,[+h\nu] \rightarrow M + M + h\nu\,[+h\nu]$

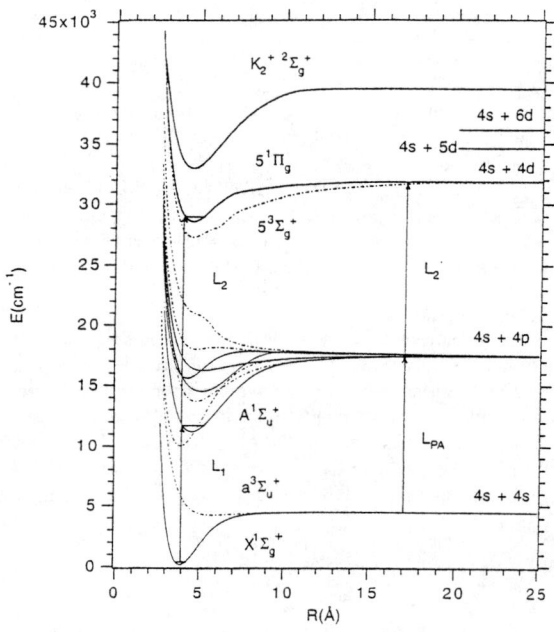

FIGURE I. Selected potential energy curves and atomic asymptotes of $^{39}K_2$.

Here we will focus especially on potassium, which our group has studied in collaboration with the groups of Profs. Phillip Gould and Edward Eyler at the University of Connecticut, and which illustrate much of the promise of the ultracold photoassociation technique. Extensive assigned spectroscopic results are also available from groups at Rice University (Li), NIST (Na), Maryland (Na and Rb), Texas (Rb), Lab. Aimé Cotton (Cs) and Utrecht (Na) (2–4).

PHOTOASSOCIATION

It has long been realized that the absorption of light by colliding atoms results in molecular continua (12-14). In particular, such spectra at ordinary temperatures are broad and continuous whether the upper electronic state is bound (photoassociation) or continuous. This breadth is due not only to the initial thermal energy. The wide range of detunings from atomic resonance over which the absorption can occur provides significant broadening, e.g. when the excited asymptote has states with strong $\pm C_3/R^3$ long range interactions as in the alkali dimers. Finally the broad range of rotational quantum numbers/angular momenta of collision provides additional broadening.

In the case of K_2 photoassociation, the 575 nm diffuse band, corresponding to free to bound $2^3\Pi_g \leftarrow a^3\Sigma_u^+$ absorption, was first observed in 1928 (15) and has been extensively studied since (16-19). It is clear that the information content in this thermal photoassociation spectrum is low; nevertheless it does serve to fairly precisely ($\pm$ 10cm^{-1}) determine the T_e value of the $2^3\Pi_g$ state of K_2, which has not been observed except via this diffuse band. It might be noted that the simulation in (18) involved calculating the free $\rightarrow$ bound Einstein B coefficients for absorption to thousands of bound $2^3\Pi_g$ levels (v'$\leq$ 50, J'$\leq$ 300) from the thermal distribution of atoms colliding on the lower $a^3\Sigma_u^+$ potential, since kT $>>$ $\Delta G_{v'+1/2}$, the upper state vibrational spacing. This is in contrast to the ultracold photoassociation described below, where not only is kT$<<$ $\Delta G_{v'+1/2}$, but also kT $<<$ $B_{v'}$ and only J" $\leq$ 2 collisions penetrate inside 100 Å.

Recent advances in atom cooling and trapping have provided samples of isotopically-selected atomic gases at densities of $> 10^{11}$ / cm^3 and ultracold temperatures (without using cryogenics) below 1 mK. Here the thermal kinetic energy [kT/h = 21 MHz at 1 mK] is comparable to or smaller than many terms in the Hamiltonian, including, for example, the natural energy linewidth of excited states! For example, at a collision energy of ~3kT (1mK) in our trap (20), only J = 0 (s-wave) and J = 1 (p-wave) collisions reach short distances. J = 2 (d-wave) collisions are reflected at ~80Å unless tunneling occurs (which has significant probability because of the long DeBroglie wavelength) and J = 3 (f-wave) collisions are reflected at ~120Å. Thus our photoassociation spectra are dominated by lower s, p and d-wave free states, giving very simple rotational spectra.

This provides new (and relatively simple and inexpensive) opportunities to study high resolution long range molecular spectroscopy by free $\rightarrow$ bound photoassociation of ultracold atoms as first pointed out in (21) and recently reviewed (2-4). This is because, compared to kT, the vibrational splittings are large (kT $<<$ $\Delta G_{v'+1/2}$) and even the rotational splittings are large (kT $<<$ $B_{v'}$). Moreover, the 4p fine structure splitting (57.7 cm^{-1}) is large and even the ^{39}K ground state hyperfine splitting (462 MHz) is large. Only the excited state hyperfine splittings, natural linewidths, Zeeman splittings and AC Stark shifts and widths are less than 10kT in our experiments.

Such ultracold photoassociation provides a simple way to determine long range interactions between atoms and precise binding energies [and dissociation energies (e.g. 22)], complementary to ordinary short range molecular spectroscopy. A side benefit is improved determination of atomic properties (e.g. radiative lifetimes (e.g. (23) and references therein).

Ultracold dynamics (photodissociation, predissociation, autoionization, energy transfer) can also be studied using related techniques. A particularly interesting topic will the low temperature limit of atomic resonant line broadening, since ultracold photoassociation provides complete information on all the potential curves near dissociation.

The use of ultracold photoassociation to form ultracold molecules is a major long term goal, discussed further in (24). Such ultracold molecules could be used for trapping, for the molecular analog of atom optics, for study of highly quantum-mechanical, resonance-dominated ultracold collisions, for fundamental nucleation studies, and for formation of molecular Bose-Einstein condensates (BECs) and molecule lasers.

The ultracold photoassociative process is exemplified by the reaction

$$K + K + h\nu_1 \rightarrow K_2^*(v',J') \tag{1}$$

where the small magnitude of kT allows for excitation of a single low J' rovibrational level in a specific electronically excited state, just as in laser-induced fluorescence of K_2. The singly excited K_2^* (v',J') molecules (here assumed to be near the 4s + 4p asymptotes) then decay radiatively in bound → bound

$$K_2^*(v',J') \rightarrow K_2(v'',J'') + h\nu_2 \tag{2}$$

or bound → free emission

$$K_2^*(v',J') \rightarrow K + K + h\nu_3 \tag{3},$$

(where K_2 (v'',J'') is either the ground $X^1\Sigma_g^+$ state or the lowest triplet state ($a^3\Sigma_u^+$)).

Since significant atomic fluorescence ($4p_{3/2} \rightarrow 4s_{1/2}$) is excited by the very near resonance trap laser, the atom density is readily monitored. Process 1 yields a decrease in atomic density, dependent on the fate of the excited molecules. If process 2 occurred exclusively, trap loss would occur with maximum efficiency (two atoms lost per photoassociative photon absorbed) assuming the excited molecule cannot emit bound-bound photons with a wavelength in the narrow bandpass of the $4p_{3/2} \rightarrow 4s_{1/2}$ filter (which is very likely). However, process 2 is a relatively minor process for levels near dissociation and process 3 dominates as is well known from earlier laser-induced fluorescence studies starting from ground state molecules. If process 3 occurs nearly exclusively, the question becomes the distribution of final kinetic energies in the bound → free emission. In particular, if the identical kinetic energies of the two separating atoms are greater than the trap depth (typically ~1K for a MOT such as ours) the atoms will escape and "trap loss" will be detected by diminished atomic fluorescence. If no such "hot" atoms (KE ≥ 1K) are formed by bound-free emission, there will be no trap loss and no photoassociation detection by trap loss. Examples of levels showing negligible trap loss in a MOT are the zero-point levels of the pure long range states (O_g^- and 1_u) at 28 and 39Å, respectively (20, 25).

Alternately, the singly excited K_2^* (v',J') molecules can be further excited (see Ref. (26))

$$K_2^*(v',J') + h\nu_4 \rightarrow K_2^{**}(v,J) \tag{4}$$

or single- or multi-photon ionized to form molecular

$$K_2^*(v',J') + nh\nu_5 \rightarrow K_2^+(v^+,N^+) + e^- \tag{5}$$

or atomic ions

$$K_2^*(v',J') + nh\nu_6 \rightarrow K + K^+ + e^- \tag{6}.$$

The singly excited K_2^* (v',J') molecules in some cases nonradiatively decay (predissociate) to fragments (Ref. (27))

$$K_2^*(v',J') \rightarrow K^*(4p_{1/2}) + K(4s_{1/2}) \tag{7}.$$

Finally, the doubly excited molecules K_2^{**} (v,J) can decay radiatively [as in (2) and (3)], or nonradiatively by predissociation as in (7) or by autoionization (Ref. (26))

$$K_2^{**}(v,J) \rightarrow K_2^+(v^+,N^+) + e^- \tag{8}$$

or by ion pair formation

$$K_2^{**}(v,J) \rightarrow K^+ + K^- \qquad (9).$$

The K_2^* (v″,J″) and K_2^{**} (v,J) can also be single- or multi-photon ionized as in (5). In addition, singly- and doubly-excited atomic fragments from predissociation of K_2^* and K_2^{**} undergo radiative decay such as

$$K^*(4p_{1/2}) \rightarrow K(4s_{1/2}) + h\nu_7 \qquad (10).$$

The doubly-excited fragments (e.g. K^{**} (5d)) undergo associative ionization as well

$$K^{**}(5d) + K(4s_{1/2}) \rightarrow K_2^+(v^+, N^+) + e^- \qquad (11),$$

and ion pair formation (e.g. K^{**} (6d))

$$K^{**}(6d) + K(4s_{1/2}) \rightarrow K^+ + K^- \qquad (12).$$

Collisional energy transfer is a final possibility, e.g.

$$K^{**}(6s) + K(4s_{1/2}) \rightarrow K^{**}(4d) + K(4s_{1/2}) \qquad (13).$$

these processes are significantly constrained by the well-known energetics (2) of the K atom (e.g. IP = 35004.18 cm^{-1}) and K_2 molecule (e.g. D_0^0 (K_2) = 4404.583 ± 0.072 cm^{-1}, IP (K_2) = E_x (v^+ = 0, N^+ = 0) – E_x (v'' = 0, J'' = 0) = 32775.5 ± 0.15 cm^{-1} and D_0^0 (K_2^+) = 6633.26 ± 0.16 cm^{-1}).

The above processes suggest a wide variety of detection schemes for ultracold photoassociation, four of which have been implemented. We measure the trap loss rate (20) by monitoring the $4p_{3/2} \rightarrow 4s_{1/2}$ atomic fluorescence of trapped K atoms using a photomultiplier-filter system. A CW ring laser provides the second photon for the two color optical-optical double-resonance photoassociative spectroscopy (26) and the fragmentation atomic resonance-enhanced multiphoton ionization spectroscopy (27). Alternatively, the translationally ultracold ground state molecules formed by photoassociation followed by bound-bound emission (28) are detected with pulsed laser resonance-enhanced multiphoton ionization and similarly detected.

DETERMINATION OF LONG RANGE POTENTIAL ENERGY CURVES

At large internuclear distances, the potential energy of a diatomic molecule can be calculated accurately by perturbation theory from the properties of its separated atoms alone (29). If the overlap of the charge distributions of the two atoms is negligible, one may express the asymptotic potential energy as a sum of terms involving inverse powers of the internuclear distance R: $V(R) = -\Sigma C_k R^{-k}$ where $V(\infty)=0$ and the leading exponent depends on the states of the two atoms; for example, k = 6 for the van der Waals interactions of an ns + ns atomic asymptote, k = 3 for the resonant dipole-dipole interaction of an ns + np atomic asymptote and k = 5 for the quadrupole-quadrupole interaction of an np + np asymptote and the resonant quadrupole-quadrupole interaction of an ns + nd atomic asymptote, respectively.

A reasonable estimate for the smallest distance at which the above equations can be used with better than 10 percent accuracy (because electron exchange is improbable) is the modified LeRoy radius (30): $2\sqrt{3} [\langle z^2 \rangle_A^{1/2} + \langle z^2 \rangle_B^{1/2}]$ (z is along the internuclear axis). The R_{LR-M} values relevant to this paper are listed in Table II.

It should also be noted that the properties of diatomic vibrational levels with outer classical turning points in the attractive $C_n R^{-n}$ long-range region are dominated by C_n, the coefficient of the long range potential, and by n, the inverse power of R(31-32). These properties include binding energy [$D_e - G(v)$], vibrational spacing ($\Delta G_{v+1/2}$), kinetic energy ($\langle T_v \rangle$), potential energy ($\langle V_v \rangle$), oscillator strength (f_{ov}), density of states (dv/dG), classical vibrational period (τ_v), outer classical turning point (R_{v+}), powers of R ($\langle R^m \rangle_v$), rotational constant ($B_v \sim \langle R^{-2} \rangle_v$), and centrifugal distortion constant (D_v) (32). For example, the binding energy of these vibrational levels is related to the vibrational quantum number v (with respect to its (noninteger) value at dissociation v_D) by the relation $D_e - G(v) = a_n^\infty (v_D - v)^{2n/(n-2)}$ where a_n^∞ depends only on n, C_n and the reduced mass μ (33-34). For the n = 3 and n = 6 cases considered here, the powers of ($v_D - v$) are 6 and 3, respectively. Alternatively, a plot of vibrational quantum numbers for long range levels versus the 1/6 and 1/3 powers of the levels' binding energies is a linear plot yielding C_n, v_D and D_e. However, it is worth noting that in ultracold photoassociative spectroscopy (in contrast to traditional short range molecular spectroscopy), the binding energy is measured directly with respect to a known atomic limit (rather than the minimum of the ground state) and one does not obtain D_e unless lower short range levels optically connected to high long range levels are known (see e.g. (22)).

TABLE II. Calculation of the Modified LeRoy Radii (R_{LR-m}) in Angstroms for the Various Observed Atomic Asymptotes of K_2 Discussed Herein.

Asymptote	R_{LR-m}		
	m = 0 (σ)	$|m| = 1$ (π)	$|m| = 2$ (δ)
4s + 4s	10.82		
4s + 4p	15.43	11.20	
4s + 4d	31.87	29.34	19.23
4s + 6s	29.89		
4s + 5d	49.47	45.27	28.42
4s + 7s	48.48		
4s + 6d	71.33	65.03	39.83

However, this simple expansion in inverse powers of R breaks down as one goes to increasingly large internuclear distance if there is a significant spin-orbit splitting. In particular, there is a change in angular momentum coupling (Hund's case) when the atomic interaction is comparable to the atomic fine structure splitting (and again when the atomic interaction is comparable to the hyperfine splittings). Finally, at very long range, retardation becomes important (35). For example, consider the case of two potassium atoms shown in Fig. II for the 4s + 4p asymptotes at large internuclear distance. There are eight case (a) electronic states, including the well-known $A\ ^1\Sigma_u^+$, $b\ ^3\Pi_u$ and $B^1\Pi_u$ states which reduce to four pairs of degenerate states $^1\Sigma_g^+ - ^3\Sigma_u^+$, $^1\Sigma_u^+ - ^3\Sigma_g^+$, $^1\Pi_u - ^3\Pi_g$ and $^1\Pi_g - ^3\Pi_u$ which further correlate at very large R with 16 Hund's case (c) molecular states due to angular momentum recoupling. Among these 16 case (c) components, we have observed those that are attractive and dipole allowed from the three ground state components, O_g^+, O_u^-, and 1_u based on the Hund's case (c) selection rules.

We first considered the O_g^+ ($4p_{3/2}$) state shown in Figure II. As shown in (20), long range calculations of C_3, C_6 and C_8 for Σ^+ and Π configurations, followed by diagonalization of a 2 x 2 matrix (solution of a quadratic equation), yield theoretical predictions of the potential energy curve in very good agreement with experiment (20). It should be noted that these results involve corrections for retardation and nonadiabatic effects and that the long-range coefficients are in excellent agreement with theory (36-40) and with a very recent result from molecular spectroscopy (11), summarized in Table III.

It should also be noted that the C_n coefficients from this state (or from molecular spectra or theory) in Table III allow one to calculate the full set of potential energy curves shown in Figure II. In particular, the C_n coefficients from the O_g^- ($4p_{3/2}$) state (20) allowed us to accurately calculate and subsequently observe the 1_u (4p$_{3/2}$) (25), to accurately calculate the 1_g (4p$_{3/2}$) state which correlates with the $1\ ^1\Pi_g$ state (10, 11) and to accurately calculate the O_g^+ (4p$_{3/2}$) state which predissociates the 1_g state (27).

The O_g^+ (4p$_{3/2}$) state can be simply thought of as a state whose C_3 coefficient changes as a function of R. At very large R, where the interatomic interaction is small compared to the spin-orbit splitting, there is an interatomic attraction $V^\infty_{O_g^-} = -(C_3/3)R^{-3}$ where $C_3 = |\langle 4s|\vec{r}|4p\rangle|^2$. However, at shorter distance, the two O_g^+ states mix (the electronic angular momenta recouple), and the O_g^+ (4p$_{3/2}$) potential is approximately described by $V_{O_g^-} = +C_3 R^{-3}$ as previously shown in Figure II. Since the change from a (-1/3) to a (+1) occurs entirely in the long-range overlap-free region (R>>R$_{LR-m}$ in Table II), it is appropriate to call this a pure long range molecule state (20).

TABLE III. Results for long-range coefficients (all in atomic units) at the 4s + 4p asymptotes of ^{39}K$_2$.

	C_3^Π	C_3^Σ	C_6^Π	C_6^Σ	C_8^Π	C_8^Σ	reference
Ultracold Photoassociation:							
	8.445 (14)	16.890 (28)	6480 (94)	9675 (141)	762300[a]	1975000[a]	(23)
	8.436 (14)	16.872 (28)	6272 (94)	9365 (141)	762300[a]	1975000[a]	(20)
Molecular Spectroscopy:							
	8.433 (9)	16.866 (18)	6840 (210)	10213 (314)	762300[a]	1975000[a]	(11)
Theory:							
	8.665	17.33	6291	9393	762300	1975000	(38, 39)
	8.768	17.54	6465	9651	713200	1892000	(40)
	9.340	18.68	6868	10130	581400	2454000	(36, 37)

[a]C_8 values from (38, 39) assumed in fit to experimental data

TABLE IV. Long range C_6 coefficients (in atomic units) for observed electronic states compared with theoretical values

Asymptote	R$_{LR-m}$ (Å)	C_6(a.u.)			
		Expt	Theory		
		(26)	(42)	(43)	(44)
4s + 6s$_\sigma$	29.89	160,000 (5,000)	155,100	146,300	—
4s + 7s$_\sigma$	48.48	370,000 (20,000)	393,700	—	391,950
4s + 4d$_{3/2\Pi}$	29.34	123,000 (3,000)	—	124,900	—
4s + 5d$_{3/2\Pi}$	45.27	330,000 (3,000)	—	—	356,200
4s + 6d$_{3/2\Pi}$	65.03	1,010,000 (70,000)	—	—	808,700

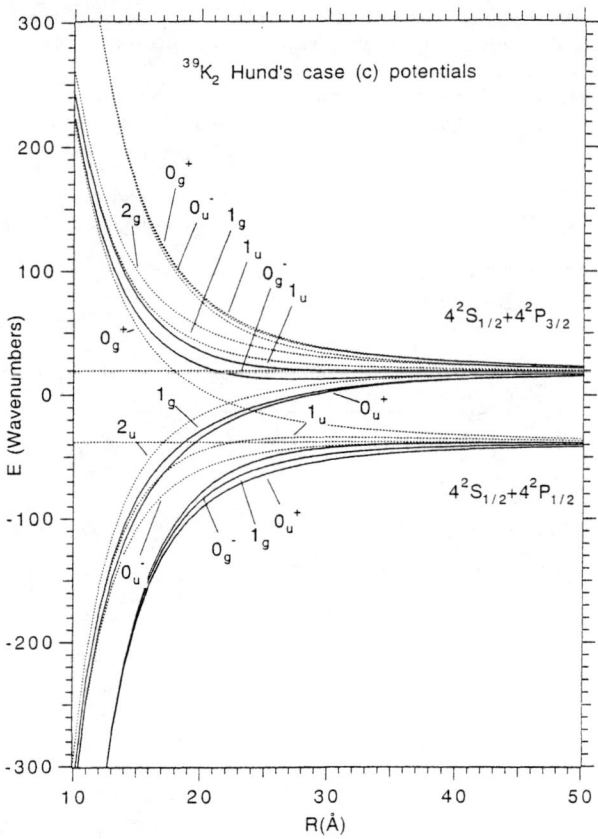

FIGURE II. The 16 adiabatic Hund's case (c) potential energy curves dissociating to the 4s + 4p$_{3/2}$ and 4s + 4p$_{1/2}$ asymptotes of K$_2$, based on the C$_3$, C$_6$ and C$_8$ values of (20). The solid curves are the seven observed states which support bound states and are accessible by dipole transitions from the 4s + 4s asymptote.

The Hund's case (a) degenerate state pairs at large internuclear distance and the correlating case (c) components can also be readily calculated for higher asymptotes. For example, the spin-orbit splitting for the 4s + 4d asymptote is very small (1.10 cm^{-1} compared to 57.71 cm^{-1} for the 4s + 4p limit). For the asymptotes we have studied thus far by two color ultracold photoassociation (4s + 6s, 7s, 4d, 5d, 6d), the long range behavior is expected to include a C$_5$R^{-5} first order resonant quadrupole term plus dispersion (C$_6$R^{-8} + C$_8$R^{-8} + C$_{10}$ R^{-10}). However, the first order term is negligible (26) and thus a linear relation between v and binding energy (D$_e$ − G(v)) to the 1/3 power is predicted and observed. The corresponding C$_6$ values in Table IV are very large, roughly two orders of magnitude larger than the 4s + 4s C$_6$ coefficient of 3813 (45) and agree well with theory (42-44).

ACKNOWLEDGMENTS

This work was carried out in collaboration with Professors Phillip Gould and Edward Eyler, Drs. He Wang and John Bahns, and Xiaotian Wang, Jing Li and Anguel Nikolov, and was supported in part by the National Science Foundation.

REFERENCES

(1) W. C. Stwalley, "Laser Manipulation of Metallic Vapors," in *Radiation Energy Conversion in Space*, K. W. Billman, Editor, Vol. **61** of Progress in Astronautics and Aeronautics, p. 593-601 (1978).

(2) W. C. Stwalley and H. Wang, "Photoassociation of Ultracold Atoms: A New Spectroscopic Technique," submitted to *J. Mol. Spectrosc.* (1998).

(3) P. D. Lett, P. S. Julienne and W. D. Phillips, "Photoassociative Spectroscopy of Laser Cooled Atoms", *Ann. Rev. Phys. Chem.* **46**, 423-452 (1995).

(4) J. Weiner, V. S. Bagnato, S. C. Zilio and P. S. Julienne, "Experiments and Theory in Cold and Ultracold Collisions," *Rev. Mod. Phys.*, in press (1998)

(5) H. E. Roscoe and A. Schuster, "Note on the Absorption Spectra of Potassium and Sodium at Low Temperatures," *Proc. Roy. Soc.* (London) **22**, 362-364 (1874).

(6) J. Heinze and F. Engelke, "The $B^1\Pi_u$ Potential Energy Curve and Dissociation Eneergy of $^{39}K_2$," *J. Chem. Phys.* **89**, 42-50 (1988).

(7) J. T. Kim, H. Wang, J. T. Bahns and W. C. Stwalley, "The $3^1\Pi_g$ and $3^1\Delta_g$ States of $^{39}K_2$ Studied by Optical-Optical Double Resonance Spectroscopy", *J. Chem. Phys.* **102**, 6966-6974 (1995).

(8) G. Jong, L. Li, T. J. Whang, A. M. Lyyra, W. C. Stwalley, M. Li and J. Coxon,, "CW All Optical Triple Resonance Spectroscopy of K_2: Deperturbation Analysis of the $A^1\Sigma_u^+$ State ($v \leq 12$) and $b^3\Pi_u$ ($13 \leq v \leq 24$) States", *J. Mol. Spectrosc.* **155**, 115-135 (1992).

(9) B. Ji, P. D. Kleiber, W. C. Stwalley, A. Yiannopoulou, A. M. Lyyra and P. S. Julienne, "Quantum State-Selected Photodissociation of K_2 ($B^1\Pi_u \leftarrow X^1\Sigma_g^+$): A Case Study of Final State Alignment in All-Optical Multiple Resonance Photodissociation", *J. Chem. Phys.* **102**, 2440-2451 (1995).

(10) A. J. Ross, C. Effantin, J. D'Incan, R. F. Barrow and J. Vergès, "The (1) $^1\Pi_g$ State of K_2 Studied by Infrared Fourier-Transform Spectroscopy," *Ind. J. Phys.* **60B**, 309-317 (1986).

(11) I. Russier, M. Aubert-Frécon, A. J. Ross, F. Martin, A. Yiannopoulou, and P. Crozet, "The (1) $^1\Pi_g$ State of $^{39}K_2$ Revisited," *J. Chem. Phys.* **109**, 2717-2726 (1998).

(12) W. Finkelnburg, "Kontinuerliche Spektren" (Springer, Berlin 1938).

(13) W. Finkelnburg and T. Peters, "Kontinuerliche Spektren" in *Spektroskopie II*, ed. by S. Flugge, Handbuch der Physik, Vol. **28**, 79-204 (Springer, Berlin 1957).

(14) G. Herzberg, "Molecular Spectra and Molecular Structure I. Spectra of Diatomic Molecules," (Van Nostrand, Princeton, 1950).

(15) J. M. Walter and S. Barratt, "The Existence of Intermetallic Compounds in the Vapour State. The Spectra of the Alkali Metals, and Their Alloys with Each Other," *Proc. Roy. Soc.* (London) **119**, 257-275 (1928).

(16) G. Pichler, S. Milosevic, D. Veza and R. Beuc, "Diffuse Bands in the Visible Absorption Spectra of Dense Alkali Vapours," *J. Phys. B* **16**, 4619-4631 (1983).

(17) D. E. Johnson and J. G. Eden, "Continua in the Visible Absorption Spectrum of K_2," *J. Opt. Soc. Am.* **B2**, 721-728 (1985).

(18) W. T. Luh, K. M. Sando, A. M. Lyyra and W. C. Stwalley, "Free-Bound-Free Resonance Fluorescence in the K_2 Yellow Diffuse Band: Theory and Experiment," *Chem. Phys. Letters* **144**, 221-225 (1995).

(19) W. T. Luh, J. T. Bahns, A. M. Lyyra, K. M. Sando, P. D. Kleiber and W. C. Stwalley, "Direct Excitation Studies of the Diffuse Bands of Alkali Metal Dimers," *J. Chem. Phys.* **88**, 2235-2241 (1988).

(20) H. Wang, P. L. Gould and W. C. Stwalley, "The Long-Range Interaction of ^{39}K (4s)+ ^{39}K (4p) Asymptote by Photoassociative Spectroscopy: Part I: The O_g^- Pure Long-Range State and the Long-Range Potential Constants," *J. Chem. Phys.* **106**, 7899-7912 (1997).

(21) H. R. Thorsheim, J. Weiner and P. S. Julienne, "Laser-Induced Photoassociation of Ultracold Sodium Atoms," *Phys. Rev. Lett.* **58**, 2420-2423 (1987).

(22) K. M. Jones, S. Maleki, S. Bize, P. D. Lett, C. J. Williams, H. Richling, H. Knöckel, E. Tiemann, H. Wang, P. L. Gould and W. C. Stwalley, "Direct Measurement of the Ground State Dissociation Energy of Na_2 ", *Phys. Rev. A* **54**, R1006-R1009 (1996).

(23) H. Wang, J. Li, X. T. Wang, C. J. Williams, P. L. Gould and W. C. Stwalley, "Precise Determination of the Dipole Matrix Element and Radiative Lifetimes of the ^{39}K 4p States by Photoassociative Spectroscopy", *Phys. Rev. A* **55**, R1569-R1572 (1997).

(24) J. T. Bahns, P. L. Gould and W. C. Stwalley, "Formation of Ultracold Molecules," to be submitted to *Advan. At. Mol. Opt. Phys.*

(25) X. Wang, H. Wang, P. L. Gould and W. C. Stwalley, "Observation of the Pure Long-Range 1_u State of an Alkali-Metal Dimer by Photoassociative Spectroscopy," *Phys. Rev. A* **57**, 4600-4603 (1998)

(26) H. Wang, X. T. Wang, P. L. Gould and W. C. Stwalley, "Optical-Optical Double Resonance Photoassociative Spectroscopy of Ultracold ^{39}K Atoms near Highly-Excited Asymptotes", *Phys. Rev. Letters* **78**, 4173-4176 (1997).

(27) H. Wang, P. L. Gould and W. C. Stwalley, "Fine Structure Predissociation of Ultracold Photoassociated $^{39}K_2$ Molecules Observed by Fragmentation Spectroscopy," *Phys. Rev. Letters* **80**, 476-479 (1998).

(28) A. N. Nikolov, E. E. Eyler, X. Wang, H. Wang, J. Li, W. C. Stwalley and P. L. Gould, "Observation of Translationally Ultracold Ground State Potassium Molecules," submitted to *Phys. Rev. Letters*.

(29) A. Dalgarno, "New Methods for Calculating Long-Range Intermolecular Forces," *Adv. Chem. Phys.* **12**, 143-166 (1967).

(30) B. Ji, C.-C. Tsai and W. C. Stwalley, "Proposed Modification of the Criterion for the Region of Validity of the Inverse-Power Expansion in Diatomic Long-Range Potentials", *Chem. Phys. Lett.* **236**, 242-246 (1995).

(31) R. J. Le Roy, in *Molecular Spectroscopy 1*, edited by R. F. Barrow *et al.*, (Chem. Soc. London, 1973), p. 113-176.

(32) W. C. Stwalley, "Long-Range Molecules", *Contemp. Phys.* **19**, 65-80 (1978).

(33) R. J. Le Roy and R. B. Bernstein, "Dissociation Energy and Long Range Potential of Diatomic Molecules from Vibration Spacings of Higher Levels," *J. Chem. Phys.* **52**, 3869-3879 (1970).

(34) W. C. Stwalley, "The Dissociation Energy of the Hydrogen Molecule Using Long-Range Forces," *Chem. Phys. Lett.* **6**, 241-244 (1970).

(35) E. A. Power, "Very Long-Range (Retardation Effect) Intermolecular Forces," *Advan. Chem. Phys.* **12**, 167-224 (1967).

(36) B. Bussery and M. Aubert-Frécon, "Multipolar Long-Range Electrostatic, Dispersion, and Induction Energy Terms for the Interaction Between Two Identical Alkali Atoms Li, Na, K, Rb, and Cs in Various Electronic States," *J. Chem Phys.* **82**, 3224-3234 (1985).

(37) B. Bussery and M. Aubert-Frécon, "Potential Energy Curves and Vibration-Rotation Energies for the Two Purely Long-Range Bound States 1_u and O_g^- of the Alkali Dimers M_2 Dissociating to M (ns $^2S_{1/2}$) + M (np $^2P_{3/2}$) with M = Li, Na, K, Rb, and Cs" *J. Mol. Spectrosc.* **113**, 21-27 (1985).

(38) M. Marinescu and A. Dalgarno, "Dispersion Forces and Long-Range Electronic Transition Dipole Moments of Alkali-Metal Dimer Excited States," *Phys. Rev. A* **52**, 311-328 (1995).

(39) M. Marinescu and A. Dalgarno, "Analytical Interaction Potentials of the Long Range Alkali Metal Dimers," *Z. Phys. D.* **36**, 239-248 (1996).

(40) F. Vigne-Maeder, "Excited Alkali Atoms: Polarizabilities and Van der Waals Coefficients for Resonant Interactions," *Chem. Phys.* **85**, 139-148 (1984).

(41) T. R. Proctor and W. C. Stwalley, "The Long-Range Interactions of S-State Alkali Atoms with Rare Gas and Hydrogen Atoms," *J. Chem. Phys.* **66**, 2063-2073 (1977).

(42) M. Marinescu, private communication.

(43) M. Marinescu, H. R. Sadeghpour and A. Dalgarno, "Dispersion Coefficients for Alkali Metal Dimers," *Phys. Rev. A* **49**, 982-988 (1994).

FAR-WING LINE-SHAPE STUDY OF THE COLLISION-INDUCED $c \leftarrow X$ TRANSITION IN Hg–RARE-GAS QUASIMOLECULES

Y. Sato, T. Kurosawa, K. Ohmori, H. Chiba, M. Okunishi, K. Ueda
A. Z. Devdariani[§], and E. E. Nikitin[†]

RISM, Tohoku University, Katahira, Sendai 980-8577, Japan
[§]Institute of Physics, St. Petersburg University, St. Petersburg, Russia
[†]Technion-Israel Institute of Technology, Haifa 32000, Israel

The far-wing excitation and probe technique is applied to observe quasimolecular absorption bands on the blue side of the Hg $6\,^1S_0 - 6\,^3P_2$ atomic forbidden line for the Hg-Ar, Hg-Kr and Hg-Xe mixtures. It is found that the excitation of these bands is followed predominantly by a rapid elastic half-collision scattering on the excited state potential yielding the nascent product state Hg($6\,^3P_2$). This gives a direct evidence for that the upper state of the absorption is the $c\,^31$ state of the Hg-RG molecules and the absorption is of purely quasimolecular nature induced by the Hg-RG collisions.

In order to determine the collision-induced-dipole moment for the transition between c state and the ground $X\,^10^+$ state as a function of the the internuclear distance R, we have made combined analyses of the $c-X$, $B\,^31-X$ and $A\,^30^+ - X$ far-wing line-shape data of the Hg-RG molecules. The resultant dipole moments show an oscillatory dependence on R due to crossings of the Π- and Σ-type electronic energies.[1]

Reference [1] T. Kurosawa *et al.* in Press, J. Chem. Phys. May 15, (1998)

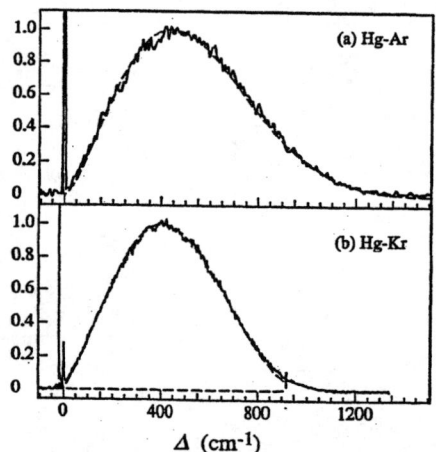

Fig. 1. Observed (solid line) and simulated (dashed line) spectra of the $c-X$ bands *vs.* the excitation wavenumber shift Δ from the Hg $6\,^1S_0 - 6\,^3P_2$ forbidden line for (a) Hg-Ar and (b) Hg-Kr systems [1].

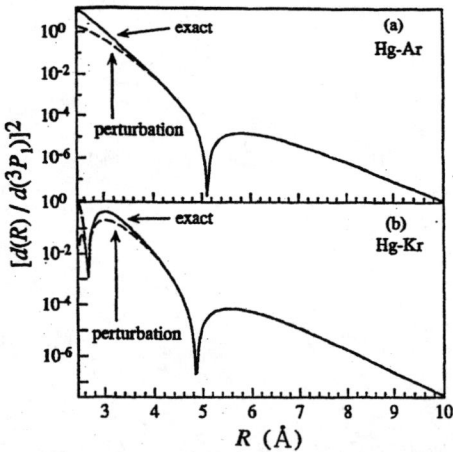

Fig. 2. Square of the $c-X$ transition dipole moment for (a) Hg-Ar and (b) Hg-Kr systems determined by the empirical potential curves of the c, A, B, and X states [1].

Combined half-collision approach to the nonadiabatic transitions in the $Hg(6s6p^3P_2)$–N_2, CO cold and thermal quasimolecules

K. Ohmori, T. Kurosawa, K. Amano, H. Chiba, M. Okunishi, K. Ueda, Y. Sato, A. Z. Devdariani*, and E. E. Nikitin[†]

RISM, Tohoku University, Sendai 980-8577, Japan
*Institute of Physics, St. Petersburg University, 198904 St. Petersburg, Russia
[†]*Department of Chemistry, Technion-Israel Institute of Technology, Haifa 32000, Israel*

We have applied a combined half-collision approach to the transit regions of Hg $^3P_2 \rightarrow {}^3P_j$ fine-structure transitions in the Hg–N_2, CO cold and thermal quasimolecules. This approach consists of (a) the far-wing excitation and probe (FEP) technique to trigger a thermal half-collision by the laser excitation of a collisional qusimolecule in the far wing of a pressure-broadened atomic-line ; and (b) the cold vdW technique to trigger a rotationally- and vibrationally-cold half-collision by the free–bound laser excitation of a relevant van der Waals complex in a supersonic-jet expansion. In both the techniques, nascent scattering products are probed by another laser pulse in a quantum-state-resolved fashion. These two types of pump and probe scheme are illustrated in Figs. 1(a) and 1(b). Excitation spectra are thus measured for the formation of the adiabatic and nonadiabatic products, $Hg(^3P_2)$, $Hg(^3P_1)$, and $Hg(^3P_0)$, by scanning the pump-laser wavelength. Inspection of the spectral line-shapes of the excitation spectra gives absolute nonadiabatic-transition-probability between the transitory quasimolecular states as a function of the angular momenta and translational energy of the half-collision pairs and have revealed for the first time that the vibrational degree of freedom in the diatomic counterparts play key roles in these nonadiabatic processes.

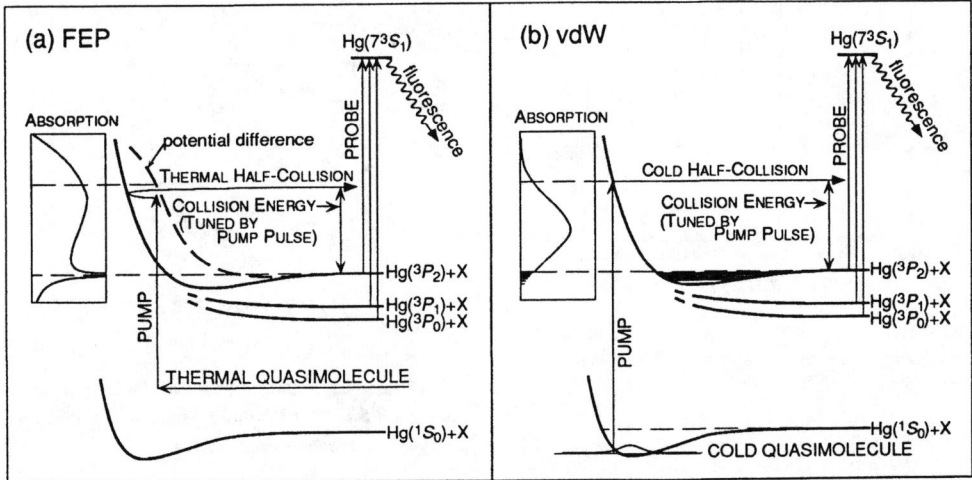

FIG. 1. Schematics of the combined half-collision approach : (a) the far-wing excitation technique to trigger a thermal half-collision ; and (b) the cold-vdW technique to trigger a rotationally- and vibrationally-cold half-collision.

First observation of the bound Hg–rare-gas complex in the dark c-state using free–bound–bound 2-step laser excitation

K. Amano, K. Ohmori, T. Kurosawa, H. Chiba, M. Okunishi, K. Ueda, and Y. Sato
RISM, Tohoku University, Sendai 980-8577, Japan

A. Z. Devdariani
Institute of Physics, St. Petersburg University, 198904 St. Petersburg, Russia

E. E. Nikitin
Department of Chemistry, Technion-Israel Institute of Technology, Haifa 32000, Israel

Figure 1 shows the $c(v' = 0 - 8) - X(v'' = 0)$ laser excitation spectrum of the Hg-Ar van der Waals complex. This corresponds to the first observation of the bound Hg–rare-gas complex in the dark c-state. In this measurement, we employ the sequence of two laser pulses ; the complex is excited by the first one to the dark bound-level in the c state and then successively excited by the second one to the bright E-state for the optical detection. The assignments of v' have been confirmed by the line shapes of the free–bound $E - c(v')$ excitation spectra shown in Fig. 2. The Birge-Sponer (BS) plot of the v' progression is well fitted by a straight line, suggesting that the c-state potential-curve of Hg-Ar is well approximated by the Morse function. The linear fitting of this BS plot gives the potential parameters for the c state.

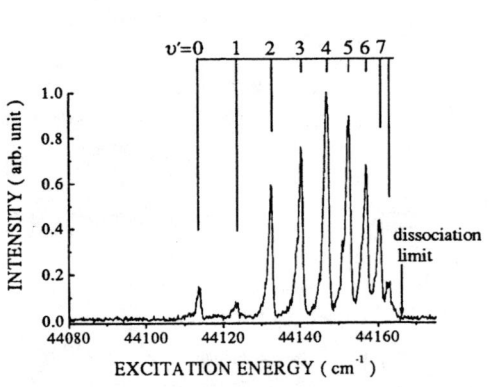

FIG. 1. $c(v') - X(v''=0)$ excitation spectrum of the Hg-Ar vdW complex plotted against the pump wave-number.

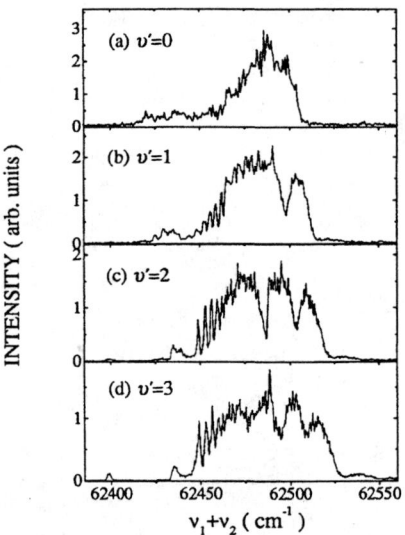

FIG. 2. $E-c(v')$ excitation spectra of Hg-Ar with (a) $v'=0$, (b) $v'=1$, (c) $v'=2$, and (d) $v'=3$ plotted against the sum of the pump and probe wave-numbers.

Far-wing line-shape study of the inter-excited-state transitions of the Hg-Ar and Hg-Ne collisional quasimolecules

K. Amano, K. Ohmori, M. Okunishi, H. Chiba, K. Ueda, and Y. Sato

RISM, Tohoku University, Sendai 980-8577, Japan

We have measured, for the first time, the far-wing spectra associated with the transitions between the 6s6p $^3P_0$ and 6s7s $^3S_1$ excited states of Hg broadened by rare gases (RG: Ar and Ne); the optical transitions are strongly forbidden between the ground $^1S_0$ state and both of those excited states. Figures 1(a) and 1(b) show the fluorescence-excitation spectra of Hg*(6 $^3P_0$)–Ar and Hg*(6 $^3P_0$)–Ne plotted against the wave-number shift Δ (in cm^{-1}) from the atomic-line center. The simulations of the spectra, based on an uniform-semiclassical treatment for the free-free Franck-Condon factor, have been carried out in order to estimate the potential-energy curves $V_E(R)$ and $V_a(R)$ for the $E\ ^31$ and $a\ ^30^-$ states, correlated to Hg*(7 $^3S_1$)+RG and Hg*(6 $^3P_0$)+RG, respectively.

We constructed a single potential-energy curve $V_E(R)$ with double minima by joining two extended Morse potentials, V_{E_1} and V_{E_2}:

$$V_E(R) = V_{E_1}(R)f_1(R) + V_{E_2}(R)f_2(R) \qquad (1)$$

where each extended Morse potential is given by

$$V_{E_i}(R) = D_{e_i}\frac{\beta_{1_i}}{\beta_{2_i} - \beta_{1_i}}\left[e^{-\beta_{2_i}(R-R_{e_i})} - \frac{\beta_{2_i}}{\beta_{1_i}}e^{-\beta_{1_i}(R-R_{e_i})}\right] + p_i, \qquad (2)$$

and $f_i(R)$s are switching functions expressed as $f_i(R) = 1/2\tanh[\pm\alpha(R - R_{hi})] + 1/2$ with the signs being $-$ and $+$ for $i = 1$ and 2, respectively. The values for the switching parameters have been chosen as $\alpha = 2.50$ Å^{-1}, $R_{h1} = 4.56$ Å, and $R_{h2} = 5.24$ Å, and the other parameters are summarized in Table I.

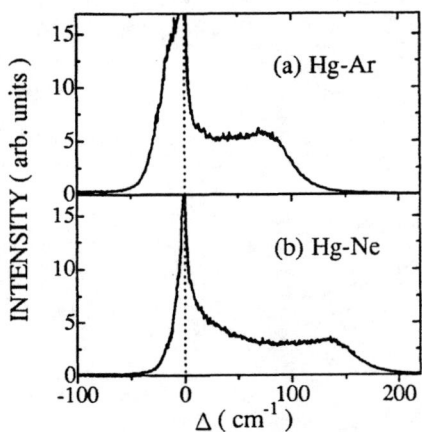

FIG. 1. Fluorescence excitation spectra observed at the $E-a$ band of the (a) Hg*(6 $^3P_0$)–Ar and (b) Hg* (6$^3P_0$)–Ne collisional quasimolecules. $\Delta = 0$ cm^{-1} denotes the 6 $^3P_0 - 7\ ^3S_1$ transition.

TABLE I. Potential parameters for the E and a states of Hg-Ar and Hg-Ne.

	Hg-Ar			Hg-Ne
	a	E_1	E_2	a
R_e (Å)	4.20	2.81	6.95	4.69
D_e (cm^{-1})	97	1560[a]	38	13
β_1 (Å^{-1})	1.03	1.94	0.40	1.25
β_2/β_1	2.00	2.10	1.01	2.00
ω_e (cm^{-1})[b]	14.4	109	3.51	8.68
$\omega_e x_e$ (cm^{-1})[c]	0.54	1.90	0.08	1.45
p_i (cm^{-1})	—	130	0	—

[a] The well depth measured from the asymptotic limit of Hg*(7 $^3S_1$)+Ar is 1430cm^{-1}.
[b] Calculated using Morse relation $\omega_e = \beta_1\sqrt{2D_e/\mu}$.
[c] Calculated using Morse relation $\omega_e x_e = \hbar\omega_e^2/4D_e$.

NEUTRAL ATOM LINESHAPES

Recent Progress in the Determination of Interatomic Potentials of Alkali - Argon Systems

Dieter Zimmermann, Markus Braune, and Dirk Schwarzhans

*Institut für Atomare und Analytische Physik, Technische Universität Berlin
Hardenbergstraße 36, D-10623 Berlin, Germany*

Abstract. The present status of determination of internuclear potentials from laser spectroscopic data is reviewed for LiAr, NaAr, and KAr. An improved potential is given for the X $^2\Sigma$ state of NaAr. For the first time, we were able to deduce the R dependence of the molecular spin-orbit operator directly from spectroscopic data for the case of the A $^2\Pi$ state of KAr. In addition, an interatomic potential has been obtained for the B $^2\Sigma$ state of KAr with $R_e = 8.0$ Å and $D_e = 9.7$ cm^{-1} as preliminary values for the equilibrium parameters.

A detailed knowledge of the interatomic potential is of crucial importance for the interpretation of line broadening and line shift experiments and may also contribute to an improved understanding of atom-atom collisions. The most reliable information on these potentials is expected from high-resolution spectroscopy in combination with a fully quantum-mechanical method of approach for inversion of the observed energy levels into the interatomic potential. The alkali-rare gas systems have been the preferred object for such studies for a long time, both experimentally and theoretically. Renewed interest in these systems stems from recent experiments studying the attachment of alkali atoms, dimers, and trimers to He clusters of nm size (see the contribution of W. E. Ernst, this volume). Unfortunately, only little information e.g. on the Na-He system is presently available from experiment. However, approximate values may be obtained from the series of the Na-rare gas systems by extrapolation.

High-resolution spectroscopy of weakly bound molecules has considerably been improved in recent years by the use of supersonic beams and of narrow-band tunable dye lasers or semiconductor laser diodes. In addition, computer-aided data recording and data analysis was an important support in handling the often large amount of spectroscopic data in due time. Moreover, distinct progress in the accurate determination of the bound part of the potential could in particular be achieved by observing <u>all</u> bound rovibrational levels for a certain electronic state.

The main objective of the present work is an accurate determination of the interatomic potential of the lowest electronic states of alkali-argon molecules. However, a second intention is to deduce information on the dependence A(R) of the molecular spin-orbit operator on internuclear separation. In this contribution we present our preliminary results on the following two subjects: i) the improved determination of the $X\,^2\Sigma$ interaction potential of NaAr. ii) the common treatment of the electronic states $A\,^2\Pi$ and $B\,^2\Sigma$ of KAr. Results for LiAr or for other electronic states of KAr have already been published (1,2) or will be published later (3).

The molecules were produced by supersonic expansion of a mixture of rare gas and alkali vapor through a nozzle of 100 μm diameter into a vacuum. The molecular beam was crossed by the laser beam under right angles and the fluorescence was collected in the direction perpendicular to both beams. By scanning the laser frequency and by observing the total fluorescence we recorded the absorption spectrum due to the transitions $A\,^2\Pi \leftarrow X\,^2\Sigma$ and $B\,^2\Sigma \leftarrow X\,^2\Sigma$. The two electronic states $A\,^2\Pi$ and $B\,^2\Sigma$ correlate with the first excited $np\,^2P$ state of the alkali atom in the limit of dissociation. About 2000 absorption lines have been

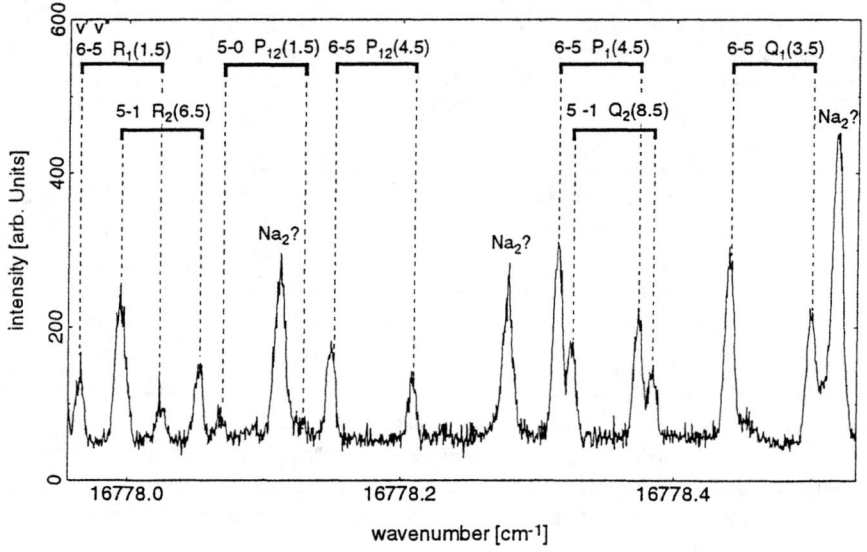

Figure 1. Section of the absorption spectrum of NaAr with absorption lines starting from v"=5 of $X\,^2\Sigma$ (upper and lower row of rotational assignment correspond to upper level with $\Omega = 1/2$ and $3/2$, respectively)

observed for each molecule and could properly be assigned. Tuning the laser to a distinct absorption line allowed us to record the spectral distribution of the fluorescence, e.g. of the $A \rightarrow X$ transition by scanning a grating monochromator being placed into the beam of the detected light.

Turning now to the $X\,^2\Sigma$ state of NaAr the vibrational levels 0....4 and the spectral distribution of the $A \rightarrow X$ fluorescence have already been investigated in previous work (4,5). Recently, we have observed the existence of two more vibrational levels v = 5 and 6 . Fig. 1 shows a section of the absorption spectrum of NaAr with molecular lines starting from v = 5 as a lower level. Altogether, 115 lines of this type could be assigned leading to 8 bound rotational levels. For v = 6 we found 3 rotational levels N = 0, 1, and 2 from 15 absorption lines. Incidentally, the level v = 6, N = 0 is bound by only 0.07 cm^{-1} and extends as far as 15 Å to its outer turning point. In Table 1 we have summarized our preliminary results for the spectroscopic parameters of v = 5 and 6. In addition, we have recorded the spectral distribution of the A → X fluorescence, once again.

Table 1: Preliminary values of spectroscopic parameters for X $^2\Sigma$ of NaAr (in cm^{-1})

v	$T_v - T_o$	$10^2 B_v$	$10^5 D_v$
5	34.22 (10)	1.72 (2)	2.3 (2)
6	35.09 (20)	0.52 (4)	---

The interatomic potential was determined by means of a standard fit procedure adjusting the free parameters of an analytical potential function until the standard deviation between input and output data reached a minimum. The output data were obtained by solving numerically the Schrödinger equation of nuclear motion. A more detailed description is given in ref. 1. For the X $^2\Sigma$ state of NaAr the input data set consisted of the 115 energy values of all existing bound rovibrational levels (including those from previous work) and 8 wavelength values corresponding to 8 maxima of the spectral distribution of fluorescence from the level v = 9, A $^2\Pi_{3/2}$. We used a Hartree-Fock-Dispersion (HFD) function with 10 free parameters

$$V(R) = A \cdot \exp(-\alpha R + \beta R^2) - \left[\sum_{n=3}^{6} C_{2n}/R^{2n}\right] \cdot F(R) - D_o \qquad (1)$$

where $F(R) = \exp[-((R_m/R)^p - 1)^2]$ is a cutting function, cutting down the long-range part of V(R) for R< R_m and being 1 elsewhere. D_o is the dissociation energy of the ground level. Our preliminary results for the parameters are given in Table 2.

Table 2: Parameter of HFD potential of the X state of NaAr

D_o = 35.159 cm^{-1}	C_6 = 6.1836·10^5 cm^{-1}Å^6
A = 6.7479·10^4 cm^{-1}	C_8 = 6.5397·10^7 cm^{-1}Å^8
α = 0.82196 Å^{-1}	C_{10} = -1.7233·10^9 cm^{-1}Å^{10}
β = -0.21313 Å^{-2}	C_{12} = 1.08816·10^{10} cm^{-1}Å^{12}
R_m = 7.47975 Å	p = 1.2966

The input data are reproduced by this HFD function with a mean deviation of $6 \cdot 10^{-3}$ cm^{-1} for the energy levels and 0.8 nm for the wavelength values. The error of V(R) is estimated to be less than ± 1 cm^{-1} for the bound part and ± 20 cm^{-1} for the repulsive part which extends up to 2000 cm^{-1}. Our result is in quite good agreement with the result of a recent theoretical calculation using the core polarisation potential method (6).

For the first excited electronic states of KAr about 160 rovibrational energies have been observed for the vibrational levels v = 6....11 of each the A $^2\Pi_{1/2}$ and the A $^2\Pi_{3/2}$ state in our previous work (1). Recently, we could deduce 52 rovibrational energies with v = 0...3 for the B $^2\Sigma$ state by recording and analyzing the B ← X absorption spectrum. The spin-orbit interaction in the A state of about 40 cm^{-1} is comparable to the difference between the A and the B potentials. In addition, the spin-orbit operator couples the two electronic states A $^2\Pi_{1/2}$ and B $^2\Sigma$. Therefore, all 3 electronic states have to be treated on a common basis.

In our approach we started from the diabatic potentials given as a Tang-Toennies function for A $^2\Pi$ (1) and as a HFD function for B $^2\Sigma$. The spin-orbit operator was represented by a modified exponential, being fixed to the atomic value of A_0=38.48 cm^{-1} in the limit of dissociation.

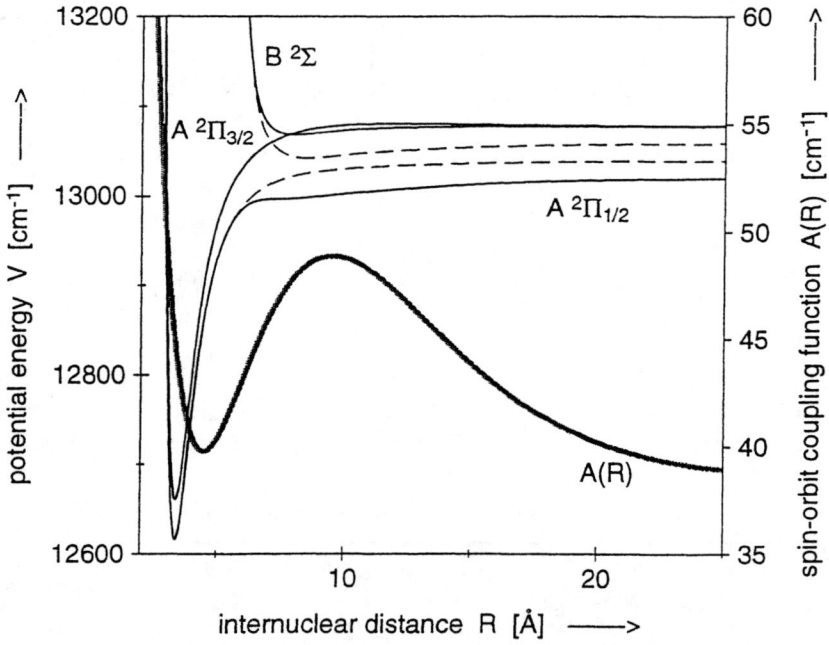

Figure 2. Adiabatic (full line) and diabatic (dashed line) potential curves of the A and the B state of KAr together with the spin-orbit parameter (grey line)

$$A(R) = [b(R-R_b)^2 + c] \exp[-a(R-R_a)] + A_0 \qquad (2)$$

The adiabatic potentials were calculated by diagonalizing the proper matrix for each value of R. Finally, the rovibrational energies were calculated from the adiabatic potentials and fitted to the input data, as described before.

In Fig. 2 we have depicted our present best result for the adiabatic and the diabatic potentials together with A(R). The input energy values are reproduced within $\pm 20 \cdot 10^{-3} \cdot cm^{-1}$ which is about 10 times the experimental error. Although we expected A(R) to increase continuously with decreasing R, only a curve showing a maximum and a minimum could reproduce the input data in a satisfactory manner (see Fig. 2). The strong effect of the non-diagonal part of the spin-orbit interaction is clearly revealed in Figure 2 by the shallow dimple in the $A\ ^2\Pi_{1/2}$ potential around R = 8.5 Å. For the equilibrium parameters of the adiabatic $B\ ^2\Sigma$ potential we get $R_e = 8.0$ Å and $D_e = 9.7\ cm^{-1}$. The equilibrium parameters of the $A\ ^2\Pi$ states agree with those given in ref. 1 within error limits. We are presently performing experiments in order to increase our data set of energy levels of the $A\ ^2\Pi$ states, in particular in the region between v = 11 and dissociation.

Acknowledgments

Thanks are due to Dipl.-Phys. Hamid Valipour for providing us with the experimental data of the $B\ ^2\Sigma$ state of KAr. The financial support by the Deutsche Forschungsgemeinschaft (contract Zi 186/5) is gratefully acknowledged.

References

1. Bokelmann, F., and Zimmermann, D., *J. Chem.Phys.* **104**, 923 - 934 (1996)
2. Brühl, R., and Zimmermann, D., *Chem. Phys. Lett.* **233**, 455 - 460 (1995)
3. Brühl, R., Dissertation, Technische Universität Berlin, 1996
4. Aepfelbach, G., Nunnemann, A., and Zimmermann, D., *Chem. Phys. Lett.* **96**, 311 - 315 (1984)
5. Tellinghuisen, J., Ragone, A., Kim, M.S., Auerbach, D.J., Smalley, R.E., Wharton, L., and Levy, D.H., *J.Chem. Phys.* **71**, 1283 - 1291 (1979)
6. Kerner, Chr., Dissertation, Universität Kaiserslautern, 1995

Experimental Study of Thermal Radiation in Dense Sodium Vapor for Broad Visible and Infrared Range

Alexei G.Leonov[*], Andrei N. Starostin[†], Dmitri I.Chekhov[*], Artem A.Rudenko[*], Andrei Yu.Sechin[†], Yuri K.Zemtsov[†]

[*]*Moscow Institute of Physics & Technology, 141700, Dolgoprudnyi, Moscow region, Russia*
[†]*Troitsk Institute for Innovation & Fusion Research, 142092, Troitsk, Moscow region, Russia*

Abstract. Purely thermal glowing of dense nonuniformly heated up to temperatures of 600-1200 K resonance medium (sodium vapor) in broad visible and IR spectral regions was investigated for the first time. Shape of obtained spectra and glowing absolute intensities in various spectral regions agree well with the numerical calculation results on the basis of the generalized theory of resonance radiation transfer.

In the recently advanced generalized theory of resonant radiation transfer [1] a number of new effects was predicted. In particular, the opportunity of essential differences in a structure of a resonant line in its central part and occurrence of an additional maximum on a far wing of a spectral lineshape was shown. For the verification of these effects a thermal glowing of resonance medium in broad visible and IR spectral regions (0.5–5 μm) was investigated for the first time. Sodium vapor heated steadily to temperatures of 600-1200 K was used as the resonance medium. The emission in the 3P-3S resonance doublet was studied. The purely thermal emission from the vapor in the absence of excitation of the vapor by an electric field or by an external radiation source was detected. The investigations were performed on setup consisting of cylindrical "heat pipe" type cell, filled with sodium and buffer gas (argon, helium at pressure $P \sim 10^4$–10^5 Pa), and recording system. To obtain the emission spectra the radiation of the paraxial zone of the pipe was directed onto input slit of an appropriate monochromator.

Obtained in the experiments specific self-inverted spectrum of thermal radiation of sodium vapor in a wavelengths range of 0.5-1.1 μm is presented in Fig.1. In the same figure the theoretical curves calculated with the generalized [1] and standard [2] theories of resonant radiation transfer also are shown. The most essential moment displayed both in the generalized theory and in experiment is the strong asymmetry of a glowing spectrum of resonant medium, which can not be obtained in the standard theory [2]. One can see from the comparison of experimental data with results of calculations, that the theoretical curve (2) well describes the experimental data near the resonance and quite satisfactorily in the field of large wavelengths. A divergence of a theoretical and experimental curves at $0.6 < \lambda < 0.8$ μm are related, apparently, with the influence of emission and absorption on electron-oscillation transitions in molecular sodium, which has been not taken into account in the calculations. The calculations under the standard theory differ significantly from experimental data in the whole range of wavelengths. Spectrum of glowing of sodium vapor in a range of wavelength 2-5 μm for various T is shown in Fig.2. It follows from the data that there is a distinctly expressed maximum in the IR region of the spectrum that confirms the theory [1].

The calculations and measurements demonstrated that the intensity of sodium vapor thermal radiation in this red maximum exceeds on some orders the

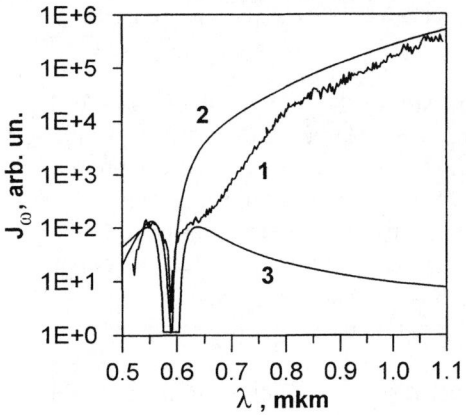

FIGURE 1. Thermal emission spectrum of sodium vapor in the visible and near IR region. 1 – experiment, 2 – calculation according to the generalized theory (Ref.1), 3 – calculation according to the standard theory (Ref.2)

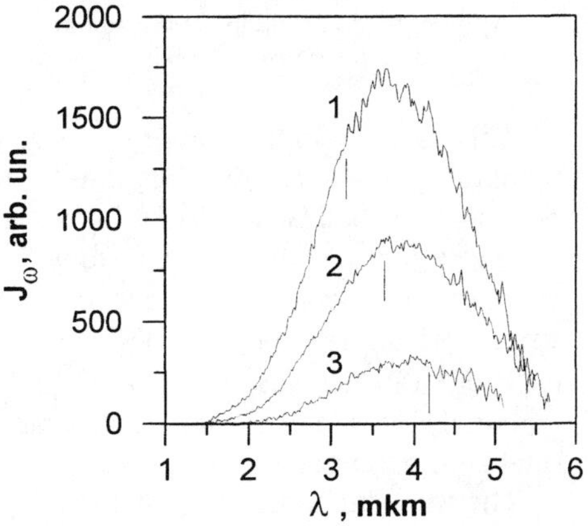

FIGURE 2. Thermal emission spectrum of sodium vapor in the middle IR region for different temperatures at the center of the heat pipe: 1 -- T= 1080 K, 2 -- 950 K, 3 -- 830 K. The vertical bars show the theoretical positions of the peaks.

value obtained from the standard theory. For this reason it is necessary to note that the red far wings can introduce sufficient contribution in total glowing intensity and for some cases may exceed essentially intensity of photorecombination and brake continuum in various types of plasma objects (discharge lamps, solar corona plasma, etc.).

REFERENCES

1. Zemtsov Yu.K.,. Sechin A.Yu and Starostin A.N, *JETP* **83**, 909 (1996).
2. Biberman L.M., Vorobjov V.S., Yakubov I.T., *Kinetics of Nonequilihrium Low-Temperature Plasmas*, New York: Plenum Press, 1987.

Quantum Width and Shift of Ar-Perturbed K Spectral Lines in Non-Impact Regime Based on a Pseudo-Potential for the Na/Ar 3s S State

W. C. Kreye* and J. F. Kielkopf[†]

*Research and Instructional Computer Center, Wright State Univ., Dayton, OH 45435 USA
[†]Department of Physics, University of Louisville, Louisville, KY 40292 USA

Abstract. The equations for a general quantum-mechanical theory for the perturbation shift and broadening of spectral lines in the non-impact regime have been numerically solved for a pseudo-potential of the Na/Ar system in the 3s S state. Spectral line shapes have been shown for low and high pressures (P). The variation with P of the linewidths for the non-impact regime and for the impact approximation is depicted in the range P = 0.5 - 20 torr.

This is a continuation of a previous study (1) in which a square-well potential (SWP) of width 0.25 Å was employed. Here, we use a 10-Å pseudo-potential for the Na/Ar system in the 3s S state (see Ref. 2 for a description of the pseudo-potentials). That study (1) and the present one determine the approximate pressure (P) above which the impact approximation is no longer valid. Baranger's theory (3) of the non-impact regime was derived on the basis of an initial wave function Ψ_{ik}, which in our case corresponds to the 3s S potential, $V_i(r')$, and a final plane-wave wave function, which corresponds to a final zero potential.

The theoretical basis for the non-impact regime is given in Baranger's Eq. (29) in Ref. 3 and in our Ref. 1. The line shape in given by

$$F(\omega)2\pi = 4\int_0^{s_{max}} ds \exp\{n \cdot R[g(s)]\} \cos\{-n \cdot I[g(s)] + \omega s\} \quad (1)$$

The correlation function g(s) consists of an impact term $g_1(s) = is<\vec{k}|V_i|\Psi_{ik}>$ plus a non-impact term $g_2(s)$ (see Refs. 1 and 3 for details of the latter). The final plane wave is expanded in the standard manner in terms of partial waves; but expansion of the initial state is unique for this pseudo-potential:

$$\Psi_{ik} = \sum_{\ell} R_\ell(\alpha r')[2\ell+1] i^\ell e^{-i\delta_\ell} P_\ell(\cos\theta) \quad (2)$$

where $R_\ell(\alpha r')$ is the solution of the radial Schroedinger equation with $V_i(r')$. The $g_2(s)$ term involves the integration over $d\varepsilon'$ of a singularity at ε. A second computational problem involved choosing the limiting value of the angular-momentum quantum number ℓ_{max}, dependent on phase shift δ_ℓ.

Figure 1A depicts a pair of impact (—) and non-impact (---) line shapes at the low pressure of 1 torr. The widths are equal (0.00168 cm^{-1}). Figure 1B

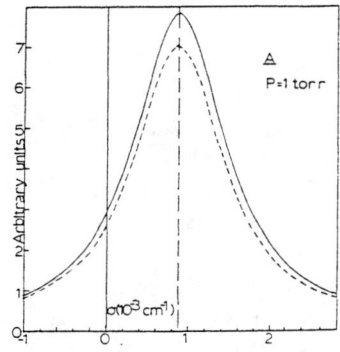
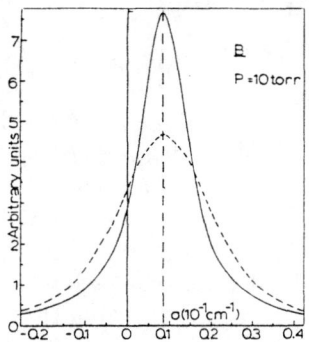

FIGURE 1. Shape vs pressure. T = 400 K, Na/Ar potential.

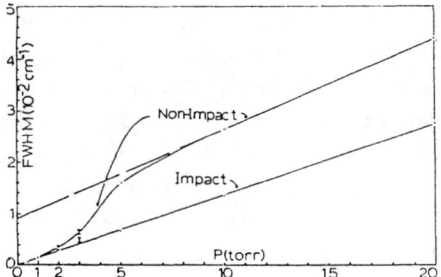

FIGURE 2. Width vs pressure.

shows the pair at a pressure of 10 torr, indicating that the difference in widths increases with P. Figure 2 depicts the dependence of the impact and non-impact widths on P. As expected, the impact curve is a straight line. The non-impact curve follows the impact curve up to ~ 2 torr (where the impact approximation holds). At intermediate pressures the curve angles sharply upward to ~ 5 torr, above which it becomes approximately linear. In a related study (4) of Ar-perturbed K-line shapes for the 7s → 4p transition, the width-vs-P curve is also approximately linear in this P region. How can we decide in which regime the experimental results should be interpreted? Our criterion is the intercept of the extrapolation of the width-vs-P curve with the width axis. Reference 4 shows that the experimental extrapolation is above the origin. Since the extrapolation of the non-impact curve in the present study is also above the origin (in contrast to the impact curve that goes through the origin), we argue that the experimental results should be analyzed with the non-impact-regime theory if P is > 4 torr. Another important feature of Figs. 1A and 1B is that the shifts of the non-impact-regime and the impact-approximation line shapes are the same.

1. Kreye, W. C., and Kielkopf, J. F., *Spectral Line Shapes*, Vol. 9, AIP Conf. Proc. 386 (Zoppi, M., and Ulivi, L., Eds.) (AIP, New York, 1996), pp. 321-322.
2. Kreye, W. C., and Kielkopf , J. F., *J. Phys. B: At. Mol. Opt. Phys.* **80**, 2075 (1997).
3. Baranger, M., *Phys. Rev.* **111**, 481 (1958).
4. Kreye, W. C., *J. Phys. B: At. Mol. Phys.* **15**, 371 (1982).

Influence of excitation processes on the shape of argon lines

A. Bielski[†], S. Brym[‡], R. Ciuryło[†] and J. Szudy[†]

[†] *Institute of Physics, Nicholas Copernicus University,*
Grudziądzka 5/7, 87-100 Toruń, Poland.
[‡] *Department of Physics, Pedagogical University,*
Żołnierska 14, 10-561 Olsztyn, Poland.

Shapes of the selfbroadened argon lines 687.1 nm ($4d_5$–$2p_{10}$), 703.0 nm ($3s_5$–$2p_9$), 750.3 nm ($2p_1$–$1s_2$) emitted from a glow discharge at low pressures were analyzed using a Fabry-Perot interferometer. The pressure broadening and shift of these lines were investigated in papers [1–3]. For the two first lines at very low pressures (below 3 Torr) appreciable departures of the observed profiles from the Voigt profile have been found. Such departures may be explained as caused by effects associated with the mechanisms by which the emitting atoms are excited. In the present study we focus our attention on the dissociative recombination of molecular ions Ar_2^+ with electrons which can cause incomplete thermalization of the excited Ar-atoms [4,5]. To analyse the profiles we used a model proposed in paper [6] for the description of similar effects in neon. We have found that the experimental profiles can be fitted well to that derived from this model. Using this model we can estimate the contribution to the line width coming from non-thermalized atoms. If we assume that the incomplete thermalization of some atoms is caused by dissociative recombination then this width (FWHM) is given by $\gamma_{DIS} = 2\omega_0\sqrt{E_D/m_E c^2}$, ($m_E$ – emitter mass, ω_0 – unperturbed frequency). Here E_D is the addi-

Table 1: Obtained values γ_{DYS}, E_D, D for argon lines.

λ [nm]	γ_{DYS} [10^{-3} cm^{-1}]	E_D [eV]	D [eV]
687.1	80 ± 9	0.28 ± 0.06	0.77 ± 0.06
703.0	86 ± 5	0.34 ± 0.04	0.58 ± 0.04

tional kinetic energy of dissociation recombination products which is equal to $E_D = E_I - E^* - D$, where E^* is the energy of given excited state of the Ar-atom produced due to the dissociative recombination, E_I is the ionisation energy of Ar and D is energy for which crossing of potential curves of the Ar*-Ar and Ar$^+$-Ar systems occurs. We evaluated E_D, D using γ_{DIS}. Results are listed in

Table 1. It should be noted that the energy D obtained from such a line shape analysis must fulfil the condition $D \leq D_0$, where D_0 is the dissociation energy of the molecular ion Ar_2^+ in its ground electronic state $^2\Sigma_u^+$. In our case this condition is fullfield with $D_0 = 1.320 \pm 0.005$ eV reported by Hall et al [7].

ACKNOWLEDGMENTS

This work was supported by a grant No 673/P03/96/10 (2 P03B 005 10) from the State Committee for Scientific Research.

REFERENCES

1. J. Wawrzyński, J. Wolnikowski, Phys. Scripta **33**, 113 (1986).
2. A. Bielski, J. Wawrzyński, J. Wolnikowski, Acta Phys. Pol. A **67**, 621 (1985).
3. D. N. Stacey, J. M. Vaughan, Phys. Lett. **11**, 105 (1964).
4. T. R. Connor, M. A. Biondi, Phys. Rev. **140**, A778 (1965).
5. L. Frommhold, M. A. Biondi, Phys. Rev. **185**, 244 (1969).
6. R. Ciuryło, A. Bielski, J. Domysławska, J. Szudy, R. S. Trawiński, J. Phys. B **27**, 4181 (1994).
7. R. I. Hall, Y. Lu, Y. Morioka, T. Matsui, T. Tanaka, H. Yoshii, T. Hayaishi, K. Ito, J. Phys. B **28**, 2435 (1995).

Speed-dependent narrowing of the 687.1 nm argon line perturbed by neon

A. Bielski[†], S. Brym[‡], R. Ciuryło[†] and J. Szudy[†]

[†] Institute of Physics, Nicholas Copernicus University,
Grudziądzka 5/7, 87-100 Toruń, Poland.
[‡] Department of Physics, Pedagogical University,
Żołnierska 14, 10-561 Olsztyn, Poland.

The pressure broadening and shift of argon line 687.1 nm $4d_5$–$2p_{10}$ perturbed by argon, neon and hel was investigated by Wawrzyński and Wolnikowski [1]. In the present work the pressure dependence of the Doppler width of argon line 687.1 nm perturbed by neon in the low-current glow discharge was investigated using a pressure-scanned Fabry-Peròt interferometer (FPI). Measurements were performed in the pressure range between 0.5 and 44.8 Torr. Assuming that the resulting line shape is given by the Voigt profile (VP) we have found, using fromula given by Ballik [2] that at higher pressures the Doppler width decreases with the increase of the perturber (neon) pressure as shown in Fig. 1.

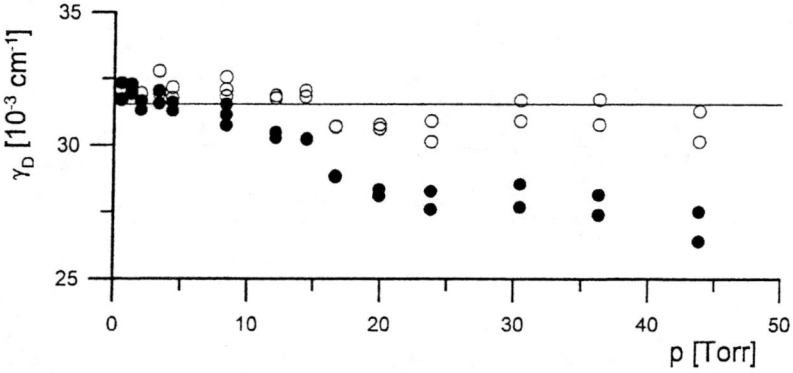

FIGURE 1: Values of the Doppler width of the argon line 687.1 nm perturbed by neon (• – VP and ∘ – SDVP analysis).

As the next step in our analysis we have assumed that the shape of emitted line is given by speed-dependent Voigt profile (SDVP) [3,4] derived by Berman. We have used an expresion derived in [5] which describes the response of FPI to the speed-dependent Voigt profile. The potential curves for Ar–Ne were assumed to be of the Lennard-Jones form. We have found that the Doppler widths obtained from such an analysis are independent of neon pressure (Fig. 1). This example shows that speed-dependent correlation effects can be also important in the case when the perturber is lighter then the emitter. In this case the the perturber-emitter mass ratio $\alpha = m_P/m_E$ is equal to 0.5.

ACKNOWLEDGMENTS

This work was supported by a grant No 673/P03/96/10 (2 P03B 005 10) from the State Committee for Scientific Research.

REFERENCES

1. J. Wawrzyński, J. Wolnikowski, Phys. Scripta **33**, 113 (1986).
2. E. A. Ballik, App. Opt. **5**, 170 (1966).
3. P. R. Berman, J. Quant. Spectrosc. Radiat. Transf. **12**, 1331 (1972).
4. J. Ward, J. Cooper, E. W. Smith, J. Quant. Spectrosc. Radiat. Transf. **14**, 555 (1974).
5. R. Ciuryło, A. Bielski, S. Brym, J. Jurkowski, J. Quant. Spectrosc. Radiat. Transf. **53**, 493 (1995).

The Effect of perturber on the pressure broadening and shift of spectral lines.

G. D. Roston and M. S. Helmi

Physics Department, Faculty of Science, Alexandria University
P.O.Box 21511 Alexandria, Egypt

The collisional broadening and shift coefficients β and δ of the Cd 326.1 nm, Hg 253.7 nm and Zn 307.6 nm intercombination lines perturbed by inert gases (Xe, Kr, Ar, Ne and He) have been calculated on the basis of the classical Lindholm-Foley impact theory for Van der waals and Lennard-Jones potentials [1,2]. Van der Waals potential parameter for the ground and excited states have been calculated using Unsold formula [3]:

$$C_6 = \alpha \; e^2 \; \langle R^2 \rangle \qquad (1)$$

Where α is the dipole polarizability of the perturbing atom (the values of α for inert gases are taken from [4,5]) and $\langle R^2 \rangle$ is the quantum-mechanical averaged value of R^2 in a given state of the radiating atom. $\langle R^2 \rangle$ was calculated using the coulomb approximation. Table I shows the obtained results of the $\langle R^2 \rangle$ [10^{-20} m] and the effective principle quantum number n^* for the ground and excited states for Hg, Cd and Zn.

Table I: The effective quantum numbers n* and the mean values of R2 [10-20m] for the ground and excited states of Hg, Cd and Zn.

Table I

Radiating Atom	Hg		Cd		Zn	
	n^*	$\langle R^2 \rangle$	n^*	$\langle R^2 \rangle$	n^*	$\langle R^2 \rangle$
Ground state	1.1416	1.37146	1.23015	1.81497	1.20367	1.67228
Excited state	1.56638	2.49664	1.6199	2.98339	1.56728	2.50861

The van der Waals potentials parameter of the ground and excited states (J m^6) of the line in case of collision of Hg, Cd and Zn with inert gas (Xe, Kr, Ar, Ne and He) atoms are illustrated in table II.

Table II: The van der Waal potential parameters (C6 J.M6) of the ground and excited stats of the resonance lines of Hg, Cd and Zn due to collisions with inert gas atoms.

Table II

Radiating atom	Hg		Cd		Zn	
Perturber	Ground state C_6	Excited state C_6	Ground state C_6	Excited state C_6	Ground state C_6	Excited State C_6
Xe	127.78297	232.62012	169.09651	277.9548	155.81197	233.73562
Kr	78.33672	142.60662	103.66378	170.39884	95.51976	143.29047
Ar	51.8811	94.44598	68.65479	112.85231	63.26114	94.89889
Ne	12.46926	22.69943	16.5007	27.12326	15.20437	22.80828
He	6.48045	11.79722	8.57562	14.0963	7.90193	11.85379

Figs (1,2) illustrate respectively the calculated values of the broadening β and shift δ based on van der Waals and L-J potential with those corresponding experimental values for Cd line [4-6], Figs (3,4) represent respectively the theoretical values of β and δ for Hg and Zn lines based on the van der Waals potential with the corresponding experimental values for Hg line only [7-8].

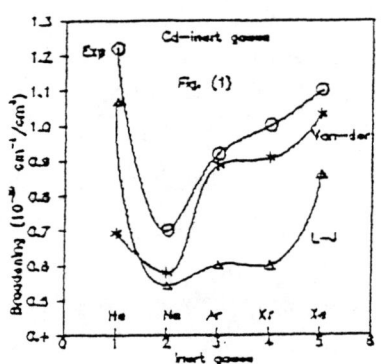
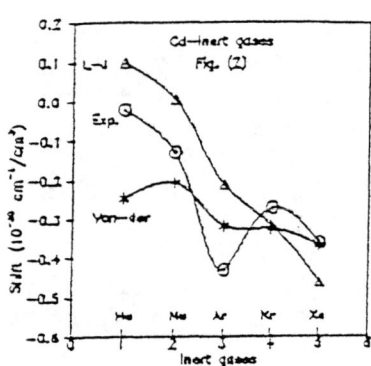

Fig.(1): Experimental and theoretical values of the broadening coefficient (Beta) for the cadmium 326.1 nm line broadened by inert gases.

Fig.(2): Experimental and theoretical values of the shift coefficient (Alpha) for the cadmium 326.1 nm line broadened by inert gas.

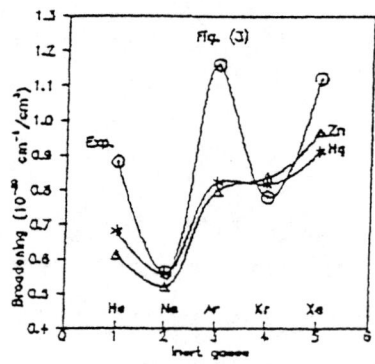 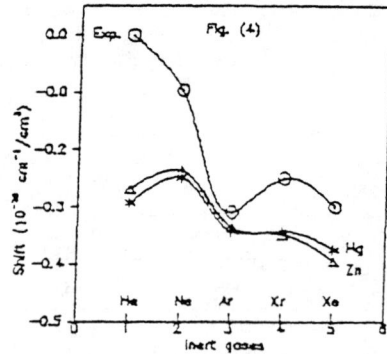

Fig.(3): Experimental and theoretical values of the broadening coefficient (Beta) for the 253.7 nm line of Hg and the 307.6 nm line of Zn broadened by inert gases.

Fig.(4): Experimental and theoretical values of the shift coefficient (Alpha) for the 253.7 nm line of Hg and the 307.6 nm line of Zn broadened by inert gases.

It is seen that the theoretical values of β and δ for Cd, Hg and Zn lines on the basis of van der Waals potential are in agreement with the corresponding experimental values especially for heavy perturbers.

REFERENCES

[1] Hindmarsh, W.R.Petford, A.D.and Smith, G.,Proc.Roy.Soc.A 297, 296,1967.
[2] R.S.Dygdata, J.phys. B: At.Mol.Opt. phys. 21 , 2039, 1988.
[3] Unsold A., Physik der Sternatmospharen, Springer, Berline, 1955.
[4] A.Bielski, S.Brym, R.Ciuryto, J.Domystawska, E.Lisicki and R.S.Trawiniski,J.phys. B:At .Mol .Opt. phys. 27,5863 1994.
[5] K.J.Dietz, P.Dabkiewic2, H.J Kluge, T.Kuhl and H.A.Schuessler, J.phys. B:At .Mol.phys.13,2749,1980
[6] S.Brym and J.Domystawska, Physica Scripta. vol. 52 , 511 , 1995.
[7] Bautax, J., and Lennuier, R., C.R. Acad. Sci., Ser. B 261 , 671 , 1965.
[8] Bautax, J. Schuller, F., and Lennuier, J.phys.33 , 635, 1972 and 35, 361, 1974.

Excitation Transfer Li(3D →3P) Occuring in Optical Collisions with Rare Gas Atoms at Thermal Energies

G.Lindenblatt[1], W.Behmenburg[1], F.Rebentrost[2], M.Jungen[3], M. Smit[3], W.Meyer[4]

[1] Institut für Experimentalphysik, Universität Dusseldorf,
   Universitätsstrasse 1, 40225 Düsseldorf, Germany
[2] Max-Planck Institut für Quantenoptik, Garching, Germany
[3] Institut für Physikalische Chemie, Basel University, Basel, Switzerland
[4] Fachbereich Chemie, Universität Kaiserslautern, 67663 Kaiserslautern, Germany

By selective optical excitation of collision pairs and observation of the reemitted fluorescence information is obtained on the potentials and dynamic couplings governing the collision. Recent results on ab initio potentials and laser spectroscopy of Li*X (X=Ne,Ar) collision molecules (1),(2) motivated the study of the optical collision process

$$\text{Li}(2P) + X + h\nu \rightarrow \text{LiX}(3D\Lambda) \rightarrow \text{Li}(3P,3D) + X$$

Characteristic for this is the branching ratio $B := p(3D\Lambda \rightarrow 3P)/(3D\Lambda \rightarrow 3D)$ for dissociation of the LiX(3DΛ) molecule into the Li(3P,3D) atomic states, its measurement as function of laser detuning $\Delta\nu$ from the Li(2P-3D) transition.giving insight into the Li(3D→3P) excitation transfer mechanism. Experiments were performed using two step cw laser excitation of gaseous mixtures of Li+X at temperatures around 600K. and monitoring the excitation of the Li(3P,3D) states by the fluorescence at 610.4 nm and 323.3 nm (fig.1)

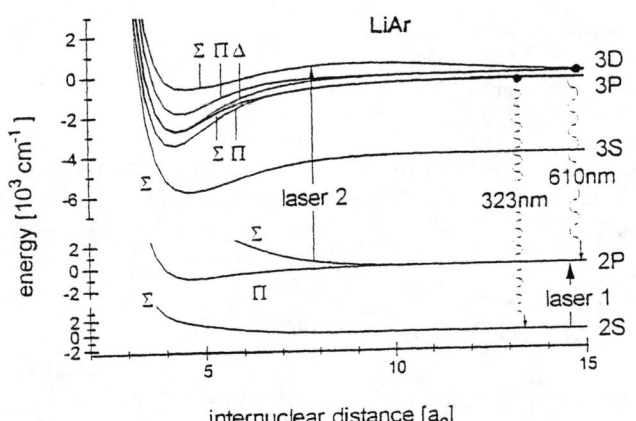

fig.1. Interaction potentials for selected LiAr molecular states from ab initio calculations after (2) and principle of experimental method.

For determination of the spectral profiles $B(\Delta\nu)$, the 3P/3D population ratio $R(\Delta\nu, p_X)$ has been measured in the detuning range $|\Delta\nu| \leq 100$ cm^{-1} of laser 2 at fixed pressures $p_{Ar}=50$ mbar, $p_{Ne}=20$ mbar. The effects of subsequent collisions Li(3D) + X following molecular excitation were eliminated, by additional measurements of the p_X dependence of $R(\Delta\nu,p_X)$ in the 2mbar to 100 mbar range at $\Delta\nu=0$.

The $B(\Delta\nu)$ profile obtained displays strong blue-red wing asymmetries both for Li*Ne (see fig.2) and Li*Ar. This reflects different dissociation probabilities from the 3DΣ or 3D(Π,Δ) states that are initially prepared by blue wing or red wing excitation, respectively.

For Li*Ne quantum coupled-channels calculations have been performed for the thermally averaged excitation transfer cross sections $\sigma(3D\Lambda^* \rightarrow 3P\Lambda)$. We have considered separately the cases of radial and angular nonadiabatic coupling. The Λ-averaged cross section $\sigma(3D-3P)$ agrees within a factor of 2 with experimental values. In addition preliminary results on the $B(\Delta\nu)$ profile are consistent with an enhanced 3P formation in the red wing.

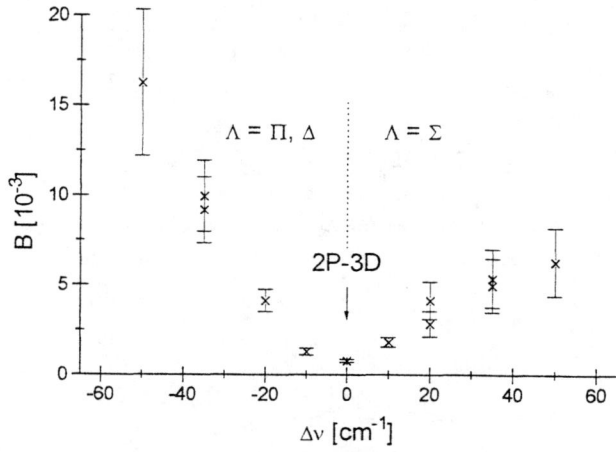

fig.2 Branching ratio for dissociation of LiNe(3DΛ) collision molecules into the Li(3P,3D) atomic states versus detuning from the Li(2P-3D) transition in the optical collision process Li(2P)+Ne+hν $\rightarrow$ LiNe(3DΛ) $\rightarrow$ Li(3P,3D)+Ne

REFERENCES

1. W. Behmenburg, A.Kaiser, F.Rebentrost, M.Jungen, M.Smit, M.Luo and G.Peach: J.Phys. B31(1998)689-708
2. W. Meyer, private commuinication

COLLISION-INDUCED SPECTRA

ATOMIC AND MOLECULAR

INTERACTIONS

Mixed vibrational and rotational excitations in liquid and solid para-hydrogen

Marco Zoppi[(*)], Lorenzo Ulivi[(*)], Massimo Moraldi[(+)], and Mario Santoro[(+)]

[(*)] Consiglio Nazionale delle Ricerche, Istituto di Elettronica Quantistica,
Via Panciatichi 56/30, I-50127 Firenze (Italy)
[zoppi@ieq.fi.cnr.it; ulivi@ieq.fi.cnr.it]
[(+)] Università di Firenze, Dipartimento di Fisica, and I.N.F.M., Sezione di Firenze,
Largo E. Fermi 2, I-50125 Firenze (Italy)
[moraldi@firenze.infn.it]

Abstract: Mixing of molecular states, in dense liquid and solid hydrogen, can give useful information on certain components of the microscopic molecular dynamics. However, since no quantum dynamic theory exists yet, the comparison can be exploited only to the level of the static quantities (integrated intensities). In this case a relevant role is played by the anisotropic components of the intermolecular potential. Since double transitions are induced by pair interactions, a purely pair theory can be applied, even in the condensed phases. The experiments were concentrated on the double rotational transition $S_0(0)+S_0(0)$ and the double rotovibrational transition $Q_1(0)+S_0(0)$, and were carried out as a function of density in order to test a relatively wide range of intermolecular distances.

INTRODUCTION

The light scattering cross section of dense hydrogen carries a wealth of information on the microscopic dynamics at the molecular level. In fact, beyond the Raman lines that are characteristic of the molecule (*intramolecular* modes), the light scattering spectrum is characterized by other interesting features that are sensitive to the *intermolecular* dynamics [1, 2]. For example, the depolarized wings of the elastic (Rayleigh) line are determined by the Collision Induced (CI) scattering and carry information on the correlation functions involving 2-, 3- and 4-particle interactions [3]. The wings of the allowed lines, in turn, are determined basically by the same phenomena, but now they involve only 2- and 3-particle correlations [4]. In addition, combination bands of the Raman lines have been observed too [5, 6]. However, differently from the low density case, where the simultaneous transitions happen on the same molecule [7], dense systems are characterized by combination bands where the *intramolecular* transitions are localized on two neighboring molecules. In this case, the dynamic process involves pair correlations only [8].

These effects are particularly interesting when the energy associated to the simultaneous transition, occurring on the same molecule, is different from that associated to the same transitions localized on distinct molecules [9]. In this case, it is easy to measure the spectral features separately. Therefore, the light scattering spectrum becomes a relatively clean tool for accessing the pair dynamics in a dense system.

There is an intrinsic importance in studying the microscopic dynamics of condensed hydrogen by means of light scattering. In fact, the microscopic probes that are usually employed to this aim, namely *X-rays synchrotron radiation*, and *neutrons*, fail in this case. The reasons are of different origin. From one side, the cross section for X-rays, which is an increasing function of the number of electrons, reaches its absolute minimum with the two electrons of molecular hydrogen and atomic helium [10]. From the other side, the ratio between the *coherent intermolecular* and the *incoherent intramolecular* neutron cross section of molecular hydrogen is extremely small, in general [11], and only a restricted region of energy can be used which is not sufficiently extended to probe satisfactorily the microscopic dynamics of the system [12]. For these reasons, not only the microscopic dynamics of dense hydrogen is still largely unknown, but also the microscopic structure factor has never been obtained with an accuracy comparable to that of other, less fundamental, biatomic molecular systems [13].

In recent years, we have carried out several experiments on condensed hydrogen using either light [8, 9, 14, 15] or neutron scattering [16-20]. In both cases, the interpretation of the experiments was possible because we could make use of a quantum mechanical simulation technique (Path Integral Monte Carlo, PIMC for short) that allows, given the pair interaction potential, the evaluation of the microscopic structure factor [21, 22]. No such possibility is allowed for dynamics quantities, yet. However, from the experimental point of view, the extended inelastic spectrum of hydrogen, which is obtained from light scattering, carries information that could be used, in principle, to access the dynamics of this quantum liquid.

In this report, we will show what kind of information can be obtained from the light scattering spectrum, and how we used them to perform some partial test on the anisotropic components of the interaction potential.

THE SCATTERING CROSS SECTION

Carrying out the investigation of a dense system using a weak probe, for example *visible photons* or *thermal neutrons*, is ruled by, basically, the same expression for the cross section [23]. For a monatomic, monoisotopic, system, the neutron scattering cross section is given by [24]:

$$\frac{d^2\sigma}{d\omega d\Omega} = N \frac{k_1}{k_0} \left[\langle b^2 \rangle S(\mathbf{k},\omega) \right], \quad (1)$$

where $S(\mathbf{k},\omega)$ is the dynamic structure factor,

$$\langle b^2 \rangle S(\mathbf{k},\omega) = \left(\frac{1}{2\pi}\right) \int_{-\infty}^{+\infty} dt \, \exp\{-i\omega t\} I(\mathbf{k},t). \quad (2)$$

and the neutron scattering intermediate function, $I_n(\mathbf{k},t)$, is defined as:

$$I_n(\mathbf{k},t) = \left(\frac{1}{N}\right) \sum_{i,j} \langle b_i^* b_j \exp[-i\mathbf{k} \cdot \mathbf{r}_i(0)] \exp[i\mathbf{k} \cdot \mathbf{r}_j(t)] \rangle. \quad (3)$$

Here, b_i is the scattering length, which depends on the structure of the target nucleus (neutrons are scattered by the nuclei) and $\mathbf{r}_j(t)$ stands for the position of the j-th nucleus at time t. The labels i and j run over the N particles that are contained in the scattering volume.

For light scattering, the intermediate scattering function becomes [23]:

$$I_l(\mathbf{k},t) = \left(\frac{1}{N}\right) \sum_{i,j} \langle [\hat{\mathbf{e}}_0 \cdot \mathbf{A}_i(0) \cdot \hat{\mathbf{e}}_1][\hat{\mathbf{e}}_0 \cdot \mathbf{A}_j(t) \cdot \hat{\mathbf{e}}_1] \exp\{i\mathbf{k} \cdot [\mathbf{r}_i(0) - \mathbf{r}_j(t)]\} \rangle \quad (4)$$

where $\hat{\mathbf{e}}_0$ and $\hat{\mathbf{e}}_1$ represent the polarization unit vectors of the incident and scattered electric field, respectively, and $\mathbf{A}_j(t)$ is the polarizability tensor of the j-th molecule at time t. Here, $\mathbf{r}_j(t)$ stands for the polarization center of the molecule that, in the following, will be assumed to coincide with the molecular center of mass.

The basic difference between neutron and light scattering resides in the fact that the neutron scattering process is localized on the nucleus, whose size is negligible with respect to the microscopic distances that are of interest here, while the light scattering process is distributed all over the molecular polarizability whose size is not too dissimilar from the intermolecular distances. In addition, the polarizability of a molecule is influenced by the presence of the neighbors, so that the scattering event is not intrinsically local. This is certainly a disadvantage for light scattering. However, this is also the reason why, in certain experimental conditions, one can *access the microscopic molecular dynamics using a probe whose wavelength is three orders of magnitude larger than the intermolecular distances of interest.*

The polarizability tensor of the *i-th* molecule, at time *t*, can be expressed as a sum of cluster terms [1, 23]:

$$\mathbf{A}_i(t) = \tilde{\alpha}^{(1)}(i;t) + \sum_{j \neq i} \tilde{\alpha}^{(2)}(i,j;t) + \sum_{(j \neq i),(k \neq i,j)} \tilde{\alpha}^{(3)}(i,j,k;t) + \cdots \quad (5)$$

where, $\tilde{\alpha}^{(1)}(i,t)$ is the polarizability tensor, at time *t*, of the *i-th* isolated molecule which depends only on the internal (*intramolecular*) normal coordinates. The next term, $\tilde{\alpha}^{(2)}(i,j,t)$, is the *pair* irreducible polarizability. This depends both on the *intramolecular* coordinates and the *intermolecular* distance r_{ij}. The following term, $\tilde{\alpha}^{(3)}(i,j,k,t)$, is the *triplet* irreducible polarizability term, and so on.

Eq. 4 can be developed, using Eq. 5, and one obtains a cluster expansion of the intermediate light-scattering function. This can be written, schematically:

$$I_l(\mathbf{k},t) = I^{(1,1)}(\mathbf{k},t) + I^{(1,2)}(\mathbf{k},t) + I^{(2,2)}(\mathbf{k},t) + \cdots \quad (6)$$

where $I^{(1,1)}(\mathbf{k},t)$ is the intermediate scattering function produced by the single molecule polarizability only (*single-single term*). Analogously, $I^{(1,2)}(\mathbf{k},t)$ is produced by the combination of a single molecule polarizability with the irreducible polarizability of a pair *(single-pair term)*, and $I^{(2,2)}(\mathbf{k},t)$ comes from the combination of two pair polarizability terms. The expansion outlined in Eq. 6 could be extended indefinitely, but we should bear in mind that the magnitude of the following terms decays very fast, so that, in general, only the first non vanishing term is important, in each spectral region that is analyzed. For example, if we consider a polarized scattering configuration ($\hat{\mathbf{e}}_0 \equiv \hat{\mathbf{e}}_1$) then the important term, in the elastic scattering region, is the *single-single* component. However, this term decays very fast as we move away from the Rayleigh line and, in the spectral region of the Rayleigh wings, the leading term becomes the *pair-pair* contribution, which is almost fully depolarized.

We report, in Fig. 1, the depolarized Raman spectrum of liquid parahydrogen, at p=17 bar and T=28 K, in the region between 0 and 800 cm^{-1}. The elastic peak has been removed, because of the spurious scattering that is always present, and only the Raman structures are visible. The most prominent feature is represented by the intramolecular rotational transition $S_0(0)$ at 352.2 cm^{-1}. At higher frequency (586 cm^{-1}) we observe the weaker rotational line $S_0(1)$ that is associated with the small residual fraction (0.5%) of ortho-hydrogen in the sample. The sharp line at 270 cm^{-1} is the $S_0(0)$ rotational line of the HD molecule, which is present as an impurity at natural concentrations in the sample.

The remaining features of the spectrum are of intermolecular origin. The region between a few cm^{-1} and ~300 cm^{-1} is dominated by the Collision Induced Light Scattering (CILS) spectrum. The observed features are similar to the depolarized

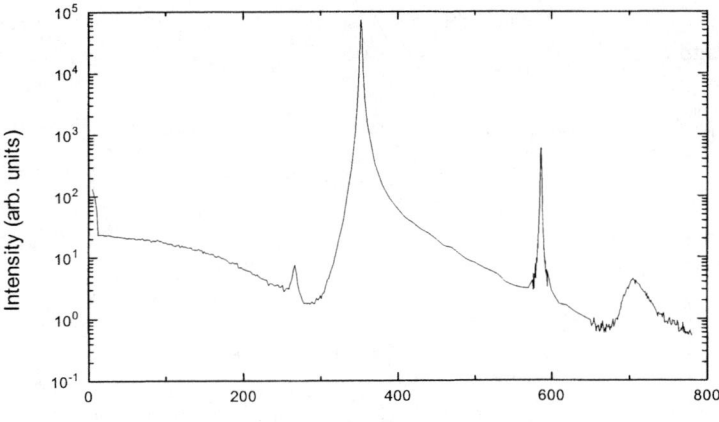

Fig. 1: Depolarized Raman spectrum of almost pure (99.5%) liquid para-hydrogen.

spectra of the condensed monatomic liquids made by the rare gases [25]. These spectral shapes are generated by exactly the same mechanism, i.e. the CI polarizability whose leading term is given by the Dipole Induced-Dipole (DID) mechanism [1]. The exponentially decaying spectral region that is observed between the intramolecular Raman line $S_0(0)$ and ~650 cm^{-1} is also partially produced by CILS phenomena. However, the leading term, here, is the pressure broadening which dominates the near wing spectrum of the intramolecular line [26]. Finally, in the region around 700 cm^{-1} we observe a broad band that is produced by the simultaneous rotational transition on two interacting molecules [5, 8, 14].

In Fig. 2 we report the rotovibrational spectrum of solid parahydrogen at p=247 bar and T=18.8 K. The most prominent feature, here, is represented by the rotovibrational transition $S_1(0)$ centered at 4483 cm^{-1}. The side peak and shoulders that are observed on this band are due to the splitting of the rotational line (lifting of the m-degeneracy) due to the anisotropic interaction [27]. The broad band, centered at ~4505 cm^{-1}, is the double transition $Q_1(0)+S_0(0)$. The origin of this band is due to the energy-leakage of the $S_1(0)$ line due to the anisotropic interaction with neigboring molecules. In fact, once a molecule has reached the (v=1, J=2) excited state, the rotational excitation can migrate on the neighboring molecules because of the presence of the anisotropic interaction potential (strictly speaking, the time consequentiality is not a necessary condition) [15]. Again, this is an effect that depends on pair correlations only.

Even in this case, there is the theoretical possibility of a simultaneous transition on two separate molecules due to the CI polarizability which produces a CILS spectrum. However, in this case, the cross section for this transition is about four orders of magnitude smaller than the other one [15].

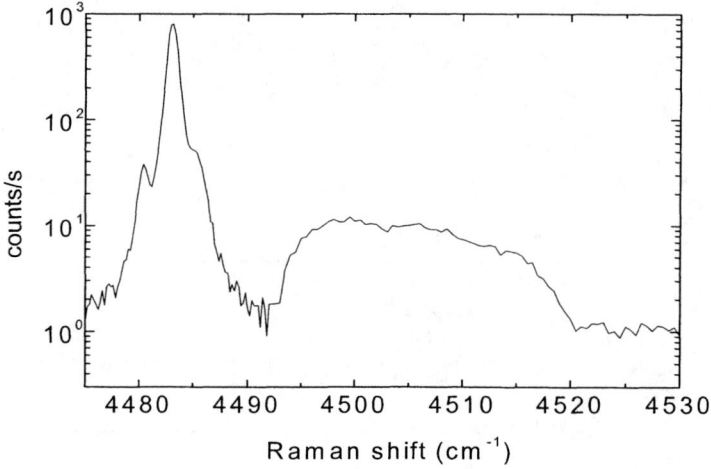

Fig. 2: The rotovibrational spectrum of solid parahydrogen. The main peak on the left is the rovibrational transition $S_1(0)$. The band on the right is the double transition $Q_1(0)+S_0(0)$

THE DYNAMIC INFORMATION

First, it is important to distinguish between the two cases of polarized and depolarized scattering. For the polarized case (quasi-elastic scattering), one can show that the leading term comes from the permanent trace component of the single isolated molecule and that, for all practical purposes, the CI contributions (pair and higher order polarizability terms) enter only as a rinormalization correction of the polarizability, and add nothing to the dynamic information [28]. Therefore, the explicit expression for polarized quasi-elastic scattering reduces to:

$$I_{l,pol}(\mathbf{k},t) = \left(\frac{(LFF)\alpha_0^2}{N} \right) \sum_{i,j} \langle \exp[-i\mathbf{k} \cdot \mathbf{r}_i(0)] \exp[i\mathbf{k} \cdot \mathbf{r}_j(t)] \rangle. \quad (7)$$

where (LFF) is a Local Field Factor, which depends on the pair induced polarizabilities, that renormalizes the single molecule trace component polarizability α_0. This term is equivalent to the neutron scattering cross section, Eq. 3. However, in this case, the momentum transfer $\mathbf{k}$ is the one characteristic of the light scattering, of the order of 10^5 cm^{-1}, i.e. probing the dynamics of the system in the hydrodynamic region.

If the depolarized spectrum is considered ($\hat{e}_0 \cdot \hat{e}_1 \equiv 0$) it is easy to show that the exponential phase factor disappears [23] and the scattering function becomes:

$$I_{l,xy} = \sum_{n,m} I_{xy}^{(n,m)}(t) = \frac{1}{N} \sum_{n,m} \sum_{i_n,j_m} \left\langle \alpha_{x,y}^{(n)}(i_n,0) \alpha_{x,y}^{(m)}(j_m,t) \right\rangle \tag{8}$$

where x and y represent the polarization unit vector directions in the laboratory reference frame, and n and m can assume the value 1 (single molecule polarizability) or 2 (pair polarizability). It is self evident that the *single-single* term ($n=1$, $m=1$) contains at most pair correlations, the *single-double* term ($n=1$, $m=2$) contains at most triplet correlations, and the *double-double* term ($n=2$, $m=2$) contains, at most, correlations of quadruplets. However, some of the correlations may vanish because of the particular symmetry or due to the presence of a random phase factor introduced by an *intramolecular* transition.

If no internal transition is excited (Rayleigh wings, CILS scattering) then the *double-double* term becomes the leading factor in the scattering function and the spectrum is composed by the combination of three different terms, namely a pair, triplet, and quadruplet correlation spectrum, depending on the choice of the four indices i_1, i_2, and j_1, j_2, in Eq. 8.

If one molecule experiences an intramolecular transition, then, in principle, all terms in Eq. 8 contribute to the scattering. There is a *single-single* term, that carries most of the intensity in the peak region. There is the *single-double* term, which produces, typically, an asymmetry of the intramolecular lines. There is, finally, the *double-double* term which contributes mostly to the wings. However, in this case, no 4-body correlation is allowed, because the molecular transition introduces a random phase factor and therefore the quadruplet correlation term vanishes identically.

If a simultaneous rotational transition occurs on two distinct molecules, also the triplet contribution vanishes and only the pair term contributes to the scattering. The observed band at ~700 cm^{-1} originates partially from this mechanism and therefore carries information on the pair dynamics of the liquid. There are, indeed, two distinct scattering mechanisms that can produce a spectral shape in the region of the $2S_0(0)$ band. One is a *single-single* term, which originates from the mixing of the rotational states of two interacting molecules induced by the anisotropic interaction. The second mechanism is a pure CILS term with $i_1=j_1$, and $i_2=j_2$. In any case, both scattering functions are driven by the dynamic correlation of the pairs in the liquid.

To conclude this section we want to show some preliminary results, that illustrate the concepts expressed above, and were obtained from the depolarized spectrum of Fig. 1. In practice, the three different portion of the spectrum, i.e. the lineshape of the Rayleigh wings, the far wings of the intramolecular $S_0(0)$ line, and the lineshape of the $2S_0(0)$ band, have been reported, normalized to 1, as a function of the shift relative to the line peak. This result is shown in Fig. 3.

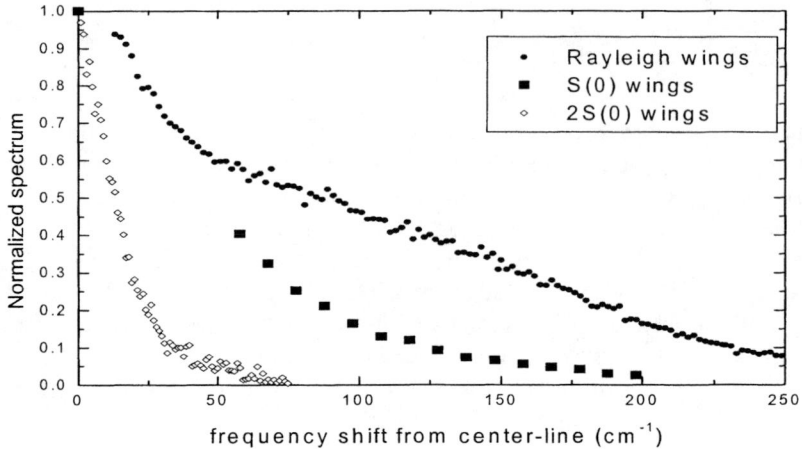

Fig. 3: Normalized Interaction-Induced spectra of molecular para-hydrogen. The full circles represent the lineshape of the total CILS spectrum (Rayleigh wings) which contains correlations involving up to four molecules. The squares are obtained from the far wings of the intramolecular $S_0(0)$ line, containing correlations up to a maximum of three molecules. The open diamonds represent the lineshape of the $2S_0(0)$ band which contains only the correlation of pairs.

Even though the results cannot be considered more than qualitative, it is apparent that the three lineshapes represent three distinct dynamic regimes in the liquid. In fact, the Rayleigh wing CILS lineshape is produced by the combination of the 2-, 3-, and 4-body correlations. The spectral shape is the most extended of the three and suggests correlations that show a rather fast decay in the time domain. The spectral shape of the $2S_0(0)$ band is produced by pair correlations only. This is the most rapidly decaying lineshape and suggests that pair correlations in the liquid are relatively long lived. Finally, in the mid-range, we find the far wings of the $S_0(0)$ line. Here, in order to decrease the effect of the line broadening produced by the *single-single* and the *single-double* contribution to the scattering function, we have cut out the first ~50 cm^{-1} of the spectral lineshape. The remaining portion, reported in the figure, is believed to be representative of the liquid dynamics including 2- and 3-molecule correlations.

What we have shown here is not new. A molecular dynamics computer simulation of the various 2-, 3-, and 4-body correlation functions in liquid argon had been reported long ago by Ladd, Montrose, and Litovitz [29]. It was shown that, because of the strong cancellation between the various contributions, the resulting total CILS time-correlation function was expected to decay much faster than each individual term. No experiment has ever been able to show directly these features. The interesting fact, inherent to Fig. 3, is that now, for the first time, such features have been observed experimentally in liquid hydrogen.

INTERPRETATION OF THE COMBINATION SPECTRA

Even though the Raman spectra of hydrogen are very rich, in terms of dynamic information, not much can be obtained at present, due to the lack of a theoretical framework for the dynamics of a quantum system. However, useful information can still be gained by analizing the combination spectra as a function of the density. In fact, due to the relatively high compressibility of hydrogen, a rather extended range of densities can be explored, without exceeding reasonable pressure limits. In this way, different intermolecular potential models can be extensively tested. In the following, we will concentrate on the interpretation of the combination bands, the $2S_0(0)$ band that has been measured in the liquid phase, and the rotovibrational band $Q_1(0)+S_0(0)$ that has been measured in the solid phase. Here, we will give a brief description on how the calculations are carried out. For a more detailed account we direct the reader to the original literature [8, 9, 14, 15].

The intensity of the $2S_0(0)$ band is produced by three contributions. There is a CILS terms, $I^{(2,2)}(t)$, with the main contribution coming from the pair induced polarizabilities. There is a MIX term, $I^{(1,1)}(t)$, where the leading contribution is given by the single molecule polarizability and the mixing of the rotational states induced by the anisotropic components of the interaction potential. Finally, there is a CROSS term, $I^{(1,2)}(t)$, produced by the interference of the other two [14]. The various terms can be evaluated theoretically using a perturbation theory whose expansion parameter is V_{anis}/E_0, where V_{anis} is the anisotropic component of the pair intermolecular potential and E_0 is the energy of the intramolecular transition J=0→2. It turns out that, to the lowest order, the MIX term is quadratic in the expansion parameter, the CROSS term is linear, and the CILS term is independent of it [14].

By assuming that a pure quadrupole-quadrupole interaction determines the anisotropic potential, V_{anis}, the MIX term is determined by the radial average of a quantity proportional to R^{-10}. This is a rather short-ranged function and turns out to be heavily affected by quantum effects. To the same approximation level, the CILS term, instead, is determined by the radial average of the pure DID contribution, i.e. R^{-6}, that is more long-ranged and therefore less affected by the quantum behavior of hydrogen. The two averages have been evaluated, using PIMC simulation, in the density range of the experiment, namely between 18.73

and 30.89 nm^{-3}. To the same level of approximation, symmetry reasons impose the vanishing of the CROSS term. Therefore, only the MIX and the CILS terms contribute to the $2S_0(0)$ band [14]. The two contributions to the scattering cross section turn out of similar size. However, it should be pointed out that, in the liquid phase, the density behavior that is obtained using a simple rigid lattice model gives incorrect results. In fact, while one would expect a $n^{10/3}$ power law for the MIX term, we find a best fitting function $n^{1.73}$. Analogously, the CILS term shoud go as n^2, while we find a best fitting function $n^{1.43}$. Morever, there no agreement between the experimental data and the calculated values, in the whole density range, with a difference that is an order of magnitude larger than the experimental error bars [14].

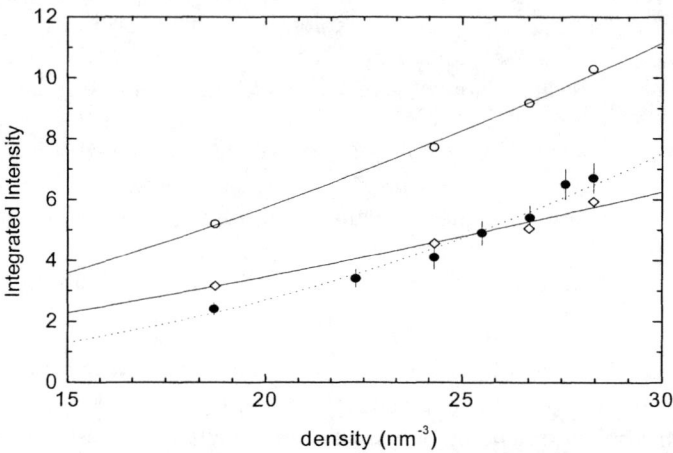

Fig. 4: Density behavior of the $2S_0(0)$ band integrated intensity. The dots with error bars are the experimental points. The open circles represent the 0-th order approximation of the thery. The diamonds are the results of an improved theoretical model (see text). The lines represent power-law fits to the data.

The theoretical model can be improved, however. This is done by extending the accuracy on the models for the anisotropic intermolecular potential and the pair CI polarizability. This task was undertaken using the hydrogen pair potential given by Norman, Watts and Buck [30] and the results of an *ab-initio* calculation for the pair polarizability by Bounds [31, 32]. The following comparison (cf. Fig. 4) is quantitatively more appealing, even though not fully satisfactory. In fact, even though the experimental data cross the theoretical function, the overall density behavior of the calculation appears different from that of the experiment. We attribute the remaining discrepancy to the still poor determination of the CI polarizability model and to a possible change in the effective size of the molecule in the explored density range.

For the second combination band, namely the rotovibrational $Q_1(0)+S_0(0)$ band, the MIX term gives most of the contribution while the CILS term appears to be negligible [9]. Again, a perturbation theory extended up to the fourth order in the expansion parameter, V_{anis}/E_0, has been developed. However, differently from the case of the $2S_0(0)$ band, here the anisotropic component of the intermolecular potential is generated by the matrix element between the ground and the first excited vibrational state. For a simple quadrupolar interaction, the relevant anisotropic potential components are:

$$V_{224}(R) = \frac{1}{3}\sqrt{\frac{14}{5}}\frac{Q_0^2}{R^5}, \tag{9}$$

for the ground vibrational state interaction [$2S_0(0)$ band], and

$$\bar{V}_{224}(R) = \frac{1}{3}\sqrt{\frac{14}{5}}\frac{Q_0 Q_1}{R^5}, \tag{10}$$

for the first excited vibrational state interaction [$Q_1(0)+S_0(0)$ band].

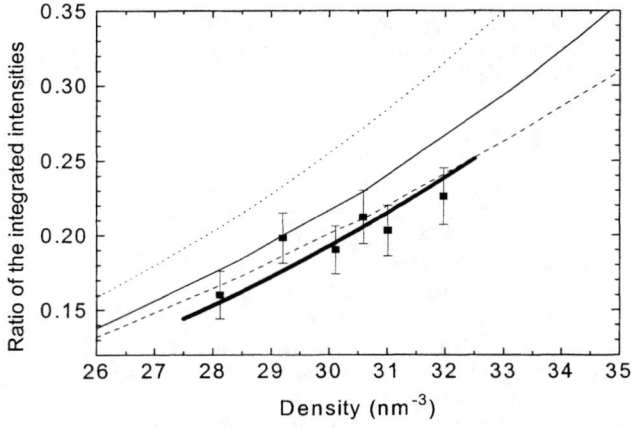

Fig. 5: Density behavior of the $Q_1(0)+S_0(0)$ band integrated intensity relative to that of $S_1(0)$. The dots with error bars are the experimental points. The dotted line represents the theory limited to 2nd order, the dashed line includes the 3rd order, and the continuous line includes the 4th order. The thick solid line represents a fitting to the data with the power law $n^{10/3}$.

In the solid phase, where this experiment was carried out, the density behavior of the combination band seems to follow the expected power law $n^{10/3}$. In fact, there is still a slight remaining systematic discrepancy between the experimental data and the results of the theoretical calculations. However, while in the previous case we could use an improved model for the anisotropic component $V_{224}(R)$ of the intermolecular potential (cf. Eq. 9), no such possibility exists, yet, for improving the model of $\bar{V}_{224}(R)$ (cf. Eq. 10). If we consider that the pure quadrupole-quadrupole interaction overestimates by some 5% the true behavior of $V_{224}(R)$ in the range of distances of interest here, namely around $6.5a_0$, we could assume that a similar correction could affect $\bar{V}_{224}(R)$. If this were the case, the numerical results would become quantitatively consistent with the experimental results of Fig. 5

REFERENCES

[1] G. Birnbaum, ed. *Phenomena Induced by Intermolecular Interactions* (Plenum Press, New York, 1985)
[2] G.C. Tabisz and M.N. Neumann, ed. *Collision and Interaction-Induced Spectroscopy* (Kluwer Academic Publishers, Dordrecht, NL, 1995)
[3] B.J. Alder, J.J. Weis, and H.L. Strauss, Phys. Rev. A **7**, 281 (1973)
[4] Y. LeDuff and A. Gharbi, Phys. Rev. A **17**, 1729 (1978)
[5] P.J. Berkout and I.F. Silvera, Comments Phys. **2**, 109 (1977)
[6] S.B. Baliga, R. Sooryakumar, K.N. Rao, R.H. Tipping, and J.D. Poll, Phys. Rev. B **35**, 9766 (1987)
[7] W. Holzer and R. Ouillon, Chem. Phys. Lett. **24**, 589 (1974)
[8] M. Zoppi, L. Ulivi, M. Santoro, M. Moraldi, and F. Barocchi, Phys. Rev. A **53**, 1935 (1996)
[9] F. Barocchi, A. Guasti, M. Zoppi, J.D. Poll, R.H. Tipping, Phys. Rev. B **37**, 8377 (1988)
[10] J.D. Jackson *Classical Electrodynamics* (Wiley, New York, 1975)
[11] J.A. Young and J.U. Koppel, Phys. Rev. **135**, A603 (1964)
[12] K. Carneiro, M. Nielsen and J.P. McTague, Phys. Rev. Lett. **30**, 481 (1973)
[13] C. Andreani, J.C. Dore, and F.P. Ricci, Rep. Prog. Phys. **54**, 731 (1991)
[14] F. Barocchi, M. Moraldi, M. Santoro, L. Ulivi, and M. Zoppi, Phys. Rev. B **55**, 12223 (1997)
[15] M. Moraldi, M. Santoro, L. Ulivi, and M. Zoppi, Phys. Rev. B **57**, 0000 (1998)
[16] M. Zoppi, U. Bafile, R. Magli, A.K. Soper, Phys. Rev. E **48**, 1000 (1993)
[17] M. Zoppi, U. Bafile, E. Guarini, F. Barocchi, R. Magli, M. Neumann, Phys. Rev. Lett. **75**, 1779 (1995)
[18] U. Bafile, M. Zoppi, M. Celli, R. Magli, A.C. Evans, J. Mayers, Physica B **217**, 50 (1996)
[19] M. Zoppi, A.K. Soper, R. Magli, F. Barocchi, U. Bafile, and N.W. Ashcroft, Phys. Rev. E **54**, 2773 (1996)
[20] U. Bafile, M. Celli, M. Zoppi, and J. Mayers, Phys. Rev. B **58**, 0000 (1998)
[21] M. Zoppi and M. Neumann, Phys. Rev. B **43**, 10242 (1991)
[22] M. Neumann and M. Zoppi, Phys. Rev. A **44**, 2474 (1991)
[23] U. Balucani and M. Zoppi, *Dynamics of the Liquid State* (Clarendon, Oxford, 1994)
[24] S.W. Lowesey, *Theory of neutron scattering from condensed matter* (Clarendon, Oxford, 1987)
[25] U. Bafile, L. Ulivi, M. Zoppi, and F. Barocchi, Phys. Rev. A **37**, 4133 (1988)
[26] A. Borysow and M. Moraldi, Phys. Rev. A **40**, 1251 (1989)
[27] J. Van Kranendonk, *Solid Hydrogen* (Plenum, New York, 1983)
[28] M. Zoppi, Unpublished
[29] A.J.C. Ladd, T.A.Litovitz, and C.J. Montrose, J. Chem. Phys. **71**, 4242 (1979)
[30] M.J. Norman, R.O. Watts, and U. Buck, J. Chem. Phys. **81**, 3500 (1984)
[31] D.G. Bounds, Mol. Phys. **38**, 2099 (1976)
[32] L. Ulivi and M. Zoppi, Mol. Phys. **90**, 971 (1997)

Interaction properties of Hg probed by Collision-Induced Raman scattering

A.Bonechi [a] and M.Moraldi [a,b]

[a] Dipartimento di Fisica, Università di Firenze ; Largo E.Fermi 2; 50125 Firenze, Italy
[b] Istituto Nazionale di fisica della Materia, Unità di Firenze;
Largo E.Fermi 2; 50125 Firenze, Italy

This paper is devoted to the study of the Collision-Induced (CI) Raman lineshapes of mercury in the density range in which only pair collisions are effective. Particularly we will be concerned with the relation between CI Raman scattering and both the pair polarizability and interatomic potential. It will be shown that the available experimental results can be reproduced if refined models for both the pair polarizability and potential are used. Particularly overlap effects (damping) on the Dipole-Induced-Dipole (DID) contributions are to be taken into account.

Introduction

CI Raman scattering has been widely used since its first observation [1] with the aim of understanding interatomic properties like CI polarizabilities. Indeed the CI Raman spectrum depends on the CI polarizability and interatomic potential, the pair polarizability being responsible for the interaction of the system with the radiation field and the interatomic potential for the dynamics of the interacting pair of atoms. Usually the measured spectra are compared with numerical calculations of lineshapes which use interatomic potential models as input data. Such a procedure has been successfully used for rare gases for which the interatomic potential could be derived from independent measurements. The pair polarizability was then obtained by comparing measured and numerically evaluated spectral lineshapes [2]. The applicability of such a procedure rests on the availability of a dependable potential, though. Moreover, a reasonable model for the pair polarizability is to be assumed. The model should depend on few parameters that can be determined by searching for a best agreement of calculated and measured CI Raman spectra. The polarizability model which has been extensively used for rare gases and which has given successful results [2,3] is composed of a long range, DID contribution plus a short range term which decays exponentially. The DID term can be derived by means of classical methods and in this case an analytical expression can be obtained as a function of interatomic distance (R) and atomic polarizability (α_0) [2]. It is known, however, that quantum (correlation) corrections to the long range contribution of polarizability can be important [4]. Quantum corrections are effective starting on second order DID terms, that is those contributions that depend on $1/R^6$. It is thus more appropriate to write the DID polarizability as a power series of $1/R^3$ and use the quantum corrected terms. Usually it is sufficient to retain only terms up to $1/R^6$. As far as the short range, exponential term, is concerned it is usually said that it should contain all overlap effects, that is all those effects that are related to the overlap of electronic charges of the two interacting atoms. In the end we write for the trace (α) and anisotropy (β) of the irreducible part of the pair polarizability

$$\alpha = \frac{C}{R^6} + B_1 \exp(-R/R_1)$$
$$\beta = \frac{6\alpha_0^2}{R^3} + \frac{D}{R^6} + B_2 \exp(-R/R_2) \tag{1}$$

with $C = 4\alpha_0^3$ and $D = 6\alpha_0^3$ in the classical limit. Corrections to C and D coefficients involving the dispersion coefficient C_6 of the potential and the atomic second hyperpolarizability γ can be estimated according to Ref.[4]

$$C = 4\alpha_0^3 + \frac{5\gamma C_6}{9\alpha_0}$$
$$D = 6\alpha_0^3 + \frac{\gamma C_6}{3\alpha_0} \tag{2}$$

However, as it will be shown in the following, the numerical evaluation of the Raman spectral lineshape according to the previous equations for the pair polarizability and the available models for the interatomic potential are not in agreement with the experimental results. This fact calls for a reexamination of the models used for both the pair polarizability and interatomic potential. It will be shown that the main problem with the previous models is related to the DID approximation for the long range part of the pair polarizability.

Preliminary numerical calculations

At first we have performed numerical evaluations of the CI spectra of mercury at the temperatures for which experimental results are available [5-7] with potentials that can be found in the literature and using the pair polarizability model described by Eq.(1).

The calculations have been performed with a Fortran code which has been widely used in the past to calculate collision induced spectral lineshapes [2]. The isotropic spectrum can be reproduced reasonably well. On the contrary the comparison of the calculated and measured spectral lineshapes shows deviations larger than experimental errors. In Fig.1 we report the depolarized spectral lineshape of mercury at 1073 K: the measured one [7] and that calculated with β from Eq.(1) and an interatomic potential derived from fluorescence data [8]. From Fig.1 it is seen that the calculated spectra show an excess of intensity at large frequency shift and a defect in the region of small frequency shift. Other calculations using different potentials (Morse [9] and 6-12 Lennard-Jones [10]) have been performed. They are not reported here but they show the same features of Fig.1.

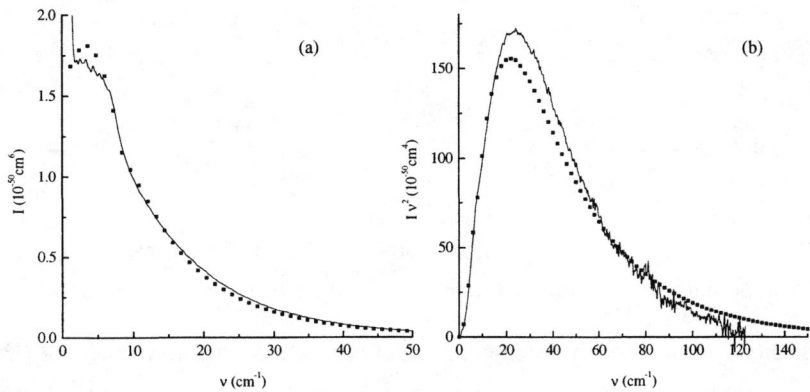

Fig.1: a) Depolarized Raman intensity. Continuos line: experimental [7]. Squares: calculated with model Eq. (1) for polarizability anisotropy and with potential model of Ref. [8]. b) Depolarized Raman intensity $\times$ frequency^2

As a first step in order to reconcile calculations and experiments we have thus considered how dependable the available potentials and what their effect on the spectral lineshapes are.

Interatomic properties of mercury have been obtained from measurements of dimeric states energies [8,9,11,12]. From those measurements we can obtain information on the properties of the potential near its minimum, particularly on the curvature, unharmonicity and the well depth. From such information a Morse [9] and a n-6 Lennard-Jones [8] potential have been derived. Such potentials are dependable only as long as the values of the potential in the well depth are important. Also, low density viscosity measurements at various temperatures are available [13]. Such measurements add some information on the value of σ, that is the interatomic distance for which the potential vanishes. Viscosity results have been reproduced by means of a 6-12 Lennard-Jones potential [10] which, however, results in a well depth which is almost twice as large as the well depth derived from dimeric state energies. Finally ab-initio calculations have been performed [14,15] which point out, among other things, the necessity to include a (damped) dispersive contribution at large interatomic distances.

A dependable potential $v(R)$ should be able to reproduce all the previous facts. Here we propose a potential model which consists of a short range and a damped dispersive contribution [16]

$$v(R) = A\exp(-bR) - \sum_{n=3}^{5} \frac{C_{2n}}{R^{2n}} f_{2n}(bR) \tag{3}$$

with [17]

$$f_{2n}(x) = 1 - \exp(-x) \sum_{k=0}^{2n} \frac{x^k}{k!} \tag{4}$$

It has been shown [16] that with appropriate values for A, b, C_6, C_8, C_{10}, the potential model of Eq.(3) is consistent with dimeric state energies and viscosity data.

The calculations of the Raman spectra by means of this new potential (Eq.(3)) and the pair polarizability model of Eq.(1) is not able to reproduce the experimental results, though. A comparison of calculated and experimental linshapes is reported in Fig. 2. It is seen that the excess intensity at large frequency shifts cannot be eliminated even though the new, more dependable potential model is used.

The next section will be devoted to a revision of the DID model employed in Eq.1.

The irreducible pair polarizability

As already explained in the introduction, the DID approximation is a useful model for the long range contribution to the pair polarizability. On the other hand, when interatomic distance vanishes it presents an unphysical singularity. Also at distances comparable with R_n (n=1,2) of Eq.1, the DID term is a sizeable contribution. On the other hand R_n represents a distance at which overlap effects are important and thus at such a distance the DID results cannot be considered valid. The situation reminds of a similar problem for the interatomic potential.

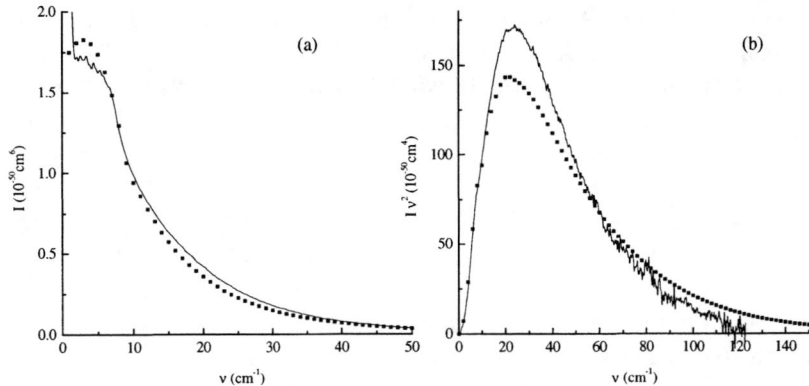

Fig.2: a) Depolarized Raman intensity. Continuos line: experimental [7]. Squares: calculated with model Eq. (1) for polarizability anisotropy and with potential model of Eq. (3). b) Depolarized Raman intensity $\times$ frequency^2

Potentials are often separated in two contributions, a short range one, usually exponential-like and a dispersive series. In such a case the dispersive series is damped at short interatomic distances [18] in order to avoid unphysical contributions due to the asymptotic divergence of the series. More or less arbitrary damping functions are used in order to suppress the large contribution of the dispersive series in the overlapping region. The most physical method [17] is based on the Drude model [19] for the interacting atoms.

Here we will use a similar method in order to study the pair polarizability. According to the model that we will describe next, each of the interacting atoms is composed of a positive localized charge (Q) and a negative one (-Q) distributed all over the space. The negative charge is assumed to have spherical symmetry and is mostly contained in a volume whose linear dimension (say $1/b$) is comparable with the distance at which overlap effects become important. The distribution of negative charge is assumed to be invariant and is bounded to the positive charge by a harmonic force. The force constant (k) of the Drude atom is related to the charge and atomic polarizability by the relation

$$k = \frac{Q^2}{\alpha_0}$$

When the two Drude atoms approach each other, the negative charges experience electrostatic (Coulomb) forces. Such forces couple the two harmonic oscillators and the coupling results in a modification of the polarizability of the pair. Also exchange forces contribute to the coupling. However, in our model we will retain only Coulomb forces and thus the effect of exchange forces on the pair polarizability is to be added to our results.

The potential energy operator of two interacting Drude atoms is written as:

$$V = \frac{Q^2}{R} + \frac{Q^2}{R_{12}} - \frac{Q^2}{R_{2A}} - \frac{Q^2}{R_{1B}} + V_{at} \tag{5}$$

where R is the distance between the two nuclei (that we have named A and B), R_{12} is the distance between the two electrons (1,2) and R_{2A} and R_{1B} are the distances of each electron with respect to the nucleus to which it is not bounded. Finally V_{at} is the potential energy of the isolated atoms

$$V_{at} = \frac{1}{2} k \left(X_{1A}^2 + X_{1B}^2 \right)$$

where we have defined by X the distance of the electron distribution center with respect to the nucleus to which it is bound.

When the atoms are very far apart they behave as 6 independent oscillators of force constant k. Their ground state energy equals $6\hbar\omega_0/2$, where ω_0 is the angular frequency

$$\omega_0 = \sqrt{k/\mu}$$

When R is much larger than $1/b$, the potential energy can be written as

$$V = V_{at} + \frac{Q^2}{R} + \frac{Q^2}{X_{12}} - \frac{Q^2}{X_{2A}} - \frac{Q^2}{X_{1B}}$$

where the generic positions of the electrons have been replaced by the positions of the centers of their charge distribution. All the X's can be written as functions of the vectors X_{1A}, X_{2B} and R. Finally the expression for V is expanded in powers of the components of X's divided by R and the expansion is truncated to quadratic terms. The result consists of 6 coupled oscillators for the 6 components of the vectors X_{1A}, X_{2B}. The oscillators can be decoupled by transforming to the new coordinates

$$R_+ = X_{1A} + X_{2B}$$
$$R_- = X_{1A} - X_{2B}$$

The six components of R_+ and R_- describe 6 independent harmonic motions that can be derived from the energy equation

$$H = \frac{(\mu/2)}{2}\left(\frac{d\mathbf{R}_+}{dt}\right)^2 + \frac{1}{2}\left(\frac{k}{2} + \frac{Q^2}{2R^3}\right)(X_+^2 + Y_+^2) + \frac{1}{2}\left(\frac{k}{2} - \frac{Q^2}{R^3}\right)Z_+^2 +$$

$$+ \frac{(\mu/2)}{2}\left(\frac{d\mathbf{R}_-}{dt}\right)^2 + \frac{1}{2}\left(\frac{k}{2} - \frac{Q^2}{2R^3}\right)(X_-^2 + Y_-^2) + \frac{1}{2}\left(\frac{k}{2} + \frac{Q^2}{R^3}\right)Z_-^2$$

where we have indicated by X_α, Y_α and Z_α (with $\alpha = +/-$) the cartesian components of the vectors $\mathbf{R}_\alpha$'s and have placed the z axis along the internuclear separation vector.
We now have 6 independent oscillators with force constants depending on R. The ground state energy of the previous hamiltonian differs from the energy of the unperturbed atoms by the quantity

$$\Delta E = -\hbar\omega_0 \frac{3}{4}\left(\frac{Q}{k}\right)^2 / R^6$$

which represents the Van der Waals energy. The new force constants can also be related to the polarizability of the pair of atoms. It is clear that only the force constants for the $\mathbf{X}_+$ motions contribute to the polarizability, the $\mathbf{X}_-$ motions leaving unaltered the electric dipole of the system. In the end we obtain

$$\alpha_{zz} = \frac{2\alpha_0}{1 - 2\alpha_0/R^3}$$

$$\alpha_{xx} = \alpha_{yy} = \frac{2\alpha_0}{1 + \alpha_0/R^3}$$

where α_{zz} and α_{xx}, α_{yy} are the polarizability components in the molecular intrinsic reference frame. The previous equation represents the DID result. The DID model is correct at large internuclear distances when the polarizability is proportional to α_0^2/R^3. Corrections to the DID polarizability are conveniently applied once the DID result is expanded in powers of α_0/R^3 [4]. For such a reason we write the anisotropy and trace of pair polarizability as

$$\beta = \alpha_{zz} - \alpha_{xx} = \frac{6\alpha_0^2}{R^3} + \frac{6\alpha_0^3}{R^6} + \ldots$$

$$\alpha = \frac{\alpha_{zz} + \alpha_{xx} + \alpha_{yy}}{3} = \frac{4\alpha_0^3}{R^6} + \ldots$$

We now inquire about the effect of the overlapping of electronic charge distributions. In this case we assume for the electronic charge distribution at position **R** an exponential function $\rho(\mathbf{R})$

$$\rho(\mathbf{R}) = Q \frac{b^3}{8\pi} \exp(-b|\mathbf{R} - \mathbf{X}|)$$

The potential energy V of Eq.(2) is now to be averaged over the distributions of the electronic charges 1 and 2

$$V(R, \mathbf{X}_{1A}, \mathbf{X}_{2B}) = V_{at} + \frac{Q^2}{R} + \int d\mathbf{R}_1 d\mathbf{R}_2 \, \rho(\mathbf{R}_1)\rho(\mathbf{R}_2) \frac{Q^2}{R_{12}} + \\ - \int d\mathbf{R}_1 \, \rho(\mathbf{R}_1) \frac{Q^2}{R_{1B}} - \int d\mathbf{R}_2 \, \rho(\mathbf{R}_2) \frac{Q^2}{R_{2A}} \quad (6)$$

V is a function of $R, \mathbf{X}_{1A}$ and $\mathbf{X}_{2B}$. It can thus be written in terms of $R, \mathbf{R}_+$ and $\mathbf{R}_-$. We notice that the electron-electron interaction (the third term in the r.h.s. of Eq.(6)) does not depend on $\mathbf{R}_+$ and thus it does not contribute to polarizability. All the other terms of Eq.(6) can be evaluated analytically. Finally we find the force constants for the $\mathbf{R}_+$ motion and obtain for anisotropy polarizability

$$\beta(R) = \frac{6\alpha_0^2}{R^3} f_3(bR) + \dots$$

with

$$f_n(x) = 1 - \exp(-x) \sum_{k=0}^{n} \frac{x^k}{k!}$$

It is interesting to observe that the same functions $f_n(bR)$ have been derived recently as damping functions for the dispersive series of the interatomic potential [17]. Damping functions can be obtained also for higher order DID terms. It is to remind however that higher order DID terms are affected by quantum effects that are not considered in the present model. In view of such a fact and because second order DID terms are not crucial for the determination of the Raman spectral lineshapes, we have preferred to use f_6 as damping functions for $1/R^6$ terms, that is the same damping function that is used to damp the van der Waals term in the interatomic potential [17]. Finally we write the pair polarizability as:

$$\alpha = \frac{C}{R^6} f_6(bR) + B_1 \exp(-R/R_1)$$
$$\beta = \frac{6\alpha_0^2}{R^3} f_3(bR) + \frac{D}{R^6} f_6(bR) + B_2 \exp(-R/R_2)$$
(7)

It should be reminded that overlap effects on Coulomb interactions have been taken into account by means of the damping functions f_n. Thus, according to the Drude model that we have used here, the exponential terms of Eqs.(7) should contain only exchange force effects.

RESULTS

Both the depolarized Raman spectral lineshape of mercury at 1073 K and the isotropic one at 973 K have been calculated by means of the interatomic potential of Ref.[16] and of the polarizability model of Eq.(7). In Fig.3 we report the comparison of the experimental [5-7] and calculated lineshapes for the depolarized spectrum.

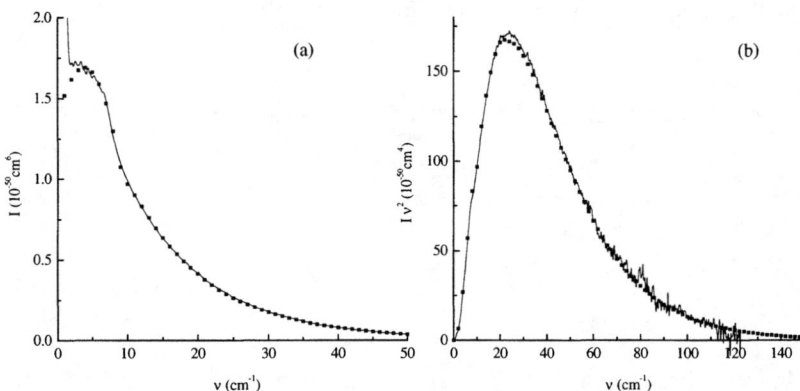

Fig. 3: a) Depolarized Raman intensity. Continuos line: experimental [7]. Squares: calculated with model Eq.(7) for polarizability anisotropy and with potential model of Eq.(3). b) Depolarized Raman intensity × frequency^2

It is now seen that the measured spectra can be reproduced within the experimental errors in the whole measured spectral range. A similar agreement is found also for the lineshape of the isotropic spectrum. We would like to stress the fact that such a result has been obtained by using a polarizability model that contains only two parameters (B_n and R_n) for both anisotropy and trace, because the quantity b

appearing in the damping functions has been assumed equal to the inverse of the range of the repulsive part of the interatomic potential ($1/b = 0.88 a.u.$ [16]).

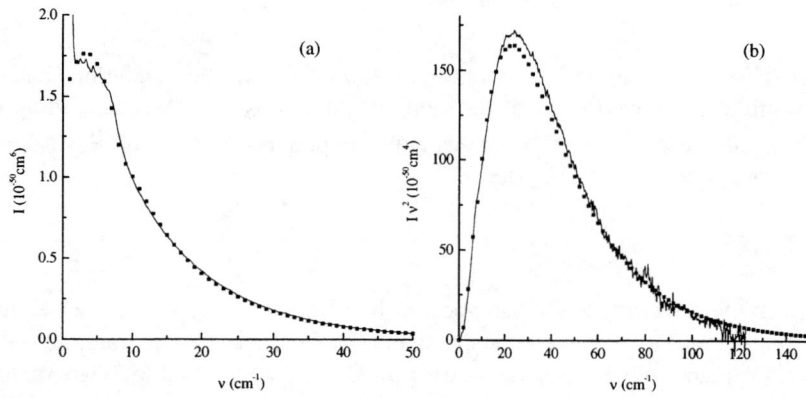

Fig. 4: a) Depolarized Raman intensity. Continuos line: experimental [7]. Squares: calculated with model Eq. (7) for polarizability anisotropy and with potential model of Ref.[8]. b) Depolarized Raman intensity $\times$ frequency^2

The values of B_n and R_n for which we find agreement are

$B_1 = 273.5\ a.u.$
$R_1 = 1.145\ a.u.$
$B_2 = 762.0\ a.u.$
$R_2 = 1.488\ a.u.$

First we observe that both R_n's are of the same order of magnitude and their values are not very different from $1/b$. That is reasonable because all those distances are related to the distance at which overlap effects become important. Moreover, the values for B_n are both positive. We remind that the exponential contributions to the polarizability of rare gases have always been found negative. Few words are necessary in order to explain such a difference which is not necessarily related to an anomalous behavior of mercury. The model that we use accounts only for linear effects in the electrostatic interaction. In reality in our equations we take account also of the second hyperpolarizability but only as far as dynamic correlation effects are concerned (dispersive-like contributions) [4]. There are also effects in the electrostatic contribution which are due to hyperpolarized electronic charge distributions [20]. Such effects are expected to be important in mercury due to the large hyperpolarizability and it has been shown that they give a positive contribution for non-bonded systems [20]. In the end we should say that the positive exponential contribution that we have derived represents both exchange and hyperpolarization effects.

Finally, in Fig. 4, we report the comparison of the experimental depolarized spectrum with a calculation that uses the model of Eq. (7) for the polarizability and the n-6 potential derived in [8]. As it is seen the agreement deteriorates though not as much as when undamped polarizabilities are used. A similar feature is observed when other potentials are used [9,10].

CONCLUSIONS

In conclusion we can say that the depolarized Raman spectrum of mercury can be reproduced only if a model for the collision induced polarizability is used that eliminates the divergence at zero internuclear separation of the classical DID contribution. In other words the DID term has to be damped at short internuclear separation. The damping we have used is derived directly from a simple Drude model of the interacting pair of mercury atoms. To our knowledge this is the first case for which the analysis of the experimental spectra shows the evidence of the damping of long range contributions to the polarizability due to overlap effects. In fact, in the case of noble gases the measured spectra could be reproduced by employing the classical (undamped) DID terms plus an exponential contribution [2,3]. The reason why in mercury a damping factor is needed is probably due both to the large atomic polarizability (as opposed to rare gases atomic polarizability) and because of the relatively high temperatures involved in the experimental determination of the CI Raman spectrum of mercury.

Moreover, in order to reproduce the Raman spectrum of mercury also the interatomic potential has to be chosen with care. In particular we have shown that either Morse or Lennard-Jones potentials are not accurate enough to explain the experimental results.

REFERENCES

[1] J.P. McTague and G. Birnbaum: Phys.Rev.Lett. **21**, 661 (1968)
[2] L. Frommhold, in Adv. Chem. Phys. **46**, I. Prigogine and S.Rice Eds. (John Wiley and Sons, Inc:, New York, 1981), p.1. See also an update: Can. J. Phys. **59**,. 1459 (1981)
[3] U. Bafile, R. Magli, F. Barocchi, M. Zoppi and L. Frommhold: Mol. Phys. **49**, 1149 (1983);
F. Barocchi, M. Zoppi, U. Bafile, R. Magli: Chem Phys. Lett. **95**, 135 (1983); N. Meinander, G.C. Tabisz, M. Zoppi: J. Chem. Phys. **84**, 3005 (1986)
[4] A.D. Buckingham: Trans. Faraday Soc. **52**, 1035 (1956); A.D. Buckingham and K.L. Clarke: Chem. Phys. Lett. **57**, 321 (1978)
[5] M. Sampoli *et al.*: Europhys. Lett. **28**, 483 (1994)
[6] M. Sampoli, F. Hensel and F. Barocchi: Phys. Rev. A **53**, 4594 (1996)
[7] A. Bonechi *et al.*: Phys. Rev. A **57**, 2635 (1998)
[8] J. Koperski, J.B. Atkinson and L. Krause: J. Mol. Spectrosc. **184**, 300 (1997)

[9] J. Koperski, J.B. Atkinson and L.Krause: Chem. Phys. Lett. **219,** 161 (1994)
[10] L.F. Epstein and M.D. Powers: J. Phys. Chem. **57**, 336 (1953)
[11] A. Zehnacker *et al.*: J. Chem. Phys. **86,** 6565 (1987)
[12] R.D. van Zee *et al.*: J. Chem. Phys. **88**, 4650 (1988)
[13] H. Braune, R. Basch and W. Wentzel: Z. Physik. Chem. **A137**, 447 (1928)
[14] W.E. Baylis: J. Phys. B **10**, L 583 (1977)
[15] C.F. Kunz, C. Hättig and B.A. Hess: Mol. Phys. **89**, 139 (1996)
[16] A. Bonechi *et al.*: submitted to J. Chem. Phys.
[17] K.T. Tang and J.P. Toennies: J. Chem. Phys. **66**, 1496 (1977); K.T. Tang and J.P. Toennies: J. Chem. Phys. **80**, 3726 (1984)
[18] G.C Maitland, M. Rigby, E.B. Smith and W.A. Wakeham, *Intermolecular Forces - Their Origin and determination* (Clarendon Press, Oxford, 1981)
[19] P.K.L. Drude, *The theory of Optics* (Longmans, Green, London, 1933)
[20] K.L. Clarke, P.A. Madden and A.D. Buckingham: Mol. Phys. **36**, 301 (1978); D.W. Oxtoby: J. Chem. Phys. **69**, 1184 (1978)

NEW FORMULATION FOR FAR-WING LINE SHAPES: APPLICATION TO CO_2 AND H_2O

Q. Ma[†] and R. H. Tipping[‡]

[†]Department of Applied Physics, Columbia University, and Institute for Space Studies, Goddard Space Flight Center, New York, NY 10027
[‡]Department of Physics and Astronomy, University of Alabama, Tuscaloosa, AL 35487

A new theoretical formalism for the calculation of the far-wing line shape within the binary-collision, quasistatic framework has been developed.[1,2] In this formalism, the eigenfunctions of the orientation, not the state functions, are chosen as the complete set of basis functions. The sums over initial and final states are transformed into multidimensional integrals over the orientational variables. Using this theory with a realistic interaction potential having an isotropic Lennard-Jones part and an anisotropic part consisting of the leading, long-range multipolar attractive term and a short-range, site-site repulsive term, we have carried out calculations for the ν_3 band of CO_2. We first verified that the integrations were converged by using two different resolutions; these results are shown in Fig. 1 where the solid curve has a resolution approximately 42 times that of the dashed curve. We then optimized the only two free parameters in the theory, the magnitude and the range of the site-site potential, so as to obtain good agreement with experimental data at 296 K. Then using the same potential, we carried out calculations for the far-wing line shape over a wide range of temperatures and the results are shown in Fig. 2. Using these line shapes, we calculated the absorption coefficient and compare the theoretical results with experimental data in Figs. 3 and 4. The + and ▲ are the data from Refs. 3 and 4, respectively; the dashed curves are the results obtained assuming a Lorentzian line shape. Using the new method, we have calculated similar results for the rotational band of H_2O for both self- and N_2 broadening. These results are compared with the experimental data of Burch et al.[5] in Figs. 5 and 6, respectively. From these comparisons, one can conclude that using the present theory with a realistic potential model, one can obtain accurate far-wing line shapes and absorption coefficients for a variety of systems that are in good agreement with experimental measurements. Work on other spectral regions and for other molecular systems is currently in progress.

REFERENCES:

1. Q. Ma and R. H. Tipping, J. Chem. Phys. **108**, 3386 (1998).
2. Q. Ma, R. H. Tipping, C. Boulet, and J.-P. Bouanich, Appl. Optics, submitted for publication.
3. R. Le Doucen, C. Cousin, C. Boulet, and A. Henry, Appl. Opt, **24**, 897 (1985).
4. J.-M. Hartmann and M.-Y. Perrin, Appl. Opt, **28**, 2550 (1989).
5. D. E. Burch and D. A. Gryvnak, AFGL-TR-0054 (1979); D. E. Burch, SPIE Proc. **277**, 28 (1981); D. E. Burch and R. L. Alt, AFGL-TR-84-0128 (1984); D. E. Burch, AFGL-TR-85-0036 (1985).

Figure captions

Fig. 1. The theoretical far-wing line shape for CO_2-CO_2 (in units of cm^{-1} atm^{-1}) as a function of frequency ω (in units of cm^{-1}) for T = 296 K calculated with two different resolutions for the multidimensional integrations; the solid curve has a resolution approximately 42 times greater than that of the dashed curve.

Fig. 2. The theoretical far-wing line shape for CO_2-CO_2 (in units of cm^{-1} atm^{-1}) as a function of frequency ω (in units of cm^{-1}) for a range of temperatures T.

Fig. 3. Comparison of the calculated absorption coefficient $\alpha(\omega)$ for CO_2-CO_2 (in units of cm^{-1} $amagat^{-1}$) with experimental data + from Ref. 3, and with results (dashed line) obtained using a Lorentzian line shape. (A) T = 218 K; (B) T = 296 K.

Fig. 4. Comparison of the calculated absorption coefficient $\alpha(\omega)$ for CO_2-CO_2 (in units of cm^{-1} $amagat^{-1}$) with experimental data ▲ from Ref. 4, and with results (dashed line) obtained using a Lorentzian line shape. (A) T = 291 K; (B) T = 414 K; (C) T = 534 K; (D) T = 627 K; (E) T = 751 K.

Fig. 5. Theoretical far-wing calculations for H_2O-H_2O as a function of frequency ω (in units of cm^{-1}): (A) The line shape (in units of cm^{-1} atm^{-1}) for T = 296 K; (B) Comparison of the calculated absorption coefficient $\alpha(\omega)$ (in units of cm^{-1} $amagat^{-1}$) denoted by ▲ with experimental data + from Ref. 5.

Fig. 6. Theoretical far-wing calculations for H_2O-N_2 as a function of frequency ω (in units of cm^{-1}): (A) The line shape (in units of cm^{-1} atm^{-1}) for T = 296 K; (B) Comparison of the calculated absorption coefficient $\alpha(\omega)$ (in units of cm^{-1} $amagat^{-1}$) denoted by ▲ with experimental data + from Ref. 5.

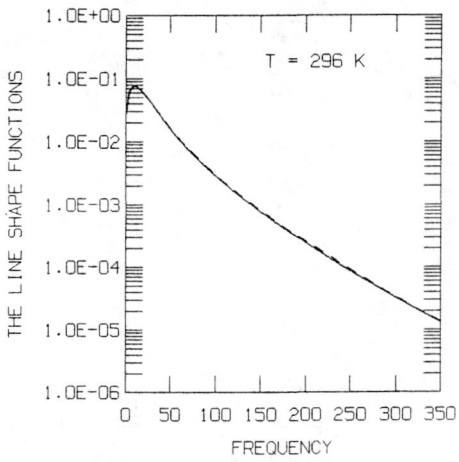

FIG. 1

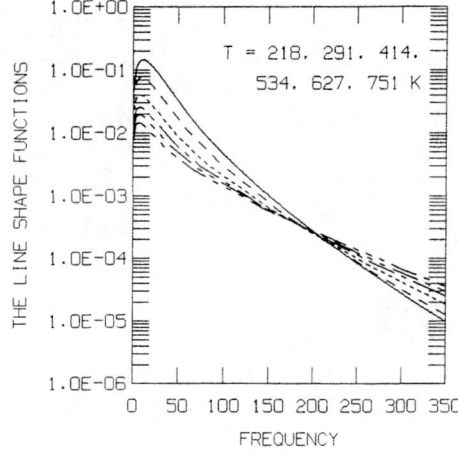

FIG. 2

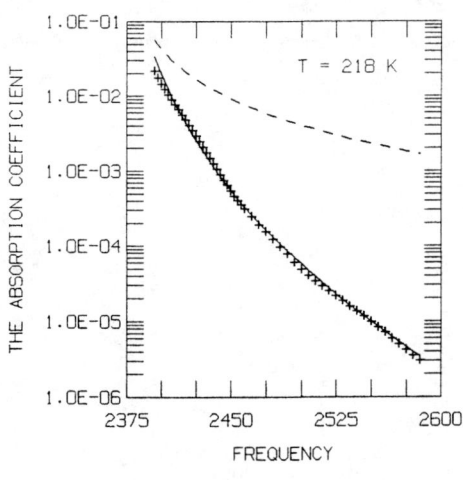

FIG. 3 (A)

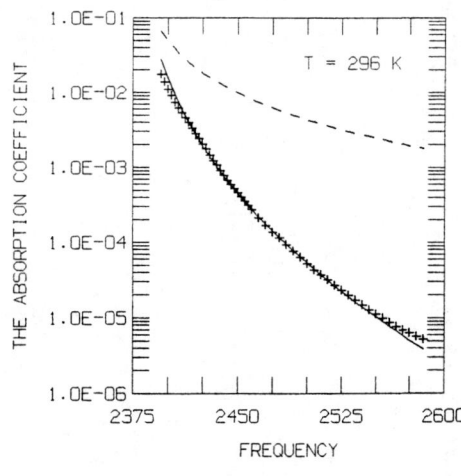

FIG. 3 (B)

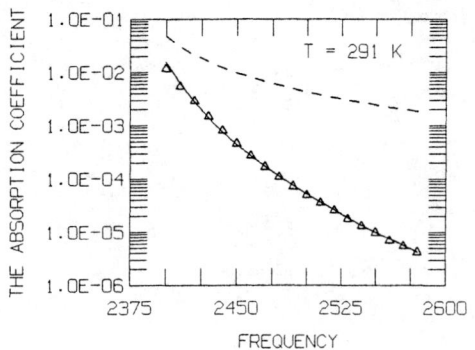

FIG. 4 (A)

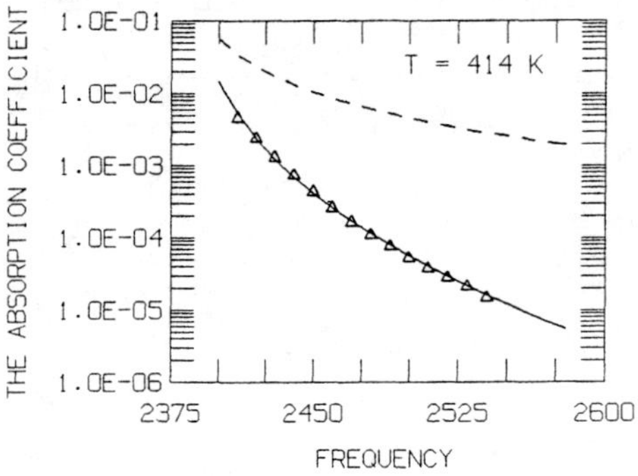

FIG. 4 (B)

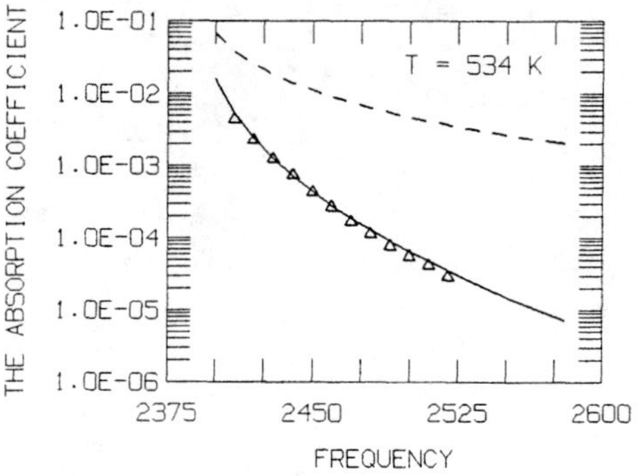

FIG. 4 (C)

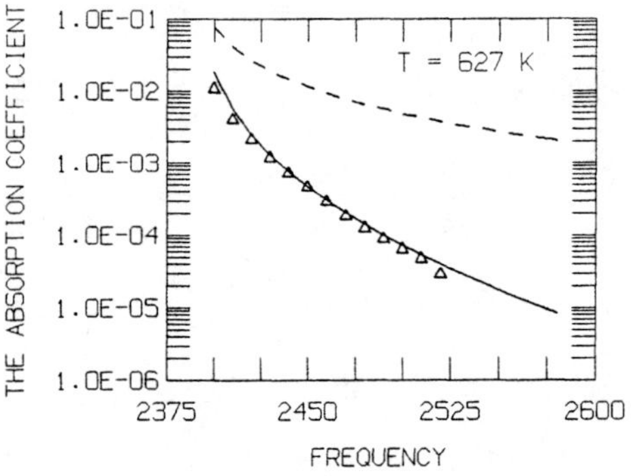

FIG. 4 (D)

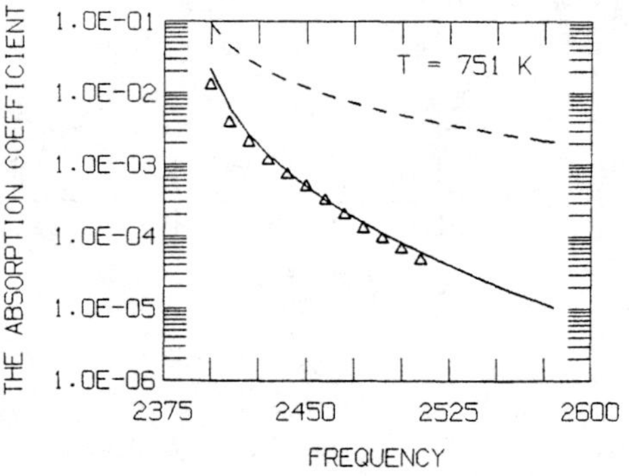

FIG. 4 (E)

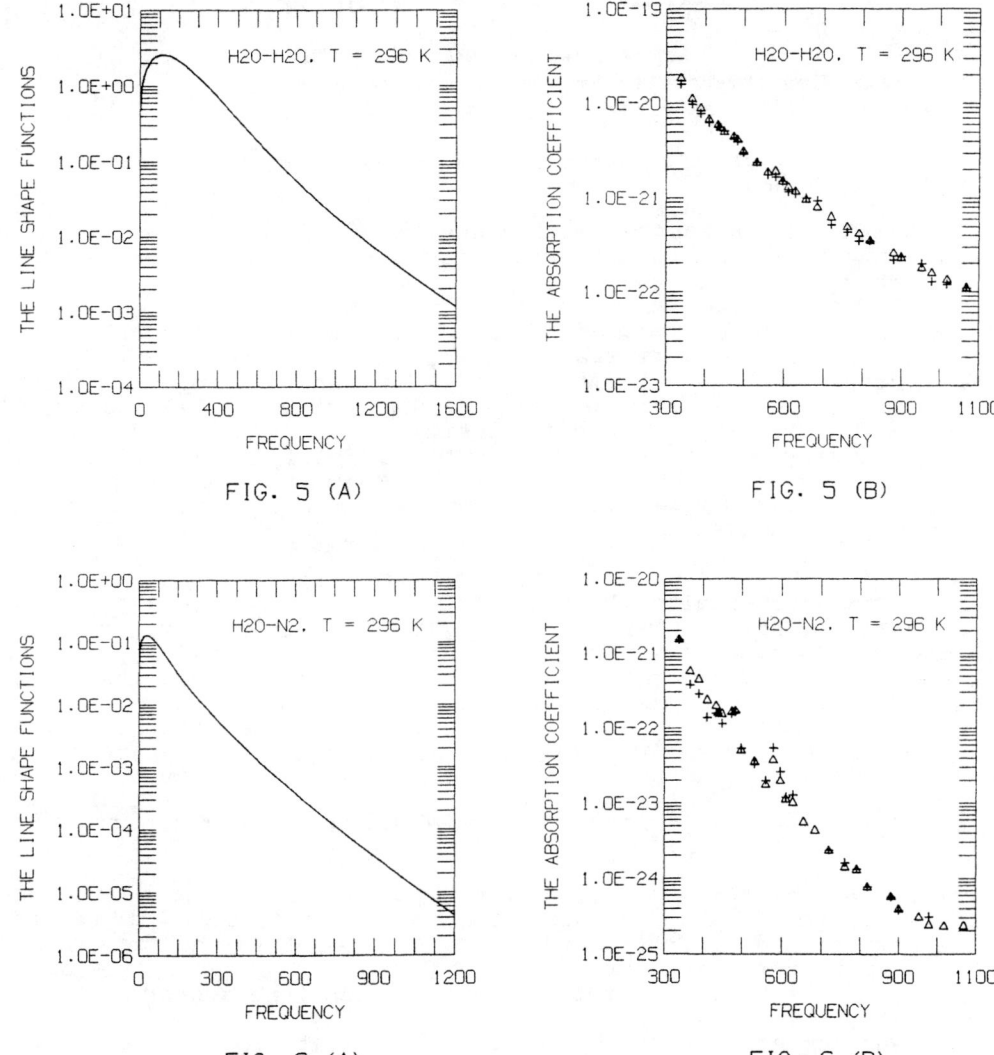

FIG. 5 (A)

FIG. 5 (B)

FIG. 6 (A)

FIG. 6 (B)

Interatomic Potentials of Cd - Kr from far-wing line profiles

G. D. Roston and T. Grycuk*

Physics Department, Faculty of Science, Alexandria University
P.O.Box 21511 Alexandria, Egypt

* Warsaw University, Institute of experimental Physics, 00-680 Warsaw, Poland

ABSTRACT: The intercombination absorption line of the Cd (326.1 nm) line broadened by pressure effects of Kr in the spectral rang extending from 800 cm^{-1} in the blue wing to 1200 cm^{-1} in the red wing was studied using a high resolution spectrometer. The temperature dependence of the studied line was analysed in the framework of the quasi-static theory. This analysis allowed us to determine the ground ($X^1 0^+$), and the excited ($^3 1$, $^3 0^+$) state potentials at the internuclear separations from 3.2 Å to 6.3 Å. The well depths with their positions for these states are respectively equal to (134±7 cm^{-1}, 3.95±0.2 Å), (72.3±4 cm^{-1}, 4.95±0.3 Å) and (471±12 cm^{-1}, 3.6 Å). These values are in close agreement with those obtained for this mixture (Cd - Kr) from other experimental methods.

1. THEORETICAL

1.1: The quasi-static theory.

The unified Frank-Condon (UFC) treatment of pressure broadening of spectral lines developed by Szudy and Baylis [1] reduces in the appropriate limits to the quasi-static theory in the far wings of the line. Using this theory, the reduced absorption coefficient (cm^3) in the line wings as a consequence of two particle interactions when normalized to the perturbing (N_p) and the radiating (N_r) atom densities is given by

$$K_m^n(\Delta v, T) = \frac{K^n(\Delta v, T)\,(cm^{-1})}{N_p N_r} = 4\pi C_m R^2 \left|\frac{dR}{d(\Delta V_m)}\right| \exp\frac{-V_1(R)}{kT} \quad (1)$$

where the symbol m = 0 and 1 are assigned to the $^3 0$ and $^3 1$ excited states respectively, R is the internuclear separation, $\Delta V_m(R)$ is the potential difference between the excited $V_m(R)$ and the ground $V_1(R)$ state potentials, given by ($\Delta V_m(R) = V_m(R) - V_1(R) = h\Delta v$) and Δv is the frequency separation from the line center. C_m is a constant.

The temperature dependence of the profile [Eq.(1)] is simply related to the boltzmann factor $\exp(-V_1/kT)$. By plotting the relation between $\ln(K_n^m(\Delta v, T))$ and (1/T) at a fixed value of Δv, we obtain a straight line whose slope is equal to V_1. Repeating this procedure for different values of Δv, the function $V_1(\Delta v)$ can be derived. The intersections of the straight lines with the ordinate axis give $\ln(K_n^m(\Delta v, T_\infty))$.

1.2: Determination of the excited state potential curves.

The integration of Eq.(1) at T ---> ∞ gives

$$\int_{\Delta v_a}^{\Delta v} K_m^n(\Delta v, T_\infty)\, d\Delta v = \frac{4\pi C_m}{h}\int_{R_a}^{R} R^2\, dR = \frac{4\pi C_m}{3h}(R^3 - R_a^3) \quad (2)$$

where R_a is the interatomic separation at $\Delta v = \Delta v_a$. The function $\Delta v(R)$ is obtained using the procedure given in [2-4].

2. EXPERIMENTAL

The spectral absorption profiles of the Cd-Kr system were measured with a high resolution spectrometer described elsewhere [2]. A high pressure XBO 150 Xenon lamp was used as a background source for absorption. The contributions affecting the spectrometer instrumental function were reduced in view of the double-beam method being used. For this purpose two identical absorption cells, 5 cm long and 3 cm in diameter, made of quartz were placed in a special oven which can be heated to temperatures ut to 1300 °K. As this work was done under the condition of unsaturated vapour the sample cell was filled with a Cd-Kr mixture and the reference one was empty. The contribution due to cd-cd interaction had been reduced by independent measurements taken by the authors [2] as was done in [3,4]. The vapour pressures were calculated using Bousquet's formula [5] and the atomic densities of Kr N_{kr} and cadmium N_{cd} were, respectively, equal to $(2.015 \pm 0.07) \times 10^{19}$ cm^{-3} and $(3.62 \pm 0.05) \times 10^{18}$ cm^{-3}.

3. RESULTS AND DISCUSSIONS

3.1: profile of Cd line 326.1 nm:

Fig. (1) shows the blue wing for both the Cd-Kr and the Cd-Cd system at the same density of cadmium (3.62×10^{18} cm^{-3}). It is seen from this figure that the effect of the Cd-Kr interaction extends only to $\Delta v = 140$ cm^{-1} from the line center. It is also seen that at $\Delta v = 55$ cm^{-1}, there is a broad shoulder which is known as the classical blue satellite. This satellite is attributed to the extremum of the potential difference $\Delta V_1(R)$, between the ground ($X^1 0^+$) and the excited ($A^3 1$) states.

The red wing profile of the Cd 326.1 nm line broadened by Kr, compared with the self broadened line profile obtained from [2], in logarithmic scale is illustrated in Fig. (2). In this figure one can see a single satellite at about 8 cm^{-1}. In the vicinity of this band two linear regions with slope ($-3/2$) are observed for the (Cd-Kr) system. The first one close to the line center may be interpreted as the quasi-static line wing which is formed by transitions between the ground state $^1 0^+$ and excited states $^3 0^+$ as well as $^3 1$, while the second corresponds to the transition to the excited $^3 0^+$ state only. The van der Waals coefficients ΔC_6^0 and ΔC_6^1 originating due to these states are respectively 37.8 ± 2 and 58.5 ± 3 (ev Å^6).

3.2 Interaction potential for the states X $^1 0^+$ and $^3 1$.

The temperature dependence of the blue wing profile of the Cd-Kr system in the interval of temperature from 960 to 1280 K is illustrated in Fig (3), which show that the maximum temperature dependence lies at about $\Delta v = 57$ cm^{-1}. This gives the position of the well depth for the ground state potential. Fig. (4) shows the linear dependence between the normalized absorption coefficient (cm^5) in logarithmic scale versus 1000/T. The slopes of these lines give the ground state potential $V_1(\Delta v)$. This is shown in Fig.(5). The well depth of this state is thus determined. After fitting the experimental results by the least square method, the well depth was estimated to be equal to 134 ± 7 cm^{-1}. To obtain the ground state potential $V_1(R)$ the relation

$\Delta v(R)$ must be obtained, as was established in the theoretical part (sec 1.2). The position of the ground state well depth R_a was taken from [6]. As the relation $\Delta v(R)$ is obtained then the ground and excited states potentials $V_g(R)$ and $V_1(R)$ for the Cd-Kr system can be obtained.

3.3. Interaction potential for the state $^3 0^+$.

As the temperature dependence of this wing is very weak, then, the difference potential $\Delta V_o(R)$ can be obtained using the ground state potential obtained in this work from the blue wing and the red wing profile data at any temperature T. the starting point R_a (Δv_a) in the integration (2) was obtained due to the fact that Kuhn's law is observed in the near red wing at $15 < \Delta v < 90$ cm^{-1} and the van der Waals potential difference $\Delta V_a = \Delta C_6^° R_a^{-6}$ can be applied. As $\Delta V_o(R)$ is obtained and $V_1(R)$ is known, then $V_o(R)$ for the excited state $^3 0^+$ is determined. Fig. (6) shows the ground ($X^1 0^+$) and the two excited ($^3 1$ and $^3 0^+$) states potentials. The parameters are illustrated in table I.

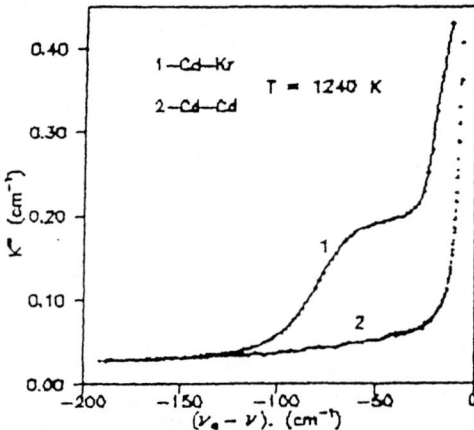

Fig.1. The blue wing for the Cd-Kr and Cd-Cd.

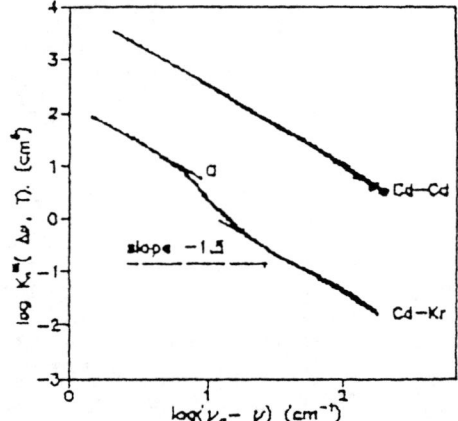

Fig. 2. Red wing profiles for the Cd-Kr and Cd-Cd in logarithmic scale (10^{-40} cm^5). The Cd-Kr profile is divided by 100.

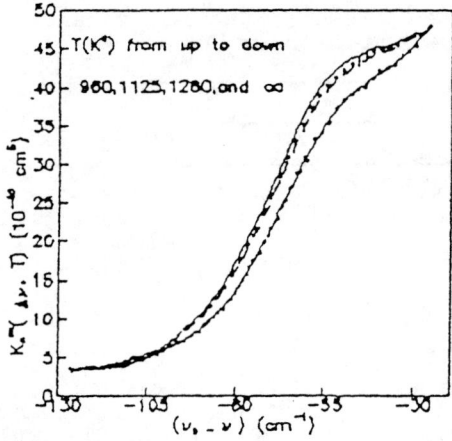

Fig.3. The temperature dependence of the blue wing of the Cd 326.1 nm line for Cd-Kr system.

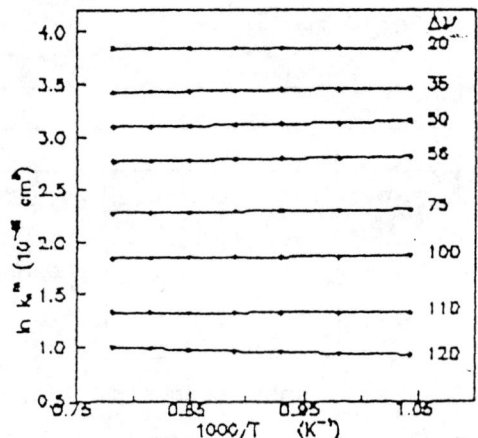

Fig. 4. Reduced absorption coefficient (10^{-40} cm^5) in the logarithmic scale against 1000/T.

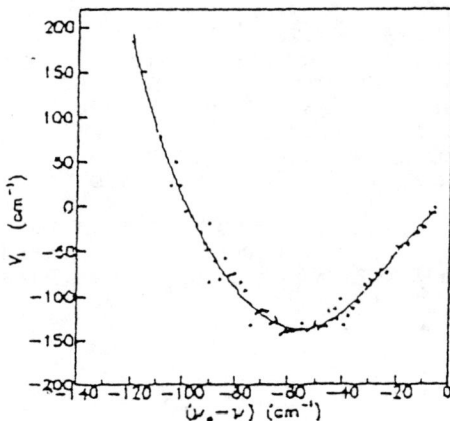

Fig.5. The ground state potential as a function of $\Delta\nu$.

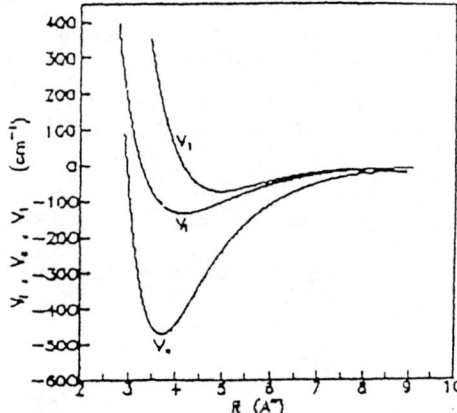

Fig.6. The ground and the two excited states potentials

Table: Comparison of our data for Cd – Kr potential parameters with others (R_m in Å and ϵ in cm^{-1}).

X^10^+		$^30^+$		^{31}		References
R_m	ϵ	R_m	ϵ	R_m	ϵ	
3.95±0.2	134±7	3.6±0.2	471±10	4.95±0.3	72.3±4	this work
3.75	130	–	–	–	–	Bousquet [5]
–	131	–	373	–	72	Kowalski [7]
4.5	114.5	3.5	498.9	5.07	60.2	Czajkowski [8]
3.63	310	3.44	407.2	3.5	–	Czuchaj [9]
–	–	–	513	–	–	kvaran [10]

It is seen from this table that the obtained ground state well depth is in good agreement with that obtained by [5,7,8] but it is very far from Czuchaj theoretical result [9]. The well depth of the excited state $^30^+$ obtained by [7] is 25 % lower than the published [9,10] and our results.

This work is supported by the Polish Committee of Scientific Research within the project 2 0261 91 01

REFERENCES

1- J.Szudy and W.Baylis,J.Quant.Spect.Radiat.Transfer,15,641,1975
2- M.S.Helmi,T.Grycuk and G.D.Roston,Spec.Chem.Acta B51, 633,1996
3- M.S.Helmi,T.Grycuk and G.D.Roston,Chem.Phys.,209, 53,1996
4- G.D.Roston,T.Grycuk and M.S.Helmi,Chem.Phys.213,365,1996
5- C.Bousquet,J.Phys.B:At.Mol.Phys.19,3859,1986
6- G.D.Roston and M.S.Helmi,to be published in 14 ICSLS USA, 1998
7- A.Kowalski,M.Czajkowski and W.H.Breckenridge,J.Chem.Phys.Lett. 121,217,1985
8- M.Czajkowski,R.Bobkowski and L.Krause.Phys.Rev A44,9,5730,1991
9- E.Czuchaj and J.Sienkiewicz,J.Phys.B:At.Mol.Phys.17,2251,1984
10- A.Kvaran,D.J.Funk,A.Kowalski and W.K.Breckenridge,J.Chem.Phys .89,6069,1988

Absolute Interaction-Induced Light Scattering Spectral Intensities from Helium Diatoms over a Large Frequency Domain

F. Rachet, C. Guillot-Noël, M. Chrysos, and Y. Le Duff

*Laboratoire des Propriétés Optiques des Matériaux et Applications, EP CNRS 130
2, bd Lavoisier, Université d'Angers, 49045 Angers, Cedex, France*

Abstract. Interaction-induced light scattering spectra from gaseous helium are measured up to 740 cm^{-1} at room temperature and for various densities. From our data, depolarized and isotropic components induced by binary interactions are extracted in absolute units over a large frequency domain unstudied as yet. Comparison of our experimental data is made with results from a quantum-mechanical computation based on a recently proposed numerical procedure. The induced polarizability model proposed by Moszynski and co-workers is found to remarkably reproduce the experimental results, allowing for a deeper comprehension of the atomic dynamics at short internuclear distances.

Experimental Setup - Measurements

The interaction-induced light scattering intensities from gaseous helium are measured with a classic experimental setup and by making use of the green spectral line (514.5 nm) of a continuous wave Ar$^+$ laser. Given that the polarization of the laser beam is perpendicular to the scattering plane, a half-wave plate associated with a glan prism is required for the registration of the signal in horizontal polarization. The power of the laser is stabilized at 2 W for all experiments held. The gaseous helium sample is contained in a high-pressure four-window cell at room temperature. The scattered signal is collected in a 90° direction with respect to the incident beam. A scrambler enables one to mix all different polarizations. As spectrometer, a double monochromator having gratings of 1800 grooves/mm is used. The detection becomes possible via a photomultiplier followed by a discriminator and a photon-counter. We underline the fact that the spectral transmission curve of the measurement device is registered by means of an etalon delivering white light.

The helium Raman scattering spectrum turns out to be extremely weak, since its intensity at frequencies close to the Rayleigh peak is weaker than that of the corresponding argon spectrum by a factor of about 10^4. We register the spectra corresponding to the two polarizations with respect to the scattering plane, namely the horizontal, I_H, and the perpendicular, I_V (Fig. 1). Taking into account the feebleness of the intensity, densities up to 300 amagat are dealt with, that is a gas pressure going up to 400 bars. For the same reason, the maximum counting time interval is fixed at 40 minutes per registered frequency. In order to extract the binary spectrum, we study the intensity at each frequency as a function of the density of the sample. Proceeding in this way also offers the advantage of eliminating the parasitic light, as well as any undesirable signal due to impurities of the gas. We stress the fact that the preparation of the cell is done with all due care in order to reduce the concentration of the impurities as much as possible. Figure 1 shows the experimental measurements, designated by symbols, whilst solid line curves simply account for smoothing. It is important to note that the spectra are calibrated in absolute intensities. The calibration is made at 76 cm^{-1}.

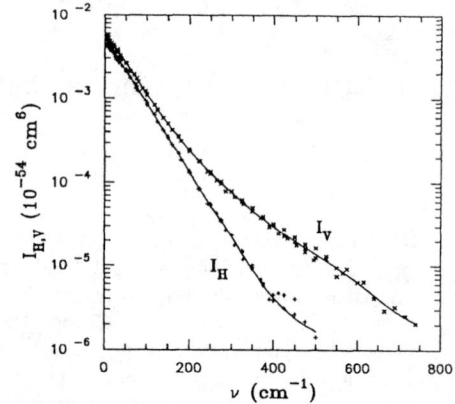

Figure 1. Experimental spectra : Perpendicular component I_V (x) and Horizontal component I_H (+). Solid line curves simply smooth the measurements.

We are able to measure the two components down to 5 cm^{-1}. This is the first time measurements this much close to the Rayleigh peak are taken for helium. We are also able to greatly extend the far Stokes wing frequency domain probed up to 740 cm^{-1} for the perpendicular component and 500 cm^{-1} for the horizontal one. This performance should be compared with the previous limit of 320 cm^{-1} by Frommhold et al (1-3). It is worth to emphasize that at the end of the seventies two other research groups studied the helium spectrum (4,5), however their spectral intensities were given in arbitrary units.

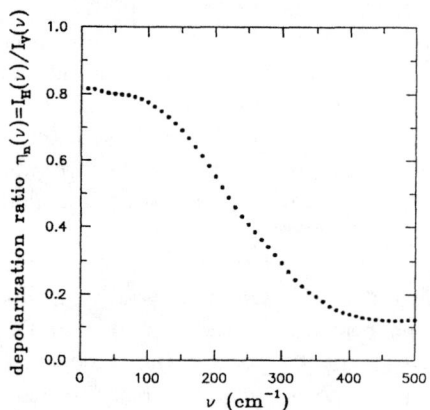

Figure 2. Depolarization ratio $\eta_n(\nu) = I_H(\nu)/I_V(\nu)$

The ratio between $I_H(\nu)$ and $I_V(\nu)$ defines what is known as the *depolarization ratio* $\eta_n(\nu)$. Our experiment reveals a fast decrease of η_n as a function of the frequency shift (Fig. 2). The very small value of approximately 0.12 is obtained already at 400 cm^{-1}, whilst beyond 400 cm^{-1} the depolarization ratio seems to form a plateau. It is noteworthy that this is the first time such a small depolarization ratio is measured for an inert gas.

Computations

From the components I_H and I_V, that both are smoothed and calibrated in absolute units, we are able to extract *depolarized*, $I_\parallel$, (Fig. 3) and *isotropic*, I_{ISO} (Fig. 4) spectra. This becomes possible through the following two formulas that take into account the aperture angle: $I_\parallel(\nu) = 1.010\ I_H(\nu) - 0.01009\ I_V(\nu)$ and $I_{ISO}(\nu) = -1.184\ I_H(\nu) + 1.017\ I_V(\nu)$. We confront our experimental results to theoretical spectra obtained with up-to-date models of both induced polarizability and interaction He-He potential. The Moszynski's trace and anisotropy (6) are elaborated, combined with the modern Janzen-Aziz potential SAPT1 (7) obtained by Symmetry Adapted Perturbation Theory. A fully quantum approach is adopted, implying Fermi-Golden rule continuum-continuum transition matrix elements between initial and final states. Their wavefunctions are built up through a step-by-step outward propagation of the wavefunction ratio defined at every pair of adjacent grid points, according to the Fox-Goodwin propagative method (8-10). A grid-size of 150 Bohr is used. Rotational quantum numbers of the quasimolecule are taken up to 700 in our computation. For each spectral frequency, integration is made on 200 energy values every 15 cm^{-1}, attaining a maximum energy of 3000 cm^{-1} where convergence of total cross sections is obtained to within 1%.

The depolarized spectrum

Figure 3 shows the helium depolarized spectrum. Points represent experiment whereas curves account for theory. The total theoretical spectrum is shown by a solid line curve (curve 1). A very good agreement is ascertained up to 400 cm^{-1}. Beyond that frequency, the two spectra are found to deviate from each other. However the numerical spectrum is still within the error bars, that attain their maximum values in this frequency domain. The signature on the spectrum of the different contributions of the Moszynski's anisotropy is also shown.

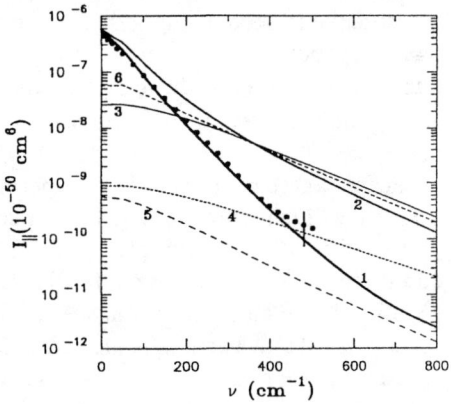

Figure 3. Depolarized spectrum $I_\parallel$; symbols: experiment; curve 1: theory using the Moszynski's anisotropy [ref. 6]; curves 2-6: partial contributions to the curve 1 (2: polarization, 3: exchange, 4: induction, 5: dispersion, 6: cross terms).

Curve 2 represents the contribution of the *polarization* alone; it turns out to be responsible for the anisotropic spectrum at low frequencies but quickly acquires very high values as frequency increases. Curve 3 stands for the *exchange* alone; it is of about 25 times weaker than polarization close to the Rayleigh peak but its contribution predominates beyond 350 cm^{-1}. This is expected since high frequencies correspond to short range interactions where electronic clouds strongly overlap. *Induction* and *dispersion* terms are indicated by curves 4 and 5 respectively. Of course there are still some missing terms, the *cross terms* (represented by curve 6), corresponding to «negative intensities». The total theoretical spectrum, found to be in remarkable agreement with the experiment, is obtained by summing all direct terms and subtracting twice the cross terms.

The isotropic spectrum

In addition to the helium depolarized spectrum we are also able to study the controversial isotropic spectrum (Fig. 4). Dark circles represent experiment in the range 200-500 cm^{-1} where the measurements are highly reliable. In the spectral range 100-200 cm^{-1}, where the depolarization ratio is large, the measurements are less reliable and are represented by open triangles. At frequencies lower than 100 cm^{-1} the measurements become highly uncertain and are not represented at all. On the other hand, beyond 500 cm^{-1}, we are able to deduce the spectrum from the perpendicular component assuming the fact that the depolarization ratio attains a small value (lying in the interval 0-0.12). Open circles correspond to the choice $\eta = 0.12$, but use of $\eta = 0$ reproduces a practically identical spectrum. Once more, one sees the excellent agreement between experiment and theory throughout the entire frequency domain probed. Again we may observe the fingerprints on this spectrum of the different contributions of the Moszynski's trace. Curve 2 represents the contribution of the polarization alone. It turns out to be too weak to play a significant role but, as one can see from figure 4, it differs from curve 1 by just a scaling constant. The exchange term (curve 3) is again parallel to the complete spectrum but far larger than it over the entire frequency domain.

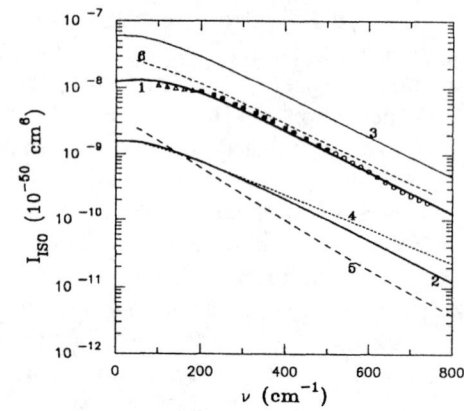

Figure 4. Same as in figure 3 but for the isotropic spectrum I_{ISO}

References

1. M. H. Proffitt, and L. Frommhold, *Phys. Rev. Lett.* **42**, 1473 (1979)
2. M. H. Proffitt, J. W. Keto, and L. Frommhold, *Can. J. Phys.* **59**, 1459 (1981)
3. P. D. Dacre, and L. Frommhold, *J. Chem. Phys.* **76**, 3447 (1982)
4. F. Barocchi, P. Mazzinghi, and M. Zoppi, *Phys. Rev. Lett.* **41**, 1785 (1978)
5. Y. Le Duff, *Phys. Rev.* **A 20**, 48 (1979)
6. R. Moszynski, T. G. A. Heijmen, P. E. S. Wormer, and A. van der Avoird, *J. Chem. Phys.* **104**, 6997 (1996)
7. A. R. Janzen and R. A. Aziz, *J. Chem. Phys.* **107**, 914 (1997)
8. D. W. Norcross, and M. J. Seaton, *J. Phys.* **B 6**, 614 (1973)
9. M. Chrysos, O. Gaye, and Y. Le Duff, *J. Phys.* **B 29**, 583 (1996)
10. O. Gaye, M. Chrysos, V. Teboul, and Y. Le Duff, *Phys. Rev.* **A 55**, 3484 (1997)

Analysis of the Collision-Induced Absorption Spectra of Double Vibrational Transitions in $H_2 - N_2$

C. Stamp, R.D.G. Prasad, P.G. Gillard, and S. Paddi Reddy

Department of Physics and Physical Oceanography
Memorial University of Newfoundland
St. Johns, NF, A1B 3X7, Canada.

Abstract. Collision-induced absorption (CIA) spectra of double vibrational transitions of $H_2(v = 1 \leftarrow 0) + N_2(v = 1 \leftarrow 0)$ were studied in their binary mixtures in the spectral region 5900 to 7100 cm^{-1}. The spectra were recorded with a 2 m high-pressure cell for partial densities of H_2 and N_2 in the range 60 to 400 and 100 to 350 amagat, respectively, at 201 K and 298 K. The observed spectra are interpreted in terms of several double transitions such as: $O_1(2) + Q_1(J)$, $Q_1(J) + O_1(J)$, $Q_1(J) + Q_1(J)$, $Q_1(J) + S_1(J)$, $S_1(0) + Q_1(J)$, $S_1(1) + Q_1(J)$, and $S_1(2) + Q_1(J)$, the first component from H_2 and the second component from N_2. These transitions occur on the high frequency wing of the H_2 CIA fundamental band. Various semi-emperical line shapes were used in the analysis of the observed spectra. Birnbaum-Cohen line-shape function was found to give satisfactory fits of the calculated profiles to the experimental profiles. Line-shape parameters and absorption coefficents obtained from the analysis are given.

Although both H_2 and N_2 do not have permanent static or vibrational electric dipole moments in their ground electronic states, they give rise to collision-induced absorption (CIA) spectra as a result of transient electric dipole moments induced in them during collisions. In 1949 the CIA spectra of H_2, O_2, and N_2 in the regions of their fundamental bands were discovered (for reference, see Welsh [1], Reddy [2], and Frommhold [3]). In the present paper we report results of our work on the CIA spectra of double vibrational transitions $H_2(v = 1 \leftarrow 0) + N_2(v = 1 \leftarrow 0)$ obtained in their binary mixtures in the spectral region 5800 to 7400 cm^{-1}. The spectra were recorded with a 2 m high-pressure cell for partial densities of H_2 and N_2 in the range 60 to 400 and 100 to 350 amagat, respectively, at 201 K and 298 K.

For a gas mixture the absorption coefficient at a given wavenumber ν (cm^{-1}) in expressed as

$$\tilde{\alpha}(\nu) = \alpha(\nu)/\nu = (1/l\nu)\ln[I_0(\nu)/I(v)], \qquad (1)$$

where l is the sample path length and $I_0(\nu)$ and $I(\nu)$ are the intensities transmitted by the cell containing the base gas H_2 and the binary mixture H_2+N_2, respectively.

A sample absorption profile of a binary mixture of $H_2 + N_2$ at 201 K, i.e., $(1/\rho_{H_2}\rho_{N_2})\tilde{\alpha}(\nu)$ against ν for $\rho_{H_2}=378.5$ amagat and $\rho_{N_2}=209.5$ amagat in the region 6000 - 7600 is given in Fig. 1.

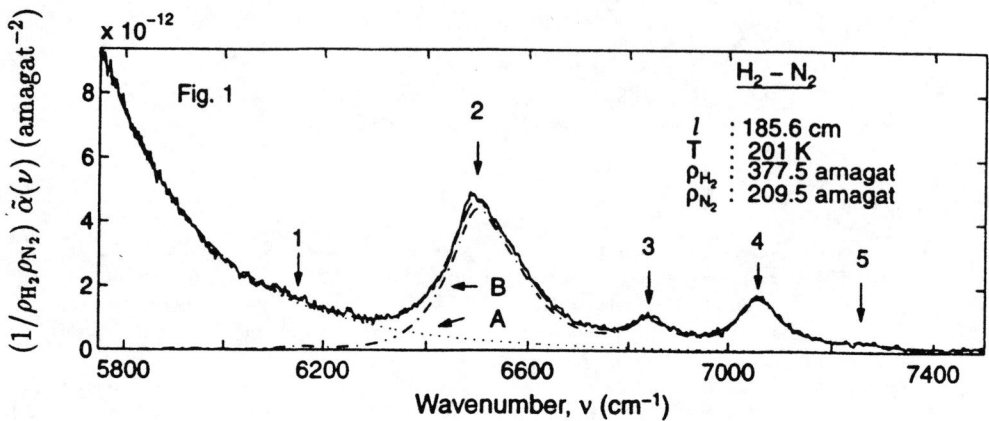

The observed absorption peaks are interpreted in terms of the double vibrational transitions $H_2(v = 1 \leftarrow 0) + N_2(v = 1 \leftarrow 0)$ and the results are summerized in Table 1. The integrated binary absorption coefficient of an induced transition in the absorption spectra can be written as (following Reddy [2] and the references therein),

$$\tilde{\alpha}_{Lm} = (1/\rho_a\rho_b)\int \frac{\alpha_m(\nu)}{\nu}d\nu$$
$$= \frac{4\pi^3 n_0^2 e^2}{3ch}a_0^5(a_0/\sigma)^{2L+1}\tilde{J}_L X_{Lm}, \quad (2)$$

where

$$\tilde{J}_L = 4\pi(L+1)\int_0^\infty x^{-2(L+2)}g_0(x)x^2 dx, \text{ and} \quad (3)$$

$$X_{Lm} = P_{J_1}P_{J_2}[C(J_1L_1J'_1;00)^2 < v_1J_1|Q_{L_1}|v'_1J'_1 >^2 \times \\
C(J_20J'_2;00)^2 < v_2J_2|\alpha_2|v'_2J'_2 >^2 + \\
C(J_2L_2J'_2;00)^2 < v_2J_2|Q_{L_2}|v'_2J'_2 >^2 \times \\
C(J_10J'_1;00)^2 < v_1J_1|\alpha_1|v'_1J'_1 >^2]. \quad (4)$$

For a detailed description of the above equations see Reddy [2].

TABLE 1. Absorption peak positions

Absorption Peak	Position (cm^{-1})	Transition (H$_2$ + N$_2$) [a]
1	6136	$O_1(2) + Q_1(J)$
2	6489	$Q_1(1) + O_1(J)$
		$Q_1(J) + Q_1(J)$
		$Q_1(J) + S_1(J)$
3	6837	$S_1(0) + Q_1(J)$
4	7060	$S_1(1) + Q_1(J)$
5	7261	$S_1(2) + Q_1(J)$

[a] For example, $O_1(2) + Q_1(J)$ represents $O_1(2)$ of H$_2$ + $Q_1(J)$ of N$_2$.

We made use of the Lewis-Birnbaum-Cohen (LBC) line-shape function in the analysis of the absorption profiles. The LBC lineshape uses the same time correlation function as the BC lineshape [4] but with Boltzmann asymmetrization to satisfy the so-called 'detailed balance'.

The dimensionless absorption coefficient $\tilde{\alpha}(\nu)$ at a given wavenumber ν of a band is the sum of the superimposed quadrupolar components and can be represented by

$$\tilde{\alpha}(\nu) = A \sum_m \tilde{\alpha}_{qm} S_L(\Delta \nu), \tag{5}$$

where A is an adjustable parameter, $\Delta\nu = \nu - \nu_m$, ν_m being the wavenumber (in cm^{-1}) of a particular transition, and the quantities ν and ω are related by $\omega = 2\pi c \nu$, and $\Delta\omega = 2\pi c \Delta\nu$. The LBC lineshape $S_L(\Delta\nu)$ ($= S_L(\Delta\omega)$) is represented by

$$S_L(\Delta\omega) = \frac{2}{1+e^{-\beta\hbar\Delta\omega}} \frac{e^{\tau_2/\tau_1}}{\pi} \frac{\tau_2}{\sqrt{1+\Delta\omega^2\tau_1^2}} K_1\left(\frac{\tau_2}{\tau_1}\sqrt{1+\Delta\omega^2\tau_1^2}\right). \tag{6}$$

Further details of the LBC lineshape can be obtained from Lewis and Stamp [5].

A superposition of individual quadrupolar components was used to construct the synthetic profile of the band. The analysis of an absorption profile is also shown in Fig. 1. In this figure the solid curve is the experimental profile, profile A represents the high wavenumber tail of the CIA spectrum of the H$_2$(v = 1 ← 0) + N$_2$(v = 0 ← 0) band, profile B represents the sum of the double vibrational transitions of H$_2$ + N$_2$, and the total synthetic profile, represented by dashes, agrees with the observed profile within experimental error. The resulting analysis at 201 K showed that the absorption depended on binary collisions only, and the time parameters τ_1 and τ_2 were density independent. The following results were obtained: $\tau_1 = 1.124$ (51) 10^{-13} s, $\tau_2 = 1.08$ (51) 10^{-14} s, and $\tilde{\alpha}_{ab} = 1.233$ (34) 10^{-9} cm^{-1} amagat^{-2} (the numbers in brackets are the standard deviations of the means).

The work presented here is supported in part from the Natural Sciences and Engineering Research Council of Canada from grant A-2440 awarded to to S. P. Reddy.

REFERENCES

1. Welsh H. L. *in MTP International Review of Science, Physical Chemistry, Vol. 3, Spectroscopy.* edited by D. Ramsay (Butterworths, London, 1972), pp 33-71.
2. Reddy S. P. *in Phenomena Induced by Intermolecular Interactions.* edited by G. Birnbaum (Plenum, New York, 1985), pp 129-167.
3. Frommhold L. *Collision Induced Absorption in Gases.* (Cambridge University Press, Cambridge, 1993).
4. Birnbaum G. and Cohen E. R. *Can. J. Phys.*, **54**,593, (1976).
5. J. C. Lewis and C. Stamp. *This Volume.*

COLLISION-INDUCED ABSORPTION IN THE FUNDAMENTAL OF O_2 AND N_2 : COMPARISON BETWEEN EXPERIMENTAL AND THEORETICAL RESULTS

G.Moreau (1), J.Boissoles (1), R. Le Doucen (1), R.H. Tipping (2) and C.Boulet (3)

(1) Laboratoire de Physique des Atomes, Lasers, Molécules et surfaces. UM.R.CNRS 6627 Université de Rennes I, Campus de Beaulieu, 35042 Rennes Cedex, France
(2) Department of Physics and Astronomy, University of Alabama, Tuscaloosa, AL 5487-0324,USA
(3) Laboratoire de Physique Moléculaire et Applications, UPR 136-CNRS, Université Paris-Sud, Bâtiment 350, Campus d'Orsay, 91405 Orsay Cedex, France

Collision induced (C.I.) spectra of pure O_2 and O_2+N_2 mixtures in the region of the fundamental of O2 have been recorded at temperatures appropriate to the Earth's atmosphere.[1] In a previous work[2], we studied theoretically the C.I. fundamental band of N_2-N_2 pairs. In particular we showed that one could achieve a reasonable agreement between theory and experiment[3] by considering only the quadrupolar and hexadecapolar mechanisms, using existing data for the various matrix elements of the moments and polarizabilities together with the quantum lineshapes given by Borysow and Frommhold[4].

In the present work, a similar modelling has been carried out for the C.I. fundamental band of O_2. The best values found in the literature for the molecular parameters are discussed in ref 2 and 5. As in our previous work[2], we neglect the anisotropic interactions between colliding molecules, leading to an absorption written in terms of additive components. To our knowledge, no quantum individual lineshapes are available for O_2. Therefore we used those of N_2. Some typical results are compared with experimental spectra in figures 1 and 2. As can be seen the theoretical spectra obtained without the introduction of adjustable parameters are in good agreement with the data, the difference being larger in the wings.

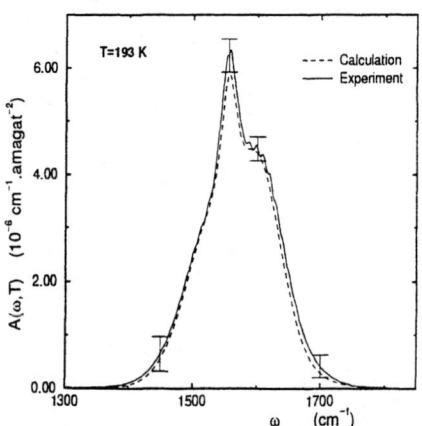
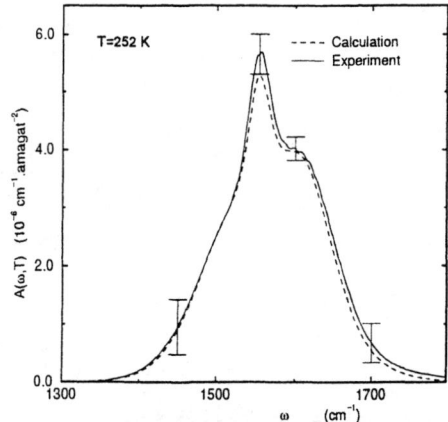

Fig 1 : Binary absorption coefficient (in units $10^{-6}cm^{-1}.amagat^{-2}$) for pure O_2 ; T=193 K

Fig 2 : Binary absorption coefficient (in units $10^{-6}cm^{-1}.amagat^{-2}$) for O_2-N_2 pair in the region of the O_2 fundamantal band ; T=252 K

By subtracting now the smooth theoretical absorption from the experimental spectra, we obtained residual spectra as shown in figure 3 and 4. Recent measurements revealed similar weak structure superimposed on the broad continuum for N_2[6] as well as for O_2[7], even at room temperature.

CP467, Spectral Line Shapes: Volume 10, 14th ICSLS,
edited by Roger M. Herman
© 1999 The American Institute of Physics 1-56396-754-5/99/$15.00

From the present work, it appears that the "rotational structure" of these residual spectra is almost identical to that shown by Mc Kellar [in figure 2 of ref 8] for the low temperature [77 K] N_2-N_2 absorption and attributed to dimers.

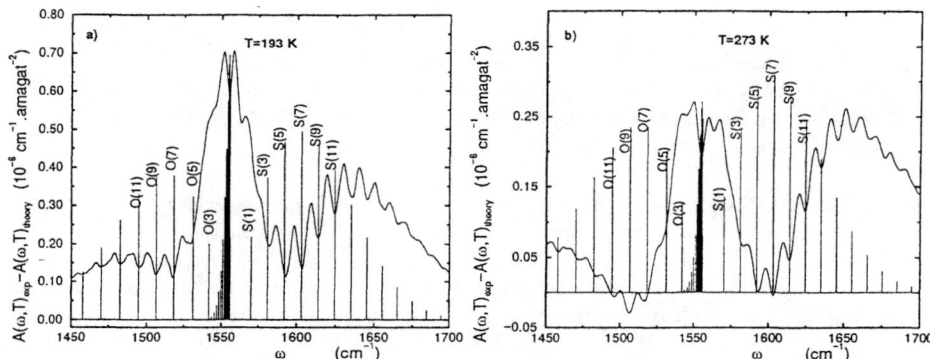

Fig 3 : Spectra obtained by subtracting the theoretical C.I.spectra from the observed spectra.
Pure O_2 (density : 6.8 Am)   (a) T=193 K   ;   (b) T=273 K
The vertical lines constitute a stick spectrum of O_2 O(J), Q(J) and S(J) transitions.

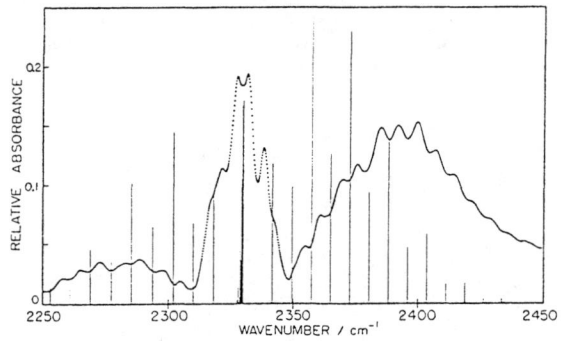

Fig 4 : Figure 2 of Ref.8 ; Pure N_2 at T=77 K

References

1 F.Thibault, R. Le Doucen, V. Menoux, L. Rosenmann, J.M. Hartmann and C.Boulet, Appl. Opt. 36 1 (1997)
2 J.Boissoles, R.H. Tipping and C.Boulet, J. Quant. Spectrosc .Radiat. Transfer. 51 615 (1994)
3 V. Menoux, R. Le Doucen, C.Boulet, A. Roblin and A.M. Bouchardy, Appl. Opt. 32 263 (1993)
4 A. Borysow and L. Frommhold, Astrophys. J. 311 1043 (1986)
5 G.Moreau, J. Boissoles, C. Boulet, R.H. Tipping and Q. Ma, submitted to J. Quant. Spectrosc. Radiat. Transfer.
6 W.J. Lafferty, A.M. Solodov, A. Weber, W.B. Olson and J.M. Hartmann, Appl. Opt. 35 5911 (1996)
7 J.J. Orlando, G.S. Tyndall, K.E. Nickerson and J.G. Calvert, J. Geophys. Res. D96 755 (1991)
8 A.R.W. Mc Kellar, J. Chem. Phys. 88 4191 (1988)

Collision-induced absorption by H_2 pairs in the second overtone band: Comparison between experimental and theoretical results

Aleksandra Borysow

Niels Bohr Institute, Juliane Maries vej 30, DK-2100 Copenhagen, Denmark

Yi Fu, Michigan Technological University, Houghton, MI 49931, USA

Claude Brodbeck, Jean-Pierre Bouanich, and Nguyen-van-Thanh

Laboratoire de Physique Moléculaire et Applications, CNRS, Bâtiment 350, Campus d'Orsay, 91405 Orsay Cedex, France

We report new quantum mechanical computations [1], and new laboratory measurements [2] of the collision induced absorption spectra of molecular hydrogen in the second (3-0) overtone band. Due to the extreme weakness of the second overtone band, it is equally difficult to measure, as to compute its intensity with acceptable accuracy. Until now, only one experimental set of data existed for the (3-0) band at low temperature (85 K) and low density [3]. Other measurements [4] were performed at the 500-1000 amagat region, but were concerned with three body effects only.

Our work focuses strictly on binary CIA spectra. New measurements have been performed at Orsay, at 77.5 K and 298 K for normal H_2. The spectra, obtained in a range of densities from 100 to 800 amagat, have been extrapolated to zero density [2]. Independently, we extended an existing [5] *ab initio* induced dipole database. This allowed us, for the first time, to compute quantum mechanical CIA spectra of H_2 pairs in that band [1], based on the first principles. The agreement between the theory and experiment for this very weak band, is found to be within 15-30%. In the figure we present the comparison between our measurements (markers) and theoretical results (solid line), where we plot binary absorption coefficient per density squared.

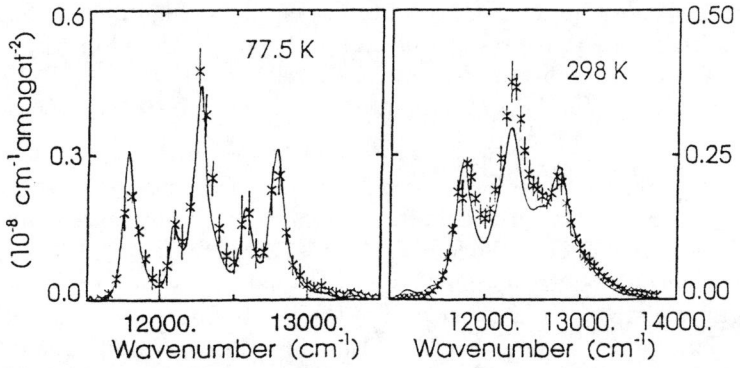

[1] Y. Fu, C. Zheng, A. Borysow, *J.Q.S.R.T.*, submitted, 1998.
[2] C. Brodbeck, J. P. Bouanich, Nguyen-van-Thanh, Y. Fu, A. Borysow, *J. Chem. Phys.*, submitted, 1998.
[3] A. R. W. McKellar and H. L. Welsh, *Proc. Roy. Soc.* (London) A, 322:421, 1971.
[4] S. P. Reddy, F. Xiang, and G. Varghese, *Phys. Rev. Lett.*, 74:367, 1995.
[5] W. Meyer, A. Borysow, and L. Frommhold, *Phys. Rev.*, A 40:6931, 1989.

A. Borysow acknowledges support of NASA, Division for Planetary Atmospheres.

The internuclear distance of molecular hydrogen in the fluid and solid phases at room temperature.

Lorenzo Ulivi,[†] Marco Zoppi,[†] and Gabriele Pratesi[‡]

[†]*Istituto di Elettronica Quantistica, CNR, via Panciatichi 56/30, I-50127 Firenze, Italy.*
`ulivi@ieq.fi.cnr.it, zoppi@ieq.fi.cnr.it`
[‡]*Dipartimento di Fisica dell'Università and INFM, largo Enrico Fermi 2, I-50125 Firenze, Italy.*
`pratesig@fi.infn.it`

In molecular gases and crystals, single molecule properties are increasingly modified rising pressure, due to interactions with neighbors. In this work we study spectroscopically the internuclear distance of the hydrogen molecule, as a function of pressure, in the compressed fluid up to the solidification pressure (5.4 GPa) and in the solid up to 20 GPa, at room temperature. [1] Previous results in the fluid at room temperature were limited to a density of about 60 mol/l [2], while in the solid have been derived, with strong simplifying assumptions, only at low temperature [3].

Hydrogen molecules are almost free rotors, even in the high density fluid and in the hcp solid up to the pressures of this study. Their rotational energy levels (for $v = 0$) may be described by the free molecule relation [4] $E(J) = B_0 J(J+1) - D_0 J^2 (J+1)^2 + H_0 J^3 (J+1)^3$. B_0 is related to the average internuclear distance. In fact, we have $r_m^2 \equiv \left\langle \frac{1}{r^2} \right\rangle_{v=0}^{-1} = \frac{h}{8\pi^2 c\mu} \frac{1}{B_0}$ with clear meaning of the symbols.

To reach pressures up to 20 GPa, we have used a diamond anvil cell, where the pressure is measured by the usual method of the ruby fluorescence shift [5]. The measured Raman frequency of the four $S_0(0)$, $S_0(1)$, $S_0(2)$, and $S_0(3)$ lines has been fitted, at each density, as a function of the rotational quantum number, using only two free parameters, B_0 and D_0, because of the limited number of rotational lines considered. The fluid data are reproduced in this way with quite good accuracy, except for the highest fluid density, while, in the solid, the position of the $S_0(0)$ line has not been considered in the fit, as discussed in Ref. [1].

In Fig. 1 we report our and previous results [2,3,6] for r_m. At low density, a good agreement is found among our and earlier data, but in the fluid at higher density, a marked decrease of the internuclear distance with a strong negative curvature becomes apparent, that could not have been predicted by an extrapolation of the

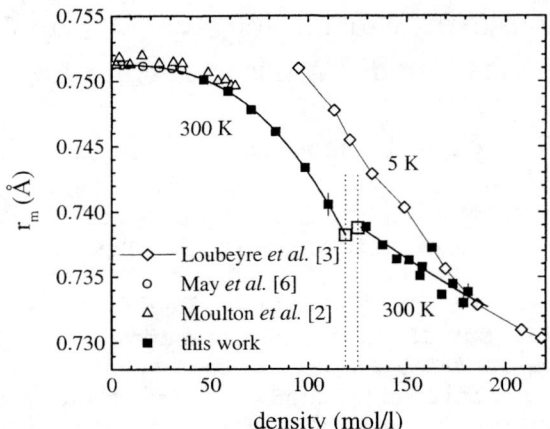

FIGURE 1. Internuclear distance of the hydrogen molecule: squares, our results; empty triangles: data of Moulton et al. [2], empty circles: results by May et al. [6], diamonds: Loubeyre et al. [3] data (5 K, solid). The two empty squares are the extrapolated values at melting, by means of a power-law equation (fluid) and by a polynomial (solid). A stretching of the molecules upon solidification is evident.

low density data. This decrease can be qualitatively attributed to the effect of the short range repulsive forces. A quantitative analysis could be of extreme importance for the characterisation of the dependence of the intermolecular potential on the internuclear distance. In the solid phase the rate of decrease of r_m is much smaller than that in the fluid, and the curvature becomes here slightly positive. The agreement with the data of Loubeyre et al. [3] is satisfactory, considering the temperature difference and the strong assumptions in both cases.

The change of the internuclear distance at melting has been derived extrapolating the value of r_m at the melting density both from the solid and the fluid side. The difference ($\simeq 0.5 \times 10^{-3}$ Å) is quite small, and may be affected by experimental errors, but it has the same sign of that found at low temperature ($\simeq 1 \times 10^{-3}$ Å). Also at room temperature, then, the ordered structure is the cause, on average, of a less intense repulsive force on the nuclei of the molecules.

REFERENCES

1. L. Ulivi, M. Zoppi, L. Gioè, and G. Pratesi, Phys. Rev. **B58**, to appear (1998).
2. N. E. Moulton, G. H. Watson, W. B. Daniels, et al. Phys. Rev.**A37**, 2475 (1988).
3. P. Loubeyre, M. Jean-Louis, and I. F. Silvera, Phys. Rev. **B43**, 10191 (1991).
4. G. Herzberg, *Spectra of Diatomic Molecules*, New York (USA), (1950)
5. H. K. Mao, P. M. Bell, J. V. Shaner, et al. J. Appl. Phys. **49**, 3276 (1978).
6. A. D. May, V. Degen, J. C. Stryland, and H. L. Welsh, Can. J. Phys. **39**, 1769 (1961).

Determination of the Ground State Well Depth Position R_m of Cd-inert Gases Van der Waals Molecules Experimentally

G. D. Roston* and M. S. Helmi

Department of Physics, Faculty of Science
Alexandria University, Egypt

The ground state well depth position R_m represents a serious problem for finding the interatomic potentials of the Van der Waals Molecules [1-4]. The aim of this work is to obtain R_m by a modified method depending on the analysis of the pressure broadening profiles of spectral lines using the UFC treatment developed by Szudy and Baylis [5]. Using this theory, the reduced absorption coefficient (cm^5) in the line wings normalized to the perturbing (N_p) and the radiating (N_r) atom densities is given by

$$K^R(\Delta v, T) = \frac{K(\Delta v, T)\,(cm^{-1})}{N_p N_r} = 4\pi c R^2 \left| \frac{dR}{d(\Delta V)} \right| \exp\frac{-V_i(R)}{kT} \quad (1)$$

Where R is the internuclear separation, $\Delta V(R)$ is the potential difference between the excited $V_s(R)$ and the ground $V_i(R)$ state potentials, Δv is the frequency separation from the line center which is related to the value of R by the following relation:

$$h\,\Delta v(R) = \Delta V(R) \quad (2)$$

T is the absolute temperature, k is the Boltzmann's constant and c is a constant.

To obtain the relation $\Delta V(R)$ and the interatomic potentials of the ground and excited states, at least one value of $R(\Delta v)$ must be known. This value is generally chosen to be the position of the ground state well depth R_m. Beside the well known methods published elsewhere [3-5] for obtaining the relation $R(\Delta v)$ the authors modified a method which gives an excellent agreement for R_m when compared with other methods. This method is summarized as follows: Let $\Delta V = C_n R^{-n}$, then $d\Delta V/dR = -n\,C_n\,R^{-(n+1)}$ where c_n is a constant and n is an integer, and as the well depth of the ground state is located at the maximum temperature dependence of the spectral line profiles, where $\Delta v=\Delta v_a$, $R=R_a$ and $V_i = \epsilon$, then Eq.(1) gives at this point the formula:

$$R_a^3 = \frac{n\, K^a(\Delta v_a, T)}{4 \pi C_a} \exp\left(\frac{\epsilon}{kT}\right) \Delta v_a \qquad (3)$$

Since $K^a(\Delta v_a, T)$ and ϵ are known experimentally, then R_a can be obtained from Eq.(3) if n is known. Different values of n were tested on (Cd-inert gas) systems using the Cd line 326.1 nm, and it was found that n=3 is the more reasonable value which gives an agreement with the values of R_a obtained by other methods. Substituting $\Delta V = C_3 R^{-3}$ in Eq.(1) we obtain $K^a(\Delta v, \infty) = A\, \Delta v^{-2}$, where A is a constant. This is shown in the figure. The obtained values of R_a with the corresponding published values are illustrated in the given table.

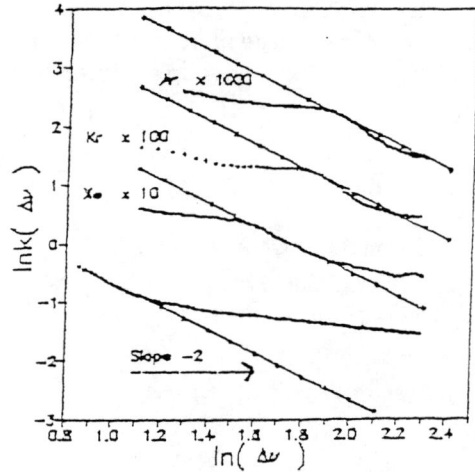

Fig: The blue wing profiles of the cadmium line for cd-inert gases quasimolecules in logarithmic scale.

Table: The obtained results of Rm(Angstrom) for Cd-inert gases with the corresponding published values.

Mol.	This work R_m (Å)	Ref. R_m (Å)
Cd-Cd	3.7 ± 0.25	3.3 [6]
Cd-Xe	3.65 ± 0.2	3.73 [7]
Cd-Kr	3.98 ± 0.15	3.75 [6]
Cd-Ar	4.25 ± 0.2	4.3 [8]

It is seen that the obtained results are in close agreement with the corresponding experimental values obtained by [7-8], but a little far from the values obtained by [6].

References

1- M.S.Helmi, T.Grycuk and G.D.Roston.Spect.Chem.acta B 51 633 1996
2- M.S.Helmi, T.Grycuk and G.D.Roston.Chem. Phys. 209, 53 1996
3- G.D.Roston, T.Grycuk and M.S.Helmi, Chem. Phys. 213, 365 1996
4- G.D.Roston and T.Grycuk, to be published.
5- J.Szudy and W.E.Baylis,j.Quant.Spectrosc.Radiat.Transfer. 15 641 1975
6- C. Bousquet, J. Phys. B: At. Mol. Phys.,19 3859 1986
7- E. Czuchaj and J. Sienkiewicz, J. Phys. B 17 2251 1984
8- R.Bobkowski, M.Czaikowski and L.Krause, Phys. Rev. A 41 234 1990

Collision-Induced Emission Spectra of H_2-He and D_2-He Pairs at Temperatures of Thousands of Kelvin

Dominik Hammer[*], Lothar Frommhold[*], and Wilfried Meyer[†]

[*] *University of Texas at Austin, Physics Department, Austin, TX 78712-1081*
[†] *Lehrstuhl Theoretische Chemie, Universität Kaiserslautern, Kaiserslautern, Germany*

The supramolecular dipole moment of the H_2-He collisional system is computed from first principles for twenty vibrational spacings of the H_2 molecule from 0.6 to 4 Bohr, eleven separations of the H_2-He pair from 2 to 6 Bohr, and five angles subtended by the inter- and intramolecular axes. From this dipole moment surface we obtain for all rotational quantum numbers J with $J' = J$ dipole matrix elements for all possible vibrational transitions $v \to v'$, with v and $v' = 0, \ldots, 14$ for H_2 and v and $v' = 0, \ldots, 20$ for D_2. The matrix elements are used to compute line shapes of individual rotovibrational transitions for temperatures of several thousands of Kelvin using a quantum line shape code [2,1]. By superposition of the individual profiles supramolecular emission spectra are determined for frequencies from 100 to 100,000 cm^{-1}, at temperatures of thousands of Kelvin. The results are applicable to the light emission, for example, of 'cool' stars or shock waves. The work shows that for the higher temperatures considered the collision-induced spectra involving large $\Delta v = v - v'$ extend into the near infrared and visible part of the spectrum. The collision-induced spectra of D_2-He are also obtained for comparison with the H_2-He spectra.

ACKNOWLEDGMENTS

The support of the R. A. Welch Foundation, grant F1346, is gratefully acknowledged.

REFERENCES

1. Frommhold, L., *Collision-induced Absorption in Gases*, Cambridge, New York: Cambridge University Press, 1993.
2. Frommhold, L., elsewhere in this volume.

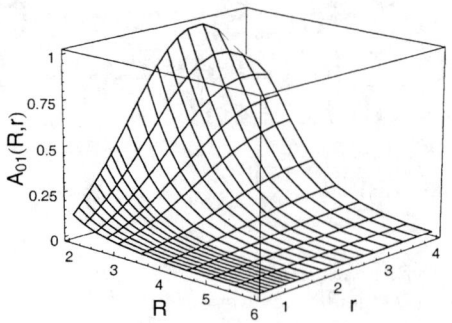

FIGURE 1. The isotropic dipole component $A_{\lambda L}(r, R)$ ($\lambda L = 0\,1$) as a function of the vibrational spacing r and the collisional separation R of the H_2-He pair, in atomic units.

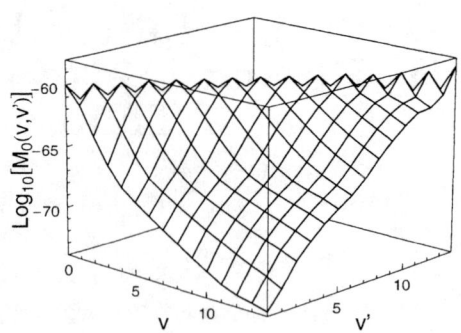

FIGURE 2. Logarithm of the integrated vibrational line strengths, as function of upper and lower vibrational quantum number, v, v', of the H_2 molecule, at 10,000 K, of the $\lambda L = 0\,1$ component, $J = J' = 0$. The intensity decreases roughly exponentially with Δv.

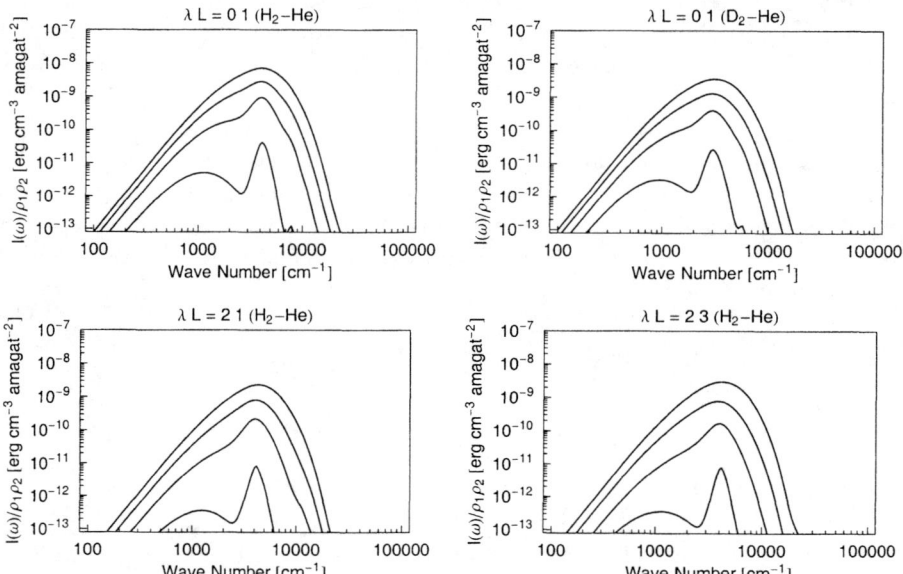

FIGURE 3. Spectra calculated for different λL-components of H_2-He and D_2-He for temperatures of, from bottom to top in each graph, 2,000 K, 6,000 K, 10,000 K, and 15,000 K. With increasing temperatures the higher λL-components become more and more important and the spectrum extends into the visible. The D_2-He spectrum is shifted to lower frequencies relative to the H_2-He spectrum and is less broad due to the higher mass of D_2.

MOLECULAR LINESHAPES

Line Mixing in CO_2 Infrared Q-Branches. A Test of the Energy Corrected Sudden Approximation

Jean-Michel Hartmann

Laboratoire de Physique Moléculaire et Applications
Université Paris-Sud, 91405 Orsay Cedex, France.

Abstract. Theoretical and experimental results on absorption by CO_2 infrared Q-branches are presented. Studies have been made for various pressures, temperatures, and the perturbers N_2, O_2, Ar, and He. Bands of different symmetries, including parallel and perpendicular transitions have been investigated for symmetric and asymmetric isotopomers. The measured spectra are analyzed using a line-mixing model based on the Energy Corrected Sudden approximation. The relaxation operator for each perturber has been constructed from 4 parameters only. Comparisons between measured and computed spectra demonstrate the quality of the approach. The more remarkable characteristics which have been pointed out and analyzed are the following: The first is that collisions with He lead to significantly narrower profiles than those obtained for the CO_2-(N_2, O_2, or Ar) systems. The second point is that vibrational angular momenta have a significant influence on the Q-branch shape through dipole matrix elements and the contributions of parity changing and conserving collisions. The last is that absorption by all isotopomers can be modeled with a single set of parameters. Finally, Consequences on absorption and emission in the atmosphere are discussed.

INTRODUCTION

The present paper is a summary of a vast study of line-mixing effects in CO_2 infrared Q branches. The latter has involved many different groups and a large number of measured spectra has been collected and used. They include absorption by various isotopomers and bands, recorded in both the laboratory and the atmosphere for a large range of conditions (pressure, temperature, collision partner). These data have been analyzed in detail using a model based on the Energy Corrected Sudden approximation. The parameters of this approach have been determined from some (very few) of the experimental data. Due to the number of results obtained, only an overview of the main points is made here but more information can be found in Refs. (1-5).

MEASURED SPECTRA

A large number of laboratory spectra has been used, which originate from different groups and instrument setups [see (4, 5)]. Twenty different Q branches have been studied, which include ten transitions for $^{12}C^{16}O_2$ and the v_2 and v_1-v_2 Q branches of the six isotopomers $^{16\ to\ 18}O-^{12}C-O^{16\ to\ 18}$. Among these data are perpendicular bands with levels from Σ to Δ (1, 2) and a $\Pi-\Pi$ parallel transition (3). In most cases the 200-300 K temperature range was covered for numerous pressures and mixtures with He, Ar, N_2, and O_2. Altogether these data represent more than 200 spectra.

Many atmospheric spectra have also been used (4, 5), which have been recorded by two high resolution Fourier balloon borne instruments and give stratospheric transmissions and emissions. Nine Q-branches have been recorded for different lines of sight and the number of spectra is about 300. They include perpendicular transitions and the v_2 Q branch of $^{16}O-^{12}C-O^{16}$, $^{16}O-^{13}C-O^{16}$, $^{16}O-^{12}C-O^{17}$, and $^{16}O-^{12}C-O^{18}$.

THEORETICAL MODEL

Within the impact and binary collision approximations the absorption coefficient α at wave number σ of a CO_2-X mixture with partial pressures $p_{CO_2} \ll P$ at temperature T is given by (6):

$$\alpha^{LM}(\sigma, p_{CO_2}, P, T) = \frac{8\pi^2}{3hc} \times \sigma \times [1 - \exp(-hc\sigma/k_B T)] \times p_{CO_2}$$
$$\times \sum_{\ell} \sum_{\ell'} \rho_\ell(T) \times d_\ell \times d_{\ell'} \times Im\left\{ \langle\langle \ell' | [\Sigma - L_0 - iPW^{CO2-X}(T)]^{-1} | \ell \rangle\rangle \right\}, \quad (1)$$

where the sum extends over all CO_2 lines ℓ and ℓ' and $Im\{...\}$ denotes the imaginary part. ρ_ℓ and d_ℓ are respectively the population of the initial level of line ℓ and the dipole matrix element of the optical transition. Σ, L_0, and W^{CO2-X} are Liouville space operators. The first two are diagonal and associated with wavenumbers (i.e.: $\langle\langle \ell' | \Sigma | \ell \rangle\rangle = \delta_{\ell,\ell'} \times \sigma$ and $\langle\langle \ell' | L_0 | \ell \rangle\rangle = \delta_{\ell,\ell'} \times \sigma_\ell$). The non diagonal relaxation operator W is complex and accounts for the effects of collisions on the absorption profile. In the absence of line mixing, it is diagonal, governed by the pressure broadening and shifting of lines ($\langle\langle \ell | W^{CO2-X} | \ell \rangle\rangle = \gamma_\ell^{CO_2-X} - i\delta_\ell^{CO_2-X}$) and the usual Lorentzian shape is obtained.

Within the Energy Corrected Sudden approach of Ref. (7), the element of $Re\{W\}$ transferring intensity from the $Q_{J'}$ line to the Q_J line within the $(v_{1f} v_{2f}^{kf} v_{3f}) \leftarrow (v_{1i} v_{2i}^{ki} v_{3i})$ band is given by:

$$\langle\langle Q_J | Re\{W^{CO2-X}(T)\} | Q_{J'} \rangle\rangle = \sum_{ki'=kf-1}^{kf+1} \theta_{J,J}^{ki',kf} \times \theta_{J',J'}^{ki',kf} \times {}^{ki,ki'}\Gamma_{J,J'}^{CO_2-X}(T), \quad (2)$$

where the dipole reduced elements $\theta_{J,J}^{k',k}$ are given in terms of 3J symbols, by:

$$\theta_{J,J}^{k,k} = (-1)^{J+k}\sqrt{2J+1}\begin{pmatrix} J & 1 & J \\ k' & k-k' & -k \end{pmatrix}. \quad (3)$$

For downward transitions (J←J'≥J), the state to state rates $^{k,k'}\Gamma_{J,J'}^{CO2-X}(T)$ are (7):

$$^{k,k'}\Gamma_{J,J'}^{CO2-X}(T) = \frac{\delta_{J,J'}}{^{k,k'}\tau_J^{CO2-X}(T)}(-1)^{k+k'}(2J+1) \times \Omega^{CO2-X}(J',T)$$
$$\times \sum_{L\text{ even}}(2L+1)\begin{pmatrix} J & L & J' \\ k & 0 & -k \end{pmatrix}\begin{pmatrix} J & L & J' \\ k' & 0 & -k' \end{pmatrix}\frac{Q^{CO2-X}(L,T)}{\Omega^{CO2-X}(L,T)} \quad \text{(for J≤J')} \quad (4)$$

whereas upward transitions (J←J'<J) are deduced from:

$$^{k,k'}\Gamma_{J,J'}^{CO2-X}(T) = [\rho_J(T)/\rho_{J'}(T)] \times {}^{k,k'}\Gamma_{J',J}^{CO2-X}(T). \quad (5)$$

The Ω^{CO2-X} and Q^{CO2-X} factors in Eq. (4) are energy corrections to the Infinite Order Sudden (IOS) approximation and the basis rates associated to the collisional transfer from rotational level J=L to J=0, respectively. In the present approach these quantities are modeled through analytical laws. $\Omega^{CO2-X}(J,T)$ is given by (8):

$$\Omega^{CO2-X}(J,T) = \left\{1 + \frac{1}{24} \times \left[\frac{\omega_{J,J-2} \times d_c^{CO2-X}(T)}{\overline{v}^{CO2-X}(T)}\right]^2\right\}^{-2}, \quad (6)$$

where $\omega_{J,J-2}$, $\overline{v}^{CO2-X}$, and d_c^{CO2-X} are the frequency spacing between level J and J-2, the mean relative velocity in CO_2-X collisions, and a scaling length. The basis rates $Q^{CO2-X}(L,T)$ are described by the widely used Exponential Power (EP) law, i.e.:

$$Q^{CO2-X}(L,T) = A^{CO2-X}(T) \times [L(L+1)]^{-\lambda^{CO2-X}(T)} \times \exp\left[-\beta^{CO2-X}(T)\frac{hc \times E_L}{k_B T}\right], \quad (7)$$

where E_L is the rotational energy of level L. The perturber and temperature dependent quantities d_c^{CO2-X}, A^{CO2-X}, λ^{CO2-X}, and β^{CO2-X} are the final parameters of the ECS-EP model. Knowledge of their values enables construction of the off-diagonal part of $Re\{\mathbf{W}\}$.

The collisional terms $^{k,k'}\tau_J^{CO2-X}$, which are related to the relaxation of the rotational angular momentum and of its associated higher order tensors (7), can be directly computed from the parameters d_c^{CO2-X}, A^{CO2-X}, λ^{CO2-X}, and β^{CO2-X} when k=k'; on the other hand, for k≠k', they should be determined from other sources, such as absorption spectra, spin relaxation, viscomagnetic effect, ..., [see Refs. (9, 10)]. Nevertheless, the results of Ref. (10) indicate that they can be satisfactory approximated by:

$$^{k,k'}\tau_J^{CO2-X} \approx {}^{k,k}\tau_J^{CO2-X}, \quad (8)$$

where $^{k,k'}\tau_J^{CO2-X}$ can be easily computed from knowledge of the ECS-EP parameters using the relation (7):

$$\sum_{J'} {}^{k,k'}\Gamma_{J,J'}^{CO2-X} = 0 \quad . \tag{9}$$

The imaginary part of **W** has been neglected in the present computations. This is justified by its small effect at the pressures considered here and the lack of data on the pressure shifts of Q lines. Note that Ref. (10) shows that $Im\{\mathbf{W}\}$ can have significant influence at elevated density.

DATA USED

The data required for computations using Eq. (1) and the ECS-EP model are, for each Q line ℓ, the line identification (rotational and vibrational quantum numbers), position σ_ℓ, dipole transition moment d_ℓ, energy of the lower level E''_ℓ, and the real part of the relaxation operator for CO_2-X collisions. The latter can be computed provided that the parameters d_c^{CO2-X}, A^{CO2-X}, λ^{CO2-X}, β^{CO2-X} are known as a function of temperature.

The line identification and the spectroscopic parameters σ_ℓ, d_ℓ, E''_ℓ have been taken in the 1996 version of the HITRAN database (11).

Two sets of d_c^{CO2-X}, A^{CO2-X}, λ^{CO2-X}, and β^{CO2-X} parameters have been determined, both adjusted in order to obtain agreement between computed and measured spectra of the $(10^00)_{II} \leftarrow (01^10)_I$ Q-branch near 618 cm^{-1}. The first, given in Ref. (1) was obtained from calculations where the diagonal part of **W** is computed from Eqs. (2)-(9). The second, which is more adapted for atmospheric applications, was adjusted (4) imposing the values of the diagonal part of **W** to some measured line-broadening data.

Two important remarks must be made at this step: (i) The first is that Eq. (4) contains an approximation for asymmetric isotopomers; indeed, for such species, odd L values should be accounted for in the sum (5). Nevertheless, since the molecules are almost symmetric the interaction potential remains dominated by even rank contributions and assuming a single set of parameters for all isotopomers is likely a satisfactory approximation. (ii) The second is that only about 4 spectra were used for the determination of the model parameters for each collision partner. Those used correspond to two pressures for the temperatures 200 and 300 K in the $(10^00)_{II} \leftarrow (01^10)_I$ band of $^{12}C^{16}O_2$. The numerous comparisons between measured and computed spectra presented thereafter for other conditions are thus a true test of the model consistency.

RESULTS

Laboratory Spectra

Effect of Pressure

Figure 1 illustrates results obtained for different pressures. As is well known, line-mixing results in transfers of intensity from regions (and lines) of small absorption towards those of intense absorption. As a consequence, neglecting line-coupling leads to underestimated values near the Q-branch peak whereas the wing is overestimated. This effect clearly increases with pressure (i.e. the ratio of line broadening and of the spectral separation between transitions) and the entire Q branch becomes similar to a single Lorentzian at elevated density.

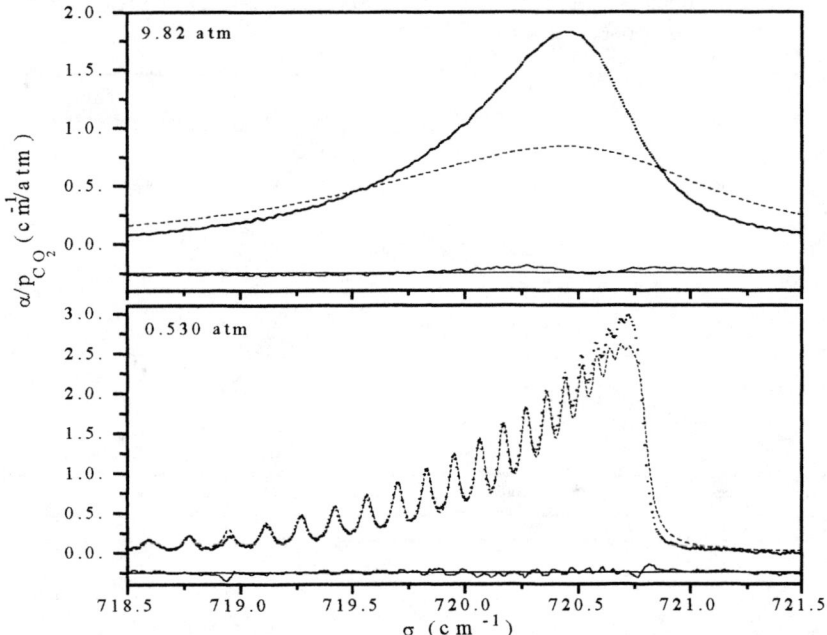

Figure 1: Influence of total pressure on the shape of the $(\nu_1-\nu_2)_I$ Q branch for CO_2-He at room temperature. ● are measured values. - - - - have been computed neglecting line-mixing. , —— are deviations between measurements and computations accounting for line-mixing.

Effect of Temperature

The effect of temperature has been studied in Ref. (1) and is expectable. Indeed, the average coupling between transitions behaves basically as the line-broadening coefficients. As for the lines, the Q branch width for constant density slightly increases with increasing temperature whereas it decreases (as $T^{-0.8}$) for constant pressure.

Effect of the Collision Partner

Figure 2 presents typical results obtained for mixtures of CO_2 with N_2, Ar, and He. It is clear that the Q branch is significantly narrower when Helium is considered although the individual line-broadening coefficients for both perturbers are quite similar (12). This behavior, analyzed in detail in (1), results from two successive effects. (i) The first is that the basis rates $Q^{CO2-X}(L)$ decrease more rapidly with increasing L in the case of He [$\lambda^{CO2-He} \approx 1.06$ and $\lambda^{CO2-Ar} \approx 0.85$ in Eq.(7)]; analytical developments and direct computations (1) show that this is due to differences in the interaction potential which is mostly repulsive for CO_2-He whereas it includes a significant mid-range attraction for CO_2-Ar. (ii) The second step, which leads to the differences observed in Fig. 2 is that low L values contribute to intrabranch (Q-Q) mixing whereas interbranch (Q-P and Q-R) couplings are favored by large L values (1). The direct result of (i) is thus that coupling between Q lines is larger for CO_2-He, leading to a narrower profile. Note that similar effects have been observed for N_2O (13).

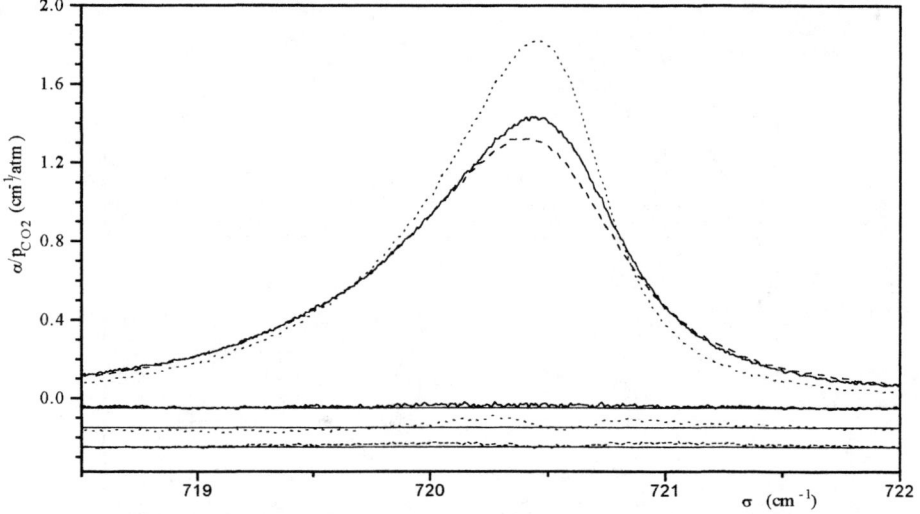

Figure 2: Influence of the perturber on the shape of the $(v_1-v_2)_I$ Q branch for about 10 atm at room temperature. ...., ———, and - - - - are measured spectra and the associated deviations with computations for He, Ar, and N_2, respectively.

Effect of the Band Symmetry

The influence of vibrational angular momenta is explicitly contained in the model, through the sum in Eq. (2) and the 3J symbols in Eqs. (3-4). The results obtained can be summarized by considering the half-widths of the Q branches, whose values are plotted in Fig. 3. They demonstrate that the band symmetry has a significant influence that is correctly accounted for by the ECS model. It also shows a number of interesting

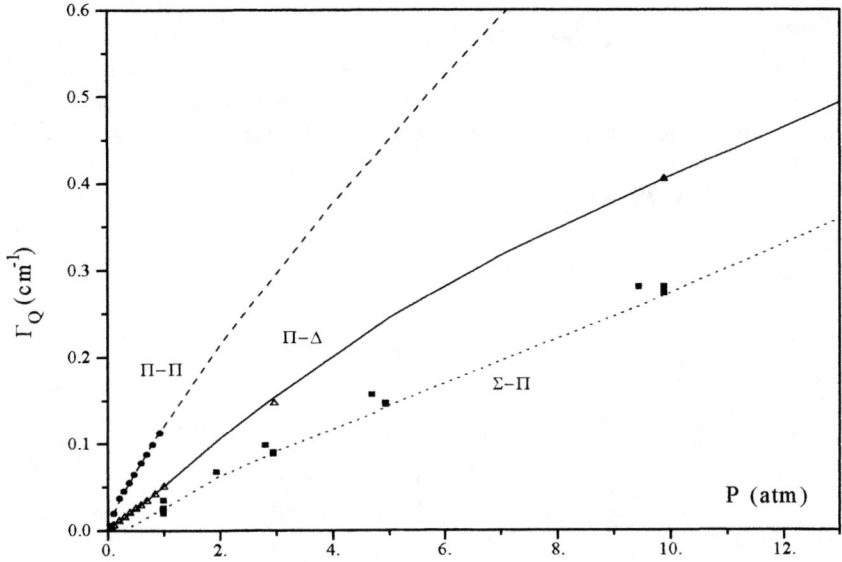

Figure 3: Room temperature CO_2-N_2 Q-branch half widths Γ_Q in different bands. Symbols and lines are values deduced from measured and computed spectra respectively. ● and - - - - are for a $\Pi \leftarrow \Pi$ Q branch [$11^11_{II} \leftarrow 01^10_I$]; Δ and ——— are for the a $\Pi \leftarrow \Delta$ Q branch [$11^10_{II} \leftarrow 02^20_I$]; ■ and ····· are for a number of $\Sigma \leftarrow \Pi$ and of $\Pi \leftarrow \Sigma$ Q branches.

characteristics which are: (i) $\Sigma \leftarrow \Pi$ and $\Pi \leftarrow \Sigma$ bands thus behave similarly validating the approximation in Eq. (8); indeed, the first involve only $^{11}\tau_J$ which can be deduced from Eq. (9) whereas the second require knowledge of $^{00}\tau_J$, $^{01}\tau_J$, and $^{02}\tau_J$ which were assumed equal to $^{00}\tau_J$. (ii) The width of the considered $\Pi \leftarrow \Delta$ band shows different behaviors at low and elevated pressure. This is the result (2) of the structure of this Q branch which involves both even and odd J lines. Indeed, at low pressure, the even J sub-branch is highly mixed but still isolated from the much sparser odd J lines. Its shape is dominated by even-even couplings. As pressure increases, absorption becomes sensitive to parity changing collisions and the broadening coefficient of the branch is lower. Note that the quality of calculations shows that the contributions of parity changing and parity conserving collisions are correctly accounted for. (iii) The Q branch of the considered parallel transition is much broader than those of perpendicular bands. This difference results (3) from the dipole transition moments which decrease very rapidly with increasing J for // transitions whereas they are practically constant for ⊥ bands. The absorption shape is thus mainly governed by the elements of the relaxation operator for low rotational quantum numbers. Since the associated Q-Q coupling elements are smaller than for the larger (J≈16) values involved in perpendicular transitions, the resulting Q branch is significantly broader.

Effect of the Isotopomer

A spectrum of enriched CO_2 in the 15 μm region is plotted in Fig. 4 where the v_2 Q branches of six isotopomers clearly appear. The quality of computations validates the hypothesis that the basis parameter of the model are the same for all species. As a result, line mixing effects are basically the same for all isotopomers leading to similar collisional effects on the Q branch shapes (5).

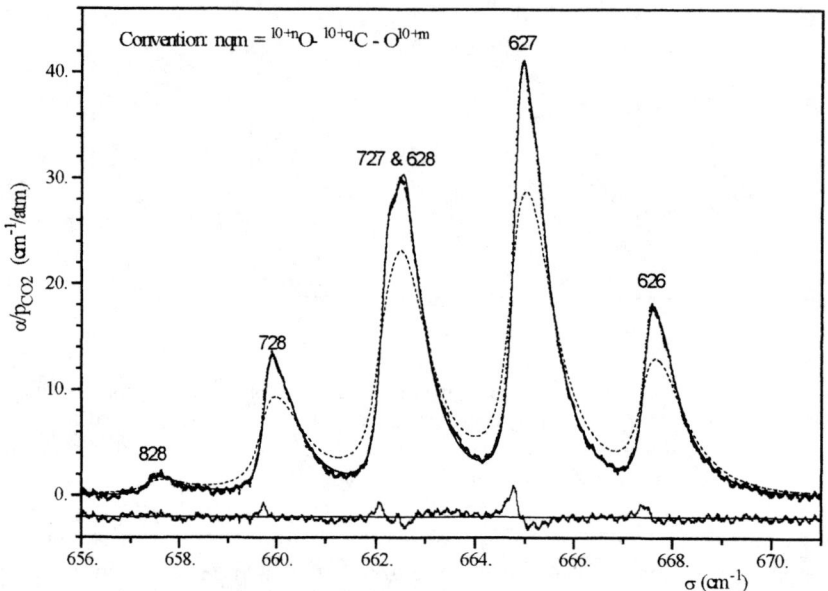

Figure 4: Room temperature CO_2-N_2 spectrum for the pressure 2.9 atm showing absorption by the v_2 Q branch of various isotopomers. ● are measured values. - - - -, ——— have been computed neglecting and accounting for line-mixing, respectively.

Overall View

For an overall view of the model quality, two quantities have been determined from each spectrum. These are the Q branch width Γ and peak absorption α^{Max}. The relative differences between values deduced from measured and calculated spectra are plotted in Fig. 5 where all CO_2-N_2 and CO_2-O_2 laboratory spectra have been considered. They show that the present ECS model leads to very satisfactory agreement with experiment since errors remain with ±5% in most cases, most of them being likely due to measurement uncertainties. On the other hand, neglecting line mixing results in errors of a factor of about two for the perturbers considered here (up to about 4 in the case of He). Note that the influence of mixing appears at pressures which depend on the line spacing so that large effects appear below 1 atm in the very narrow $11^1 0_{II} \leftarrow 02^2 0_I$ Q branch near 597 cm^{-1}.

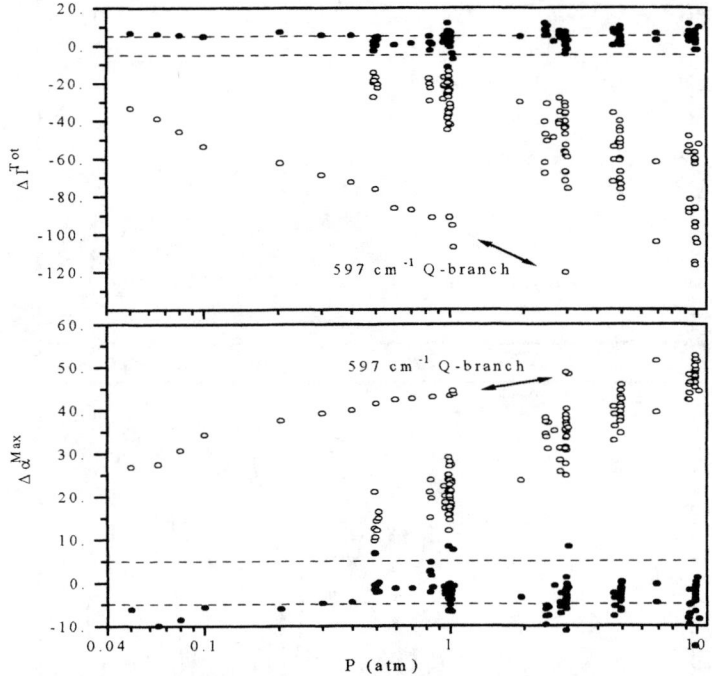

Figure 5: Relative deviations (%) between measured and computed Q-branch widths and peaks. Each point represents a spectrum (CO_2-N_2 and CO_2-O_2, all bands, pressures and temperatures). ● and ○ have been obtained accounting for and neglecting line-mixing, respectively.

Atmospheric Spectra

Examples

Typical comparison between measured and computed atmospheric transmission and emission are plotted in Fig. 6. Again, they confirm the quality of the approach and the need to account for line-mixing. Neglecting these effects leads to a strong overestimation of the absorption and emission in the near wing of the Q-branch, as a direct result of Fig. 5. These conclusions are confirmed by numerous spectra for other bands and isotopomers (4, 5). Recall that line-mixing effects on emission spectra depend on the geometry of the line of sight. Indeed, neglecting line-mixing may overestimate or underestimate radiance, depending on the optical thickness and temperature/pressure profiles (4). The quality of computations in Fig. 6 should have interesting consequences in terms of the retrieval of T/P vertical distributions from atmospheric spectra. Indeed, use of absorption in the Q-branch wing should lead to improved accuracy since it varies with P^2 whereas the dependence of spectral features on pressure is less important.

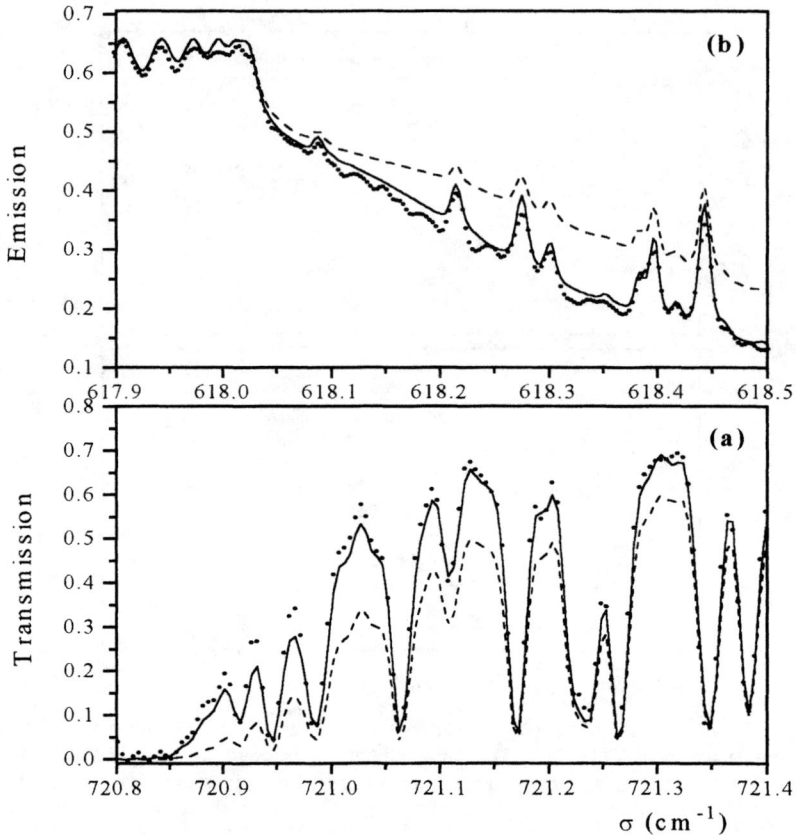

Figure 6: Stratospheric spectra in the wing of two Q branches. ● are measured values. - - - -, ——— have been computed neglecting and accounting for line-mixing, respectively. (a) transmission in the $(v_1-v_2)_I$ band for a balloon altitude of 15 km looking up ($\phi=8.5°$) at (67°N, 22°E). (b) emission in the $(v_1-v_2)_{II}$ band for a balloon altitude of 36 km looking down (tangent height=24 km) at (34°N, 254°W).

Overall View

For an overall view of the model quality, a single quantity was deduced from each of the stratospheric transmission spectrum: it is the absolute difference Δt^{Max} between measured and computed values at the spectral point where line-mixing has the most influence on transmission. The values obtained (each point is a spectrum) for three Q branches are plotted in Fig. 7 vs. a parameter (4) which quantifies the thickness of the optical path. Again, the quality of the model is clear and most results within the noise level. Neglecting line mixing leads to errors of 0.25 on transmission for thick paths. This results from the overestimation of the absorption coefficient by a factor of about two shown in Fig. 5. Note that use of emission spectra confirm the quality of the model and the need to account for line-mixing effects. (4, 5).

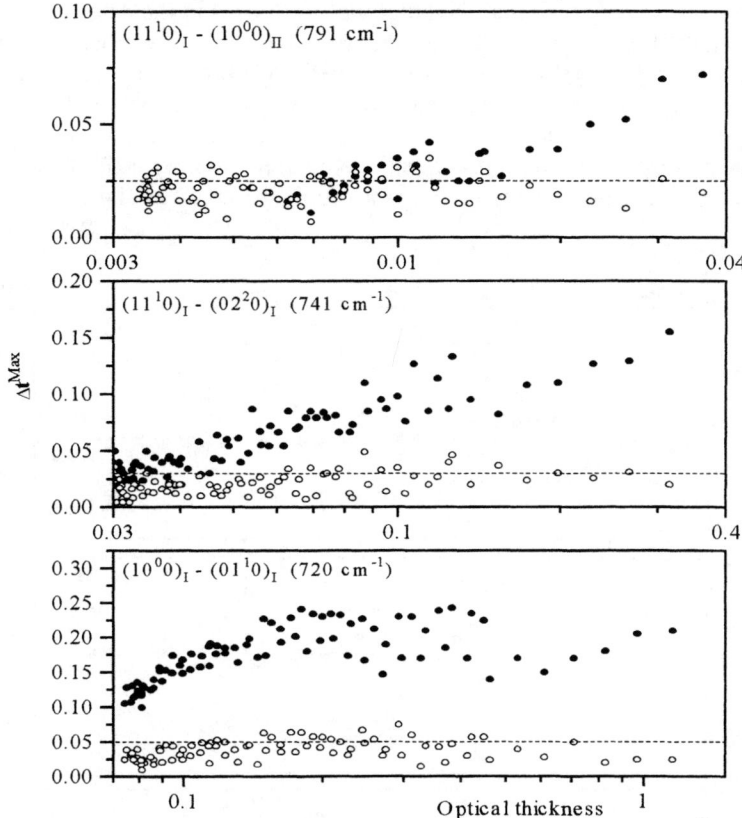

Figure 7: Absolute errors on stratospheric transmission spectra in the wing of three Q branches. ○ and ● have been obtained accounting for and neglecting line-mixing, respectively. Values are plotted vs. a parameter which quantifies the optical thickness of the path in the considered Q branch.

DISCUSSION - CONCLUSION

The results obtained on line-mixing in CO_2 infrared Q-branches have demonstrated the quality of the ECS model. Indeed, absorption is correctly computed under all considered conditions. This shows that the theoretical approach satisfactory accounts for the influence of pressure and temperature as well as that of the band symmetry. The different behaviors observed in $\Sigma-\Pi$, $\Pi-\Delta$, and $\Pi-\Pi$ transitions is a successful test of the model consistency, showing that the prediction of the coupling of angular momemta is correct.

Nevertheless, one problem remains when line-broadening coefficients are considered. Indeed, as shown in Fig. 8, expect for CO_2-He the agreement between measured and computed line-widths is not as satisfactory as that obtained when absorption is considered [recall that the latter was privileged in the determination of the model parameters (1)]. This result is likely due to the energy corrections that have been introduced [through Eq. (6)] which are negligible for He but have significant influence for the other collision partners. Furthermore, use of a molecule-atom model for CO_2-N_2 and CO_2-O_2 is clearly questionable. These problems are "hidden" when the model is applied to spectra, by the use of adjusted parameters; on the other hand the independent test using line-broadening somehow shows a problem (although slight) in the model consistency.

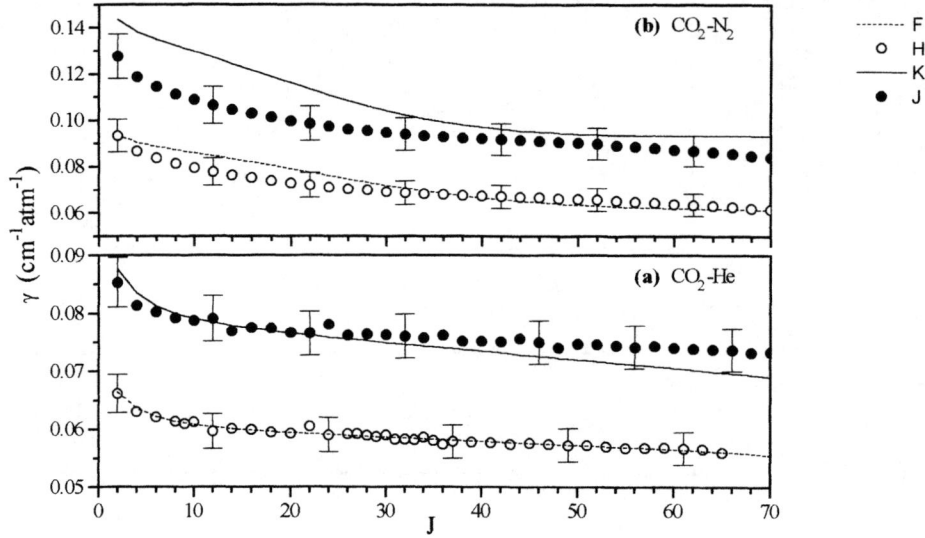

Figure 8: Individual line-broadening parameters. Symbols are measured values and lines are computed with the ECS-EP model with: upper results for 200 K and lower values for 300 K.

ACKNOWLEDGMENTS

J.M. Hartmann is grateful to the many people who have contributed to the study of line mixing in CO_2 Q-branches whose results are summarized here. This work has been supported by the Centre National d'Etudes Spatiales.

REFERENCES

1. R. Rodrigues, B. Khalil, R. Le Doucen, J.M. Hartmann, and L. Bonamy, *J. Chem. Phys.*, **107**, 4118-4132 (1997).
2. R. Rodrigues, Gh. Blanquet, J. Walrand, B. Khalil, R. Le Doucen, F. Thibault, et J.M. Hartmann, *J. Mol. Spectrosc.*, **186**, 256-268 (1997).
3. J.P. Bouanich, R. Rodrigues, J.M. hartmann, J.L. Domenech, a,d D. Bermejo, *J. Mol. Spectrosc.* **186**, 269-275 (1997).
4. R. Rodrigues, K.W. Jucks, N. Lacome, Gh. Blanquet, J. Walrand, W.A. Traub, B. Khalil, R. Le Doucen, A. Valentin, C. Camy-Peyret, L. Bonamy, et J.M. Hartmann,, *J. Quant. Spectrosc. Radiat. Transfer* (in press).
5. K.W. Jucks, R. Rodrigues, R. Le Doucen, W.A. Traub, C. Claveaux, et J.M. Hartmann,, *J. Quant. Spectrosc. Radiat. Transfer* (submitted).
6. A. Ben-Reuven, *Phys. Rev.* **145**, 7-22 (1966).
7. L. Bonamy and F. Emond, *Phys. Rev. A* **51**, 1235-1240 (1995).
8. A.E. DePristo, S.T. Augustin, R. Ramaswamy, and H. Rabitz, *J. Chem. Phys.* **71**, 850-865 (1979).
9. B. Lavorel, G. Fanjoux, G. Millot, L. Bonamy, and F. Emond, *J. Chem. Phys.* **103**, 9903-9906 (1995).
10. R. Rodrigues, C. Boulet, L. Bonamy, and J.M. Hartmann, *J. Chem. Phys.* (in press).
11. L.S. Rothman, ASA meeting, Reims, France, 4-6 Sept. (1996).
12. F. Thibault, J. Boissoles, R. Le Doucen, J.P. Bouanich, Ph. Arcas, and C. Boulet, *J. Chem. Phys.* **96**, 4945-4953 (1992).
13. J.P. Bouanich, J.M. Hartmann, G. Blanquet, J. Walrand, D. bermejo, and J.L. Domenech, *J. Chem. Phys.* (submitted).

Egelstaff Time and the Birnbaum-Cohen Line Shape

John Courtenay Lewis
and
Clifford Stamp

Department of Physics and Physical Oceanography
Memorial University of Newfoundland
St. John's, NF
Canada A1B 3X7.

Abstract

Two approaches to constructing line shape functions for fitting collision-induced spectra are examined. Starting with the symmetric time correlation function of Birnbaum and Cohen, we compare the lineshape obtained by including detailed balancing using a Boltzmann factor with that obtained using Egelstaff's complex time (the Birnbaum-Cohen lineshape). The lineshape obtained using Boltzmann-factor asymmetrization is found to be slightly superior in quality of fit to the Birnbaum-Cohen lineshape, and is faster to compute.

1 Quantal Lineshapes from Classical Models

A collision-induced spectral band can be written as

$$S(\omega) = \frac{1}{2\pi} \int_{-\infty}^{\infty} d\tau e^{-i\omega\tau} M(\tau), \ M(\tau) \equiv \frac{1}{Z} tr \left[e^{-\beta H} \vec{\mu}(\tau) \cdot \vec{\mu}(0) \right] \quad (1)$$

and $\vec{\mu}(\tau)$ is some component of the induced dipole moment. Symmetrized and anti-symmetrized correlations function for $\vec{\mu}(\tau)$ can be defined by

$$M_{\pm}(\tau) \equiv \frac{1}{2Z} tr \left[e^{-\beta H} \{ \vec{\mu}(\tau) \cdot \vec{\mu}(0) \pm \vec{\mu}(0) \cdot \vec{\mu}(\tau) \} \right].$$

A model of the system, expressed in terms of a symmetrical time correlation function $C(\tau) \equiv \kappa(\tau^2)$ can be related to the observed spectrum in different ways. One widely used approach is to assume that

$$M_{+}(\tau) \simeq C(\tau) = \kappa(\tau^2), \quad (2)$$

or, equivalently,

$$S(\omega) = \frac{2G(\omega)}{1+e^{-\beta\hbar\omega}}, \quad G(\omega) \simeq \frac{1}{2\pi}\int_{-\infty}^{\infty} d\tau\, e^{-\iota\omega\tau} C(\tau)$$

where $G(\omega)$ is the symmetrised line shape function. This satisfies detailed balancing exactly. Levine and Birnbaum[1] followed this approach in developing a lineshape which has been used successfully in fitting collision-induced spectra, in particular the Q branch. It can be shown, in this scheme, that

$$\iota M_-(\tau) = \beta\hbar\tau\kappa'(\tau^2) + O(\beta^3\hbar^3). \tag{3}$$

Another approach, developed by Egelstaff[2] in the theory of neutron scattering, and applied by Birnbaum and Cohen[3] to collision-induced absorption, is to replace τ in $C(\tau)$ with complex "Egelstaff time", and then to assume that the result approximates the autocorrelation function $M(\tau)$:

$$M(\tau) \simeq C\left(\sqrt{\tau^2 - \iota\beta\hbar\tau}\right), \quad S(\omega) \simeq \frac{1}{2\pi}\int_{-\infty}^{\infty} d\tau\, e^{-\iota\omega\tau} C\left(\sqrt{\tau^2 - \iota\beta\hbar\tau}\right).$$

This also satisfies detailed balancing exactly. It can be shown that

$$M_+(\tau) = \kappa(\tau^2) + O(\beta^2\hbar^2), \quad \iota M_-(\tau) = \beta\hbar\tau\kappa'(\tau^2) + O(\beta^3\hbar^3). \tag{4}$$

It is evident from comparing Eq. (2) with Eq. (4) and Eq. (3) with eq. (4) that the two approaches to obtaining a line shape function which satisfies detailed balancing from a classical model agree only through terms of first order in $\beta\hbar$.

Birnbaum and Cohen obtain their widely used and successful lineshape by taking

$$C(\tau) = \exp\left\{\tau_1^{-1}\left[\tau_2 - \sqrt{\tau_2^2 + \tau^2}\right]\right\} \tag{5}$$

to obtain

$$S(\omega) = \frac{1}{\pi} e^{\eta\tau_2} e^{\beta\hbar\omega/2} \frac{a\eta}{\sqrt{\omega^2+\eta^2}} K_1\left(a\sqrt{\omega^2+\eta^2}\right), \quad a = \sqrt{\tau_2^2 + (1/4)\beta^2\hbar^2}, \quad \eta = \tau_1^{-1}.$$

Using eq. (5) in the first approach gives

$$S(\omega) = \frac{2}{1+e^{-\beta\hbar\omega}} \frac{e^{\frac{\tau_2}{\tau_1}}}{\pi} \frac{\tau_2}{\sqrt{1+\omega^2\tau_1^2}} K_1\left(\frac{\tau_2}{\tau_1}\sqrt{1+\omega^2\tau_1^2}\right). \tag{6}$$

Table 1: Lineshape Parameter Correlation

	Lewis-Birnbaum-Cohen Lineshape				Birnbaum-Cohen Lineshape			
	S	τ_1	τ_2	dv	S	τ_1	τ_2	dv
S	1.00	0.04	0.52	-0.18	1.00	-0.70	0.95	0.18
τ_1	0.04	1.00	-0.79	-0.46	-0.70	1.00	-0.84	-0.46
τ_2	0.52	-0.79	1.00	0.25	0.95	-0.84	1.00	0.28
dv	-0.18	-0.46	0.25	1.00	0.18	-0.46	0.28	1.00

2 Comparisons with Experiment

We have fitted collision-induced spectra for pure H_2, H_2-Ar, H_2-N_2, and H_2-D_2 both with the Birnbaum-Cohen lineshape and with Eq. (6). In all cases eq. (6) fits slightly better than the Birnbaum-Cohen lineshape. For example in the first overtone band of H_2 at 77 K, the average decrease in standard error of 0.1. Furthermore, its parameters show less statistical correlation than the parameters of the Birnbaum-Cohen lineshape, and it is faster to compute. The decrease in correlation is shown in Table 1.

3 Conclusion

We conclude that the usefulness of the Birnbaum-Cohen lineshape in fitting collision-induced spectra comes from the underlying symmetric lineshape, not from any particular virtue of "Egelstaff time". In fact, calculating a symmetric spectrum from the Birnbaum-Cohen *correlation function*, Eq. (5), and then multiplying by a Boltzmann factor gives a lineshape which is slightly better in quality of fit, and significantly better in terms of interpretation of the results of fits, than the Birnbaum-Cohen lineshape itself.

4 References

1. Levine, H.B. and G. Birnbaum, Phys. Rev. **154**, 72 (1967).

2. Egelstaff, P.A., *An Introduction to the Liquid State. 2nd Ed.*, Oxford University Press, 1994. Egelstaff introduced "Egelstaff time" in 1961.

3. Birnbaum, G. and E.R. Cohen, Can. J. Phys. **54**, 593 (1976).

ENERGY CORRECTED SUDDEN CALCULATIONS OF LINE WIDTHS AND LINE SHAPES BASED ON COUPLED STATES CROSS SECTIONS: THE TEST CASE OF CO_2-ARGON

F. THIBAULT, J. BOISSOLES
U.M.R. C.N.R.S. 6627, Université de Rennes I, Campus de Beaulieu, 35042 RENNES CEDEX, France.

C. BOULET, L. OZANNE and J. P. BOUANICH
Laboratoire de Physique Moléculaire et Applications, (Laboratoire associé aux Universités Paris-Sud et P. et M. Curie), UPR 136 du CNRS, Université de Paris-Sud, Centre d'Orsay, Bât. 350, 91405 ORSAY CEDEX, France.

C. F. ROCHE and J. M. HUTSON
Department of Chemistry, University of Durham, Durham, DH1 3LE, England.

The accuracy of the energy corrected sudden (ECS) formalism for line mixing calculations has been investigated using coupled states calculations for CO_2-Ar collisions on the recently developed "single repulsion" potential of Hutson et al [1].

In the ECS approach, the real part of the relaxation matrix may be expressed in terms of the so-called "basic rates" $Q_L(T)$ which are actually thermally averaged inelastic cross sections for downward transitions to the ground state, $\sigma^0(L \to 0; E)$. In the past, the basic rates have been usually modelled through some simple analytical expression, based on some adjustable parameters deduced from measurements through an "ECS – inversion". In the present work, the basic rates have been calculated, with the coupled states approximation using the MOLSCAT program [2]. The basis set used included all rotor functions up to J≈38, which should give reasonable convergence for basic rates up to about L≈30.

The reader will find a résumé of the ECS formalism for the calculation of the real part of the relaxation matrix in ref.[3].

The ECS approach does not provide the imaginary part of the relaxation matrix which cannot be neglected at high perturber pressures. A suitable model has been proposed and discussed in ref. [4] and will be used here for Σ-Σ bands at high perturber density, together with effective number density of the perturber, which takes excluded volume effects into account as explained in ref.[5]. Finally it has been shown [6] that the adiabaticity factor of Bonamy et al [7] gives better agreement with experiment and direct calculations than that of De Pristo et al [8]. Therefore we will consider here only results obtained with the "Bonamy" factor.

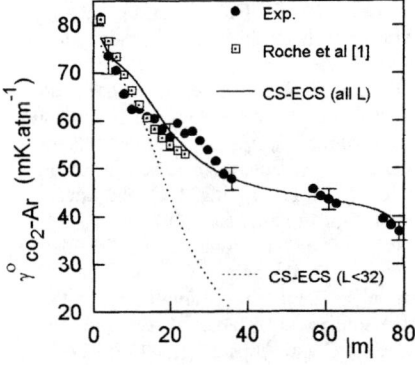

Fig.1: Line widths

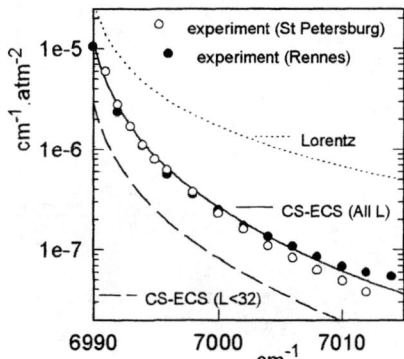

Fig.2 : Absorption in the band wing of the 00^03-00^00 band

From Fig.1, it appears that the ECS line widths for low initial J_i, $J_i \leq 16$, are sensitive only to the low L basic rates for which the CS calculations are converged ; comparing them with directly calculated CS line widths thus gives a stringent test of the ECS model and it works well (~10%).

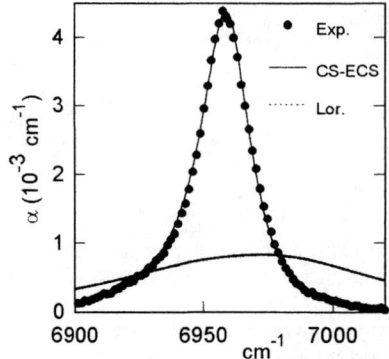

Fig.3 : absorption in the $00^03\text{-}00^00$ band for P[Ar]=592 Atm.

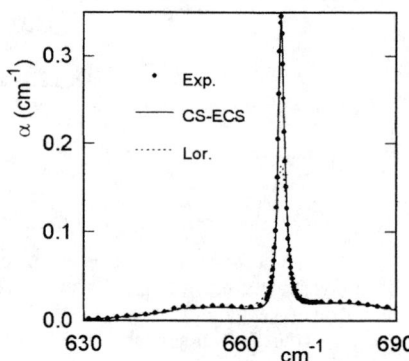

Fig.4 : absorption in the $01^10\text{-}00^00$ band for P[Ar]=20 Atm

However, for higher J_i lines and wings calculations (see Fig.2), basic rates for higher L ($L \geq 30$) are needed for convergence. These have been obtained by an extrapolation procedure based on the following experimental data : line widths and absorption in the near wing of the $3\nu_3$ band [6]. ECS calculations using the resulting basic rates are designated "extrapolated CS-ECS calculations" and are found to give accurate results for high J line widths (Fig.1), near wing absorption (Fig.2), high density band profiles (Fig.3) and absorption in Q branches of any symmetry ($\Pi\text{-}\Sigma$: Fig.4 ; $\Delta\text{-}\Pi$: Fig.5).

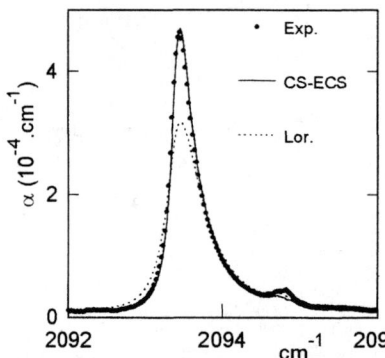

Fig.5 : absorption in the $12^201\text{-}01^101$ band for P[Ar]=2.4 Atm.

REFERENCES

1. C . F. Roche, A. S. Dickinson, A. Ernesti and J. M. Hutson, J. Chem. Phys. **107**, 1824 (1997).
2. J. M. Hutson and S. Green, MOLSCAT computer programm, version 14, distributed by Collaborative Computational Project N° 6 of the UK Science and Engineering Research Council, 1994.
3. J. Boissoles, F. Thibault and C. Boulet, J. Quant. Spectrosc. Radiat. Transfer **56**, 835 (1996).
4. R. Rodrigues, C. Boulet, L. Bonamy and J. M. Hartmann, J. Chem. Phys., to be published.
5. L. Ozanne, Q. Ma, Nguyen-Van-Than, C. Brodbeck, J. P. Bouanich, J. M. Hartmann, C. Boulet and R. H. Tipping, J. Quant. Spectrosc. Radiat. Transfer **58**, 261 (1997).
6. F. Thibault, J. Boissoles, C. Boulet, L. Ozanne, J.P. Bouanich, C. Roche and J.M. Hutson, J. Chem. Phys. Submitted.
7. L. Bonamy, J.M. Huet, J. Bonamy and D. Robert, J. Chem. Phys. 95, 3361 (1991).
8. A. E. DePristo, R. Ramaswamy, S. D. Augustin and H. Rabitz, J. Chem. Phys. **71**, 850 (1979).

Pressure Broadening and Saturation Lineshapes of CO at Pressures between 10^{-2} Pa and 10^2 Pa

Peter Palm, Dirk Hanke, Manfred Mürtz, Bertold Frech and Wolfgang Urban

Institut für Angewandte Physik der Universität Bonn, Wegelerstraße 8, 53115 Bonn, Germany
Email: manfred@iap.uni-bonn.de

Abstract.
Saturated absorption lineshapes in the CO $v = 1 \leftarrow 0$ band were measured at pressures between 10^{-2} Pa and 10^2 Pa in pure CO and CO-He mixtures with a spectral resolution of better than $\Delta\nu/\nu = 3 \times 10^{-10}$. The CO saturation signals mainly consist of narrow dips with a homogeneous width that nonlinearly depends on pressure due to dominant contributions of velocity changing collisions (VCC) and a broader background that can also be attributed to VCC. Preliminary evaluation shows a change of the CO self broadening coefficient from 96 kHz/Pa for $p < 3$ Pa to 38 kHz/Pa for 15 Pa $< p <$ 40 Pa. The CO-He broadening coefficient was found to be 47 kHz/Pa for 0.2 Pa $< p <$ 16 Pa.

INTRODUCTION

The long natural life time of rovibrational transitions in the infrared spectral region leads to collision induced line broadening even down to very low pressures ($p < 1$ Pa). Saturation spectroscopy can be used to overcome the dominant Doppler broadening at these pressures by selecting only a narrow longitudinal velocity class within the Maxwellian distribution. In this case the time of coherent interaction between field and molecule is strongly effected by velocity changing collisions (VCC) that can remove molecules from the resonant velocity class. This leads to a much higher and pressure dependent broadening coefficient at low pressures compared to the high pressure case. In contrast to earlier measurements on relatively complex systems like CH_4, CO_2 [1], our new data for the CO-CO and CO-He systems could be used to test modern ab-initio calculations starting with well known intermolecular potentials [2].

EXPERIMENTAL SETUP

Figure 1 shows the experimental setup of our newly developed ultra high resolution spectrometer. It is based on a liquid nitrogen cooled, flowing gas CO laser. To frequency stabilize the CO laser about 0.1% of its output power is transfered into a sideband produced by an electro optical modulator (EOM). The sideband is frequency modulated and tuned into resonance with an OCS transition. The OCS saturation signal is used to lock the sideband frequency to the OCS frequency by tuning the laser cavity length. Thus the laser frequency is stabilized but can be tuned within the gain profile of each laser line by tuning the microwave frequency applied to the EOM. To detect saturation lineshapes of CO the unmodulated carrier is split into a strong pump beam and a weak probe beam, passing the absorption cell anticollinear. Only the pump beam is amplitude modulated using a chopper. After passing the absorption cell part of the probe beam is extracted by a beam splitter and focussed onto a liquid nitrogen cooled InSb detector. Background free saturation signals are obtained by phase sensitive detection at the chopper frequency. Depending on pressure absorption cells with path lengths between 0.04 m and 24 m, saturation powers between 270 μW and 8 mW, and beam diameters between $2\omega_0 = 1$ mm and $2\omega_0 = 25$ mm are used.

PRELIMINARY RESULTS

We measured saturation lineshapes of the P(15) transition in the CO $v = 1 \leftarrow 0$ band at pressures between 10^{-2} Pa and 10^2 Pa. This extends our earlier measurements [3] to lower as well as higher pressures by one order of

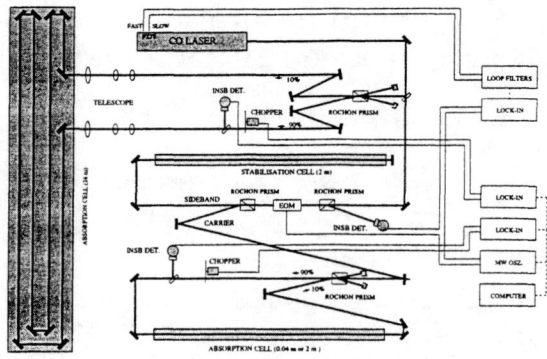

FIGURE 1. Experimental setup.

magnitude. A resolution of better than $\Delta\nu/\nu = 3 \times 10^{-10}$ was obtained, which represents the highest resolution at $\lambda = 5$ μm reported so far and improves our earlier results by one order of magnitude.

For a first evaluation Lorentzian lineshapes were fitted to the measured saturation dips. Figure 2c shows the half widths at half maximum obtained versus pressure. No corrections for power broadening, residual Doppler broadening, and transit time broadening were applied. Further analysis will be published shortly. The CO self broadening coefficient was found to be strongly nonlinear. A linear fit of the halfwidths at pressures below 3 Pa gives a broadening coefficient of 96 kHz/Pa, changing to 38 kHz/Pa for a linear fit at pressures between 15 Pa and 40 Pa. For CO-He the broadening coefficient was found to be 47 kHz/Pa for a linear fit at pressures between 0.2 Pa and 16 Pa. A closer look at the lineshapes showes that the wings are much wider than those of Lorentzian profiles (Figure 2b). We attribute this to an additional background due to velocity changing collisions from longitudinal velocity class v_z to velocity class $-v_z$, transfering saturation from the velocity class resonant with the pump beam to the velocity class resonant with the probe beam, leading to a saturation signal outside of the velocity class $v_z = 0$ that is in resonance with both beams even without VCC.

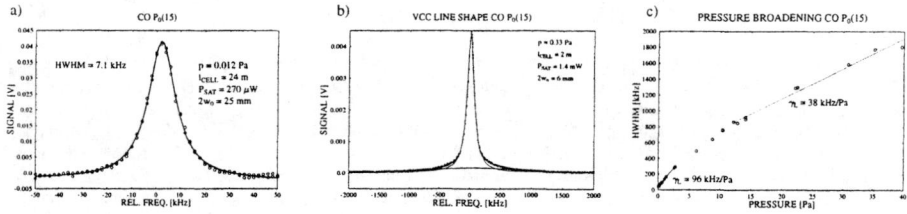

FIGURE 2. a) Saturation lineshape at very low pressure. b) Lorentzian and background used to fit non-Lorentzian wings. c) Pressure dependent CO self broadening.

REFERENCES

1. S.N. Bagayev, In *Laser Spectroscopy IV*, Berlin: Springer, 1979.
2. S. Green, *Proceedings of the International Conference on Spectral Lineshapes*, New York: AIP Press, 1995
3. M.H. Wappelhorst, M. Mürtz, P. Palm, W. Urban, *Appl. Phys. B* **65**, 25 (1997).

The 2093 and 2130 cm^{-1} CO$_2$ Q-Branches Revisited: Line Mixing Effects

Adriana Predoi-Cross, R. Berman, J. R. Drummond and A. D. May

Department of Physics, University of Toronto
60 St. George St., Toronto, Ont. M5S 1A7, Canada

In the past few years the 2093 and 2130 cm^{-1} Q-branches have been the focus of detailed investigations for pure CO$_2$ [1], for CO$_2$ perturbed by N$_2$, O$_2$ and Ar [2] and for CO$_2$ perturbed by He [3]. We have re-investigated the line mixing and band collapse of CO$_2$ Q-branch spectra in the 4.7 µm spectral region, covering a pressure range from 1 kPa to 30 atm.

Berman et al. [4] have recently reported direct measurements of line mixing coefficients in the 2076 cm^{-1} v_1+v_2 Q-branch. The same approach has been used for the present analysis. For pressures below 70 kPa the spectra have been recorded using a tunable difference frequency laser spectrometer and a temperature controlled cell. The spectral resolution was 2 MHz and the signal to noise ratio was in excess of 1500 for a 1 sec integration time. Our first experiments covered the 0.5 to 15 kPa pressure range at room temperature. The spectra were fitted successfully using the Rosenkranz first order theory of line mixing. Translational effects were accounted for using the hard collision model modified to include an asymmetry component due to line mixing. The narrowing parameters were found to be strongly non-linear with pressure, a consequence of the speed dependent effects unaccounted for in our model. Additional spectra, up to 70 kPa, were fitted including higher order line mixing effects. To analyse these spectra, an empirical speed dependent profile was formed from the sum of two hard collision profiles (double hard collision model).[5] The spectra displayed mixing effects non-linear in density. The intensity transfer between lines was directly measured and found to agree qualitatively with that predicted by an EPG scaling law calculation.

As pressure increases one Q-branch line becomes the dominant line and acquires the strength of the entire band. The individual line structure of the Q-branches collapse towards a single Lorentzian line with a width that decreases with pressure. This effect is due to line mixing [6]. Based on the results from the lower pressure CO$_2$ spectra, predictions were made for the line mixing strengths, broadening coefficients and symmetric strength components in high pressure spectra. New spectra have been recorded at room temperature and 75°C in the 1 to 30 atm pressure range using a Bomem FTIR instrument. In this pressure range line mixing effects are very strong; however, the bands are not fully collapsed. We have analysed these spectra using a simple Lorentzian model and a more complex line by line calculation. Our results highlight the rich effects of line mixing in these bands.

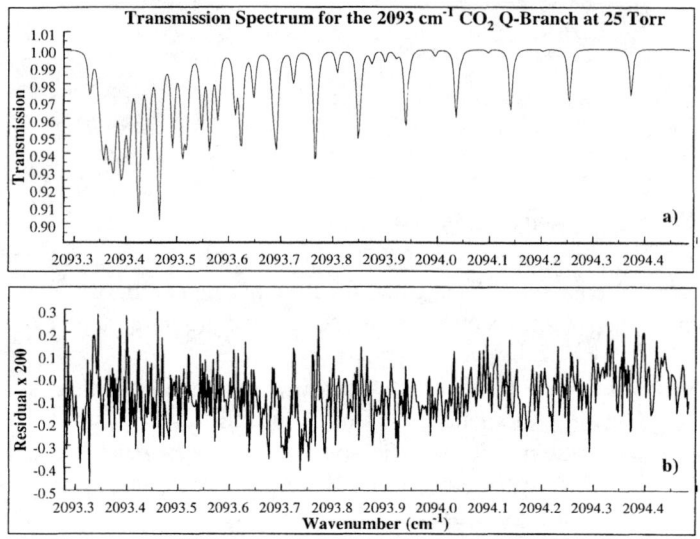

Fig. 1. a) Transmission spectrum for the 2093 cm^{-1} CO_2 Q-branch recorded at room temperature and 25 torr. (b) Residual from the fit for the line strengths using the double hard collision model.

ACKNOWLEDGMENTS

This work was supported in part by the Natural Sciences and Engineering Research Council of Canada and the University of Toronto Research Fund.

REFERENCES

1. M. Margottin-Maclou, F. Rachet, C. Boulet, A. Henry, and A. Valentin, *J. Mol. Spectrosc.* **172**, 1 (1995).
2. F. Rachet, M. Margottin-Maclou, A. Henry, and A. Valentin, *J. Mol. Spectrosc.* **175**, 315 (1996).
3. J. Boissoles, F. Thibault, F. Rachet, A. Valentin, and C. Boulet, *J. Quant. Radiat. Transfer* **57**, 519 (1997).
4. R. Berman, P. Duggan, P.M. Sinclair, A.D. May, and J.R. Drummond, *J. Mol. Spectrosc.* **182**, 350 (1997).
5. R. Berman, P.M. Sinclair, A.D. May, and J.R. Drummond, submitted to *Applied Optics*.
6. J.M. Hartmann and C. Boulet, *J. Phys. Chem.* **94**, 6406 (1991).

Evidence of Inhomogeneous Broadening and Shifting in the Raman Q Branch of D_2 and D_2–He at Low Temperatures

S. H. Fakhr–Eslam, G. D. Sheldon, J. R. Drummond and A. D. May

Department of Physics, University of Toronto
Toronto, Canada, M5S 1A7

Using high resolution Raman gain spectroscopy, we have measured the width and shift of the Q(0), Q(1) and Q(2) lines in pure D_2 and D_2–He mixtures at 100.7, 150.0, 200.0 and 250.0 K. As an example, Fig.1 shows a plot of the frequency shifts of the Q(2) line at 250.0 K as a function of the density, for

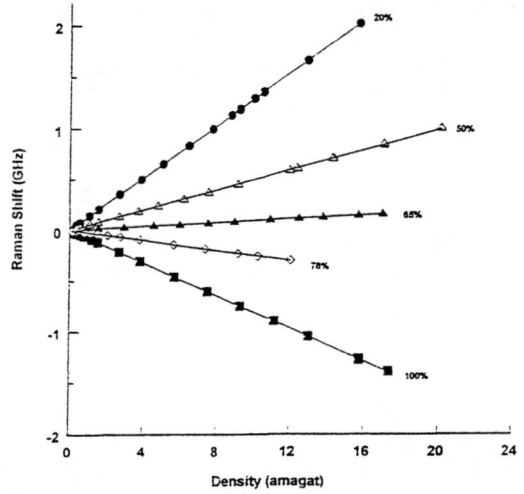

Figure 1 Line shift versus density for the Q(2) line in D_2–He mixtures at 250 K.

different mixture ratios. Here, as in most experimental set ups, only relative frequencies are measured. To determine shifts, the raw measurements, ν_i, for each series of densities, are fitted to an equation of the form $\nu = \nu_{ins} + \delta^1 \rho$ and the results, $\nu_i - \nu_{ins}$, presented as experimental shifts, Δ_i. (Conceptually, the instrumental constant, ν_{ins}, converts the measurements to absolute frequency relative to the free molecule value.) Clearly a plot of the "shift", Δ_i, versus density, ρ, must pass through the origin and have the slope (shifting coefficient) δ^1. This gentle "massaging" of the raw data is theoretically justifiable for low density gases, provided the relaxation of the optical coherence is speed independent (homogeneous). For precise measurements or higher densities it may be necessary to add terms in the fitting routine that are quadratic or cubic in the density. Such nonlinear terms are expected and observed at very high densities. Visually, Fig.1 supports a low density view and to the best of our knowledge, so do all measurements of shifts in low density gases.

To determine broadening coefficients, the raw data must also be gently "massaged", but for a different reason. Here the both translational motion and the relaxation of the optical coherence play a role. At not too low a density the widths, $\Delta\nu_i$, would be fit to an equation of the form, $\Delta\nu = a/\rho + \gamma^1\rho$, where the

$1/\rho$ term describes the effect of Dicke narrowing. In a parallel treatment, one could present the results, $\Delta\nu_i - a/\rho$, as the collisional widths, Γ_i. A plot of Γ_i would be proportional to density with a slope (broadening coefficient), γ^1.

In summary, conventional analysis of spectral profiles of isolated lines in low density gases, extracts a broadening and a shifting coefficient by treating the collisional widths and shifts as <u>proportional</u> to density. In the past, small deviations from a proportional relationship have been attributed to the presence of terms of higher order in density. They have been detected only by plotting residuals, the difference between measured widths and shifts and best fit lines. Departures from a proportional behaviour were not visually apparent in plots like Fig.1. Here we report on departures from a proportional behaviour that is visually apparent in a plot of both the width and shift of a line. However, in contrast with previous interpretations[1], the departures can not be ascribed to weak terms quadratic or cubic in density.

As our clearest example of this new behaviour, Fig.2 shows a plot of the width of the Q(1) line in pure D_2 at 100.7 K, as a function of density. We see

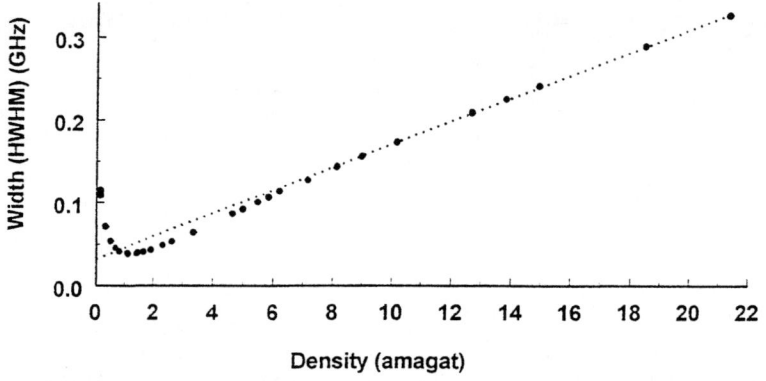

Figure 2 Width (HWHM) of the Q(1) in pure D_2 at 100 K along with a straight line fit to data above 10 amagat.

from the dotted line that the measured widths (raw data) from about 10 up to 22 amagat units of density fall on a straight that does not pass through the origin. In this region of density the collisional width can be represented by the equation, $\Gamma = \Gamma^o + \gamma^1\rho$, i.e. a width <u>linear</u> in, as opposed to <u>proportional</u> to, the density. The Dicke narrowing contribution is evident at lower densities but the entire curve cannot be represented by, $\Gamma = a/\rho + \gamma^1\rho$, or by a similar curve with weak terms, quadratic or cubic in the density. Measurements of shifts show a similar linear behaviour at intermediate densities (a few tens of amagat). Terms nonlinear in density are weak and only appear in the neighbourhood of 100 amagat. We[2] have interpreted the constant offsets (Γ^o for the width, Δ^o for the shift) as inhomogeneous contributions arising from a speed dependence to the relaxation of the optical coherence. Complete details of these new measurements are given in reference 2.

1   P. M. Sinclair, P Duggan, J. W. Forsman, J. R. Drummond and A. D. May, Can. J. Phys. **72**, 891–896 1994).
2   S. H. Fakhr–Eslam, Ph.D. Thesis, University of Toronto, to be submitted.

Direct Measurements of Line Mixing in Pure CO_2

R. Berman, P. M. Sinclair, J. R. Drummond, and A. D. May

Department of Physics, University of Toronto
Toronto, Canada, M5S 1A7

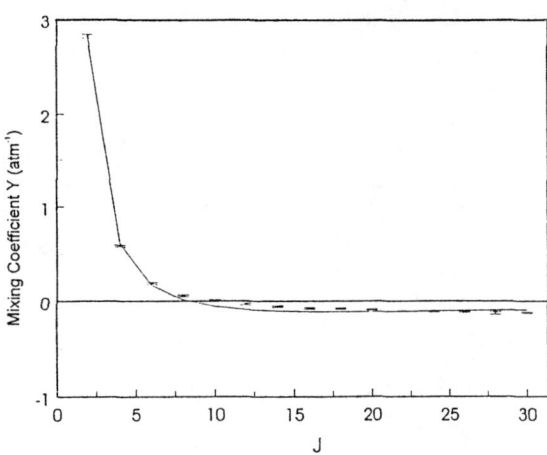

Fig.1 Mixing coefficients as a function of J.

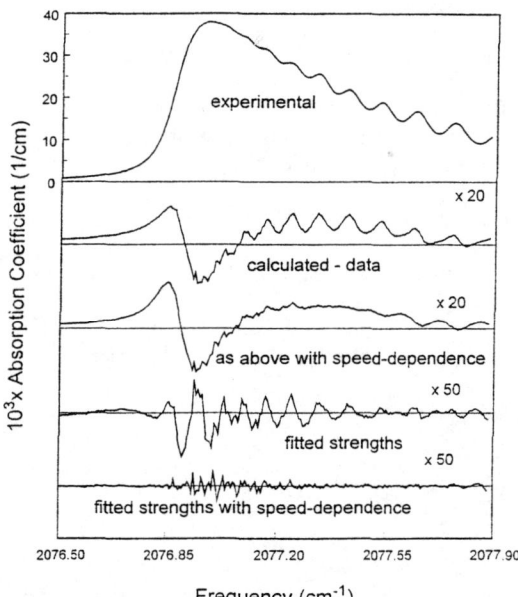

Fig.2 Spectral profile at 32.5 kPa and residuals to fits.

We have made high resolution (2 MHz) and high signal to noise (3000:1) measurements of the $\nu_1+\nu_2$ Q-branch of pure CO_2. Line mixing effects have been measured directly from spectra at low density[1] using a line by line fitting and the weak mixing expression of Rosenkranz[2]. Fig.1 shows the measured mixing coefficients. We have fitted the broadening coefficients (also measured at low density) to a PEG[3] law and then used this law to predict the line mixing coefficients. The result is shown as a solid line in Fig.1.

Fig.2, top panel, shows a spectral profile at 32.5 kPa where there is severe mixing. We have modified the usual expression for strong mixing to allow for the direct contribution of the translational motion to the spectrum by convoluting the standard profile with a profile generated using the hard collision model. The second panel in Fig.2 shows the difference between the measured profile and the profile calculated using the PEG law parameters determined from the low density measurements. Considerable structure is evident in the residual. This structure is attributed to a neglect of speed dependence in the broadening. We have included speed dependent broadening[4] using the speed dependent expression given by Ward et al[5], and a force law with q = 6 in their terminology. As is evident from the third panel in Fig.2, much of the fine structure in the residual has disappeared The coarse structure in the residual can be attributed to a failure in the PEG law, specifically a failure to predict the intensities of the spectral components in the strongly mixed spectrum. The fourth panel shows the residual when the strengths are fitted, but no allowance is made for speed dependent broadening. Finally the last panel shows that all structure in the residual has vanished when both effects are included in the model spectrum.

Fig.3 shows the fitted values of the line strengths versus those predicted by the PEG law as a function of density. We believe this to be the first direct

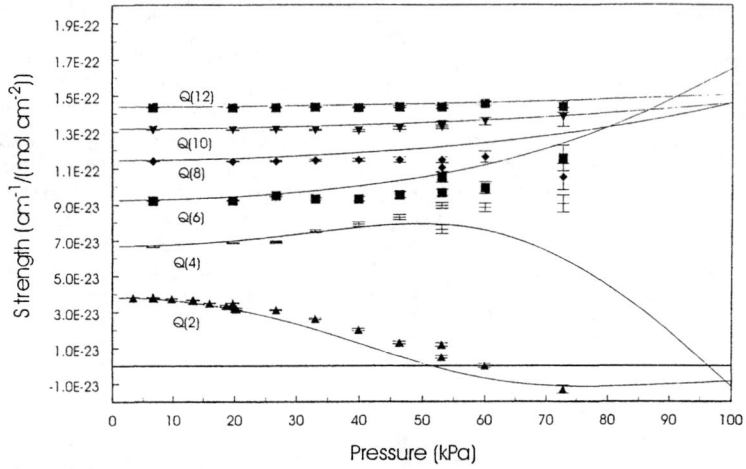

Fig.3 Fitted line strengths (point) and values calculated from the PEG law.

observation of any intensity transfer due to line mixing. Note, as predicted by the strong mixing theory, the Q(2) component acquires a "negative strength" at the highest density. Further details may be found in reference 4.

1   R. Berman, P. Duggan, P. M. Sinclair, A. D. May, and J. R. Drummond, J. Mol. Spec. **182**, 350 (1997).

2.   P. W. Rosenkranz, IEEE Trans. Anten. Propagat. **AP–23**, 498 (1975).

3   D. P. Edwards and L. L. Strow, J. Geoph. Res. **96**, 20859 (1991)

4   R. Berman, Ph. D. Thesis, Dept. of Physics, University of Toronto (1998).

5   J. Ward, J. Cooper and E. W. Smith, J. Quant. Spectrosc. Radiat. Transfer, **14**, 555 (1974).

Line Mixing in HD: Bridging Density Regimes

G. D. Sheldon, J. R. Drummond and A. D. May

Department of Physics, University of Toronto
Toronto, Canada, M5S 1A7

Using a high resolution, shot noise limited, Raman gain spectrometer, we have measured the width and mixing parameter of the Raman Q branch lines in HD from 1 to 7 amagat at 304.6 K. Fig.1a shows a plot of the width of the symmetric component of the spectral profiles as a function of density. The slope of the lines (broadening coefficients) agree with previous measurements[1] but are an order of magnitude more precise. Of greater interest are the mixing parameters or the ratio of the amplitude of the asymmetric component to the amplitude of the symmetric component of the spectral profiles.

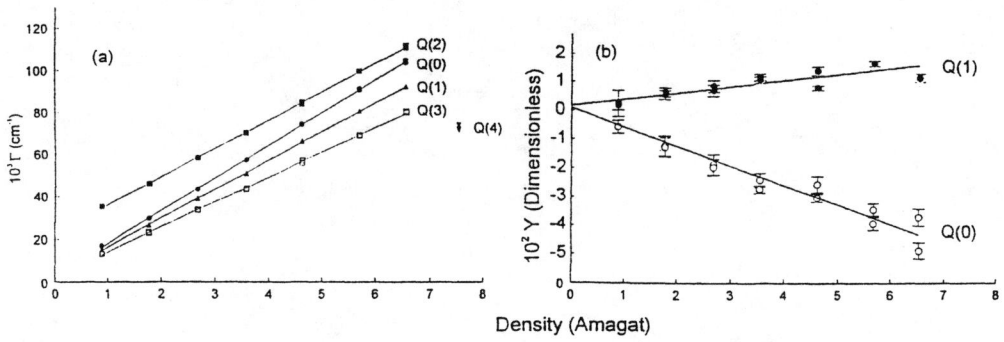

Figure 1 (a) Line width (HWHM) versus density for the Q(0) through Q(4) lines. The Q(2) data are offset by 2×10^{-3} cm^{-1}. (b) Line mixing parameter versus density for the Q(0)) and Q(1) lines.

As an example Fig.1b shows the mixing parameter of the Q(0) and Q(1) line as a function of density. The Q(0) mixing coefficient (slope of the line) is in agreement with a previous measurement[1]. The Q(1) through Q(3) mixing coefficients have not been measured previously.

The broadening coefficients were fitted to a variety of empirical energy gap laws, after the calculated[2] constant contribution for vibrational dephasing was subtracted. Variations of the exponential gap law (EGL) and the modified exponential gap law (MEGL) were found to fit both the broadening and the mixing coefficients. The variation for both gap laws consisted of using a different prefactor for an odd or even change in the rotational quantum number, J. A preference for ΔJ odd was found. The quality of fit was equal for both gap laws. Since the EGL required one less fitting parameter than the MEGL it is a preferred representation of the relaxation process. It was used as a bridge between the present results and earlier measurements of the band profile at high densities[3].

Using the fitting parameters from the low density data one can easily calculate the entire relaxation matrix at any density, provided one assumes that the rates continue to scale linearly with density. Then, using the well known expression of Baranger[4], it is a simple matter to calculate the band spectrum in the severe overlapping regime. In this manner we can predict a spectrum at high densities and we can compare it with earlier results. Of course one must add back to the relaxation matrix, a constant times the unit matrix to account for the pure vibrational broadening. As the shifting is known to be nonlinear in density at high density, one can add an imaginary fitting parameter times the unit matrix to the relaxation matrix or one can simply shift the entire calculated spectrum to match the high density data.

Fig.2 shows a comparison of the profiles, computed as outlined above, and the results of reference 3. The agreement is remarkable given that we have linearly extrapolated (except for the isotropic shift) measurements below 7 amagat, to near liquid densities. The level of agreement is superior to the agreement found by Bonamy et al. between their theoretical calculation[5] and the high density spectra[3].

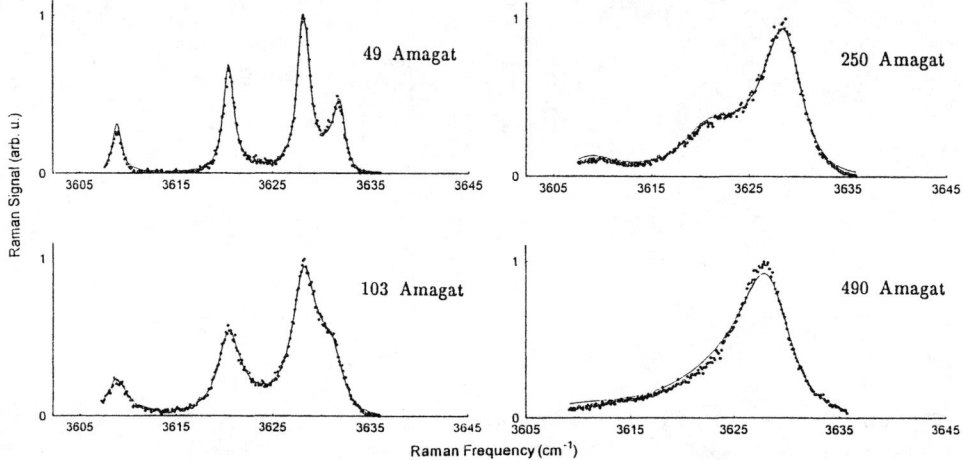

Figure 2 A comparison, at several densities, between spectra computed from our low density results and the high density results of reference 3.

The fitted values of the shifts, while not highly accurate, are nevertheless consistent with the low density measurements[3] and with theoretical calculations[5]. Complete details of the present work have been submitted for publication[6].

1 G. J. Rosasco, A. D. May, W. S. Hurst, L. B. Petway and K. C. Smyth, J. Chem. Phys. **90**, 2115—2124 (1989).
2 J. Bonamy and D. Robert, Chem. Phys. Lett. **57**, 22—28 (1978).
3 T. Witkowicz and A. D. May, Can. J. Phys. **54**, 575—583 (976), P. Dion and A. D. May, Can. J. Phys. **51**, 36—39 (1973).
4 M. Baranger, Phys. Rev. **111**, 48—493 (1958), Phys. Rev. **111**, 494—504 (1958), Phys. Rev. **112**, 855—864 (1958).
5 J. Bonamy, L. Bonamy and D. Robert, J. Chem. Phys. **67**, 4441—4453 (1977).
6 accepted for publication in J. Mol. Spectrosc.

On the Form of Rotational Relaxation Matrix in the Infinite-Order Sudden Approximation Corrected for Energy and Frequency

A. P. Kouzov

Institute of Physics, Saint Petersburg University, Peterhof,
Saint Petersburg 198904, Russia

Based on the Fano-Mori formalism, the Energy-and-Frequency Corrected infinite-order Sudden Approximation (EFCSA) is developed to incorporate the adiabatic and nonmarkovian effects into the IOSA scheme. The EFCSA rotational relaxation matrix is shown to obey the basic relations derived from the first principles. It allows self-consistent modelling of any rotational spectral characteristics based on the unique set of parameters characterizing the binary collisions. The EFCSA extends far the scopes of the existing markovian models applicable to the case of scalar photon-molecule coupling. The proposed model is successfully checked against a number of the relaxation data known for the room-temperature nitrogen.

Derivation of the relaxation matrix $\Gamma(r,\omega)$ in the form accounting both for the rank r of the tensor quantity involved in the relaxation process and the duration τ of collision is a question at issue for many spectroscopic and kinetic gas-phase studies. Rapid accumulation of spectroscopic data made obvious the limitations of the scattering theory approaches ($\tau\omega \to 0$). However, except the perturbation theory (PT) [1] whose feasibility is confined to few molecular systems, no theoretical model of Γ was so far developed to treat the r-th rank relaxation induced by noninstantaneous rotator-buffer particle collisions. This general case is presently investigated by using the Fano-Mori technique. It allowed us to correct strictly the Infinite-Order Sudden Approximation (IOSA) Γ-matrix for ω and the rotational energy E_J without the use of the common semiempirical routine [2]. Remarkably, the form of the symmetric EFCSA matrix derived (for brevity, we give only its off-diagonal elements)

$$\Gamma_{if}^{i'f'}(r,\omega) = -(1/2 n_{if} n_{i'f'}) \Pi_{J_i J_f J'_i J'_f} \sum_L \Pi_{LL} \begin{pmatrix} J'_i & L & J_i \\ 0 & 0 & 0 \end{pmatrix}\begin{pmatrix} J'_f & L & J_f \\ 0 & 0 & 0 \end{pmatrix} \times$$

$$\begin{Bmatrix} J_i & J_f & r \\ J'_f & J'_i & L \end{Bmatrix} \int_0^\infty \exp(-i\omega t)[\rho_i \exp(i\omega_{f'i}t)F_L(t) + \rho_{i'}\exp(i\omega_{fi'}t)F_L^*(-t) +$$

$$\rho_f \exp(i\omega_{fi'}t)F_L^*(t) + \rho_{f'}\exp(i\omega_{f'i}t)F_L(-t)]dt$$

coincides with that of the PT one [1] differing only in the expressions for the intracollisional time correlation functions $F_L(t)$; the latter are fundamental IOSA/EFCSA characteristics to be modeled or calculated from the basic equations. The coefficients n_{if} and $n_{i'f'}$ are given via the rotational Boltzmann factors ρ_k:

$n_{kl} = \sqrt{(\rho_k + \rho_l)/2}$. The derived Γ was shown to be positively defined and to obey both the sum rules and the Ben-Reuven symmetry relations. In the marcovian IOSA limit, all exponential factors in the integrand should be set equal to unity. Similarly to the IOSA case, the real part of the EFCSA matrix can be given in terms of the basic rates (i.e. the $0 \to L$ transition rates for the scalar case $r=0$) defined by

$Q_L = \Phi_L(\omega_{0L}) \equiv 1/2 \int_{-\infty}^{\infty} \exp(-i\omega_{0L}t) F_L(t) dt$. Due to the thermal equilibrium, the even functions $\tilde{\Phi}_L(x) = (1 + \exp(-\hbar x / kT)) \Phi_L(x)/2$ can be introduced. When the imaginary part of Γ is negligible, their knowledge is sufficient to obtain any characteristics of rotational relaxation. The developed model ($\tilde{\Phi}_L(\omega) \sim \Phi(\omega/\kappa_L); \Phi(x) = \exp(-\sqrt{\gamma^2 + (x\tau)^2}); \kappa_L = \sqrt{1 + \chi L(L+1)}$) is specified by three parameters: τ, the mean collision duration; χ, the squared ratio of the anisotropic core range to the Lennard-Jones diameter, and γ. It was applied to the room-temperature nitrogen for which the experimental basic rates [3] can be fitted by the dependence $Q_L = Q_0 \exp(-hE_L / kT)$ with $h = 2.0$. Their use as the input data simplifies the procedure and excludes the amplitude factors $\Phi_L(0)$ from the treatment. The calculation was performed at the fixed values $\gamma = 1.5$ and $\tau^{-1} = 100$ cm^{-1}, $\chi = 0.01$ compatible with the existing data on the N_2-N_2 potential. The amplitude $Q_0 = 20.9$ mK/atm was found by equating the simulated broadening coefficient $\Gamma_{JJ}^{JJ}(0,0)$ at $J = 8$ to its experimental value in the isotropic Raman Q_1-branch. A good accord with the variety of the isotropic Raman and kinetic measurements was thus achieved, approximately same as that of the multiparameter ECSA model [4]. The calculated energy relaxation cross section $\sigma_E = 9.7 \text{Å}^2$ agrees well with the measurements ($\sigma_E = 9(3) \text{Å}^2$). In the depolarized Rayleigh (DPR) case, the EFCSA gives a realistic value of the broadening cross section $\sigma_Q = 32.8 \text{Å}^2$ ($\sigma_Q(\exp) = 34.4(6) \text{Å}^2$ [5]) and accurately predicts the deviations of the DPR shape from the dispersion profile [1,5]. The obtained angular momentum relaxation cross section $\sigma_J = 16.1 \text{Å}^2$ is close to the measured one (14.5Å^2) [6]. Besides, a reasonable agreement was found for the anisotropic Raman halfwidths.

References

[1] A. P. Kouzov and J. V. Buldyreva, Chem. Phys. **221**, 103 (1997).
[2] A. E. DePristo, S. D. Augustin, R. Ramaswamy, and H. Rabitz, J. Chem. Phys. **71**, 850 (1979).
[3]. G.O. Sitz and R.L. Farrow, J. Chem. Phys., **93**, 7883 (1990).
[4]. S. Temkin, J. M. Thuet, L.Bonamy, J. Bonamy, and D. Robert, Chem. Phys., **158**, 89 (1991).
[5]. F.R.W.McCourt, J.J.M.Beenakker, W.E.Köhler, and I.Kuscer, *Nonequilibrium Phenomena in Polyatomic Gases*, Clarendon Press, Oxford, 1991.
[6]. C.J. Jameson and A.K. Jameson, J. Chem. Phys., **93**, 3237 (1990).

Line mixing effect on IR line clusters and line wings: relaxation matrix and applications.

M. V. Tonkov and N. N. Filippov

Institute of Physics, St.Petersburg University, Peterhof, St.Petersburg, 198904 Russia

IR band shapes prove to be defined by the spectral function which have a symmetrized representation

$$\Phi(\omega) = \frac{1}{\pi} \mathrm{Re} \sum_{m,m'} A_m A_{m'} \left. \frac{1}{i(\omega - L_0) + \Gamma(\omega)} \right|_{mm'}, \qquad (1)$$

where ω is a frequency detuning from the band origin, m, m' are the line indices, $A_m = \sqrt{P_m} d_m$, P_m is the population of initial state, d_m is the reduced matrix element of the dipole moment, and L_0 gives the diagonal matrix of unperturbed line frequencies ω_m. In this notation, the line strength is $S_m = A_m^2$. $\Gamma(\omega)$ is the generalized (frequency dependent) relaxation matrix in a symmetrized representation: $\Gamma_{mm'}(\omega) = \Gamma_{m'm}(\omega)$. Its diagonal elements determine the shape of uncoupled lines and off-diagonal elements are responsible for non-additive effects (line mixing).

When the rotational perturbation play the major role in the band shape forming, the elements of $\Gamma(\omega)$ matrix are connected by the following **sum rule**:

$$\sum_m A_m \Gamma_{mm'}(\omega) = 0; \quad \sum_{m'} A_{m'} \Gamma_{mm'}(\omega) = 0, \qquad (2)$$

which is a consequence of the commutation properties of the dipole moment and perturbation operators. This rule can be used for the examination of calculated relaxation matrices and for the band shape asymptotic estimation.

For examples, in the line wing region, when $|\omega - \omega_m| \gg |\Gamma(\omega)_{mm'}|$, the spectral function can be expressed by Ben-Reuven formula

$$\Phi(\omega) = \frac{1}{\pi} \mathrm{Re} \sum_{m,m'} \frac{A_m A_{m'} \Gamma_{mm'}(\omega)}{(\omega - \omega_m)(\omega - \omega_{m'})}, \qquad (3)$$

For Ben-Reuven formula at $|\omega| \gg |\omega_m|$ one can expand Eq. (3) in a power series of ω_m / ω and $\omega_{m'} / \omega$. Due to the sum rule (2) the first three terms of the expansion go to zero, and we have

$$\Phi(\omega) = \frac{1}{\pi \omega^4} \mathrm{Re} \sum_{m,m'} A_m A_{m'} \omega_m \omega_{m'} \Gamma_{mm'}(\omega). \qquad (4)$$

In the Markov approximation $\Gamma(\omega) = \Gamma(0) \equiv \Gamma$ we obtain

$$\Phi(\omega) \propto \omega^{-4}. \qquad (5)$$

We calculated profile (4) for CO_2+He and CO_2+Xe mixtures in the region of ν_3 CO_2 band wing. Our calculations show that the major source of shape deviations from Lorentzian contour, even in the wing region, is the line mixing effect.

In the case of separated lines the band shape in Markov approximation can be approximated by Rosenkranz formula

$$\Phi(\omega) = \frac{1}{\pi} \sum_m \frac{A_m^2 [\Gamma_{mm} + (\omega - \omega_m) Y_m]}{(\omega - \omega_m)^2 + \Gamma_{mm}^2}; \quad Y_m = 2 \sum_{m' \neq m} \frac{A_{m'}}{A_m} \frac{\Gamma_{m'm}}{\omega_m - \omega_{m'}}. \qquad (6)$$

For Rosenkranz shape at $|\omega| \gg |\omega_m| \gg |\Gamma_{mm}|$ we can neglect the value Γ_{mm}^2 in the denominator in Eq. (6). Expressing Eq. (6) as a power series in ω_m/ω and $\omega_{m'}/\omega$ and using the relation

$$\sum_m A_m^2 Y_m = 0 \qquad (7)$$

we find for Rosenkranz shape

$$\Phi(\omega) = \frac{1}{\pi\omega^2}\left[\sum_m A_m^2(\Gamma_{mm} + Y_m\omega_m) + \frac{A_m^2(2\Gamma_{mm} + Y_m\omega_m)\omega_m}{\omega} + \frac{A_m^2(3\Gamma_{mm} + Y_m\omega_m)\omega_m^2}{\omega^2} + \ldots\right].$$

One can show with the sum rule (2) that the first and second terms in brackets vanish. As the third term is not generally equal to zero, we also have $\Phi(\omega) \propto \omega^{-4}$ for Rosenkranz shape asymptotic.

To construct the relaxation matrix we started from IOSA model. For a Σ-Σ band of linear molecule we get

$$\Gamma_{mm'}^{IOSA} = n_B \overline{v} \sqrt{\frac{p_{j_i}}{p_{j_i'}}} \sum_l Q_l\, C(j_i j_f; j_i' j_f'; l); \quad m = (j_i \to j_f); \quad m' = (j_i' \to j_f'), \qquad (8)$$

where $C(j_i j_f; j_i' j_f'; l)$ is the conventional combination of 3J- and 6J-symbols, and Q_l is the dynamic cross-section. The defined matrix Γ^{IOSA} is not a symmetric one:

$$\Gamma_{mm'}^{IOSA} = \exp(-\hbar\omega_{mm'}/kT)\, \Gamma_{m'm}^{IOSA}, \qquad \omega_{mm'} = \hbar^{-1}(E_{j_i} - E_{j_i'}) \qquad (9)$$

To correct this disadvantage and to take into account an adiabatic factor, we introduced resonance function corrected (RFC) matrix:

$$\Gamma_{mm'}^{IOS/RFC} = R(\omega_{m'm})\Gamma_{mm'}^{IOSA}; \qquad m' \neq m. \qquad (10)$$

with the resonance function $R(\omega)$ of the form

$$R(\omega) = \frac{2}{1+\exp(-\hbar\omega/kT)} \cdot \frac{k_c}{1+(\omega/\omega_c)^2/24}. \qquad (11)$$

The function $R(\omega)$ should satisfy the detailed balance relation and matrix (10) is a symmetric one: $\Gamma_{mm'}^{IOS/RFC} = \Gamma_{m'm}^{IOS/RFC}$. We obtained the diagonal elements of $\Gamma^{IOS/RFC}$ from the off-diagonal ones using the sum rule (2).

For CO_2+He system the cross-sections Q_l were calculated using our semiclassical method. We found that the short-range anisotropic repulsive forces are the main cause of the rotational perturbation of CO_2 in this system. Calculating Q_l, we described the time dependence of the absorber-perturber distance r using a classical trajectory $r(t)$ which is formed by the isotropic part of the intermolecular potential. Then we adjusted the parameters of the resonance function (11) using broadening coefficients data. With this matrix we calculated the wing shape of $3\nu_3$ band CO_2, the result is in a good accordance with experimental data.

This model was also extended to calculate the profiles of Π-Σ bands. The influence of line mixing effect on the Q branch shape was examinated in the CO_2+He spectra in the 15 μm regon, the calculated shape is quite close to the experimental one.

Spectrally Resolved Determination of the Linear Dipole-Polarizability of Molecular Iodine in the Range of the B ← X Transition Between 11500cm^{-1} and 17800cm^{-1}

Uwe Hohm

Institut für Physikalische und Theoretische Chemie, Hans-Sommer-Str. 10, D-38106 Braunschweig, FRG

There is currently a renewed interest in the physicochemical behaviour of iodine, I_2. Recent efforts include a study of the I_2-Ne van der Waals system, the study of I_2 in solid Kr, and I_2 in rare-gas solvents as well as detailed spectroscopic investigations of I_2, which are helpful e.g. for realizing frequency standards. Moreover, the study of the collisional shift and broadening of iodine spectral lines in air brought forth the need for accurate dipole-polarizability values for this molecule [1]. In our contribution we present an experimental and theoretical study of the mean dipole-polarizability $\alpha(\sigma)$ of molecular iodine in the wavenumber range between 11500cm^{-1} and 17800cm^{-1} [2]. This is the region of the B ← X transition of iodine in which strong absorption occurs. Therefore, the objective of our work is to measure the polarizability spectrum $\alpha(\sigma)$ and to compare the experimental results with calculations which take into account the Franck-Condon factors of the transitions between different vibrational levels of the ground and excited state. This work extends previous efforts by Abraham et al. [3], who have measured and analyzed the refractive index spectrum in the visible inside a small wavenumber range of $\Delta\sigma \approx 1$cm^{-1} inside a single rotational line.

In order to obtain $\alpha(\sigma)$ we have measured the refractive index spectrum $[n(\sigma, \varrho, T) - 1]$ of iodine at a temperature of $T = 296.3$K and a density of $\varrho = 0.014797$mol m^{-3}. At our experimental conditions $[n - 1]$ amounts to $\approx 7 \times 10^{-7}$. We use the technique of asymmetric white-light interferometry with a spectral resolution of ≈ 5cm^{-1} and a sensitivity of $\Delta n \approx 2 \times 10^{-8}$. The dipole-polarizability is obtained from the measured refractivity via the Lorentz-Lorenz equation.

In our calculations of $\alpha(\sigma)$ we use an expression of the polarizability emerging from perturbation theory.

$$\alpha(\omega) = \frac{2}{\hbar} \frac{1}{Z} \sum_{v'',J''} g_N e^{-E(v'',J'')/kT} |\mu_e(\bar{R})|^2 | <J'|J''> |^2 \times$$

$$\times \left[\sum_f \frac{\omega_{if} |<v'|v''>|^2 (\omega_{if}^2 - \omega^2)}{(\omega_{if}^2 - \omega^2)^2 + \gamma_{if}^2 \omega^2} + \int_{\omega' > \epsilon'} \frac{|<\omega'|v''>|^2 (\omega'^2 - \omega^2) d\omega'}{(\omega'^2 - \omega^2)^2 + \gamma^2 \omega^2} \right] , \quad (1)$$

ω_{if} and ω' are transition energies into bound and unbound states, respectively. $|v>$ and $|J>$ are the vibrational and rotational wavefunctions, $|\mu_e(\bar{R})|^2$ is the squared electronic transition dipole-moment and Z is the partition function. The other symbols have their usual meaning. The first term in the brackets denotes the contributions from bound-bound-transitions, the second one is due bound-free transitions. We take into account the $B^3\Pi_g^+ \leftarrow X$, $^1\Pi_u \leftarrow X$, and $A^3\Pi_u \leftarrow X$ transitions. The relevant Franck-Condon factors are calculated by usual methods from the corresponding potential curves which are obtained via the Rydberg-Klein-Rees method from spectroscopic input data [4].

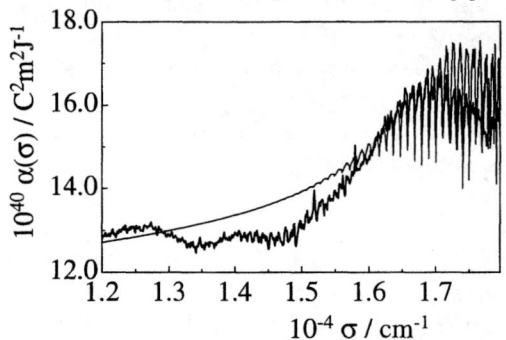

The measured (thick line) and calculated (thin line) polarizability spectra of iodine are compared in the figure on the left. Overall, an acceptable agreement between the two curves can be seen.

$\alpha(\sigma)$ in the visible can change at least by 15% inside a single vibrational line. This change in the polarizability is a large effect and often not negligible for a number of optical processes which are linear in $\alpha(\sigma)$ (e.g. refractivity $(n-1)$) or even depend on higher powers of α (e.g. scattering of light or dispersion interaction energy). In the low density range used in our experiments we assume the influence of intermolecular interactions on the polarizability spectrum to be negligible. However, despite of experimental difficulties future work should extend to higher pressures and it will be interesting to observe the effect of collisions on $\alpha(\sigma)$ and the corresponding change of the position and the shape of the spectral lines.

REFERENCES

1. Fletcher D.G., and McDaniel J.C., *J. Quant. Spectrosc. Radiat. Transfer* **54**, 837 (1995).
2. Maroulis G., Makris C., Hohm U., and Goebel D., *J. Phys. Chem. A* **101**, 953 (1997).
3. Abraham R.G., Booth L.J., and Dalby F.W., *Can. J. Phys.* **68**, 81 (1990).
4. Hohm U., *Mol. Phys.* (1998), accepted for publication.

Investigation of Line Broadening in the $\nu_1 + 3\nu_3$ Band of Acetylene

Hamid Valipour and Dieter Zimmermann

Institut für Atomare und Analytische Physik, Technische Universität Berlin
Hardenbergstraße 36, D-10623 Berlin, Germany

Abstract. Line broadening and line shift coefficients have been determined for 21 rotational lines of the $\nu_1 + 3\nu_3$ vibrational band of acetylene. Results are presented for the case of self-broadening and of broadening by N_2.

A detailed knowledge of line broadening coefficients is an important prerequisite for a quantitative determination of the amount of gaseous pollutants in remote sensing or in in-situ measurements under conditions of atmospheric pressure. Line shift coefficients, on the other hand, are of particular interest, if molecular absorption lines are taken to serve as accurate secondary frequency standards, as recently demonstrated for the wavelength region around 1.5 µm using acetylene (1). In the present contribution we report on an investigation of the line shift and the line broadening for several rotational transitions of the vibrational band $\nu_1 + 3\nu_3$ of acetylene in the wavelength range around 789 nm. The line strength of this combination-overtone vibrational band is considerably lower compared to the fundamental. However, this disadvantage is at least partially compensated by the use of inexpensive tunable cw semiconductor laser diodes as a powerful light source as well as by the reduced opacity of the atmosphere in the near-infrared range. Therefore, this absorption lines are quite well suited to serve as monitoring transitions for the detection of small amounts of acetylene, as recently demonstrated experimentally (2).

As an absorption cell we used a multi-pass cell of the Herriott-type with a physical length of 0.5 m allowing to get an effective absorption length of 35 m. The laser beam was provided by a semiconductor laser diode (Hitachi HL 7851 G) with 50 mW output power at 789 nm. The temperature of the laser diode was held constant to about 0.005 K. Tuning of the laser frequency was accomplished by changing the laser diode current being provided by a well-stabilized power supply. In order to reduce the frequency jitter the laser diode was locked to an external Fabry-Perot Resonator (FPR), the reference cavity of a Coherent 599-21 dye laser, being well stabilized against changes in temperature and pressure. Changing the frequency of the FPR by electronically tilting a glass plate inside the FPR allowed a scanning of the laser diode over a wavenumber range of 0.8 cm^{-1}. As the laser power was not constant over a scan, it was separately recorded in order to allow a normalization of the absorption signals.

In addition to the strong absorption lines due to the $v_1 + 3v_3$ band weaker lines occurred in the absorption spectrum which were assigned to different vibrational bands, e.g. the $v_2 + 3v_3 + 2v_4$ band. Only lines being not effected by overlapping with these weak lines even under conditions of pressure broadening have been selected. Usually, a single scan extended over a range of 0.8 cm^{-1}. The observed absorption curves were fitted to a Voigt-profile with the Doppler-contribution to the line width fixed to its theoretical value of 0.0305 cm^{-1} for a temperature of (295 ± 2) K. All data have been taken at this temperature.

The line width obtained for different gas pressures showed a linear dependance within limits of error in all cases. Our results for the broadening coefficients γ and

Table 1. Preliminary values for self and N_2 broadening and line shift coefficients
(in cm^{-1}/bar)

Line	v_o [cm^{-1}]	γ_{self}	δv_{self}	$\gamma(N_2)$	$\delta v(N_2)$
P (4)	12665.962(5)	0.330 (11)	-0.0119 (7)	0.190 (5)	-0.0122 (3)
P (6)	12660.798(5)	0.302 (12)	-0.0104 (13)	0.177 (9)	-0.0142 (8)
P (7)	12658.145(5)	0.327 (22)	-0.0094 (37)	0.171 (16)	-0.0142 (20)
P (8)	12655.436(5)	0.277 (29)	-0.0133 (38)	0.160 (17)	-0.0166 (23)
P (9)	12652.670(5)	0.307 (21)	-0.0133 (16)	0.167 (9)	-0.0163 (13)
P(10)	12649.868(5)	0.280 (19)	-0.0117 (20)	0.160 (9)	-0.0165 (12)
P(11)	12647.010(5)	0.281 (7)	-0.0125 (9)	0.169 (3)	-0.0161 (2)
P(17)	12628.829(5)	0.263 (6)	-0.0155 (7)	0.157 (4)	-0.0165 (2)
R (1)	12680.231(5)	0.364 (12)	-0.0050 (9)	0.202 (6)	-0.0090 (5)
R (3)	12684.584(5)	0.333 (10)	-0.0086 (8)	0.189 (5)	-0.0112 (5)
R (5)	12688.731(5)	0.304 (10)	-0.0123 (16)	0.175 (8)	-0.0137 (9)
R (6)	12690.726(5)	0.294 (9)	-0.0122 (7)	0.169 (5)	-0.0138 (4)
R (7)	12692.672(5)	0.318 (15)	-0.0105 (14)	0.168 (11)	-0.0134 (15)
R (8)	12694.566(5)	0.303 (8)	-0.0168 (9)	0.167 (4)	-0.0157 (4)
R (9)	12696.414(5)	0.294 (19)	-0.0147 (28)	0.151 (16)	-0.0141 (25)
R(10)	12698.207(5)	0.287 (18)	-0.0148 (29)	0.159 (15)	-0.0146 (18)
R(11)	12699.951(5)	0.286 (17)	-0.0145 (14)	0.155 (11)	-0.0162 (14)
R(12)	12701.645(5)	0.280 (15)	-0.0175 (16)	0.152 (9)	-0.0163 (13)
R(13)	12703.285(5)	0.286 (7)	-0.0176 (6)	0.161 (2)	-0.0163 (2)
R(15)	12706.426(5)	0.262 (8)	-0.0196 (25)	0.152 (6)	-0.0170 (10)
R(22)	12715.882(5)	0.228 (12)	-0.0243 (21)	0.138 (7)	-0.0183 (9)

the line shift coefficients δv are compiled in the table for self and N_2 broadening. Both coefficients refer to the FWHM. Altogether, 8 lines of the P branch and 13 lines of the R branch have been studied. The error bars of γ and δv are mainly due to the uncertainty of ±1 mbar in measuring the pressure. We used a Piezovak (Leybold PV 111) which was calibrated by means of a Baratron (MKS 270 C). Additional contributions to the error come from the variations of the baseline and from the statistical error being taken as 3σ. For values being underlined in the table, results are also available from previous work (2,3), which are, however, in good agreement with our values in most cases. As can be seen from the table, the values of γ and δv with the same lower level J are close to each other. The broadening coefficients decrease with increasing value of J for both self and N_2 broadening, whereas $|\delta v|$ shows the opposite behavior. We are presently applying existing theoretical models in order to get an interpretation of this observations. In addition, values of γ and δv have been obtained using O_2, air, and Ar as perturbers, which will be published in a separate paper.

References

1. Nakagawa K., de Labachelerie, M., Awaji, Y., and Kourogi, M., *J. Opt. Soc. Am.* **B 13**, 2708 - 2714 (1996)
2. Cancio, P., and Pavone, F.S., *Physica Scripta* **T58**, 86 - 93 (1995)
3. Lucchesini, A., De Rosa, M., Pelliccia, D., Ciucci, A., Gabbanini, C., and Gozzini, S., *Appl. Phys.* **B 63**, 277 - 282 (1996)

SYMPOSIUM IN HONOR OF THE 100th ANNIVERSARY OF THE BIRTH OF ALEKSANDER JABŁOŃSKI

Aleksander Jabłoński and the Atomic and Molecular Physics

Józef Szudy

Institute of Physics, Nicholas Copernicus University,
87-100 Toruń, ul. Grudziądzka 5, Poland

I. LIFE AND PERSONALITY

Aleksander Jabłoński was born on February 26, 1898 in Voskresenovka near Kharkov, Ukraine which at that time was part of Russia. In 1916 he entered the University of Kharkov to study physics. His study at Kharkov was interrupted by his military service first in Russia and later, during World War I, in the newly organized Polish Army. At the end of 1918, when an independent Poland was recreated after more than 120 years of occupation by neighbouring powers, Jabłoński left Kharkov and arrived in Warsaw where he entered the Warsaw University to continue his study of physics. His study in Warsaw was again interrupted in 1920 by his military service during the Polish-Bolshevik war.

An enthusiastic musician, Jabłoński played the first violin at the Warsaw Opera from 1921 to 1926 in parallel with his studies at the University under Stefan Pieńkowski for his doctorate which he received in 1930 with a thesis *On the influence of the wavelength of excitation light on the fluorescence spectra*. Although Jabłoński left Opera in 1926 and devoted himself entirely to scientific work, music remained his great passion until the last days of his life.

After receiving his doctorate, Jabłoński spent two years (1930-31) as a fellow of the Rockefeller Foundation in Germany working first with Peter Pringsheim in Berlin and later with W.I. Lenz in Hamburg. While in Berlin he attended the famous colloquia at the Physics Institute chaired by Walther Nernst. These colloquia were attended by such scholars as Albert Einstein, Max Planck, Max von Laue, Erwin Schrödinger and Lise Meitner who were working at that time in Berlin. The greatest scientists of the *Golden Age of Quantum Physics* were among the invited speakers who delivered seminars. There is no doubt that such an atmosphere had to have a great influence on the personality of Jabłoński as a physicist.

In 1934 he acquired his habilitation from the Warsaw University with the thesis *On the influence of intermolecular interactions on the absorption and emission of light.* Throughout the 1920's and 30's the Physics Department at the Warsaw University was an active centre for studies on luminescence phenomena. During most of this period Jabłoński worked both experimentally and theoretically on fundamental problems of photoluminescence of liquid solutions as well as on pressure effects on atomic spectra in gases.

In 1938 Jabłoński accepted a faculty appointment at the Stefan Batory University in Wilno (Vilnius), where he developed experimental studies on spectral line broadening. These studies were interrupted by the outbreak of World War II on September 1, 1939 when Poland became attacked from the West and North by the Nazi Germany. Again in the field service Jabłoński went through the Polish-German September campaign. On September 17, 1939 when due to the Ribbentrop-Molotov agreement Poland was attacked from the East by the Soviet Army, Jabłoński with his unit crossed the Polish-Lithuanian border and was sent by Lithuanian authorities to an internment camp. At the end of 1939 he was released from the camp and came to Vilnius. In the meantime Lithuania became occupied by Soviet Union and in July 1940 Jabłoński was arrested by Soviet authorities and sent to Kozielsk, a camp in which a few months earlier several thousands of Polish Army officers were confined until April 1940, who were all murdered in a nearby Katyn forest.

In June 1941 after the attack of the Nazi Germany against Soviet Union Jabłoński was conveyed from Kozielsk to another internment camp in Griazowiec from where he was eventually released to join the Polish Army organized by the Polish government in exile in the Soviet territory. With the Polish Army he left Soviet Union and then through the Middle East he finally arrived in the summer of 1943 in Great Britain. Being on leave from the army he became a lecturer of physics at the Polish School of Medicine affiliated to the University of Edinburgh, Scotland until the end of the war. In Scotland he was able to return to scientific work and devoted his attention to the line shape theory. At Edinburgh at that time was Max Born and Jabłoński attended Born's Physical Colloquia where he delivered a seminar on his theory of pressure effects on spectral lines.

After the war in November 1945 Jabłoński returned to Poland and started to work again in the Warsaw University. Soon, however, he moved to Toruń, where in the fall of 1945 a new university bearing the name of Nicholas Copernicus who was born in that town was established by the professors of the former Stefan Batory University who left Vilnius. On January 1, 1946 Jabłoński was nominated a full professor of Copernicus University. As the chairman of the Physics Department from its very beginning in 1946 to his retirement in 1968 Jabłoński created in Torun a centre of studies in atomic, molecular and optical physics, and in particular, in molecular luminescence and photophysics. Aleksander Jabłoński died on September 9, 1980.

II. CONTRIBUTIONS TO MOLECULAR LUMINESCENCE AND PHOTOPHYSICS

II.1. Photoluminescence of liquid solutions

Aleksander Jabłoński was one of the pioneers in the development of the field of molecular luminescence and photophysics. His early work at Warsaw included measurements of absorption spectra in liquid solutions and the experimental proof that the intensity distribution in the fluorescence spectra in typical cases is independent of the excitation light wavelength [1]. He introduced then the concept of a *luminescent centre*, i.e. a system composed of the excited molecule and its closest neighbourhood [2]. Using the Franck-Condon principle (FCP) generalized to such centres, he explained the main features of the fluorescence phenomena in liquid solutions [3,4]. In particular, he explained the Stokes shift of the fluorescence and absorption spectra and the independence of the intensity distribution in the fluorescence band of the excitation light frequency in liquid solvents. Jabłoński's early work on luminescence of liquids [1-4] was the basis to formulate the so-called Kasha's rule [5,6] according to which luminescence emission is observed almost exclusively from the lowest excited states of any given multiplicity. This results from the very high rate constants of non-radiative deactivation for polyatomic molecules within each multiplicity manifold, compared with the rate constants of radiative transitions.

II.2. The Jabłoński diagram

A characteristic feature of luminescence phenomena observed in liquid solutions of polyatomic dye molecules is the existence of a long-lived component in the fluorescence (called today *delayed* or *slow fluorescence*) and the *phosphorescence*, i.e. an emission which occurs at higher wavelengths than ordinary fluorescence. Although these phenomena were the subjects of numerous experiments since the classic work by Stokes in 1852 [7] several attempts to explain their mechanisms failed.

The first step in a good direction was made in 1929 by Francis Perrin [8] who introduced the concept of a metastable level lying below the fluorescence level of a dye molecule. In 1931-32 H. Kautsky [9] demonstrated experimentally the existence of metastable levels of dyes adsorbed on gels as sensitizers. Kautsky's experiments were immediately analyzed by Jabłoński, who in 1933 wrote a short note to *Nature* [10], giving the first presentation of his famous diagram for luminescent molecules. In this diagram a metastable level M as an *intrinsic electronic state* of a dye molecule was introduced. This level lies below the fluorescence level F and it was assumed that the distance between levels F and M was small enough for the thermal excitation M $\to$ F to occur. In 1935 Jabłoński [11] presented a theoretical analysis of the consquences of his diagram. Starting with a simple kinetic equation for the depopulation of dye molecules in the metastable state M, he derived formulae for the lifetime of M-state molecules in the presence and absence of thermal excitation to state F. He also derived formulae for the fluorescence and phosphorescence quantum yields as well as for the phosphorescence exponential decay law.

In later years the Jabłoński diagram was generalized to include several radiationless processes in dye molecules such as *intersystem crossings* [12,13] and *intramolecular transitions* [14]. The diagram has withstood the test of time and now it serves as the starting point of all modern textbooks on photochemistry and photophysics (see e.g. [6]). In his comprehensive overview of various aspects of the Jabłoński diagram Michael Kasha [13] has written: "...*the Jabłoński 1933 note and 1935 paper stand as giant steps in the transformation, of a long history of qualitative luminescence observation, into the beginnings of a quantitative science of molecular photophysics*". Jabłoński appropriately ends his 1935 paper [11] with the statement: "...*by these deliberations the phenomena of photoluminescence have lost some of their mysterious character*".

II.3. Polarization of photoluminescence

The problem that intrigued Jabłoński for many years was the polarization of photoluminescence of isotropic and anisotropic systems [4,15-27]. In 1935 he formulated [16] a general theory of polarization based on the model of spatial oscillators. To explain some discrepancies between experiment and theory he analyzed various factors responsible for the depolarization of luminescence. In 1950 he suggested [20] a hypothesis that the fluorescence depolarization is due to torsional vibrations of luminescent molecules. It is worth to mention that at the end of 1950's Jabłoński introduced a term *emission anisotropy* [22,23] instead of the commmonly used degree of polarization. The emission anisotropy, unlike the degree of polarization, is an additive quantity thus leading to much simpler calculations.

In 1934 Jabłoński proposed [15] a method for orienting dye molecules in anisotropic matrices which is still widely applied in photochemical and biophysical studies as a tool for determining the directions of the absorption electronic transition moments.

The last two papers of Jabłoński [27] published in the late 1970's once more dealt with the influence of torsional vibrations on the polarization of photoluminescence in rigid media and with the time dependence of the emission anisotropy of the fluorescence of liquid solutions excited by a short pulse of linearly polarized light.

II.4. Decay and quenching of photoluminescence

Throughout his career, Jabłoński was interested in the problem of the decay of photoluminescence. In 1935 he described [18] the principles of phase-modulation fluorometry and derived a simple formula which is the basis for a method of analyzing the decay curves obtained from phase-modulation fluorometers. Using such an analysis he found [18] that the decay times of the fluorescence components polarized parallel and perpendicular to the excitation light direction are different.

In a series of papers published in 1950's [22,28-30] Jabłoński developed a theory of concentration quenching of photoluminescence of liquid solutions based on the so-called *multilayer shell model* of the luminescent centre. In this model the luminescent centre consists of an excited molecule surrounded by monomolecular layers of solvent molecules, including quenchers, i.e. disturbing molecules. Assuming the probability of different configurations of the disturbing molecules to be given by the Smoluchowski distribution, Jabłoński [28,29] derived a general formula for the photoluminescence quantum yield in liquid solutions which in limiting cases transforms into well-known Perrin's formula for the self-quenching or into the Stern-Volmer law. In 1958 Jabłoński [30] analyzed the quenching of photoluminescence of solutions caused by the excitation energy transfer assuming the Foerster law according to which the quenching probability is inversely proportional to the sixth power of the distance between the quencher and the luminescent molecule.

III. CONTRIBUTIONS TO ATOMIC SPECTROSCOPY

Jabłoński first became involved in atomic spectroscopy research even before his doctorate when in 1927 he performed experimental studies of the absorption and fluorescence of cadmium vapours and found several continuous bands in the UV region [32]. The most fruitful period of his work in atomic physics occured between 1930 and 1939, encompassing collaborations with many of the most distinguished scientists of that period. In 1930 he came to Berlin to work in the *Physikalisches Institut der Universitaet* in the laboratory of P. Pringsheim. His first work in Berlin dealt with the quenching and polarization of the D-line emission from the sodium vapours irradiated by a *yellow light*, and included measurements of the quenching of sodium fluorescence by nitrogen as well as the effect of the magnetic field on the polarization of the Na-D lines. Results of these studies were published in 1931 in two papers co-authored with Pringsheim [33].

In the middle of 1931 Jabłoński published [34] his pioneering paper in which he made the first attempt at a quantum theory of collisional effects on spectral lines. It was that paper in which he proposed his so-called *quasimolecular model* of the pressure broadening and shift phenomena. Jabłoński was probably the very first person to introduce in that paper the notion of the Franck-Condon principle (FCP) into atomic spectroscopy as a starting point of the analysis of pressure-broadened spectral lines.

The significance of this model was recognized immediately. In the middle of 1930's it had attracted much attention by theoreticians like Victor Weisskopf [35], Henry Margenau [36,37] and M. Kulp [38] as well as experimental physicists such as A.C.G. Mitchell and M.W. Zemansky [40], H.G. Kuhn and O. Oldenberg [41]. Using this model Margenau [36,37] and Kuhn and London [39,41] developed the celebrated *quasistatic* theory of pressure broadening which was later used by many researches as a poweerful tool for investigating in detail various aspects of interatomic interactions on the basis of an analysis of spectral line wings [51,62,63].

In a series of papers published between 1937 and 1945 Jabłoński formulated [42-44] the mathematical foundations of his quantum-mechanical theory of pressure broadening. The system of radiating atom plus a perturber-bath was treated by him as a huge (N+1)-atom quasimolecule, where N is the total number of perturbers. In order to calculate the shape of a broadened line he applied the Born-Oppenheimer approximation and FCP in its quantum-mechanical formulation [45] and expressed the line shape in terms of the *one-perturber free-free Franck-Condon factors* [44]. To obtain analytic results he used the semiclassical JWKB wave functions to calculate the Franck-Condon factors and showed that in the classical limit, the line shape at a far wing of a line becomes identical with that resulting from the quasistatic theory due to Margenau and Kuhn. This significant result was later discussed in numerous monographs and review articles [46-54] and regarded as the quantum-mechanical justification of the quasistatic theory.

When in the spring of 1938 Jabłoński came to Vilnius to the Stefan Batory University, he soon established there a laboratory for a research in atomic spectroscopy. Together with H. Horodniczy he began systematic experimental study of the influence of temperature on the pressure broadening of spectral lines. Their experiments were performed in absorption on the Hg 253.7 nm line perturbed by He and Ar [55]. They found that the broadening by He did not depend appreciably upon temperature. On the other hand their experiment on the broadening by Ar indicated that temperature influence really seemed to exist in this case although it was much smaller than that predicted by the impact theories of Lorentz and Weisskopf.

The last two years of World War II Jabłoński spent in Edinburgh. Although teaching duties consumed most of his time he was also involved in research work. In March 1945 he published in *Nature* a note [57] in which he had explained the long duration of the emission of hydrogen Balmer lines excited by an electrodeless discharge in a tube under a pressure of 0.2 Torr [56] as due to effects caused by the recombination of H$^+$ ions.

Most of his time in Edinburgh Jabłoński devoted to work on the further extension of his earlier quantum-mechanical theory of pressure broadening and gave a new, more consistent, form to this theory which was presented in his well-known paper published in 1945 [44]. In the next three decades that paper became a source of inspiration of many theoreticians working in the field of spectral line shapes. In particular, that paper was used by M. Baranger [58] and I.I. Sobelman [49,59] as a starting point to develop a theory of line broadening by electrons in plasmas. They have shown that although Jabłoński in his papers [42-44] restricted his considerations to cases of heavy perturbers only, his treatment appears to be so general that it can be applied to light perturbers such as electrons.

In 1968 F.H. Mies [60] performed a very thorough study concerned with a reexamination of Jabłoński's theory. First of all, he tested the validity of Jabłoński's main approximations and emphasized the astounding accuracy of both the JWKB and stationary-phase approximation for the case of broadening by neutral perturbers. He thus demonstrated the applicability of the classical FCP for free-free transitions. The characteristic feature of the Jabłoński-JWKB stationary-phase line shape is that it contains the factor $\cos^2 \Phi(R)$, where $\Phi(R)$ is the phase-shift difference (at the interactomic separation R) between wave functions of the upper and lower state. This factor is a quantum-mechanical interference effect with no classical interpretation. Mies [60] has shown that this factor leads to the oscillatory structure in the wings of pressure broadened spectral lines when the interaction potential in the initial state is deeply attractive compared to the thermal energy and the interaction in the final state is repulsive. Such *quantum oscillations* have been observed in the intensity distribution of the far-UV continuous radiation at 60 nm emitted by metastable He atoms during collision with ground-state He atoms. At high enough temperatures these oscillations tend to be averaged out if the Jabłoński-JWKB stationary phase-phase formula is integrated over distribution of collision energies and impact parameters. This is equivalent to the use of the random phase approximation which consists of replacing the $\cos^2 \Phi(R)$ term by its mean value 1/2. In this approximation the Jabłoński-JWKB formula yields then the quasistatic line shape which depends on temperature through the Boltzmann factor only. The quasistatic profile becomes singular as R approaches a position of an extremum in the potential difference. This singularity can be removed if the potential difference in the Jabłoński-JWKB formula is expanded around its extremum [64,65]. In such a case the resultant line shape may be expressed in terms of an Airy function in full analogy to the Ford-Wheeler theory of the rainbow scattering [66] and the quasistatic singularity becomes then a famous *rainbow satellite*.

The classic example of the rainbow satellite is an intensity maximum located at 162.3 nm in the red wing of the self-broadened Lyman-α line of hydrogen [67] and its shape was first theoretically described on the basis of the Jabłoński-JWKB formula by Sando and Wormhoudt [64] in 1973. Another rainbow satellite (at 140.5 nm) should also appear on the red wing of this line if perturbations by protons are taken into account [68,69]. Such satellite features on the red wing of the Lyman-α line were found first in the spectra of DA white dwarf stars [70,71] and recently in laboratory conditions [72].

IV. CONCLUSION

Aleksander Jabłoński has had a profound and lasting influence on the development of molecular photophysics and atomic and molecular spectroscopy. He gained worldwide recognition for his pioneering work in the field of photoluminescence and pressure broadening of spectral lines, particularly for his contributions to the understanding of the mechanisms of molecular luminescence and collisional effects on atomic spectra. Some of Jabłoński's papers have influenced other fields such as condensed matter physics, molecular spectroscopy and scattering theory [73,74]. For instance, his general approach proposed in 1931 [75] to interpret broad band spectra of dye solutions was extended [76] to explain the spectra of impurity ions and molecules in condensed matter. There is a sum rule due to Jabłoński [77,73] that says that the sum of the transition strengths from a single vibrational level to all other levels (of the other electronic state) is a constant independent of the single level chosen.

There is one more important area of atomic physics which was also influenced by Jabłoński. At the end of 1970's and in 1980's many laboratories focused their attention on profiles associated with *collisional redistribution of radiation,* i.e. the effect that the collisions have on the near-resonance scattering of laser light in dilute gases [78-82]. The theory of collisional redistribution has been developed in a series of papers by John Cooper and his coworkers at JILA, Boulder, Colorado and it is reviewed in the article by Keith Burnett [82]. It appears that in spite of some differences there are also smimilarities between the profiles associated with collisional redistribution and those of ordinary pressure broadened spectral lines, especially in cases when the incident light in a redistribution experiment is detuned into the far line wings. In a paper [81] on collisional redistribution of radiation in the non-impact region of spectral lines dedicated to Aleksander Jabłoński on the occasion of his 80th birthday which was published in a *Festschrift* issue of *Acta Physica Polonica* John Cooper has written: "The pioneering work of Jabłoński on the description of line wing radiation has made it possible to approach this complicated problem with a high degree of insight and understanding".

REFERENCES

[1] A. Jabłoński, Comp. Rend. de la Soc. Polon. Phys. **7**, 1 (1926).
[2] A. Jabłoński, Z. Phys. **73**, 460 (1931); Phys. Z. Sowjetunion **8**, 105 (1935).
[3] A. Jabłoński, Acta Phys. Polon. **2**, 97 (1933).
[4] A. Jabłoński, Acta Phys. Polon. **5**, 271 (1936).
[5] M. Kasha, Faraday Soc. Discuss. **9**, 14 (1950).
[6] P. Suppan, *Chemistry and Light*, Royal Soc. of Chemistry, Cambridge 1994.
[7] G. Stokes, Philos. Trans. Roy. Soc. London **142**, 463 (1852).
[8] F. Perrin, Ann. Phys. (Paris) **12**, 169 (1929).
[9] H. Kautsky, Ber. Deutsch. Chem. Ges. **64**, 2053 and 2677 (1931); **65**, 401 (1932).
[10] A. Jabłoński, Nature **131**, 839 (1933).
[11] A. Jabłoński, Z. Phys. **94**, 38 (1935).
[12] M. Kasha, J. Chem. Phys. **20**, 71 (1952).
[13] M. Kasha, Acta Phys. Polon. A**71**, 661 (1987).
[14] S. Leach, Acta Phys. Polon. A**71**, 671 (1987).
[15] A. Jabłoński, Acta Phys. Polon. **3**, 421 (1934).
[16] A. Jabłoński, Z. Phys. **96**, 236 (1935).
[17] A. Jabłoński, Acta Phys. Polon. **4**, 311 (1935); **4**, 389 (1935).
[18] A. Jabłoński, Z. Phys. **95**, 53 (1935); **103**, 526 (1936).
[19] A. Jabłoński, Acta Phys. Polon. **7**, 15 (1938).
[20] A. Jabłoński, Acta Phys. Polon. **10**,33 (1950); **10**,193 (1950).
[21] P. Drzewiecki, A. Jabłoński, A. Kawski, and M. Kryszewski, Acta Phys. Polon. **12**, 149 (1953).
[22] A. Jabłoński, Acta Phys. Polon. **14**, 295 (1955); **14**, 497 (1955); **16**, 471 (1957).
[23] A. Jabłoński, Bull. Acad. Polon. Sci. **8**, 259 (1960); **10**, 55 (1962).
[24] A. Jabłoński, Z.Naturforsch. **16a**,1 (1961).
[25] A. Jabłoński, in: *Luminescence of Organic and Inorganic Materials*, ed. H. Kallman and L. Spruch, John Wiley and Sons, New York 1962, p. 110.
[26] A. Jabłoński, Acta Phys. Polon. **26**, 427 (1964); **26**, 717 (1965).
[27] A. Jabłoński, Bull. Acad. Polon. Sci. **25**, 603 (1977); **27**, 1 (1979).
[28] A. Jabłński, Acta Phys. Polon. **13**, 175 (1954).
[29] A. Jabłoński, Acta Phys. Polon. **15**, 263 (1956).
[30] A. Jabłoński, Bull. Acad. Polon. Sci. **6**, 663 (1958).
[31] A. Jabłoński, Acta Phys. Polon.A **38**, 453 (1970); **39**, 87 (1971).
[32] A. Jabłoński, Z. Phys. **45**, 878 (1927).
[33] A. Jabłoński and P. Pringsheim, Z. Phys. **70**, 593 (1931); **73**, 281 (1931).
[34] A. Jabłoński, Z. Phys. **70**, 723 (1931).
[35] V. Weisskopf, Z. Phys. **75**, 287 (1934); Phys. Z. **34**, 1 (1933).
[36] H. Margenau, Phys. Rev. **40**. 387 (1932); **43**, 129 (1933).
[37] H. Margenau and W.W. Watson, Phys. Rev. **44**, 92 (1933); Rev. Mod. Phys. **8**, 22 (1936).
[38] M. Kulp, Z. Phys. **79**, 495 (1932).
[39] H.G. Kuhn and F. London, Phil. Mag. **18**, 983 (1934).

[40] A.C.G. Mitchell and M.W. Zemansky, *Resonance Radiation and Excited Atoms*, University Press, Cambridge 1934.
[41] H.G. Kuhn, Phil. Mag. **18**, 987 (1934); Proc. Roy. Soc. A**158**, 212 (1937); H.G. Kuhn and O. Oldenberg, Phys. Rev. **41**, 72 (1932).
[42] A. Jabłoński, Actan Phys. Polon. **6**, 371 (1937); **7**, 196 (1938); **8**, 71 (1939).
[43] A. Jabłoński, Physica **6**, 541 (1940).
[44] A. Jabłoński, Phys. Rev. **68**, 78 (1945).
[45] E.U. Condon, Phys. Rev. **32**, 825 (1928).
[46] S.Y. Ch'en and M. Takeo, Rev. Mod. Phys. **29**, 20 (1957).
[47] H.G. Kuhn, *Atomic Spectra*, Academic Press, New York 1969.
[48] R.G. Breene, jr., Rev. Mod. Phys. **29**, 94(1957); *The Shift and Shape of Spectral Lines*, Pergamon Oxford 1961.
[49] I.I. Sobelman, *Introduction to the Theory of Atomic Spectra*, Pergamon, Oxford 1972.
[50] G. Peach, Adv. Phys. **30**, 367 (1981).
[51] N.F. Allard and J. Kielkopf, Rev. Mod. Phys.**54**, 1103 (1982).
[52] V.S. Lisitsa, Sov. Phys. Uspekhi **20**, 603(1977).
[53] A. Gallagher, in: *Atomic, Molecular and Optical Physics Handbook*, ed. G.W.F. Drake, AIP Press, Woodbury, NY 1996, p.220.
[54] G. Peach, in: *Atomic, Molecular and Optical Physics Handbook*, ed. G.W.F. Drake, AIP Press, Woodbury, NY 1996, p.669.
[55] H. Horodniczy and A. Jabłoński, Nature **142**, 1122 (1938); **144**, 594 (1939).
[56] Lord Rayleigh, Proc. Roy. Soc. A**183**, 26 (1944); Nature **155**, 84 (1945).
[57] A. Jabłoński, Nature **155**, 397 (1945).
[58] M. Baranger, Phys. Rev. **111**, 481 (1958); **111**, 494 (1958); **112**, 855 (1958).
[59] I.I. Sobelman. Opt. Spektrosk. **1**, 617 (1956).
[60] F.H. Mies, J. Chem. Phys. **48**, 482 (1968).
[61] Y. Tanaka and K. Yoshino, J. Chem. Phys. **39**, 308 (1963).
[62] R.E. Hedges, D. Drummond and A.C. Gallagher, Phys. Rev. A**6**, 1519 (1972); A.C. Gallagher, Acta Phys. Polon.A **54**, 761 (1978).
[63] W. Behmenburg, Z. Naturforsch. **27a**, 31 (1972); J. Losen and W. Behemenburg, Z. Naturforsch. **28a**, 1620 (1973).
[64] K.M. Sando and J.C. Wormhoudt, Phys. Rev. A **7**, 1889 (1973).
[65] J. Szudy and W.E. Baylis, J. Quant. Spectrosc.Radiat. Transfer **15**, 641 (1975); Phys. Rep. **266**, 127 (1996).
[66] K.W. Ford and J.A. Wheeler, Ann. Phys. **7** 259 and 287 (1959).
[67] K.M. Sando, R.D. Doyle and A. Dalgarno, Astrophys. J. **157**, L143 (1969).
[68] J. Stewart, J.M. Peek and J. Cooper, Astrophys. J **179**, 983 (1973).
[69] N.F. Allard, D. Koester, N. Feautrier and A. Spielfiedel, Astron. Astrophys. Suppl.Ser. **108**, 417 (1994).
[70] J.L. Greenstein, Astrophys. J. **241**, L87 (1980).
[71] D. Koester, N.F. Allard and G. Vauclair, Astron. Astrophys. **291**, L9 (1994).
[72] J.F. Kielkopf and N.F.Allard, Astrophys. J. **450**, L75 (1995).
[73] G. Herzberg, *Molecular Spectra and Molecular Structure*, vol. 1, Van-Nostrand Reinhhold, New York 1950.

[74] R.P. Futrelle, Phys. Rev. A **5**, 2152 (1972).
[75] A. Jabłoński, Z. Phys. **73**, 460 (1931).
[76] D. Berry and F. Williams, Acta Phys. Polon. **54**, 705 (1978).
[77] A. Jabłoński, Acta Phys. Polon. **6**, 350 (1937).
[78] J.L. Carlsten, A. Szöke, and M.G. Raymer, Phys. Rev. A**15**, 1029 (1977).
[79] A. Corney and J.V.M. Ginley, J. Phys. B **14**,3047 (1981).
[80] W.J. Alford, N. Andersen, K. Burnett, and J. Cooper, Phys. Rev. A **30**, 2366 (1984).
[81] J. Cooper, R.J. Ballagh, and E.W. Smith, Acta Phys. Polon. A **54**, 729 (1978).
[82] K. Burnett, Phys. Rep. **118**, 339 (1985).

Polarization Processes in Electron-Heavy Ions Collisions

Astapenko V.A.*, Bureyeva L.A.**, Lisitsa V.S.***

*Moscow Institute for Physics and Technology, Dolgoprudnyi, Moscow region 141700, Russia
**Scientific Council on Spectroscopy of the RAS, Leninski Pr.53, Moscow, 117924, Russia
**RRC «Kurchatov Institute», Kurchatov Sq., 46, Moscow, 123182, Russia

A polarization channel for radiative transitions (free-free, free-bound, bound-bound) of electrons on heavy ions with a complex core is under investigation. The channel is connected with a dynamic polarization of an ion's core which results in radiation of the core and an inelastic transition of electron. This channel is estimated in the frame of statistical model of the complex ion. It is shown that a contribution of polarization radiation may be comparable or exceed the standard radiation contribution. The significance of interference effects is demonstrated as well.

1. Introduction.

Polarization radiative transition have been extensively investigated for free-free transitions being considered as polarization Bremsstrahlung [1]. The essence of such kind of processes is as follows. The electron moving in an atomic potential can radiate a light quantum in two ways, namely 1) due to it's own acceleration in the potential or 2) due to a dynamic polarization of the atomic core by it's variable electric field (Fig.1).

In the first case one deals with Bremsstrahlung radiation whereas in the second case we face with the polarization radiation. In both cases the colliding electron loses it's initial energy due to radiation of the quantum, but such radiation provides by the colliding electron itself in ordinary Bremsstrahlung case and by the atomic core in the case of the polarization Bremsstrahlung. The last process can be considered as a special kind of scattering of the electron energy by the atomic core with a transformation into the light quanta in analogous with the scattering of light. The conception of polarization radiative transitions are very close to the conception of Max Born radiation theory (see [2]). Really M.Born considered a radiation process as a radiation of a quantum by the whole interacting system without separation between transitions of external colliding electrons and bounded atomic electrons. In the frame of his conception one must operate with a interaction of radiation field with a total dipole momentum of the whole system being a sum of dipole momenta of all interacting particles.

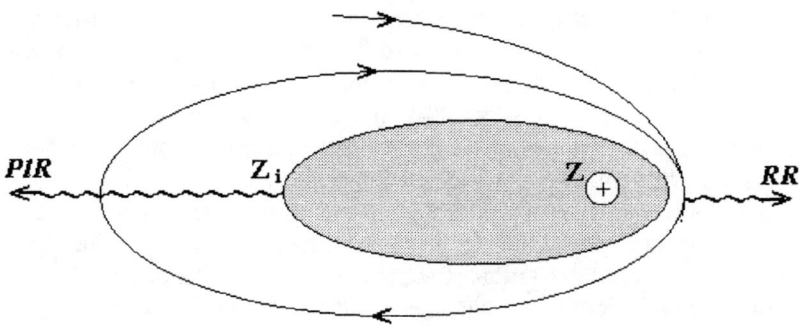

Fig.1 A scheme of standard radiation (RR) and polarization radiation (PIR) of an electron in the field of heavy ion.

Alexander Jablonski [3] has generalized the Born conception on the case of bound-bound transitions. He has developed the general spectral line broadening theory considering the broadening process as a radiation of the light quantum by the compound system: «radiating atom +surrounded perturbing particles». The broadening phenomena in such a picture look like an energy exchange between the perturbing particles and the radiating atomic electron. It looks like a some kind of polarization radiation where the polarized atomic electron provides a radiation of the quanta whereas the energy losses (or increase) are possessed by colliding particles.

As a conclusion of the discussion above we arrive to a following general conception of polarization radiation and accompanied processes. Considering the radiation of light quanta by the whole interacting system one must take into account the induced dipole momenta due to polarization of different parts of the radiating system due to the interaction. The effect is well known in the broadening theory in the case of the radiation of forbidden components of spectral lines due to induced dipole momentum at the forbidden transition in interacting system «plasmas + radiating atom» [4,5]. In the present case we'll be interested in polarization radiative transitions due to a polarization of the atomic core in colliding with electrons. It is obvious that different types of electron-atomic processes can take place in polarization radiative transitions, namely free- free (pointed above) , free-bound (polarization recombination), and polarization bound-

bound transitions. The possibility of such types of transitions has been pointed out in [6] and specific calculations for bound-bound transitions has been performed in [7] (note that polarization phenomena in the bound-bound transitions in alcali elements was taken into account many years ago, see corresponded references in [8] and the recent work [9]).

In all the cases pointed the one electron (resonance type) transitions were taken into account. At the same time in calculations of polarization Bremsstrahlung the broad peak in spectral distribution of emitted quanta was obtain theoretically and observed experimentally [10, 11]. It points on a possibility of some types of collective effects due to emission of atomic electrons strongly correlated within a given electron shell. So it is of interest to estimate a contribution to polarization effects such types of multiparticle phenomena. In the frame of such collective conception the radiating atom is considered as a cloud of electrons polarized by the colliding electron and radiading as a whole at some collective frequencies connected with virtual excitation of many electrons simultaneously [12, 13]. Such types of polarization radiative phenomena will be considered lower.

2. General approaches.

The estimation of polarization radiation (PlR) rates is based on the comparison of spectral radiation intensities determined by Fourier components of the dipole moment d_ω of the incident electron and by a dipole $D_\omega = \alpha(\omega) F_\omega$ induced in the ion's core by the electric field $F(t) = e\, r(t)/r^3(t)$ of the colliding electron, $\alpha(\omega)$ being a dynamical polarizability of the core, F_ω is a Fourier component of the electric field. The relationship between Fourier components of electron acceleration determining the classical radiation intensity and Fourier components of the electric field determining the polarization of the atomic core one can find by taking into account the equation of electron motion in an atomic potential ($d = e\, r$ below atomic units are used):

$$\ddot{\mathbf{r}}(t) = -Z_{eff}(r)\mathbf{r}/r^3 \qquad (1)$$

where $Z_{eff}(r)$ is an effective charge of the ion's core changing between the ion charge Z_i and the nuclear change Z.

When taking the Fourier transformation of eq.(1) we'll use for simplicity an approximate version of this relationship introducing an effective charge $Z_{eff}(r_0^\omega)$ for electron penetration into the core at the radius r_0^ω. The radius r_0^ω responsible for the radiation of the frequency ω is determined by the relationship [14-16]

$$E + |U(r_0^\omega)| = \omega^2 (r_0^\omega)^2 / 2 \qquad (2)$$

where E is the initial electron energy, $U(r_0^\omega) = Z_{eff}(r_0^\omega)/r_0^\omega$. The relationship (2) expresses the identity of the emitted frequency ω and the angular velocity of electron rotation at it's trajectory in the atomic potential. Using this approximation we arrive to the following expression for the Fourier component of eq.(1):

$$\ddot{\mathbf{r}}_\omega = \omega^2 \mathbf{r}_\omega \approx Z_{eff}\left(r_0^\omega\right) \mathbf{F}_\omega \qquad (3)$$

Finding further the ratio $R(\omega)$ of spectral intensities (or corresponding radiation cross section $d\sigma/d\omega$) of polarization $I^{PIR}(\omega)$ and conventional radiative transitions $I^{RR}(\omega)$ we arrive to:

$$\frac{I^{PIR}}{I^{RR}} = \frac{|D_\omega|^2}{|d_\omega|^2} = \left|\frac{\omega^2 \alpha(\omega)}{Z_{eff}(\omega)}\right|^2 \equiv R(\omega) \qquad (4)$$

the result (4) has been obtained in [1] for the case of the atomic potential $U(r)$ close to Coulomb one when the effective charge $Z_{eff}(\omega)$ is close to the ion's charge Z_i.

Substituting into eq.(2) instead of $U(r)$ the statistical ion's Thomas-Fermi potential one can determine the value r_0^ω and consequently the effective charge $Z_{eff}(\omega)$. Practically for collisions of slow electrons the value of $Z_{eff}(\omega)$ is close to the ion charge Z_i.

The relationship (4) is of universal type. It expresses the ratio $R(\omega)$ of two channels of radiative transitions in terms of the core polarizability $\alpha(\omega)$ and effective charge $Z_{eff}(\omega)$. The value $\alpha(\omega)$ consists of a lot of resonances connected with the excitation of the core. We'll be as pointed beyond interested also in a regular (smooth) dependence of $\alpha(\omega)$ on the frequency ω which is supposed to be responsible for the increased background in the radiative rates. Such a dependence can be found for example from the statistical Thomas-Fermi model of atoms.

3. Calculations of dynamic polarizabilities.

Problems connected with calculations of dynamic polarizabilities of complex ions and atoms are subjects for a lot of papers, see [10,12,13,17]. We only point here at two general cases in such calculations where some general properties can be extracted for polarization radiation. The first one is associated with one or two electron resonances in atomic structures where the resonance approximation for electron polarizabilitiy can be applied. It is the case also for Rydberg atomic states where a lot of resonances can be taken into account in Coulomb field [17].

The second case is based on statistical (Thomas-Fermi) model of heavy atoms where polarizabilities of the systems are connected with a distribution of electron density inside the atom [12,13]. The resonant polarizability (or one in the Coulomb field) can be calculated directly according the general formulae:

$$\alpha(\omega) = \frac{e^2}{m} \sum_n \frac{f_n}{\omega_n^2 - \omega^2} \qquad (5)$$

where f_n is the oscillator strength of corresponding transition at a frequency ω_n, e, m, electron charge and mass (atomic units are used lower).

In resonant approximation only one (resonance) term must be stay in the sum (5), which result in Lorentz-type dependence with sharp maximum at the resonance transition.

In general the polarizability of a ground atomic state is a functional of electron density of the ground state of the atom. If one takes into account the only change in the atomic Hamiltonian connected (in the frame of dipole approximation) with the change of orbital momentum of the state then the Fourier transform of the atomic dipole momentum leads to a relation between the emitted/absorbed frequency ω and the change in centrifugal potential. So the atomic photoionization cross-section is expressed in integration of squared of atomic electron wave function (that is electron density) with delta function [18]:

$$\operatorname{Im}\alpha^{Rost}(\omega) = \frac{137}{4\pi\omega}\sigma(\omega)^{Rost} \approx \frac{\pi Z^2}{3\omega^2}\int d\mathbf{r}|\Psi(\mathbf{r})|^2 \delta\left(\omega - \frac{1}{r^2}\right) \qquad (6)$$

The result (6) suggested in [18] gives an excellent results for the photoionization cross-sections in the cases of one and two- electron atoms.

One can apply the result (6) to the case of many electron Thomas-Fermi atom by substitution the statistical density $n(r)$ instead of squared wave function into eq.(6). The results are in reasonable correspondence with other approximations pointed lower.

From Fig.2 one can see that considered approximation gives good static limit of ion polarizability and correct frequency dependence.

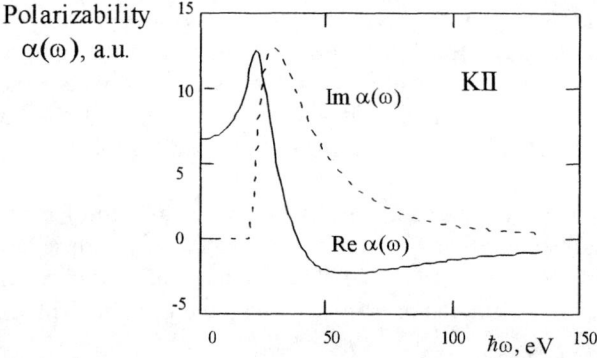

Fig.2 Dynamical polarizability of *KII* ion according to the Rost approximation [18]

One of the interesting approximation is connected with plasma approaches to the calculation of atomic polarizability. In the frame of these approximations the polarized core radiate at a plasma frequency $\omega_p(r)=\sqrt{4\pi n(r)}$ so that the photoionization cross-section is determined by [12]:

$$\operatorname{Im}\alpha^{B-L}(\omega)=\frac{137}{4\pi}\sigma^{B-L}(\omega)=\frac{\pi}{2}\int dr\, n(r)\delta(\omega-\omega_p(r)) \qquad (7)$$

the real part of the polarizability is determined with the help of Kramers-Kronig relationship [19].

A generalization of plasma statistical theory with account of nonlocal interelation between radiated and observed frequencies was done in [13].
The idea of radiation of an electron at some specific frequency during it's motion in an atomic potential is not new as has been pointed out in connection with eq.(2). It has been already applied to the Bremsstrahlung radiation of electrons in atomic potentials [14,16,20] (so called rotation approximation - RA).

In the frame of RA all spectral distribution are obtained with the help of substitution of delta-function containing differences between emitted ω and rotation ω_{rot} frequencies into equations determining the power emitted at the specific radius r. Such substitution seems to be of the same nature as the substitution of delta-function at plasma frequency in eq.(7). Really if one considers the motion of one atomic electron in a self-consistent potential of other electrons it is clear that the situation is just the same as in the case of colliding electron with the energy $E=0$. So one can determine the photoabsorption cross-section with the help of relationship given by eq.(7) with delta-function containing $\omega-\omega_{rot}$. Practically all types of frequencies substituted into the argument of delta-function are rather close to one another; all discrepancies are inside discrepancies between statistical theories themselves.

4. Polarization of Thomas-Fermi atoms and ions.

Many electron atom in the frame of statistical model is supposed to be an electron cloud with the density $n(r)$ determined by well known Thomas-Fermi function $n(r)=(32Z^2/9\pi^3)$ $(\chi(r/a_{TF})a_{TF}/r)^{3/2}$ (a_{TF}=0.885 $Z^{1/3}$ being a Thomas-Fermi length) [21]. The ion radius is equal to $X(q)a_{TF}$, where $X(q)$ is a dimensionless ion radius depending on the ion ionization stage $q=Z_i/Z$.

The ratio $R(\omega)$ with accounting of eq. (11) takes the form

$$R(\omega)=(0.885\,X)^6\beta^4 F_X(\beta)(Z/Z_{eff})^2 \qquad (8)$$

where $\beta=\omega/Z$ (a.u.) and we introduced the function

$$F_X(\beta)=\phi_X^2(\beta)+\psi_X^2(\beta) \qquad (9)$$

To determine $R(\omega)$ from eq.(12) it is necessary to substitute parameters $Z_{eff}(\omega)$, X and β.

5. Free-free and free-bound transitions

The most calculations and measurements are performed for free-free transitions [1, 10-11,22]. We'll consider both types of the transition mentioned in the frame of universal statistical theory [23].

The value of $R(\omega)$ as well as the specific form of the function $F_X(\beta)$ from eq. (13) are determined by the structure of specific ions, namely by nuclear charge Z and ionization parameter $q=Z_i/Z$.

The function $F_X(\beta)$ for $q=0.3$, $X=5$ is directly obtained from numerical data [13]. The values of $Z_{eff}(\omega)$ and β are determined by the initial electron energy E and a final energy E' or an energy E_n of the captured electron given correspondingly by

$$\beta \equiv \omega/Z = \begin{cases} (E - E')/Z & \text{free} - \text{free} \\ (E + |E_n|)/Z & \text{free} - \text{bound} \end{cases} \quad (10)$$

The factor $R(\omega)$ is expressed in simple analytical forms in limiting cases $\omega \to 0$ and $\omega \to \infty$ using static and high frequency polarizabilities of the ion correspondingly. Thus putting $F_X(\beta) \approx \phi_X^2(\beta) = 1$ for $\beta \to 0$ one obtains from eq.(8):

$$R(\omega) \approx (0.885 X)^6 \beta^4 (Z/Z_i)^2 \quad (11)$$

where we have put $Z_{eff} \approx Z_i$ for $\omega \to 0$ according eq. (2). In opposite limiting case $\alpha(\omega)$ is determined for the high frequency domain $\omega \to \infty$ by the polarization of free core electrons that is

$$-\omega^2 \alpha(\omega) \underset{\omega \to \infty}{\to} N = Z(1-q) \quad (12)$$

where N is a number of bound electrons in the ion ($N=Z-Z_i$). So in this domain of electron energies we arrive to

$$R(\omega) = \left[\frac{2N}{Z+Z_i}\right]^2 = 4\left[\frac{1-q}{1+q}\right]^2 \quad (13)$$

where we have put $Z_{eff}=(Z+Z_i)/2$ which is in accordance with calculations [24] (see also [14-16]) a good approximation for wide domain of electron energies.

The boundary value β_X^* is determined by equating expressions for $|\alpha(\omega)|$ for limiting cases $\omega \to 0$ and $\omega \to \infty$. Putting $\alpha(0)=(X\, a_{TF})^3$ for $\omega=0$ and $|\alpha(\omega)|=N/\omega^2$ for $\omega \to \infty$ we obtain for the boundary value of β_X^* (a.u.):

$$\beta_X^* = \frac{\sqrt{1-q}}{(0.885\,X)^{3/2}} \approx \frac{\sqrt{1.5(1-q)}}{X^{3/2}} \tag{14}$$

The value of β_X^* from eq. (14) is in good correspondence with numerical data [13], namely, for $X=5$, $q=0.3$ it follows $\beta_5^* \approx 2.5$ eV which is close to the minimum value of the curve $Re\,\alpha(\omega)$ in [13].

The scaling of the type given by eq.(14) may be obtained from the local plasma frequency model. Really, if one analyses selfconsistently a couple of equations for the total electron number in the ion and photoionization cross section $\sigma_{ph}(\omega)$ expressed in terms of local electron density $n(r)$ than it is seen that $\sigma_{ph}(\omega)$ is a function of reduced frequency $\tilde{\omega} = \omega\, X(q)^{3/2}/Z\sqrt{1.5(1-q)}$ that is just the scaling (19).

The estimation of the factor $R(\omega)$ in eq.(8) for uranium ion U^{+28} in the case of intermediate domain of frequencies gives it's value near the factors from 3 to 4, while for KII and RbII R value may be about 100 for sufficiently high frequencies. Thus polarization recombination may spectrally dominate over radiative one. In this sense the situation is analogous to polarization Bremsstruhlung.

6. Polarization bound-bound transitions.

The first calculations of the core polarization effects on radiative transitions were done many years ago, see references in [8,9]. The calculations were performed for oscillator strengths in alkali atoms where the radiated frequencies are small enough to produce a dynamic polarization of the cores having the closed shells and high excitation frequencies. So it is enough to take into account the polarization effects in the frame of static approximations for atomic polarizabilities [9]. Nevertheless theoretical estimations and experimental observations demonstrate a strong dependence on frequency ω at a lot of transitions [8].

Let us consider polarization bound-bound transitions in more details. An excited electron inside an atom can reach it's ground state by two ways: firstly, by standard radiative transitions and, secondly, by exciting of the atomic core accompanied by it's energy loss. The second type of transitions is just polarization bound-bound transitions. As it has been already pointed out such types of transitions have been observed experimentally and considered theoretically [8,9] for electron transitions in alkali atoms (Na, Cs, Rb). It is clear that these polarization transitions are due to polarization of atomic cores with close electron shells which radiative frequencies are relatively small, that is of the order of ionization potentials of alkali atoms. So one can account the polarization effects in a static approximation for a dynamic polarizability of the cores ($\alpha(\omega) \approx \alpha(0)$). The calculation models [8,9] use a combined electron dipole momentum instead of the dipole momentum of external electron, that is

$$d_{if} = \int r dr P_i(r) P_f(r) \left(1 - \alpha(0)/r^3\right) \tag{15}$$

where $P_{i,f}(r)$ are radial wave functions.
Such models are in reasonable agreement with experimental observations. At the same time the dynamic effects in polarization transitions result in considerable disagreement between the theory and experiment, see [8].

The essential contribution of dynamic polarizabilities for bound electron transitions has been demonstrated in [7] for electron transitions from highly excited (Rydberg) atomic states in multicharged ions with complex cores. The essence of the effect is as follows: the ion's states with the excited core (that is double excited states) can lie lower than one-electron ionization potential. In this case the double excited states occurs to be put in a resonance with highly excited one-electron (Rydberg) states. It is just the case for *NeIV* ion where the double excited state *2s2p³3l* is near resonance for the Rydberg states *2s²2p²9l, 10l*. If one compare the radiative transition probabilities at the same frequencies for standard and polarization channels it occurs that polarization channel dominates over standard one about 30 times [7].

The bound-bound polarization transitions result in energy exchange between radiating and core electrons. These types of transitions are known as Oge-effects [2], where the energy exchange results in production a hole in the atomic core. What is the contribution of nonresonant (collective) polarization processes is not clear at present time. An interesting observations of polarization Oge-type transition in collisions of fast heavy ions with metal films were done in [25]. Such observations demonstrate an appearance of X-ray spectral line due to an excitation of M-shell in heavy projectile ions. At the same time a relative velocity of heavy projectile is too small to produce any electron excitation from M-shell. The explanation of the effect observed is connected with a channel leading to a capture of an electron from the target to an excited state of the projectile. Further stabilization of the system is due to the energy exchange between excited and core electron resulting in a hole in the M-shell of the heavy ion.

7. Stimulated polarization Bremsstrahlung including multiphoton processes. Interference effects.

Total stimulated Bremsstrahlung cross-section including both static and polarization channels as well as their interference term were calculated at the first time [26] in the framework of the fixed current approximation for the scattering of quasiclassical electrons on highly charged ions (HCI). The photon emission is considered as a transition between the states of the field oscillator under the influence of the electromagnetic field of the fixed radiating current, see for example [27,28]. Total radiating current is supposed to be a sum of two terms: fixed current of the incident particle in the field of HCI and the polarization current of the HCI core electrons induced by the incident particle $j_\omega^{total} = j_\omega^{projectile} + j_\omega^{polariz}$. The polarizability of HCI depends upon the stimulating field amplitude E_0 for sufficiently strong and/or near resonant field. Expression for the probability of n-photon total Bremsstrahlung including polarization and interference effects has the form:

$$W_\Sigma(n) = J_n^2 \left\{ 2(n_{k\lambda}^{Las} n_{k\lambda}^{stat})^{1/2} \left[1 + \frac{m\omega^2 \alpha(\omega, E_0)}{Z_i e^2} \right] \right\} \quad (16)$$

here n^{Las} is the photon occupation number of the stimulating field mode (k,λ) and n^{stat} is the photon number spontaneously emitted via static channel in the mode (k,λ), J_n is the n-order Bessel function.

The Bremsstrahlung cross-section of the one-photon absorption, integrated over Maxwellian velocity distribution and scattering angles of incident electron, as a function of frequency detuning Δ from the closest resonance transition $2s$-$3p$ in the Li-like oxygen ($Z=5$, $\omega=3.037$ a.u., $T=0.625$ a.u., $f_0=0.26$) is presented in Fig.3. Separate contributions of static and polarization channels are shown as well.

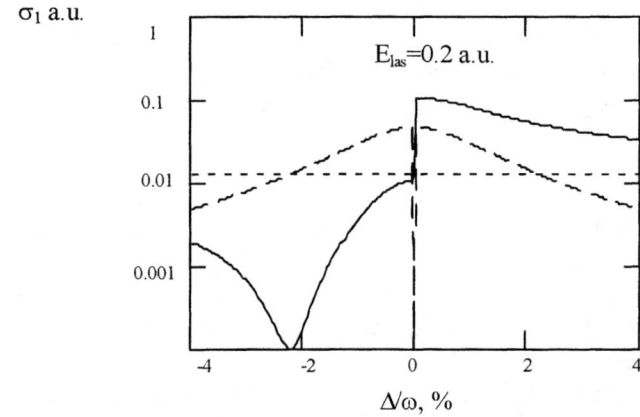

Fig.3 Spectral line shape of the near-resonant stimulated Bremsstrahlung.

From this figure one can see the spectral line asymmetry due to the interference between polarization and static channels in stimulated near-resonant Bremsstrahlung.

The interference effects are also of importance for standard and polarization radiation channels in the case of low charges of ions with many electron cores. Such interference effect is shown in Fig.4 for *KII* ion where a ratio of squared modulus of the sum of both channels to the sum of their modulus is presented as a function of the frequency. A dip in such the ratio demonstrates a strong compensation of contributions of these two channels for a specific frequency domain.

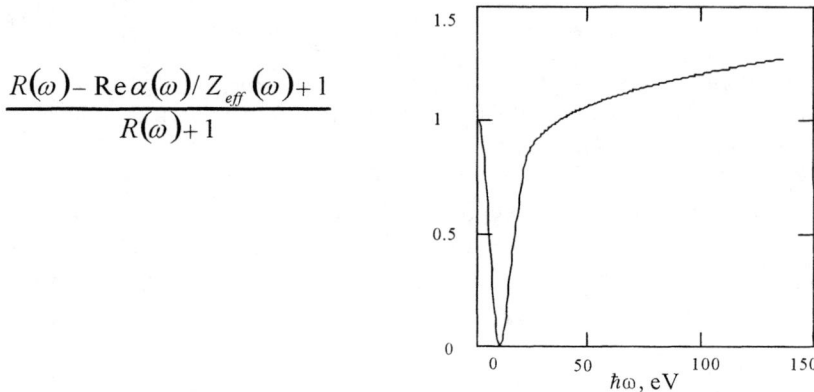

$$\frac{R(\omega) - \operatorname{Re}\alpha(\omega)/Z_{eff}(\omega) + 1}{R(\omega) + 1}$$

Fig.4 The spectral dependence of the ratio of radiative transition rate with and without account for the interference effects between standard and polarization channels for *KII* ion calculated in the frame of Rost approximation for core polarizability

8. Conclusions

Considered polarization processes in electron-heavy ion collisions show the importance of new (polarization) channels in radiation phenomena for various types of quantum transitions of the incident particle: free-free, free-bound and bound-bound. The significance of the interference effects between polarization and ordinary channels is demonstrated as well.

Applications of the considered theory comprise the cases of low temperature plasmas, astrophysical plasmas, laser plasmas and storage rings experiments. The latter can provide the most detailed information on the polarization channel. In plasma experiments the presence of new recombination channel results in the shift of ionization equilibrium. In the case of laser plasma stimulated polarization absorption/emission processes can considerably influence on the energy exchange between laser radiation and plasma.

9. Acknowledgments

This work was performed with the support of the Russian Ministry of Sciences (project «Theoretical spectroscopy of Rydberg states of atoms and ions») and the RFBR (project 98-02-16763).

REFERENCES

1. Polarization Bremsstrahlung, Edited by Tsytovich V.N. and Oiringel I.M., Plenum, 1992.
2. Mott N.F., Massey H.S.W. The theory of atomic collisions, Oxford, 1965.
3. Jablonski A. Phys. Rev., **68** (1945) 78.
4. Griem H.R. Plasma spectroscopy, McGraw-Hill C., N.Y., 1965.
5. Sobelman I.I. Introduction to the theory of atomic spectra, Pergamon, Oxford, 1973.
6. Kukushkin A.B., Lisitsa V.S. Ch.11 in [1].
7. Kukushkin A.B., Lisitsa V.S. Phys. Lett.A **159** (1991) 184.
8. Chichkov B.N., Shevelko V.P. Phys. Scripta **23** (1981) 1055.
9. Migdalek J. J.Phys.B **31** (1998) 1947.
10. Amusia M.Ya. Ch. 7 in [1].
11. Korol A.V., Solovyov A.V. J.Phys.B **30** (1997) 1105.
12. Brandt W., Lundqvist S. Phys.Rev. **139** (1965) A612.
13. Vinogradov A.V., Tolstikhin O.I. Sov.Phys. JETP **69** (1989) 683.
14. Kogan V.I., Kukushkin A.B. Sov.Phys.JETP **60** (1984) 665.
15. Lisitsa V.S. Atoms in Plasmas, Springer Verlag, Berlin-Tokyo 1994.
16. Bureyeva L.A., Lisitsa V.S. A Perturbed Atom, (In Russian) IZDAT, 1997.
17. Beigman I.L., Bureyeva L.A., Pratt R. J.Phys.B **27** (1994) 5833.
18. Rost J.M. J.Phys.B **28** (1995) L601.
19. Landau L.D., Lifshitz E.M. Electrodinamika sploshnykh sred, M., «Nauka», 1992.
20. Kogan V.I., Kukushkin A.B., Lisitsa V.S. Phys. Rep. **213** (1992) 1.
21. Landau L.D., Lifshitz E.M. Quantum Mechanics, Pergamon Press, New York-Oxford, 1974.
22. Astapenko V.A., Buimistrov V.M., Krotov Yu.A., et al. Sov.Phys.JETP **61** (1985) 930.
23. Bureyeva L.A., Lisitsa V.S. J.Phys.B **31** (1998) 1477.
24. Kim L., Pratt R. Phys.Rev.A **36** (1987) 45.
25. Baird S. et al. Phys.Lett.B **361** (1995) 184.
26. Astapenko V.A., Kukushkin A.B., JETP **84** (1997) 229.
27. Berson I. Sov.Phys. JETP **53** (1981) 63.
28. Lisitsa V.S., Savel'ev Yu.A. Sov.Phys. JETP **65** (1987) 273.

Optical Transitions in Excited Lithium+Rare Gas Collision Molecules and Related Interatomic Potentials

W. Behmenburg

Institut für Experimentalphysik
Heinrich-Heine-Universität Düsseldorf, D40225 Düsseldorf

Abstract. Experimental collision induced spectra of Li*He and Li*Ne associated with the transitions $2P\Lambda \rightarrow 3(D,P)\Lambda^*$ as well as $3S\Sigma \rightarrow 2S\Sigma$ are analysed by means of quantum mechanical simulations, using *ab initio* as well as *semiempirical* potentials and transition moment functions as input. By comparison, the observed spectral features can be understood in terms of the contributions from free-free and bound-free electronic transitions between the different molecular states, and the accuracy of the input data is tested.

1. INTRODUCTION

This paper reports on recent studies of electronic transitions involving the excited $3(D,P,S)\Lambda$ states of Li*-X collision molecules with X=He, Ne. Because of their simple electronic structure the spectra of such systems are well suited for a quantitative testing or determining the interaction potentials involved. In particular, for LiHe(2PΠ, 3DΔ) and LiNe(2PΠ, 3SΣ) accurate potentials have been extracted from rotationally resolved $2P\Lambda \rightarrow 3(D,P,S)\Lambda^*$ excitation spectra (1), (2). Complementary, from rainbow satellite bands, observed in the far wings of the Li(2P-3D) transition, barrier heights and positions of the long range repulsive $3(P,D)\Sigma$ potentials have been derived (3). Disagreement between these spectroscopically determined potentials and calculated ones (4), (5), (6), (7) has stimulated the development of new calculational techniques based on *ab initio* as well as semiempirical methods. Using these, interaction potentials and transition moment functions of much improved accuracy have been obtained for the LiHe and LiNe systems. The improvement was demonstrated by quantummechanical spectrum simulations, using the new interaction data as input and comparison with the observed spectra (8), (9).

Apart from testing potential calculations spectrum simulations also show, how the observed structures arise from the contributions due to the different molecular transitions involving free and bound states. Based on the results reported in (8) and (9), I shall illustrate in my talk both aspects at 2 example cases, namely the excitation spectrum $2P\Lambda \rightarrow 3(D,P)\Lambda^*$ of Li*He and the emission spectrum $3S\Sigma \rightarrow 2S\Sigma$ of Li*Ne. In the case of the Li*He spectrum I shall also illustrate the improved agreement obtained with experiment when bound-bound transitions are included in the simulation.

2. GENERATING THE SPECTRA

2.1. General experimental method

The excitation spectra on the 2PΛ→3(P,D)Λ*transitions are generated by 2 step cw laser excitation in gaseous mixtures of Li+X at temperatures around 600 K (Fig. 1). Laser 1 excites Li atoms resonantly to the 2P level, where they subsequently collide with rare gas atoms forming molecular Σ or Π states. A second laser scans over the manifold of electronic transitions shown in Figure 1. to excite the collision molecules into the higher 3(D,P)Λ states, that subsequently dissociate into Li3P or 3D atoms. Using 3P-3D mixing by rare gas collisions during the excited state life time the atomic Li 3D→2P emission at 610.3 nm is monitored as a measure of molecular absorption. The detection sensitivity is mainly limited by the considerable 610 nm background fluorescence produced in blue wing of the 670 nm resonance line and from Li3D-population due to energy pooling collisions of Li2P atoms.

For comparison with quantummechanical simulations, the 610 nm signal rate $S_{3D}(\nu)$, depending on scan laser frequency ν is normalized to the rare gas density n_X and the integrated signal rate $\int S_{3D}(\nu)d\nu$. The letter was obtained by scanning laser 2 across the center and into the wings of the 610 nm line. Under the experimental conditions this normalized signal rate equals the normalized molecular absorption cross section in eq. (2) of sect. 3 of this paper.

Excimer emission spectra 3(P,S)Σ→2SΣ were generated by means of 2-step resonant cw laser pumping of Li atoms into the upper 3D level. The Li3D atoms are then transferred by rare gas collisions to the lower 3P and 3S levels and by secondary collisions

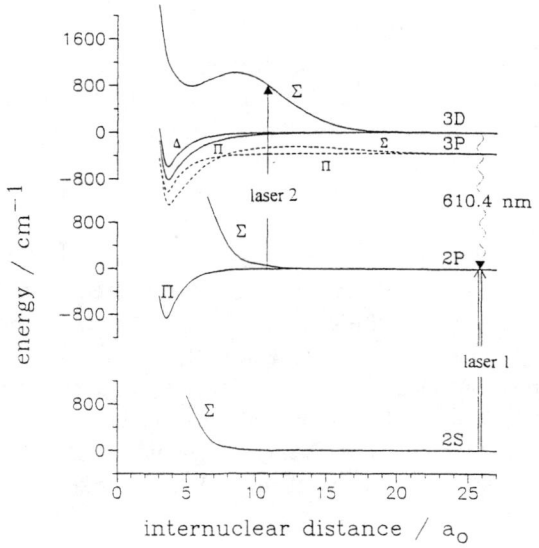

FIGURE 1. Interaction potentials for LiHe system (8) and the general experimental scheme.
Li atoms are laser-excited to the 2P state and a second laser scans over the manifold of electronic transitions 2P→3(P,D). Atomic Li emission at λ = 610.4 nm is monitored as a signature of the molecular LiHe absorption.
From ref. (8)

with rare gas atoms Li3(P,S)X molecules are produced. These molecules emit fluorescence spectra on transitions to the repulsive 2SΣ ground state in the 330 - 420 nm region, that is scanned by a grating monochromator.

2.2. Experimental Details

The schematic diagram of the apparatus used for the two types experiments is seen in Figure 3. A stainless steel cell with cooled windows contains lithium vapour-rare gasmixtures at around 650 K and densities of the order of 10^{11} cm^{-3} for Li and 10^{18} cm^{-3} for

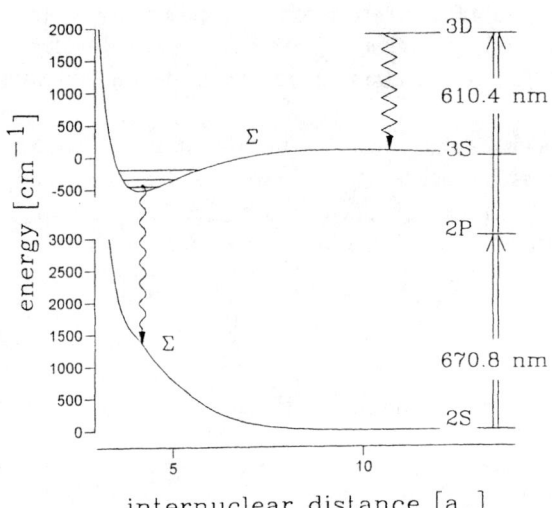

FIGURE 2.
Interaction potentials for the LiNe system (9) and the experimental scheme for excitation and observation of the emission spectrum 3SΣ→2SΣ. From ref. (9)

the rare gas. The two dyelasers used for fluorescence excitation are simultaneous pumped by an Ar⁺laser and multimode operated. Fluorescence is excited within the overlap region of the two laser beams passed antiparallel through the cell. The fluorescence detection chain for both experiments consists essentially of a monochromator M with photomultiplier PM3 and photon counting system

For the excitation spectrum an additional 610 nm interference filter (not shown in Figure 2) in front of M is used to block atomic Li 670 nm fluorescence scattered from the cell windows. Also a spatial filter is introduced into the scan laser beam path to suppress broadband ASE background that would otherwise produce extra 610 nm fluorescence background. For registration of the emission spectrum an additional short-pass cut-off filter CF is used to block instrumental 610 nm and 670 nm radiation. In both types of experiments the Li2P or Li3D population is monitored by detecting fluorescence from these levels by filtered multipliers in a separate reference channel. Further experimental details are described in refs. (3) and (17).

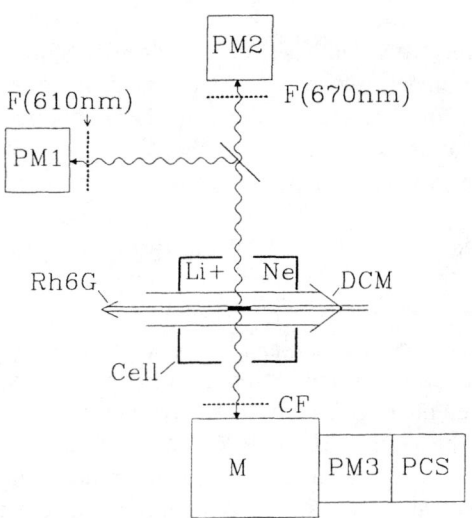

FIGURE 3.
Schematic diagram of the experimental apparatus for recording excitation and emission spectra:

M: 1m Czerny-Turner monochromator, Spex 1704;
PM1, PM2: photomultiplier EMI 9783;
PM3 photomultiplier Hamamatsu R943-02;
PCS: Photo counting system.
F(610): interference Filter MAZ, Schott;
F(670): colour glass filter RG630, Schott;
CF: lowpass cut-off filter LO 595;
Rh6G, DCM: dyelasers CR-599.

3. QUANTUM CALCULATIONS OF THE SPECTRA

3.1. Excitation spectra 2PΛ→3(P,D)Λ* of LiX

For testing the different approaches to the LiX interactions quantum calculations of the free-free spectra 2PΛ→3PΛ*, 3DΛ* were performed with the corresponding potential and transition moment sets as input. The resulting thermal absorption cross sections σ (per active atom) under the conditions of binary collisions is given by (10)

$$\sigma(\bar{v}) = \frac{8\pi^3 c}{3} \cdot \frac{\bar{v}}{Q(T)} \cdot \frac{n_X}{g_i} \cdot \sum_{\Lambda\Lambda^*} \int dE \sum_J (2J+1) \left| \left\langle \varphi_{\Lambda E}^J \left| d_{\Lambda\Lambda^*}(R) \right| \varphi_{\Lambda^* E^*}^J \right\rangle \right|^2 e^{-E/kT}, \quad (1)$$

where the φ' s are the energy normalized wave functions for the nuclear motion related to the molecular channels of the free-free transitions $\Lambda \to \Lambda^*$ in AX and $E^* = E + hc\Delta\bar{v}$. $Q(T) = (2\pi\mu kT/h^2)^{3/2}$ is the translational partition function. Similar expressions pertain to bound-free and bound-bound transitions (10).

Eq. (1) is related to the result of the early Jablonski quasimolecular approach to optical collisions (11), (12) in two respects. It represents, on the one hand, the single perturber approximation of the multi perturber interaction with the radiating atom, originally considered. On the other hand it is an extension of the early model to real systems with several molecular channels and where the dipole transition moment may vary considerably with internuclear separation.

The actual evaluation of eq. (1) was performed here by the theoretical scattering treatment of optical collisions (13). The variation of the transition moments $d_{\Lambda\Lambda^*}(R)$ is accounted for in the calculation of the profiles. However all nonadiabatic effects due to mixing of electronic levels during a collision were ignored, and the total absorption due to free-free transitions is expressed as a sum of independent contributions from pairs of adiabatic potentials. As long as no long lived quasibound levels are involved the calculated intensities also refer to the excitation of the atomic 3D and 3P states of Li caused by optical collisions, and are therefore directly comparable to fluorescence rates from these levels. Furthermore, the profiles are thermally averaged, so that direct comparison with cell experiments is possible.

The spectral profiles obtained in this way are, within the adiabatic limit, valid over the whole spectral range, including the impact region, except in the very narrow range of the central line core. Also it should be noted, that such quantum calculations, other than some of the semiclassical approaches, lead to correct rainbow satellite shapes for any given potential set.

As in (3) the experimental profiles are given in terms of a normalized excitation cross section

$$\sigma^*(\bar{\nu}) := \frac{\sigma(\bar{\nu})}{n_{He} \cdot \int \sigma(\bar{\nu}) d\bar{\nu}}, \qquad (2)$$

which involves also an integration over the line core region. In the binary collision regime σ^* is obtained from the calculated absorption cross section by using in eq. (2) the relation

$$\int \sigma(\bar{\nu}) d\bar{\nu} = \frac{8\pi^3}{3h} \cdot \frac{\bar{\nu}_o}{g_i} \cdot \sum_{\Lambda\Lambda^*} d^2_{\Lambda\Lambda^*}(\infty). \qquad (3)$$

For the Li*(2P)X excitation spectra the molecular states involved in the vicinity of the Li2P→3D transition are $\Lambda = \{2P\Sigma, 2P\Pi\}$ and $\Lambda^* = \{3P\Sigma, 3P\Pi, 3D\Sigma, 3D\Pi, 3D\Delta\}$.

Figure 4 displays the calculated Li*(2P)He spectrum using the potentials and transition moment functions from the AI method. From the contributions of the different molecular channels it is seen, that the far blue wing region is dominated by transitions 2PΣ →3DΣ and 2PΠ→3DΣ. The satellites S1 and S2 are due to the maxima in the respective difference potentials, that inturn originate from the 3DΣ potential barrier around 8 a_o.

The red wing region, on the other hand is composed of two different contributions: A continuously decreasing part formed by "allowed" 2PΛ→3DΛ* transitions and a superimposed single satellite structure S3 caused by the asymptotically dipole forbidden transitions 2PΛ→3PΣ near the maximum of the 3PΣ potential around 13 a_o; in this range the transition moments in Figure 6 reach their maxima too, thus giving rise to a pronouced intensity peak.

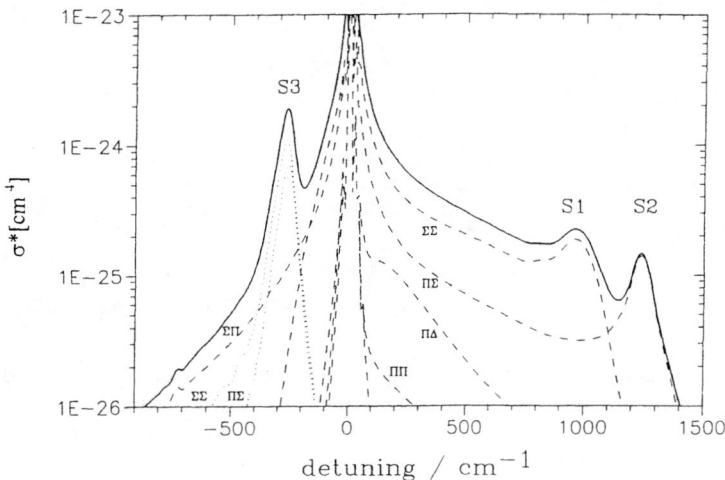

FIGURE 4: Normalized excitation cross section of Li*(2P)He versus detuning from the Li transition 2P–3D at T = 720 K. Quantum calculations according to eq. (1) using AI potentials and transition moment functions as input. The contributions from various molecular channels are indicated.
-----2P→3D; ·····2P→3P; ———— total. From ref. (8).

3.2. Excimer emission spectra 3DΣ→2SΣ LiX

The thermally averaged emission rate for free-free transitions was calculated from the absorption cross section for optical collisions σ(v) given by eq. (1). Introducing the spectral emission rate $k_{ff}(\bar{v})$ per collision pair (cm³/(cm⁻¹ s)) as defined in (10) and the absorption cross section $q_{ff}(\bar{v})$ per collision pair (cm⁵), one obtains according to (13)

$$k_{ff}(\bar{v}) = \frac{8\pi}{h} \bar{v}^2 \frac{g_\ell}{g_u} \exp(-hc\Delta\bar{v}/kT) \, q_{ff}(\bar{v}) \, ,$$

where $hc\Delta\bar{v} = V_u(R) - V_\ell(R)$ is the detuning from the asymptotic resonance and g_ℓ/g_u is the ratio of the statistical weights for the lower and upper molecular state.

The thermally averaged emission rate for bound-free transitions was calculated from the spectral emission rate $A_{bf}(\bar{v})$ per molecule (1/(cm⁻¹ s)) by means of the computer program given in ref. (14). This quantity is related to the spectral bound-free emission rate coefficient $k_{bf}(\bar{v})$ (cm³/(cm⁻¹ s)) of ref. (10) by

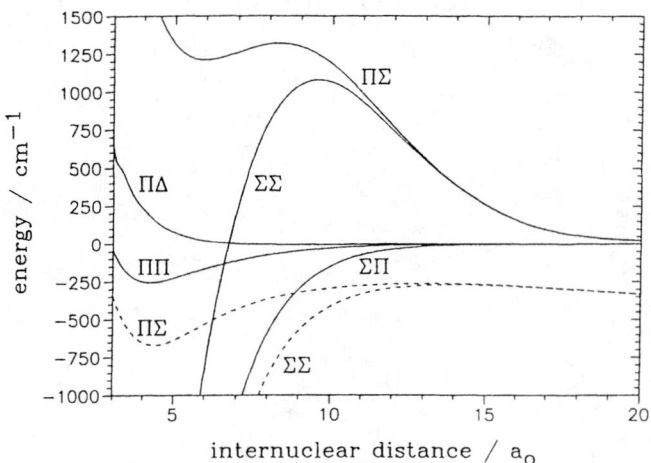

FIGURE 5: Difference potentials of LiHe for different Li transitions. The curves are obtained from the AI potentials. Energy values with reference to the asymptotic energy difference Li 2P–3D. ——— 2P–3D; ----- 2P–3P. From ref. (8)

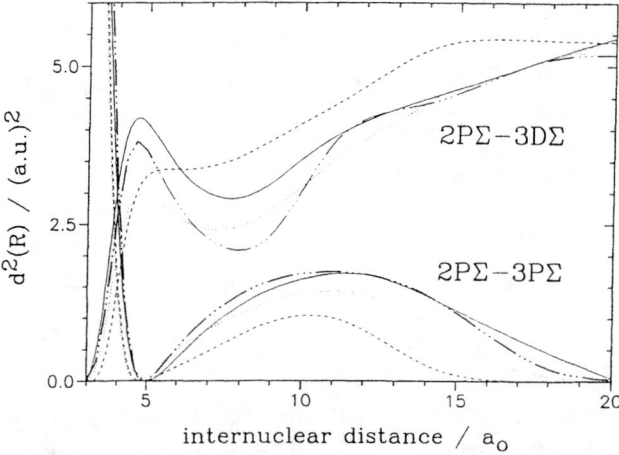

FIGURE 6: Squared transition moments of the molecular transitions 2PΣ→3PΣ,3DΣ for LiHe, calculated using different methods.——— (AI); —··· (PP); – – – (MFP); ··· (MP). From ref. (8)

$$k_{bf}(\overline{v}) = \frac{Q_b}{Q_f} \cdot A_{bf}(\overline{v}) \qquad (7)$$

Here, $Q_f = (2\pi \mu\, kT/h^2)^{3/2}$ is the translational partition function and Q_b, the partition function for the bound states, is approximately given by (15)

$$Q_b(T) = \frac{kT}{B_e} \sum_v \exp(-E_v/kT), \qquad (8)$$

where B_e denotes the usual rotational constant and E_v is the rotation-free binding energy in vibrational state v with reference to the asymptotic electronic energy.

For LiNe $3S\Sigma$, $B_e = 0.69$ cm^{-1} and with the E_v values calculated from the $3S\Sigma$ potential one obtains at 650 K for $Q_b/Q_f = 1.86 \times 10^{-22}$ cm^3.

From the results, included in Figure 10, the bound-free (bf) contribution dominates over that from free-free (ff) transitions. It is also seen how the bf fraction is composed of the partial spectra of the various $3S\Sigma$ vibrational states, that reflect the shapes of the respective wavefunctions $\Psi_v(R)$. The main peak of the total bf spectrum stems from the inner turning point maxima of $\Psi_v(R)$, whereas the short-wavelength modulation is due to the maxima of the $v \geq 1$ states at larger R. This modulation, however, is attenuated by the rapid drop of the transition moment function (Fig. 6).

4. CALCULATED VERSUS EXPERIMENTAL SPECTRA

4.1. Absorption spectrum of LiHe in the wings of the Li(2P-3D) transition

Figure 7 shows the absorption spectrum $2P\Lambda \rightarrow 3(D,P)\Lambda^*$ of Li(2P)He measured on absolute intensity scale in the vicinity of the Li(2P-3D) transition. For comparison are included the simulated free-free spectra, denoted by (AI, MP, MFP), that are obtained with the corresponding potential and transition moment sets from the different approaches in ref. (8) as input. The experimental spectrum displays an overall blue wing asymmetry, two far blue wing satellites S1,S2 followed by a steep cutoff and a single red wing satellite S3 superimposed on a continuous intensity decrease. These overall features are qualitatively reproduced by the simulated spectra. In detail however noticable discrepancies remain, that will be discussed further below.

In the detuning range up to ± 50cm^{-1}, the different molecular channels $2P\Lambda \rightarrow 3D\Lambda^*$ contribute, according to the simulations, with comparable intensities to the spectrum. This impact region is rather insensitive to the detailed potential shapes and thus well suited for a consistency test of the evaluation procedures for the experimental and theoretical cross sections. $\sigma^*(v)$. From Figure 8 the good agreement between experimental and calculated values in this region shows that the evaluation procedures have been correctly performed.

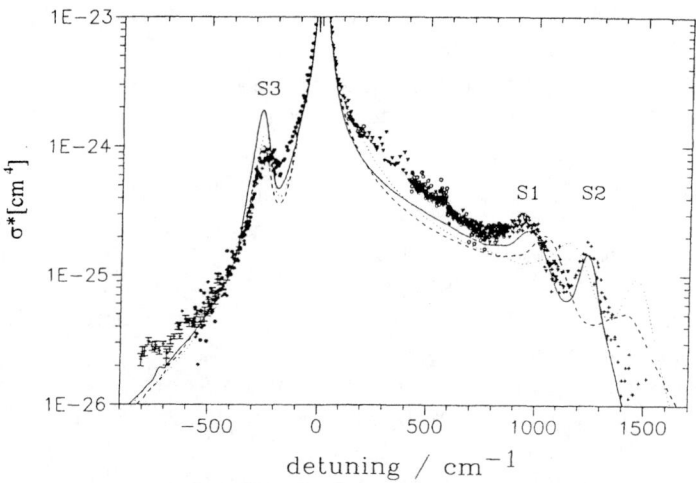

FIGURE 7: Excitation spectrum of Li*(2P)He at T = 720 K. The normalized atomic Li fluorescence signal at 610.4 nm is plotted versus the scan laser detuning from the Li 2P–3D transition. Experimental results (+++) and quantum calculations using potentials and transition moment functions from different methods.
——— (AI);   – – – (EH);   ····· (MP).   From ref. (8).

Blue wing region

The difference potentials in Figure 5 and the simulations show, that the whole blue wing spectrum beyond +50 cm^{-1} is dominated by 2P$\Lambda \rightarrow$3DΣ free-free transitions. In the near quasistatic region +50 cm^{-1} to +200 cm^{-1} shown in Figure 8 these transitions occur at internuclear separations 16-19a_0. There the lower 2P(Σ,Π) potential energies are still unshifted from their asymptotic value, so that the 2PΛ-3DΣ difference potentials are nearly identical with the upper 3DΣ potential. Thus different intensities at a given detuning in this spectral range reflect mainly differences in the slope of the 3DΣ energy curve in the 16-19 a_0 region.

Figures 7 and 8 show, that the calculated intensities between 100 and 700 cm^{-1} tend to be systematically smaller than the measured ones. However, the calculations neglect bound states, that may contribute noticeably to the total intensities. In fact pronounced rovibronic bands in this detuning range due to bound-bound transitions 2P$\Pi \rightarrow$3D(Π,Δ) have been observed and studied by rotationally resolved laser spectroscopy in ref. (1)

In the satellite region the AI spectrum reproduces the S2 satellite position within the experimental error bounds of ±15 cm^{-1}, whereas the S1 position is slightly shifted. This means, that the calculated 3DΣ barrier height is accurate within 30 cm^{-1} but the barrier position R_o is significantly too large (see table). Moreover, the good agreement with the experimental spectrum beyond 700 cm^{-1} indicates, that the whole AI shape of the 3DΣ potential barrier seems to be very accurate. The S1, S2 positions and cutoff in the MP and MFP spectra, however, are shifted towards larger detunings of the order of 100 cm^{-1} showing, that the 3DΣ barriers in the MP and MFP potentials are too high by the corresponding amounts (see table). Furthermore the S2 satellite in the MFP spectrum is much less pronounced than observed, demonstrating that the predicted shoulder in the 2PΠ–3DΣ difference potential is not realistic.

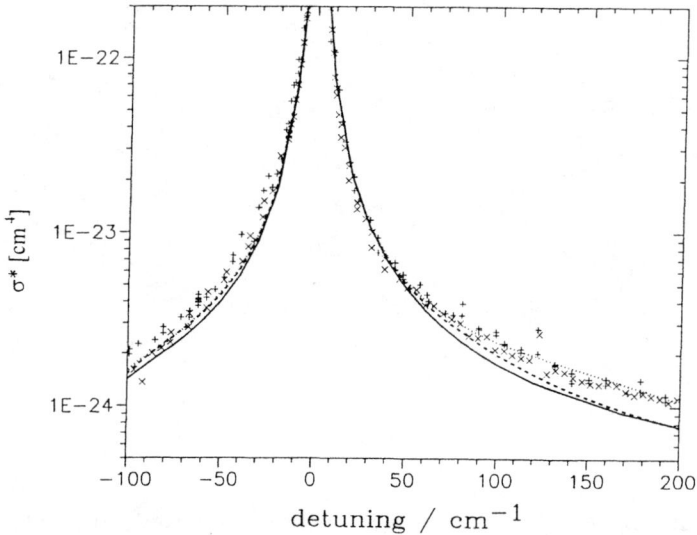

FIGURE 8: Excitation spectrum of Li*(2P)He at T = 720 K. Experimental results (+,×) from independent runs and quantum calculations using potentials and transition moment functions from different methods. ——— (AI); ----- (MFP); ····· (MP). From ref.(8)

TABLE: Binding energies/ barrier heights (cm^{-1}) and equilibrium distances (a_o) for LiHe in adifferent molecular states. From ref. (8).   a) ref. (1);   b) ref. (3);   c) shoulder

LiHe	2PΠ		3PΣ		3DΣ		3DΔ	
	R_o	V_o	R_o	V_o	R_o	V_o	R_o	V_o
AI	3.38	−1005	12.4	114	8.2	1244	3.58	−613
PP	3.55	−678	13.3	135	9.2	2070	–	–
MP	3.32	−920	12.7	120	8.1	1444	3.50	−557
MFP	3.62	−785	12.5	112	7.0	1396^c	4.0	−579
exp.	3.37(3)^a	−1020(20)^a	–	123(10)^b	7.9(3)^b	1270(20)^b	3.52(2)^a	−610(20)^a

With the aim to reduce the noticable discrepancies between simulated and experimental spectrum in the 100 - 700 cm^{-1} region, calculations were performed using the numerical routine described in (17) and assuming that the angular momentum coupling for the LiHe molecular states is well described by Hund's case b. The results of the calculations (19), using the AI-data of (8) as input, are presented in Figure 9. It is seen, that taking into account bound states considerably improves the agreement with experimental spectrum.

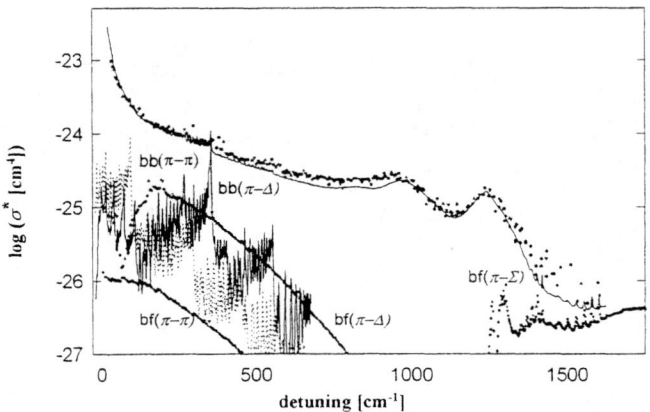

FIGURE. 9: Normalized excitation cross section of Li*(2P)He versus detuning from the Li transition 2P-3D at 720 K. The calculation (——) of ref.(19) is compared with experiment (·····) of ref.(3). The contributions from the bound-bound and bound-free transitions are indicated.

Red wing region

From the channel specific simulations (Fig. 4) and difference potentials (Fig. 5) the continuous part of the total experimental spectrum beyond -100 cm^{-1} is dominated by 2PΣ →3DΠ free-free transitions in the R-range 7-12 a_o. In the near red wing region -100 to -30 cm^{-1} also 2PΠ→3DΠ transitions contribute to the spectrum in the same R-range. The good agreement of simulation with experiment shows, that in this range the corresponding difference potentials are very realistic.

Good agreement is also found concerning existence and position of the S3 satellite band, superimposed to the background continuum. According to the simulations and difference potentials this satellite is formed by asymptotically forbidden 2P(Σ,Π)→3PΣ free-free transitions. around 12 a_0, where the transition moments maximize (Fig. 6). The good reproduction of the S3 position demonstrates that the height of the 3PΣ potential barrier is predicted by the AI method within about 15 cm^{-1}. However, discrepancy remains concerning intensities in the S3 region, indicating inaccuracies of the AI transition moments. The origin of the observed red wing oscillations beyond -500 cm^{-1} is not yet understood. The absence of such oscillations in the quantum profile (Fig. 4) shows, that the observed structure is not due to interferences between the ingoing and outgoing Condon transitions occurring during an optical collision (Stückelberg oscillations). The potentials and difference potentials indicate that bound-bound transitions 2PΠ→3PΣ and free-bound transitions 2PΣ→3PΣ, 3DΠ may contribute to this red wing region. However, from the vibrational band positions and relative intensities calculated with the AI potentials, a 2PΠ→3PΣ contribution seems to be very unlikely, since additional structures expected in the -500 cm^{-1} to -200 cm^{-1} range are not observed. In order to improve agreement with experimental red wing spectrum new calculations including also transitions between bound states are in progress.

4.2 Excimer emission spectrum 3SΣ→2SΣ of LiNe

In Figure 10 the emission spectrum Li*Ne in the 410 - 370 nm region is presented. The good matching of experiment with simulation shows, that the spectrum is formed by collision-induced asymtotically dipole forbidden transitions 3SΣ→2SΣ. According to the potentials in Figure 2 the transitions occur mainly from the well region of the upper 3SΣ potential, where the ground states 2SΣ are already highly repulsive.

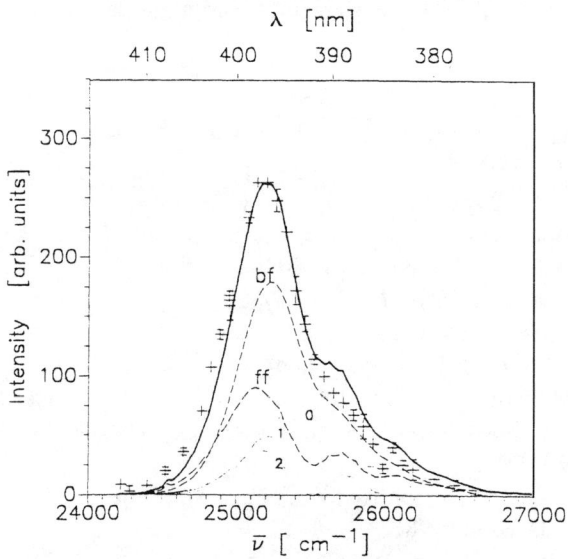

FIGURE 10: Emission spectrum of LiNe 3SΣ→2SΣ at T = 650 K, p_{Ne} = 508 mbar.
+++ experimental results ----- quantum calculation with the 2SΣ potential and transition moment function from the AI-method and the experimental 3SΣ potential from ref. (2)
- - - - Contributions from free-free and bound-free transitions.—— Partial bound-free spectra from 3SΣ vibrational states v as indicated. From ref. (9)

The experimental spectrum is roughly reproduced by the simulation. In detail, however, the experimental position of the intensity peak is shifted by about 50 cm^{-1} towards smaller values, and there is only a slight indication for the blue-wing hump predicted. The shift of the peak position is consistent with the experimental value of the 3SΣ binding energy, which is by 25 cm^{-1} smaller than the AI value. A quantum calculation that replaces the 3SΣ potential from the AI set by the experimental one of ref. (2) reduces the shift to about 30 cm^{-1}. Accounting for an $\pm$ 7 cm^{-1} uncertainty in the experimental 3SΣ potential minimum it follows that the 2SΣ energy at 4.11 a$_o$, the 3SΣ equilibrium distance, is smaller by about 40 cm^{-1}. This deviation, however, is within the limits of confidence of 20 - 100 cm^{-1} of the AI calculation in (9).

The missing hump in the experimental spectrum indicates a deviation from the Boltzmann population distribution among the rovibrational levels in the 3SΣ potential. For Neon densities of 4x10^{18} cm^{-3} at 650 K, the rate for elastic collisions of 4x10^{8} s^{-1} may be estimated. Comparing this to the spontaneous decay rate of the Li 3S state 3.49x10^{7} s^{-1} (16), we have about 10 collisions during the excited state lifetime, so that an equilibrium distribution may not be completely established.

ACKNOWLEDGEMENT

The author is indebted to H.Wenz for reading the manuscript and technical assistence.

REFERENCES

(1) Lee,C.J., Havey,M.D., Meyer,R.P.: Phys. Rev. A **43**, 77 (1991)
    : Rev. A **43**, 6066 (1991)
(2) Lee C.J., Havey,M.D.,: Phys. Rev. A **43,** 6066 (1991)
(3) Makonnen,A., Kaiser,A., Behmenburg,W.: Z. Phys. D **36**, 325 (1996)
(4) Jungen,M. , Staemmler,V.: J. Phys. B **21**, 463 (1988)
(5) Peach, G.: Comm. At. Mol. Phys. **11**, 101 (1982)
(6) Czuchaj,E., Rebentrost,F., Stoll,H., Preuss,H.: Chem. Phys. **136**, 79 (1989)
(7) Zagrebin,A.L., Tserkovnyi,S.I.: Sov. J. Chem. Phys. **10**, 907 (1992)
(8) Behmenburg,W., Makonnen,A., Kaiser,A., Rebentrost,F., Staemmler,V., Jungen,M., Peach,G., Devdariani,A., Tserkovnyi,S. Zagrebin,A.,Czuchaj,E.: J.Phys. B **29**, 3891 (1996)
(9) Behmenburg,W., Kaiser,A., Rebentrost,F., Jungen,M., Smit,M., Luo,M. and Peach,G.: J.Phys. B**31**,689 (1998)
(10) Julienne,P.S., Krauss,M., Stevens,W.: Chem. Phys. Lett. **38**, 374 (1976)
(11) Jablonski,A.: Z. Phys. **70,** 732(1931); Acta Phys. Pol. **6**, 371(1937)
(12) Jablonski,A.: Phys.Rev. **68,** 78 (1944);
(13) Kulander,K.C., Rebentrost,F. : J. Chem. Phys. **80**, 5623 (1984)
(14) LeRoy,R.J.: Comput. Phys. Comm. **52**, 383 (1989)
(15) Herzberg,G.: Molecular Spectra and Molecular Structure,Vol. I, van Nostrand Company., 2nd ed. (1963)

(16) Wiese,W.L., Smith,M.W., Glennon,B.M.: Atomic transitions probabilities, Vol.I NSRDS-NBS 4, Washington (1966)
(17) Kaiser, A.: Dissertation Düsseldorf (1997)
(18) Findeisen,M. and Grycuk,T.: J.Phys. B**22**, 1583 (1989)
(19) Grycuk,T., Behmenburg,W. and Staemmler,V.: Jablonski Centennial Conf., Torun Juli 23-27, (1998)

Collision Perturbations in the Spectra and Incremental Polarizabilities of Inert Gas Atoms

M. O. Bulanin

Institute of Physics, St Petersburg University,
Peterhof, 198904 St Petersburg, Russia

Abstract. The changes in the dipole polarizabilities of interacting atoms observed in high-density gases are interpreted in terms of collisional perturbations of the absorption spectra. Contributions to the second refractivity virial coefficients due to the collision-induced low-energy shifts of the photoionization continua are calculated for the Ne, Ar, Kr, and Xe atoms and are shown to be significant.

Optical studies of dense gases provide an important information on the electric properties of the colliding species. Interactions between atoms modify their dipole polarizabilities, so that the mean dynamic polarizability $\alpha_{12}(\omega, r)$ of a colliding pair at a distance r deviates from the sum of polarizabilities for the isolated atoms. This leads to a nonlinear density dependence of the Lorentz-Lorenz function, which for neat moderately dense gases can be written in the form a truncated virial expansion [1]

$$\frac{n^2 - 1}{n^2 + 2} = A_R(\omega)\rho + B_R(\omega, T)\rho^2 + \cdots$$

$$= A_R(\omega)\left(\frac{p}{RT}\right) + \left[B_R(\omega, T) - B_P(T) A_R(\omega)\right]\left(\frac{p}{RT}\right)^2 + \cdots, \quad (1)$$

where $n = n(\omega, \rho)$ is the refractive index at the frequency ω, ρ is the molar density, A_R and B_R are the first and the second refractivity virial coefficients, T is the temperature, p is the pressure, R is the molar gas constant, and $B_P(T)$ is the second thermodynamic virial coefficient.

The term linear in density dominates the Lorentz-Lorenz function and depends on the polarizability α_0 of non-interacting atoms: $A_R(\omega) = (4\pi/3)N_A\alpha_0(\omega)$, where N_A is Avogadro's constant. The coefficient B_R describes the effect due to two-body interactions and depends on the binary incremental polarizability trace

$$\Delta\alpha(\omega, r) = \alpha_{12}(\omega, r) - 2\alpha_0(\omega). \quad (2)$$

Classical statistical mechanics of binary correlations between like species yields [2]

$$B_R(\omega, T) = \frac{8\pi^2 N_A^2}{3} \int_0^\infty \Delta\alpha(\omega, r) \, exp[-\beta V_G(r)] \, r^2 dr, \quad (3)$$

where $\beta = (k_B T)^{-1}$ and $V_G(r)$ is the potential energy function in the atomic ground electronic state.

Several *ab initio* and semiempirical models have been developed revealing the nature of interactions contributing to the pair trace (2) in the static ($\omega = 0$) limit, however, the agreement between the experimental and theoretical data on B_R for heavier inert-gas atoms remains modest, at best [3] - [5]. Also the computed second virial coefficients fail to reproduce the reported strong temperature variations of B_R [4, 6].

An alternative approach to the problem was proposed [7, 8] that allows to obtain the dynamic pair trace from analysis of the collision-induced perturbations in the absorption spectrum of an atomic gas. The polarizability can be expressed in terms of the density-dependent dipole oscillator strength distribution (DOSD), $D(E) = df/dE$, or the absorption cross section, $\sigma(E)$, via the Kramers-Kronig-type integral

$$\alpha(\omega, \rho) = \frac{1}{4\pi^3} \int_{\Delta E} \frac{\sigma(E, \rho) \, dE}{E^2 - \omega^2}. \tag{4}$$

Here ΔE is the spectral range in which the information on the $\sigma(E, \rho)$ function is either available from independent experiments, or can be obtained from the theory of spectral line broadening and shift.

Strictly speaking, Eq. (4) will yield full atomic polarizability in the range ΔE from the excitation threshold to very high energies, however, only a few spectral regions with the greatest DOSD are really of practical importance. The major contribution to the polarizability of inert-gas atoms is provided by the low-energy part of the photoionization continuum [9], which also exhibits a strong dependence on the gas density (see, e. g. [10]). In the present work the integration range in Eq. (4) was taken from the low-density ionization onset, at $E = IP$, to the energy of the first minimum in the ionization continuum denoted E_{max}.

Equation (4) can be further modified using the adiabatic statistical model of the spectral line broadening and shift to explicitly account for the effect of collisions. According to this model, the dipole transitions in a pair of colliding species at any given separation r occur between the ground $V_G(r)$ and the upper $V_U(r)$ molecular potential energy curves correlating with the electronic terms of non-interacting atoms. The transition frequencies for each radiative channel $\langle U| \leftarrow \langle G|$ shift according to the difference potential

$$\Delta V(r) = [V_U(r) - V_G(r)] - [V_U(r = \infty) - V_G(r = \infty)]. \tag{5}$$

In the frequency ranges removed from single-photon resonances, the collision-induced shift of the absorption intensity towards lower energies ("red shift") makes positive contribution to the incremental trace, whereas the blue shift makes negative contribution [11].

The upper molecular terms radiatively coupled in the photoionization continuum to the atomic pair ground state $^1\Sigma_g^+$ are the *ungerade* states of the inert-gas dimer ions Rg_2^+ (Rg = Ne, Ar, Kr, Xe). The spin-orbit splitting of the Rg^+ cation ground state into $^2P_{3/2}$ and $^2P_{1/2}$ sublevels with two different ionization limits generates three relevant molecular states labeled here by $i = 1, 2,$ and 3. The Rg_2^+ states $^2\Sigma_{1/2u}^+$ and $^2\Pi_{3/2u}$ correlate with the $^2P_{3/2}$ sublevel, the $^2\Pi_{1/2u}$ state correlates with the $^2P_{1/2}$ sublevel [12]. The separation between the ionization thesholds $IP_1 = IP_2 = IP(^2P_{3/2})$ and $IP_3 = IP(^2P_{1/2})$ progessively increases in the Ne to Xe sequence, from 0.1 eV for Ne to 1.3 eV for Xe [9].

Contribution to the pair trace due to the collision-induced shift of the ionization continuum can thus be written as

$$\Delta\alpha(\omega,r) = \frac{1}{2\pi^3}\sum_{i=1}^{3}\left(\int_{IP_i}^{E_{max}}\frac{\sigma_i(E)}{[E+\Delta V_i(r)]^2 - \omega^2}dE\right) - 2\alpha(\omega,\rho \to 0), \quad (6)$$

where $\alpha(\omega,\rho \to 0)$ is the collisionless limit of Eq. (4), $\Delta V_i(r)$ are the difference potentials (5), and σ_i are the partial cross sections for each of the existing radiative channels. The branching ratios $[\sigma_1 + \sigma_2]/\sigma_3$ beyond the $^2P_{1/2}$ ionization limit were assumed constant throughout the integration range and were taken from ref. [13]. The absorption cross section distributions in the photoionization spectral ranges of the inert-gas atoms were tabulated by Chan et al. [14].

In order to calculate the difference potentials defined in Eq. (5) the knowledge of the potential energy curves is required for each of the three Rg_2^+ states that contribute to absorption in the continuum. Strongly bound lowest $^2\Sigma_{1/2u}^+$ states with the dissociation energies in the 0.98 - 1.33 eV range [15] are well studied both experimentally and theoretically for all the inert-gas dimer ions. Information on the properties of two other states $^2\Pi_{3/2u}$ and $^2\Pi_{1/2u}$ is far less reliable, because these are mostly repulsive states with shallow van der Waals minima at large interatomic separations, which limits the accuracy of the present B_R estimates based on Eq. (6). The *ab initio* potential energy curves used in the calculations of the $\Delta V_i(r)$ functions were tabulated in [16, 17]. Highly accurate ground-state potentials $V_G(r)$ for the inert-gas atoms are readily available [18].

Table 1 compares the values of B_R obtained from Eqs. (6) and (3) with most of the available experimental and theoretical data. We see that the interaction-induced shift of the ionization continuum yields a positive contribution to the pair trace. This effect is directly related to lowering of the ionization potential with density observed for the inert gases [10], which it turn can be traced to the properties of the potential energy curve for the lowest dimer ion state. The state $^2\Sigma_{1/2u}^+$ is much more strongly bound compared to the neutral Rg_2 ground state $^1\Sigma_g^+$. This makes the corresponding difference potential $\Delta V_1(r)$ large and negative at all separations r sampled by the radial distribution function in the integrand of Eq. (3) and, as is evident from Eq. (6), results in a large positive contribution to the pair trace. For two other essentially repulsive dimer ion states ($i = 2, 3$), the difference potentials are smaller and positive at the relevant interatomic separations, forming negative contributions, although the net effect of the collision-induced redistribution in the continuum remains positive.

Measurements performed so far indicate the dispersion of the second virial coefficients for the atomic gases in the energy range from the static limit to the visible - near UV to be smaller than the probable rather large experimental errors. The difficulty in the experimental determination of B_R stems from the product $B_P(T) A_R(\omega)$ in Eq. (1) being orders of magnitude greater than the values of B_R themselves, thus

Table 1. Second refractivity and dielectric virial coefficients (cm^6 mol^{-2}) of atomic gases

T/K	ω^{-1}/nm	$B_R(\omega,T)$	Reference
		Neon	
323	∞	-0.12 ± 0.06	[19]
77	∞	$+0.07 \pm 0.02$	[19]
303	633	-0.11 ± 0.02	[20]
300	∞	-0.13	[4] (calc.)
323	∞	$+0.10$	This work
77	∞	$+0.11$	This work
		Argon	
298	632.8	1.57 ± 0.58	[21]
303	∞	1.22 ± 0.09	[19]
303	633	1.75 ± 0.05	[20]
322	∞	0.48	[3] (calc.)
300	∞	0.0615	[4] (calc.)
300	∞	0	[5] (calc.)
303	∞	3.73	This work
		Krypton	
298	632.8	6.23 ± 1.55	[21]
323	∞	4.3 ± 0.7	[19]
243	∞	8.2 ± 0.4	[19]
303	633	5.96 ± 0.06	[20]
296	325	8.4 ± 0.4	[22]
322	∞	4.68	[3] (calc.)
300	∞	1.366	[4] (calc.)
323	∞	8.48	This work
		Xenon	
298	632.8	25.5 ± 2.85	[21]
323	∞	32 ± 2	[19]
243	∞	35 ± 2	[19]
348	633	28.5 ± 0.5	[20]
322	∞	21.6	[3] (calc.)
300	∞	16.28	[4] (calc.)
323	633	24.5	This work

even small uncertainty in the $B_P(T)$ (typically, a few percent) may result in an unacceptably large error in B_R. A differential technique described in detail, e. g. in [1, 20] was developed to overcome this difficulty. However, a remarkably strong correlation found in [6] between $B_R(T)$ and $[-B_P(T)]$ showing the same Boyle point may indicate the effect of the thermodynamic virial coefficient not to be entirely compensated by this technique. The accuracy of the existing experimental data for Ar was recently questioned [5].

The results collected in Table 1 show that at similar conditions our estimates for

Ne and Ar are too positive with respect to the experimental data. For Kr and Xe our values are close enough to experiment. One should emphasize at this point that the values of B_R calculated here are *not* the total second refractivity virial coefficients, but only contributions generated by the interaction-induced absorption cross section redistribution in the ionization continua. Additional contributions, that can be of either sign depending upon the pattern of the spectral intensity redistribution, must be provided by the collisional perturbations in the discrete range of the absorption spectra as well. A positive contribution of 4.2 cm^6 mol^{-2} at 633 nm and room temperature was found from analysis [7] of the effect due to asymmetric broadening of the first resonance line at 147 nm in the Xe gas [23]; combined with the value listed in Table I makes the total B_R (633 nm) = 28.7 cm^6 mol^{-2}, in excellent agreement with the result reported for Xe by Achtermann *et al.* [20]. Our results for Ne and Ar suggest that negative contributions to the pair traces are to be expected from the collisional effects in the discrete spectra of these gases.

ACKNOWLEDGMENT

The support of the Deutsche Forschungsgemeinshaft and of the Russian Foundation for Basic Research is gratefully acknowledged.

References

[1] T. K. Bose, in *Phenomena Induced by Intermolecular Interactions*, edited by G. Birnbaum (Plenum Press, New York, 1985), p. 49.

[2] A. D. Buckingham and J. A. Pople, *Trans. Faraday Soc.* **51**, 1029, 1173 (1955); A. D. Buckingham, *ibid.* **52**, 1035 (1956); A. D. Buckingham and C. Graham, *Proc. Roy. Soc. Lond.* A, **336**, 275 (1974).

[3] P. D. Dacre, *Molec. Phys.* **45**, 1, 17 (1982); *ibid.* **47**, 193 (1982); P. D. Dacre and L. Frommhold, *J. Chem. Phys.* **76**, 3447 (1982).

[4] N. Meinander, *Chem. Phys. Lett.* **228**, 295 (1994).

[5] C. G. Joslin, J. D. Goddard, and S. Goldman, *Molec. Phys.* **89**, 791 (1996).

[6] M. O. Bulanin, U. Hohm, Yu. M. Ladvishchenko, and K. Kerl, *Z. Naturforsh.* **49a**, 505 (1994).

[7] M. O. Bulanin, *Chem. Phys. Lett.* **217**, 466 (1994).

[8] M. O. Bulanin, U. Hohm, and K. Kerl, *Molec. Phys.* **92**, 929 (1997).

[9] J. Berkowitz, *Photoabsorption, Photoionization, and Photoelectron Spectroscopy* (Academic Press, New York, 1987).

[10] R. E. Huffman and D. H. Katayama, *J. Chem. Phys.* **45**, 138 (1966); R. Reininger, V. Saile, and A. M. Köhler, *J. Phys. B*, **20**, 2239 (1987).

[11] M. O. Bulanin, *Opt. Spectrosc.* **75**, 574 (1993).

[12] R. S. Mulliken, *J. Chem. Phys.* **52**, 5170 (1970).

[13] J. A. R. Samson, J. L. Gardner, and A. F. Starace, *Phys. Rev. A*, **12**, 1459 (1975).

[14] W. F. Chan, G. Cooper, X. Guo, and C. E. Brion, *Phys. Rev. A*, **45**, 1420 (1992); W. F. Chan, G. Cooper, X. Guo, G. R. Burton, and C. E. Brion, *Phys. Rev. A*, **46**, 149 (1992).

[15] R. I. Hall, Y. Lu, Y. Morioka, T. Matsui, T. Tanaka, H. Yoshii, T. Hayaishi, and K. Ito, *J. Phys. B*, **28**, 2435 (1995); *J. Chem. Phys.* **102**, 1553 (1995).

[16] J. S. Cohen and B. Schneider, *J. Chem. Phys.* **61**, 3230 (1974).

[17] W. R. Wadt, *J. Chem. Phys.* **68**, 402 (1978).

[18] R. A. Aziz and M. J. Slaman, *Chem. Phys.* **130**, 187 (1989); *J. Chem. Phys.* **92**, 1030 (1990); A. K. Dham, W. J. Meath, A. R. Allnatt, R. A. Aziz, and M. J. Slaman, *Chem. Phys.* **142**, 173 (1990).

[19] J. Huot and T. K. Bose, *J. Chem. Phys.* **95**, 2683 (1991).

[20] H. J. Achtermann, J. G. Hong, G. Magnus, R. A. Aziz, and M. J. Slaman, *J. Chem. Phys.* **98**, 2308 (1993).

[21] R. C. Burns, C. Graham, and A. R. M. Weller, *Molec. Phys.* **59**, 41 (1986).

[22] K. Kerl, private communication.

[23] M.-C. Castex, *J. Chem. Phys.* **74**, 759 (1981).

Following in the Footsteps of A. Jabłoński: Some Considerations on Collisional Interference in the HD Fundamental Band

R. M. Herman
Department of Physics
The Pennsylvania State University
104 Davey Laboratory
University Park PA 16802 USA

It is a great pleasure for me to dedicate this paper to the memory of Professor Aleksander Jabłoński on the occasion of the 100th anniversary of his birth. While I regret not having more extensively known Professor Jabłoński, I did have the pleasure of sitting with him during an afternoon session of the conference "Optical Pumping and Line Shape (OPaLS)" which was organized by Professor T. Skalinski and held in Warsawa in 1968. I had been quite aware of the contributions of Professor Jabłoński to our subject, in part because I was a student of Professor H. Margenau whose work, as we know, was heavily influenced by that of Jabłoński.

Jabłoński showed us how to solve static (high density and/or line wing problems) using the Franck-Condon principle as applied to transitions between the optically active molecule-perturber pair stationary translational states[1]. The Franck-Condon principle remains applicable, yet is not so directly obvious in its application, in the low density (or near line centers) regime where the impact theory holds, because there the energy defects associated with the line shapes are shared between all perturbers and are manifested through the elastic scattering amplitudes in the asymptotic parts of the translational states. Thus, characteristically, the lineshape problem has the two rather incompatible limits – the impact, and the static which also includes, to a large extent, the collisionally induced spectra. Of course, some duration-of-interaction effects govern the shapes of the latter spectra - but this is represented, automatically, through the inelastic translational state transitions, as modulated by the permanent or collisionally induced dipole operator function in two body interactions, using the Franck-Condon principle in an obvious way.

Now there also exists, to the author's knowledge, a single observed example in which both limits are separately and simultaneously valid in the description of single spectral lines. This is the case of the vibration-rotation (and pure rotation) $\Delta J = \pm 1$ absorption spectra for the heteronuclear isotopomers of molecular hydrogen. Let us focus upon the (0-1) absorption band of HD under foreign gas pressures. Here, a single collisional pair gives rise to a dipole moment according to the static theory[2] as conventionally used in CIA studies. In HD there exists a dipole transition component which depends only upon a scalar function of the intermolecular displacement[3]. This component has the internal coordinate dependence of the tiny allowed HD dipole moment, and thus is phase-locked to the latter[4]. As a result, chains of collisionally induced transition amplitudes interfere among each other and with the permanent dipole, leading to corresponding long-time phase memory effects giving rise to narrow spectral features. This behavior is entirely analogous to that of the

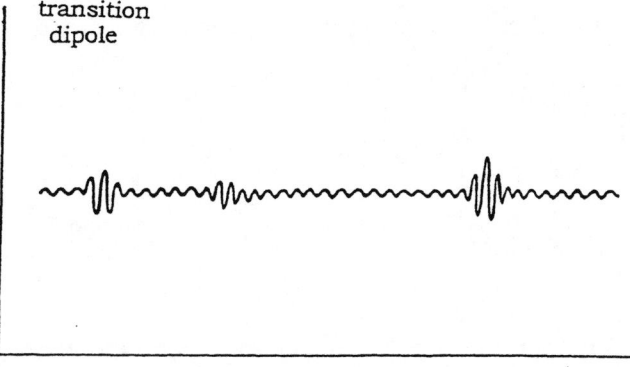

Fig. 1. Schematic illustration of the transition dipole component giving the ΔJ=±1 vibration-rotation spectrum of HD. The perfectly phased collisional dipoles interfere with the permanent dipole and with each other over long times, producing a superposition of narrow spectral features.

Ramsey double-resonator experiment. In the present case, the (multiple) resonators are the many outer "surfaces" of the perturbing atoms. See Fig. 1. Because this interference persist for the coherence time of the permanent dipole, impact theories based upon the usual dephasing through well-separated intervening collisions govern the manner in which the dipole correlation is lost, giving rise to spectral features characterized by impact theory lineshape parameters. The actual collisional dynamics, as they enter the dipole production, nonetheless, must be analyzed through single collision dynamics, which involves the static limit type of calculation, with the time averaged dipole produced through a single encounter being perfectly represented through the Franck-Condon dipole integral with the restriction that the incoming translational state propagation vectors **k** and the phases for translational states accompanying the initial and final internal vibrational states be equal. (That the **k** vectors remain unchanged is a condition for the collisional interference; the inelastic Franck-Condon transitions result in an (unresolved) broad background associated with the dipole component of interest. In HD, the absence of large collisionally-induced phase shifts in the dipole-producing collisions allows the interference effects to exist.)

The lineshape is, as always, proportional to the dipole-dipole correlation function,

$$C(\tau) = \{\mu_{if}(0)\mu_{fi}(\tau)\} , \qquad (1)$$

with { } representing the ensemble average, and μ containing the permanent and collisionally induced dipole components mentioned above.

While the statistical properties of $\mu_{fi}(\tau)$ appear to be clearly understandable, through the fact that phase accumulations as collisions progress in time can be translated into the quantum dynamical language of matrix elements associated with translational states, the situation is not so clear with the dipole production at the earlier time. In that case there is a sort of role-reversal in time coordinates relating the peak of the dipole-producing collision to other times during the dipole producing collisions. Probably the best solution to this dilemma is to write

$$C(\tau) = \{\mu_{if}(-\tau/2)\mu_{fi}(\tau/2)\} \tag{2}$$

which then can be recast into the form

$$C(\tau) = \frac{\{\mu_{if}(-\tau/2)\mu_{fi}^A(0)\mu_{if}^A(0)\mu_{fi}(\tau/2)\}}{|\mu_{if}^A(0)|^2} \tag{3}$$

where $\mu_{if}^A(0)$ is the (allowed) permanent dipole component, which for the purposes above, simply carries a reference phase at time $\tau=0$. Because of the statistical independence of the intervals $(-\tau/2 \rightarrow 0)$ and $(0 \rightarrow \tau/2)$, these intervals can be averaged separately, giving

$$C(\tau) = \frac{\{\mu_{if}(-\tau/2)\mu_{fi}^A(0)\}\{\mu_{if}^A(0)\mu_{fi}(\tau/2)\}}{|\mu_{if}^A|^2} \tag{4}$$

In view of the fact that in all such correlation functions, $C(-\tau)=C^*(\tau)$, this becomes

$$C(\tau) = \frac{\{\mu_{if}^A(0)\mu_{fi}(\tau/2)\}^2}{|\mu_{if}^A|^2} \tag{5}$$

From impact theory, we know that $\{\mu_{if}^A(0)\mu_{fi}(\tau/2)\}$ varies as $\exp[(i(\omega_{fi} + \Delta\omega_{fi}) - \frac{\gamma_{fi}}{2})(\tau/2)]$, $\Delta\omega_{fi}$ and γ_{fi} being the impact-theory line shift and FWHM. At the same time, the dipole production at time $\tau/2$ leads to the allowed dipole amplitude (which can be regarded as being real) together with complex collisionally induced transition amplitudes. $C(\tau)$ can then be represented, in the notation of Gao, Cooper and Tabisz[5] in the form

$$C(\tau) = |\mu_{fi}^A|^2 [1 + (N_b/2)(a+ic)]^2 \exp\left[\left(i(\omega_{fi} + \Delta\omega_{fi}) - \frac{\gamma_{fi}}{2}\right)\tau\right], \tag{6}$$

N_b being the perturber number density with $\Delta\omega_{fi}$ and γ_{fi} also being proportional to N_b. The resulting line shape function therefore is given by

$$L(\omega) = \frac{1}{\pi}\text{Re}\left[\frac{\left(1+(N_b/2)(a+ic)\right)^2}{i(\omega - \omega_{fi} - \Delta\omega_{fi}) + \frac{\gamma_{fi}}{2}}\right], \tag{7}$$

which represents the McKellar line profile, consisting of pressure dependent normal and anomalous dispersion components. The quantities a and c now can be calculated using the translational states, much as envisioned by Jabłoński. For the combination HD-He, λ is too large ($\cong 0.035$ nm at 77K) for classical path dynamics to be valid, but large enough that the number of partial translational waves required appears to be reasonably small (≤ 12 for T $\leq$ 300K) in a quantum dynamical treatment. The WKB approximation appears not to be good, except for larger intermolecular spacings, where it is useful in establishing translational state normalization.

According to Cooper, Tabisz and coworkers[5,6,7], the collision-induced transition amplitudes (a+ic) result from three interfering paths. For example, for the $R_1(0)$ line, in which one makes transitions $(v,J)=(0,0)\rightarrow(1,1)$, in addition to a direct collisionally induced term, represented by $B_1^{01}(0;R)$ in ref. 2, having internal coordinate dipole symmetry and scalar (rotationally invariant) dependence upon the external displacement coordinate $\vec{R}$, there are two more or less equivalently participating paths of the following description: A term of the intermolecular potential arising from the displacement of the center of mass from the geometric center in HD, $V_{||}(R) P_1 (\vec{R}\cdot\vec{r})$, causes the (01) state to slightly admix into the (0,0) state. The resulting state then can reach the (1,1) state through the operation of the relatively large collisionally induced dipole component $B_0^{01}(1;R)$ of ref. 2, which has external dipole symmetry and internal scalar symmetry. The other more or less equivalent type path arises through slightly admixing some (1,0) into the upper (1,1) state, together with the $B_0^{01}(1;R)$ operator causing the transition from (0,0) to (1,1) through the (1,0) admixture component. While Gao, Cooper and Tabisz[5] and Tabisz and McQuarrie[6] have specifically calculated a four-parameter expression for the N_b dependence of the line shapes, the above expression emphatically indicates a dependence upon only two independent parameters - a fact that appears to be entirely supported by the substantial data of McKellar and Rich[8] for the $R_1(0)$ and $R_1(1)$ lines perturbed by a wide variety of foreign gases.

Bibliography
1. See, for example, the summary of Jabłoński's work in R. G. Breene, Jr., The Shift and Shape of Spectral Lines (Pergamon Press, Oxford, 1961).
2. A. Borysow, L. Frommhold and W. Meyer, J. Chem. Phys. **88**, 4855 (1988).
3. R. M.Herman, Phys. Rev. Lett. **42**, 1206 (1979); R. M. Herman, R. H. Tipping and J. D. Poll, Phys. Rev. A**20**, 2006 (1979).
4. R. M. Herman in Spectral Line Shapes, R. Exton, ed. (Deepak, Hampton, VA, 1987).
5. B. Gao, J. Cooper and G. C. Tabisz, Phys. Rev. A **46**, 5781 (1992).
6. B. McQuarrie and G. C. Tabisz, J. Mol. Liq. (Netherlands) **70**, 159 (1996).
7. B. Gao, G. C. Tabisz, M. Trippenbach and J. Cooper, Phys. Rev. A**44**, 7379 (1991).
8. A. R. W. McKellar and N. Rich, Can. J. Phys. **62**, 1665 (1984).

APPENDIX

Minutes of the ICSLS Program Committee

The Committee meeting was held June 25, 1998.

It was confirmed that the next Conference, ICSLS-15, will be held at the PTB, Berlin, Germany, in the Summer of the year 2000, with exact dates to be chosen so as to optimally interface with other conventions and events at that time. Joachim Seidel will serve as organizing chairman.

It was decided to hold ICSLS-16 in the year 2002 in Livermore, CA, USA. Christina Back will serve as organizing chairman, for which the Committee expressed its gratitude. The Committee also expressed its appreciation for the kind offer of G. Tabisz (University of Manitoba) to serve as organizing chairman of ICSLS-16.

It was agreed that Christian Boulet (Orsay) would replace Hoe Nguyen as a Committee member and thanked Dr. Nguyen for his service in past years. In addition, Kenji Ohmuri (Tohoku University) and William Stwalley (University of Connecticut) agreed to serve as representatives of the newly emerging area of low temperature lineshapes. The Program Committee will consist of the following members:

C. Back	tinaback@llnl.gov
C. Boulet	christian.boulet@lpma.u-psud.fr
A. Z. Devdariani	ponik@devdar.spb.su
M. S. Dimitrijevic	mdimitrijevic@aob.bg.ac.yu
N. Feautrier	Nicole.Feautrier@obspm.fr
R. M. Herman	rmh@phys.psu.edu
J. Kielkopf	jfkiel01@nimbus.physics.louisville.edu
V. S. Lisitsa	LISITSA@qq.nfi.kiae.su
A. D. May	dmay@physics.utoronto.ca
M. Moraldi	moraldi@firenze.infn.it
K. Ohmori	ohmori@rism.tohoku.ac.jp
E. Oks	goks@physics.auburn.edu
G. Peach	g.peach@ucl.ac.uk
G. Pichler	pichler@its.hr
J. Seidel	seidel@chbrb.berlin.ptb.de
R. Stamm	rstamm@piima1.univ-mrs.fr
W. C. Stwalley	stwalley@uconnvm.uconn.edu
J. Szudy	szudy@phys.uni.torun.pl
G. C. Tabisz	tabisz@cc.umanitoba.ca
R. H. Tipping	rtipping@ualva.ua.edu

A discussion was held regarding procedures to be followed in deciding upon the holding of symposia honoring a present or former member of the spectral lineshape scientific community. While no firm decisions were made, the Committee expressed the concern that this should be done in a deliberate and timely fashion, with full consultation of the entire Committee.

Respectfully submitted

Roger M. Herman
Organizing Chairman, ICSLS-14
25 June 1998

LIST OF PARTICIPANTS BY NATIONALITY

Berman, R.	Canada	rberman@atmosp.physics.utoronto.ca
Cho, C-W.	Canada	cho@kelvin.physics.mun.ca
Dolbeau, S.	Canada	Stephanie.Dolbeau@france-mail.com
Lewis, J.C.	Canada	court@kelvin.physics.mun.ca
May, A.D.	Canada	dmay@physics.utoronto.ca
Predoi-Cross, A.	Canada	adriana@atmosp.physics.utoronto.ca
Reddy, S.P.	Canada	spreddy@kelvin.physics.mun.ca
Stamp, C.	Canada	stamp@kelvin.physics.mun.ca
Tabisz, G.C.	Canada	tabisz@cc.umanitoba.ca
Pichler, G.	Croatia	pichler@bobi.ifs.hr
Borysow, A.	Denmark	aborysow@nbi.dk
Polzik, E.S.	Denmark	polzik@dfi.aau.dk
Behmenburg, W.	Deutschland	behmenbu@uni-duesseldorf.de
Hohm, U.	Deutschland	u_hohm@tu-bs.de
Kunze, H.-J.	Deutschland	H.-J.Kunze@ep5.ruhr-uni-bochum.de
Schmidt, M.	Deutschland	martina.schmidt@ptb.de
Seidel, J.	Deutschland	joachim.seidel@ptb.de
Sorge, S.	Deutschland	stefans@darss.mpg.uni-rostock.de
Steiger, A.	Deutschland	andreas.steiger@ptb.de
Voslamber, D.	Deutschland	voslam@drfc.cad.dea.fr
Zimmermann, D	Deutschland	dz@kallium.physik.tu-berlin.de
Ghatass, Z.F.	Egypt	GDANIEL@alex.eun.eg
Helmi, M.S.	Egypt	GDANIEL@alex.eun.eg
Roston, G.D.	Egypt	GDANIEL@alex.eun.eg
del Val, A.J.	Espana	juanval@hp9000.cpd.uva.es
Meinander, N.	Finland	meinander@phcu.helsinki.fi
Audebert, P	France	patrik@greco2.polytechnique.fr
Boulet, C.	France	christian.boulet@lpma.u-psud.fr
Brillant, S.	France	stephane.brillant@obspm.fr
Calisti, A.	France	annette@piima1.univ-mrs.fr
Ferri, S.	France	sandy@piima1.univ-mrs.fr
Hartmann, J.-M.	France	jean-michel.hartmann@lpma.u-psud.fr
Laboucher-Dalimier, E.	France	lebda@moka.ccr.jussieu.fr
Lesage, A.	France	lesage@obspm.fr
Mosse, C.	France	caro@piima1.univ-mrs.fr
Rachet, F.	France	rachet@babinet.univ-angers.fr
Stamm, R.	France	rstamm@piima1.univ-mrs.fr
Stehle, C.	France	chantal.stehle@obspm.fr
Talin, B.	France	btalin@piima1.univ-mrs.fr
Vitel-Lepinay, Y.	France	yv@ccr.jussieu.fr
Ben-Reuven, A.	Israel	abram@post.tau.ac.il
Yurovsky, V.A.	Israel	volodia@post.tau.ac.il
Barocchi, F.	Italia	barocchi@fi.infn.it
Moraldi, M.	Italia	moraldi@firenze.infn.it
Zoppi, M.	Italia	zoppi@ieq.fi.cnr.it
Ohmori, K.	Japan	ohmori@rism.tohoku.ac.jp
Sato, Y.	Japan	satoyuki@rism.tohoku.ac.jp
Takami, M.	Japan	mtakami@postman.riken.go.jp

Szudy, J.	Polska	szudy@phys.uni.torun.pl
Bulanin, M.O.	Russia	bulanin@niif.spbu.ru
Bureyeva, L.A.	Russia	bureyeva@sci.lebedev.ru
Devdariani, A.Z.	Russia	ponik@devdar.spb.su
Gavrilenko, V.P.	Russia	gavrilenko@fpl.gpi.ru
Kouzov, A.P.	Russia	alex@aph.usr.pu.ru
Leonov, A.G.	Russia	leonov@post.mipt.rssi.ru
Lisitsa, V.S.	Russia	LISITSA@qq.nfi.kiae.su
Shapiro, D.A.	Russia	249@okibox.iae.nsk.su
Starostin, A.N.	Russia	staran@fly.triniti.troitsk.ru
Tonkov, M.V.	Russia	tonkov@nicf.spb.ru
Peach, G.	U.K.	g.peach@ucl.ac.uk
Wark, J.	UK	justin.wark@physics.ox.ac.uk
Back, C.	USA	back2@llnl.gov
Bernheim, R.A.	USA	r5b@psuvm.psu.edu
Betz, A.	USA	betz@spot.colorado.edu
Birnbaum, G.	USA	birnbaum@micf.nist.gov
Bulanin, K.	USA	bulanin@che.udel.edu
Chang, E.	USA	chang@phast.umass.edu
Cole, M.W.	USA	mwc@psu.edu
Derevianko, A.	USA	andrei@atomic2.phys.nd.edu
Ernst, W.E.	USA	wee1@psuvm.psu.edu
Frommhold, L.	USA	frommhold@physics.utexas.edu
Glenzer, S.H.	USA	glenzer1@llnl.gov
Griem, H.R.	USA	griem@Glue.umd.edu
Hammer, D.	USA	dhammer@mail.utexas.edu
Heinzen, D.	USA	heinzen@physics.utexas.edu
Henry, M.E.	USA	m_henry@clinch.edu
Herman, R.M.	USA	rmh@phys.psu.edu
Huennekens, J.P.	USA	jph7@lehigh.edu
Hulet, R.	USA	randy@atomcool.rice.edu
Iglesias, E.J.	USA	iglesias@Glue.umd.edu
Julienne, P.S.	USA	paul@molphys.nist.gov
Junkel, G.C.	USA	junkel@phys.ufl.edu
Kielkopf, J.F.	USA	john@nimbus.physics.louisville.edu
Kreye, W.C.	USA	WKREYE@desire.wright.edu
Looney, J.P.	USA	jlooney@nist.gov
Ma, Q.	USA	crqxm@nasagiss.giss.nasa.gov
Oks, E.	USA	goks@physics.auburn.edu
Palm, P.	USA	palm@rclsgi.eng.ohio-state.edu
Savchenko, V.I.	USA	vsavchen@pppl.gov
Smith, M.A.	USA	m.a.h.smith@larc.nasa.gov
Steele, W.A.	USA	was@chem.psu.edu
Stwalley, W.C.	USA	stwalley@uconnvm.uconn.edu
Tipping, R.H.	USA	rtipping@ua1vm.ua.edu
Touma, J.	USA	jtouma@physics.auburn.edu
Tricinelli, R.	USA	LAMBDAUSA@aol.com
Varanasi, P.	USA	PVARANASI@ccmail.sunysb.edu
Yeazell, J.	USA	yeazell@phys.psu.edu

Author Index

A

Ahmad, I., 39
Allard, N. F., 228, 263, 264
Amano, K., 389, 390, 391
Angelo, P., 64, 139, 141
Aparicio, J. A., 203
Astapenko, V. A., 146, 520

B

Ban, T., 259
Behmenburg, W., 411, 532
Bengtson, R. D., 187
Ben-Reuven, A., 313
Berman, R., 240, 489, 493
Bielski, A., 404, 406
Błachowicz, T., 341
Blagojević, B., 189
Blair, D. W., 252
Boissoles, J., 457, 485
Bonechi, A., 427
Borysow, A., 207, 459
Bouanich, J. P., 485
Boulet, C., 457, 485
Bradley, D. K., 136
Braune, M., 395
Brillant, S., 256
Brym, S., 404, 406
Bulanin, M. O., 546
Bureyeva, L. A., 111, 520
Büscher, S., 39

C

Calisti, A., 64, 88, 115, 119, 139
Cardenoso, V., 119
Ceccotti, T., 64, 141
Chang, E. S., 252
Chekhov, D. I., 301, 400
Chenais-Popovics, C., 141
Chiba, H., 388, 389, 390, 391
Chrysos, M., 449
Ciuryło, R., 328, 404, 406

D

de la Rosa, M. I., 77
Delettrez, J. A., 136
del Val, J. A., 203
Deming, D., 252
Demura, A. V., 127
Derevianko, A., 148
Derfoul, H., 64, 139, 141
Devdariani, A. Z., 139, 325, 388, 389, 390
Djurović, S., 191, 193
Dolbeau, S., 240
Drira, I., 263, 264
Drummond, J. R., 240, 489, 491, 493, 495
D'yachkov, L. G., 183

E

El Bezzari, M., 183
El-Raey, M., 199
Elton, R. C., 143
Escarguel, A., 185

F

Fakhr-Eslam, S. H., 491
Faraggiana, R., 264
Feautrier, N., 127
Ferri, S., 115, 119
Filippov, N. N., 499
Fisch, N. J., 339
Förster, E., 141
Frank, A. G., 195
Frech, B., 487
Frommhold, L., 167, 465
Fuhr, J. R., 27

G

Gauthier, P., 64, 141
Gavrilenko, V. P., 14, 195
Gerbaldi, M., 264

Gerton, J. M., 364
Ghatass, Z. F., 197, 199
Gigosos, M., 119
Gillard, P. G., 453
Glenzer, S. H., 49
Griem, H. R., 3, 143
Grützmacher, K., 77, 155, 201
Grycuk, T., 444
Guillot-Noël, C., 449
Gunderson, M. A., 136

H

Hald, J., 267
Hammer, D., 465
Hanke, D., 487
Hartmann, J.-M., 469
Haynes, Jr., D. A., 136
Helmi, M. S., 408, 462
Henry, M. E., 286
Herman, R. M., 286, 552
Hoang-Binh, D., 336
Hodges, J. T., 275
Hohm, U., 501
Hooper, Jr., C. F., 136
Hulet, R. G., 364
Hutson, J. M., 485

I

Iglesias, E. J., 143

J

Jaanimagi, P. A., 136
Johannsen, U., 77
Jørgensen, U. G., 207
Jungen, M., 411
Junkel, G. C., 136

K

Khvorostov, E. B., 317
Kielkopf, J. F., 228, 263, 264, 402
Klein, L., 88
Kobilarov, R., 191, 193

Kondrat'eva, M. F., 332
Konjević, N., 27, 189, 191, 193
Kosarev, I. N., 127
Kotb, M. M., 199
Koubiti, M., 119
Kouzov, A. P., 497
Kreye, W. C., 402
Kunze, H.-J., 39
Kurilenkov, Y. K., 183
Kurosawa, T., 388, 389, 390
Kurucz, R., 264
Kyrie, N. P., 195

L

Leboucher-Dalimier, E., 64, 139, 141
Le Doucen, R., 457
Le Duff, Y., 449
Lee, R. W., 88, 115
Leonov, A. G., 301, 400
Lesage, A., 27, 185, 260
Lewis, J. C., 482
Lindenblatt, G., 411
Lisitsa, V. S., 111, 127, 520
Looney, J. P., 275

M

Ma, Q., 439
Mar, S., 203
Matheron, P., 260
Mathys, G., 256
May, A. D., 240, 489, 491, 493, 495
McQuarrie, B. R., 332
Meyer, W., 411, 465
Mijatović, Z., 191, 193
Mohamed, M. M., 197, 199
Moraldi, M., 415, 427
Moreau, G., 457
Mossé, C., 88
Mouret, L., 119
Mürtz, M., 487

N

Nikitin, E. E., 388, 389, 390
Nikolić, D., 191, 193

O

Ohmori, K., 352, 388, 389, 390, 391
Oks, E., 99, 123, 148, 150, 187
Okunishi, M., 388, 389, 390, 391
Osborn, T. A., 332
Ozanne, L., 485

P

Palm, P., 487
Panteleev, A. A., 339
Pichler, G., 259
Polzik, E. S., 267
Popović, M. V., 189
Poquérusse, A., 64, 141
Prasad, R. D. G., 453
Pratesi, G., 460
Predoi-Cross, A., 489

R

Rachet, F., 449
Rebentrost, F., 411
Reddy, S. P., 453
Regan, S., 136
Revalde, G., 179
Richou, J., 185, 260
Roche, C. F., 485
Roston, G. D., 408, 444, 462
Rubtsova, N. N., 317
Rudenko, A. A., 400

S

Sackett, C. A., 364
Santoro, M., 415
Sato, Y., 388, 389, 390, 391
Sauvan, P., 64, 139, 141
Savchenko, V. I., 339
Schmidt, M., 155
Schwarzhans, D., 395
Scott, H., 143
Sechin, A. Y., 301, 400
Seidel, J., 134
Seiser, C., 77
Shalaby, E. A., 199
Shapiro, D. A., 321
Sheldon, G. D., 491, 495
Sinclair, P. M., 493
Skenderović, H., 259
Skudra, A., 179
Smit, M., 411
Sørensen, J. L., 267
Stamm, R., 88, 115, 119
Stamp, C., 453, 482
Starostin, A. N., 132, 301, 339, 400
Stehlé, C., 127, 256
Steiger, A., 77, 155, 201
Stepanov, M. G., 321
Stwalley, W. C., 377
Szudy, J., 328, 404, 406, 509

T

Tabisz, G. C., 332
Takami, M., 345
Talin, B., 64, 88, 115, 119, 139
Thibault, F., 485
Tipping, R. H., 439, 457
Tonkov, M. V., 499
Touma, J., 150, 187

U

Ueda, K., 388, 389, 390, 391
Ulivi, L., 415, 460
Urban, W., 487

V

Valipour, H., 503
Van Regemorter, H., 336
van Zee, R. D., 275
Vasilenko, L. S., 317
Vitel, Y., 183
Vollbrecht, M., 141
Voslamber, D., 134

W

Welling, M., 364
Wrubel, T., 39

Y

Yakunin, I. I., 132
Yurovsky, V. A., 313

Z

Zemtsov, Y. K., 301, 400
Zimmermann, D., 395, 503
Zoppi, M., 415, 460